P9-CRE-990
SPRINGER
STUDY
EDITION

G. Pólya G. Szegö

Problems and Theorems in Analysis

Volume II

Theory of Functions · Zeros · Polynomials
Determinants · Number Theory · Geometry

Translation by C. E. Billigheimer

Springer-Verlag

George Pólya · Gabor Szegö
Department of Mathematics
Stanford University
Stanford, California 94305
USA

Claude Elias Billigheimer
McMaster University
Hamilton, Ontario, Canada
and
Maimonides College
Toronto, Ontario, Canada

Revised and enlarged translation of *Aufgaben und Lehrsätze aus der Analysis II,* 4th edition, 1971; Heidelberger Taschenbücher, Bd. 74

AMS Subject Classifications (1970): 10-01, 15-01, 15A15, 30-01, 30A06, 30A08

Library of Congress Cataloging in Publication Data
Pólya, George, 1887–.
Problems and theorems in analysis
Vol. 2 translated by C. E. Billigheimer. Rev. and enl. translation of Aufgaben und Lehrsätze aus der Analysis, 4th ed. 1970–.
Contents: v. 1. Series, integral calculus, theory of functions. v. 2. Theory of functions, zeros, polynomials, determinants, number theory, geometry.
1. Mathematical analysis—Problems, exercises, etc.
I. Szegö, Gabor, 1895– joint author. II. Title.
QA301. P64413. 515′.076. 75-189312.

Printed in the United States of America.

9 8 7 6 5 4

ISBN 0-387-90291-0 Springer-Verlag New York
ISBN 3-540-90291-0 Springer-Verlag Berlin Heidelberg

Contents

Part Four. Functions of One Complex Variable. Special Part

Chapter 1. Maximum Term and Central Index, Maximum Modulus and Number of Zeros

Chapter 2. Schlicht Mappings

Chapter 3. Miscellaneous Problems

Part Five. The Location of Zeros

Chapter 1. Rolle's Theorem and Descartes' Rule of Signs

Chapter 2. The Geometry of the Complex Plane and the Zeros of Polynomials

Chapter 3. Miscellaneous Problems

Part Six. Polynomials and Trigonometric Polynomials

Part Seven. Determinants and Quadratic Forms

Part Eight. Number Theory

Chapter 1. Arithmetical Functions

Chapter 2. Polynomials with Integral Coefficients and Integral-Valued Functions

Chapter 3. Arithmetical Aspects of Power Series

Chapter 4. Some Problems on Algebraic Integers

Chapter 5. Miscellaneous Problems

Part Nine. Geometric Problems

Appendix

Notation and Abbreviations

We have attempted to be as consistent as possible in regard to notation and abbreviations and to denote quantities of the same nature by the same symbol, at least within the same part. A particular notation may be specified for a few sections. Otherwise the meaning of every letter is explained anew in every problem except when we refer to a previous problem. A problem that is closely related to the preceding one is introduced by the remark "continued"; if it is related to some other problem the relevant number is mentioned, e.g. "continuation of **136**".

We denote parts by roman numerals, chapters (where necessary) by arabic numerals. The problems are numbered in bold-face. Within the same part only the number of the problem is given; if, however, we refer to another part its number is also indicated. For example if we refer to problem (or solution) **123** of Part IV in a problem (or solution) of Part IV we write "**123**"; if we refer to it in a problem (or solution) of any other part we write "IV **123**".

Remarks in square brackets [ ] in a problem are hints, while in a solution (particularly at the beginning of the solution) they are citations or references to other problems that are used in various steps of the proof. All other remarks appear in ordinary parentheses. A reference to a problem number indicates in general that one should consult both problem and solution, unless the opposite is explicitly stated, e.g. "solution **75**".

Almost always references to the sources are given only in the solution. If a problem has already appeared in print, this fact is indicated in the citations. If the author but no bibliography is mentioned, the problem has been communicated to us as a new problem. Problems whose number is preceded by the symbol * (as in ***206** of Part IV) or contains a decimal point (as in **174.1** of Part IV) are new, that is they are either not contained in the original German edition, or else are contained there but are essentially modified in the present English version. If the problem is the same as in the original edition but the solution has some essentially new feature, the symbol * is used only in the solution. The abbreviations of the names of journals are taken from the index of Mathematical Reviews and, if not listed there, from the World List of Scientific Periodicals Published 1900–1960, Peter Brown, British Museum, Washington, Butterworths, 1963.

The most frequently quoted journals are:

Abh. Akad. Wiss. St. Petersburg	= Akademie der Wissenschaften, St. Petersburg
Acta Math.	= Acta Mathematica, Stockholm
Acta Soc. Sc. Fennicae	= Acta Societatis Scientiarum Fennicae
Amer. Math. Monthly	= American Mathematician Monthly
Arch. Math. Phys.	= Archiv der Mathematik und Physik
Atti Acad. Naz. Lincei Rend. Cl. Sci. Fis. Mat. Natur.	= Atti dell' Accademia Nazionale dei Lincei Rendiconti. Classe di Scienze Fisiche, Matematiche e Naturali, Roma
Berlin. Ber.	= Berliner Berichte
C.R. Acad. Sci. (Paris) Ser. A–B	= Comptes rendus hebdomadaires des séances de l'Académie des Sciences, Paris, Séries A et B
Giorn. Mat. Battaglini	= Giornale di Matematiche di Battaglini
Jber. deutsch. Math. Verein.	= Jahresbericht der deutschen Mathematiker-Vereinigung
J. Math. spéc.	= Journal de Mathématiques Spécieles, Paris
J. reine angew. Math.	= Journal für die reine und angewandte Mathematik
Math. Ann.	= Mathematische Annalen
Math. és term ért	= Matematikai és természettudományi értesitö
Math. Z.	= Mathematische Zeitschrift
Münchner Ber.	= Münchner Berichte

Nachr. Akad. Wiss. Göttingen = Nachrichten der Gesellschaft der Wissenschaften Göttingen
Nouv. Annls Math. = Nouvelles Annales de mathématiques
Nyt. Tidsskr. = Nyt tidsskrift for matematik
Proc. Amer. Math. Soc. = Proceedings of the American Mathematical Society
Proc. Lond. Math. Soc. = Proceedings of the London Mathematical Society
Trans. Amer. Math. Soc. = Transactions of the American Mathematical Society

The following textbooks are quoted repeatedly and are usually cited by the name of the author only or by a suitable abbreviation (e.g. Hurwitz-Courant; MPR.):

G.H. Hardy and E.M. Wright: An Introduction to the Theory of Numbers, 4th Ed. Oxford: Oxford University Press 1960.

E. Hecke: Vorlesungen über die Theorie der algebraischen Zahlen, Leipzig: Akademische Verlagsgesellschaft 1923.

E. Hille: Analytic Function Theory, Vol. I: Boston – New York – Chicago – Atlanta – Dallas – Palo Alto – Toronto – London: Ginn & Co. 1959; Vol. II: Waltham/Mass. – Toronto – London: Blaisdell Publishing Co. 1962.

A. Hurwitz – R. Courant: Vorlesungen über allgemeine Funktionentheorie und elliptische Funktionen, 4th Ed. Berlin – Göttingen – Heidelberg – New York: Springer 1964.

K. Knopp: Theory and Applications of Infinite Series, 2nd Ed. London – Glasgow: Blackie & Son 1964.

G. Kowalewski: Einführung in die Determinantentheorie, 4th Ed. Berlin: Walter de Gruyter 1954.

G. Pólya: How to Solve It, 2nd Ed. Princeton: Princeton University Press 1971. Quoted as HSI.

G. Pólya: Mathematics and Plausible Reasoning, Vols. 1 and 2, 2nd Ed. Princeton: Princeton University Press 1968. Quoted as MPR.

G. Pólya: Mathematical Discovery, Vols. 1 and 2, Cor. Ed. New York: John Wiley & Sons 1968. Quoted as MD.

G. Szegö: Orthogonal Polynomials, American Mathematical Society Colloquium Publications Vol. XXIII, 3rd Ed. New York: American Mathematical Society 1967.

E.C. Titchmarsh: The Theory of Functions, 2nd Ed. Oxford – London – Glasgow – New York – Melbourne – Toronto: Oxford University Press 1939.

E.T. Whittaker and G.N. Watson: A Course of Modern Analysis, 4th Ed. London: Cambridge University Press 1952.

The following notation and abbreviations are used throughout the book:

$a_n \to a$ means "a_n tends to a as $n \to \infty$."

$a_n \sim b_n$ (read "a_n is asymptotically equal to b_n") means "$b_n \neq 0$ for sufficiently large n and $\frac{a_n}{b_n} \to 1$ as $n \to \infty$".

$O(a_n)$, with $a_n > 0$, denotes a quantity that, divided by a_n, remains bounded, $o(a_n)$ a quantity that, divided by a_n, tends to 0 as $n \to \infty$.

Such notation is used analogously in limit processes other than for $n \to \infty$.

$x \to a+0$ means "x converges to a from the right";

$x \to a-0$ means "x converges to a from the left".

$\exp(x) = e^x$, where e is the base of natural logarithms.

Given n real numbers $a_1, a_2, \ldots, a_n$, $\max(a_1, a_2, \ldots, a_n)$ denotes the largest (or one of the largest) and $\min(a_1, a_2, \ldots, a_n)$ the smallest (or one of the smallest) of the numbers $a_1, a_2, \ldots, a_n$. Max $f(x)$ and min $f(x)$ have an analogous meaning for a real function defined on an interval a, b, provided $f(x)$ assumes a maximum or a minimum on a, b. Otherwise we retain the same notation for the least upper bound and the greatest lower bound, respectively. Analogous notation is used in the case of functions of a complex variable.

sgn x denotes the signum (Kronecker) function:

$$\operatorname{sgn} x = \begin{cases} +1 & \text{for} \quad x > 0 \\ 0 & \text{for} \quad x = 0 \\ -1 & \text{for} \quad x < 0. \end{cases}$$

$[x]$ denotes the greatest integer that is not greater than x $(x-1<[x]\leqq x)$. Square brackets are, however, also used instead of ordinary parentheses where there is no danger of confusion. (They are also used in a very special sense restricted to Part I, Chap. 1, § 5.)

$\bar{z}$ is the conjugate to the complex number z.

For the determinant with general term $a_{\lambda,\mu}$, λ, $\mu=1, 2, \ldots, n$, we use the abbreviated notation

$$|a_{\lambda\mu}|_1^n \quad \text{or} \quad |a_{\lambda\mu}|_{\lambda,\,\mu=1,2,\ldots,n} \quad \text{or} \quad |a_{\lambda 1}, a_{\lambda 2}, \ldots, a_{\lambda n}|_1^n.$$

A non-empty connected open set (containing only interior points) is called a *region*. The closure of a region (the union of the open set and of its boundary) is called a *domain*. As this terminology is not the one most frequently used, we shall sometimes emphasize it by speaking of an "open region" and a "closed domain".

A *continuous curve* is defined as a single-valued continuous image of the interval $0\leqq t\leqq 1$, i.e. as the set of points $z=x+iy$, where $x=\varphi(t)$, $y=\psi(t)$, with $\varphi(t)$ and $\psi(t)$ both continuous functions on the interval $0\leqq t\leqq 1$. The curve is *closed* if $\varphi(0)=\varphi(1)$, $\psi(0)=\psi(1)$, and is *without double points* if $\varphi(t_1)=\varphi(t_2)$, $\psi(t_1)=\psi(t_2)$, $t_1<t_2$, imply $t_1=0$, $t_2=1$. A curve without double points is also called a *simple curve*. A simple, continuous curve that is not closed is often referred to as a *simple arc*.

A simple closed continuous curve (a *Jordan* curve) in the plane determines two regions of which it forms the common boundary.

The paths of integration of line integrals or complex integrals are assumed to be continuous and rectifiable.

(a, b) denotes the open interval $a<x<b$, $[a, b)$ the half-open interval $a\leqq x<b$, $(a, b]$ the half-open interval $a<x\leqq b$, $[a, b]$ the closed interval $a\leqq x\leqq b$. When we do not need to distinguish between these four cases we use the term "interval a, b".

"Iff" is used occasionally as an abbreviation for "if and only if".

Problems

Part Four. Functions of One Complex Variable Special Part

Chapter 1. Maximum Term and Central Index, Maximum Modulus and Number of Zeros

§ 1. Analogy between $\mu(r)$ and $M(r)$, $\nu(r)$ and $N(r)$

Let $a_0, a_1, a_2, \ldots, a_n, \ldots$ be complex numbers not all zero. Let the power series

$$f(z)=a_0+a_1 z+a_2 z^2+\cdots+a_n z^n+\cdots$$

have radius of convergence R, $R>0$. If $R=\infty$, $f(z)$ is called an entire function. Let $0 \leqq r<R$. Then the sequence

$$|a_0|, \quad |a_1|r, \quad |a_2|r^2, \quad \ldots, \quad |a_n|r^n, \quad \ldots$$

tends to 0, and hence it contains a largest term, the *maximum term*, whose value is denoted by $\mu(r)$. Thus

$$|a_n|r^n \leqq \mu(r)$$

for $n=0, 1, 2, 3, \ldots, r \geqq 0$ [I, Ch. 3, § 3].

The *central index* $\nu(r)$ is the value of n for which the maximum of $|a_n|r^n$ is attained, so that $\mu(r)=|a_{\nu(r)}|r^{\nu(r)}$. If several of the numbers $|a_n|r^n$ are equal to $\mu(r)$, then $\nu(r)$ denotes the largest of the corresponding n.

The above applies for $r>0$; for $r=0$ cf. **15**.

Denote by $M(r)$ the *maximum modulus* of the function $f(z)$ on the circle $|z|=r$. Then $M(r)$ is also the maximum of $|f(z)|$ in the circular disc $|z| \leqq r$ [III **266**]. We have

$$|a_n|r^n \leqq M(r)$$

for $n=0, 1, 2, \ldots, r>0$. We have equality if and only if all terms of the sequence $a_0, a_1, a_2, \ldots$ with the exception of a_n vanish [III **122**].

The *number of zeros* $N(r)$ is the number of zeros of $f(z)$ in the closed circular disc $|z| \leqq r$ counted with the proper multiplicity.

The above notation applies to the whole chapter.

1. Calculate $\mu(r)$ and $\nu(r)$ for the power series

$$1+\frac{z}{1!}+\frac{z^2}{2!}+\cdots+\frac{z^n}{n!}+\cdots.$$

2. Calculate $M(r)$ and $N(r)$ for

$$e^z=1+\frac{z}{1!}+\frac{z^2}{2!}+\cdots+\frac{z^n}{n!}+\cdots.$$

3. Calculate $\mu(r)$ and $\nu(r)$ for

$$\frac{1}{1!}+\frac{z}{3!}+\frac{z^2}{5!}+\cdots+\frac{z^n}{(2n+1)!}+\cdots.$$

4. Calculate $M(r)$ and $N(r)$ for

$$\frac{\sin\sqrt{z}}{\sqrt{z}}=\frac{1}{1!}-\frac{z}{3!}+\frac{z^2}{5!}-\cdots+\frac{(-z)^n}{(2n+1)!}+\cdots.$$

5. Calculate $\nu(r)$ for the geometric series

$$1+z+z^2+\cdots+z^n+\cdots.$$

6. Calculate $N(r)$ for

$$1+z+z^2+\cdots+z^n+\cdots.$$

7. In the case of a polynomial of degree n

$$a_0+a_1z+a_2z^2+\cdots+a_nz^n, \qquad a_n\neq 0,$$

we have

$$\lim_{r\to\infty}\frac{\log\mu(r)}{\log r}=\lim_{r\to\infty}\nu(r)=n.$$

8. In the case of a polynomial of degree n

$$f(z)=a_0+a_1z+a_2z^2+\cdots+a_nz^n, \qquad a_n\neq 0,$$

we have

$$\lim_{r\to\infty}\frac{\log M(r)}{\log r}=\lim_{r\to\infty}N(r)=n.$$

9. In the case of a transcendental entire function the two limits in **7** are equal to ∞.

10. In the case of a transcendental entire function the first limit in **8** is equal to ∞, the second may be finite or ∞.

11. Denote by $\mu_k(r)$ and $\nu_k(r)$ the maximum term and the central index of the series

$$a_0+a_1z^k+a_2z^{2k}+\cdots+a_nz^{nk}+\cdots, \qquad k=1, 2, 3, \ldots$$

respectively. Express $\mu_k(r)$, $\nu_k(r)$ in terms of $\mu_1(r)$, $\nu_1(r)$.

12. Denote by $M_k(r)$ and $N_k(r)$ the maximum modulus and the number of zeros of $f(z^k)$ in the circle $|z|\leqq r$, $k=1, 2, 3, \ldots$. Express $M_k(r)$, $N_k(r)$ in terms of $M_1(r)$, $N_1(r)$.

13. Calculate the limit $\lim_{r\to\infty}\frac{\nu(r)}{\log\mu(r)}$ for the two power series

$$1+\frac{z^2}{2!}+\frac{z^4}{4!}+\frac{z^6}{6!}+\cdots+\frac{z^{2n}}{(2n)!}+\cdots=\frac{e^z+e^{-z}}{2},$$

$$1+\frac{2z^2}{2!}+\frac{2^3z^4}{4!}+\frac{2^5z^6}{6!}+\cdots+\frac{2^{2n-1}z^{2n}}{(2n)!}+\cdots=\left(\frac{e^z+e^{-z}}{2}\right)^2=\frac{1}{2}+\frac{e^{2z}+e^{-2z}}{4}.$$

14. In contrast to the notation of **12**, we now denote by $M_k(r)$ and $N_k(r)$ the maximum modulus and the number of zeros of $(f(z))^k$ in the circle $|z| \leqq r$ $k=1, 2, 3, \ldots$ respectively. The quotient $\dfrac{N_k(r)}{\log M_k(r)}$ is independent of k.

15. If $a_0 = a_1 = \cdots = a_{q-1} = 0$, $a_q \neq 0$, let $\nu(0) = q$. The function $\nu(r)$ is piecewise constant. It increases at its jump points by a positive integer and is *everywhere* continuous from the right [I **120**].

16. If $a_0 = a_1 = \cdots = a_{q-1} = 0$, $a_q \neq 0$, then $N(0) = q$. The function $N(r)$ is piecewise constant. It increases at its jump points by a positive integer and is *everywhere* continuous from the right.

17. If $a_0 = 0$, $0 < r_1 < r_2 < R$ then

$$\frac{\mu(r_2)}{\mu(r_1)} \geqq \frac{r_2}{r_1}.$$

18. If $f(0) = 0$, $0 < r_1 < r_2 < R$ then

$$\frac{M(r_2)}{M(r_1)} \geqq \frac{r_2}{r_1}.$$

19. The function $\eta = \log \mu\,(e^\xi)$ is represented in a rectangular coordinate system ξ, η by a curve which is non-decreasing and convex. [Consider the aggregate of the straight lines

$$\eta = \log |a_0|, \qquad \eta = \xi + \log |a_1|, \qquad \eta = 2\xi + \log |a_2|, \ldots \eta = n\xi + \log |a_n|, \ldots,$$

omitting meaningless terms with $a_n = 0$. How can $\mu(r)$ and $\nu(r)$ be interpreted in terms of this configuration?]

20. The function $\eta = \log M(e^\xi)$ is represented in a rectangular coordinate system ξ, η by a curve which is always increasing and is strictly convex, apart from certain exceptional polynomials for which the curve degenerates into a straight line.

21. Let α be fixed, $0 < \alpha < 1$. The quotient $\mu(\alpha r)/\mu(r)$ is non-increasing with increasing r.

22. Let α be fixed, $0 < \alpha < 1$. The quotient $M(\alpha r)/M(r)$ is strictly decreasing with increasing r, apart from certain polynomials [**20**].

23. Let $R = \infty$. For fixed α, $0 < \alpha < 1$,

$$\lim_{r \to \infty} \frac{\mu(\alpha r)}{\mu(r)} = 0,$$

if the power series does not terminate, $= \alpha^n$ if the power series reduces to a polynomial of degree n.

24. Let $f(z)$ be an entire function. For fixed α, $0 < \alpha < 1$,

$$\lim_{r \to \infty} \frac{M(\alpha r)}{M(r)} = 0,$$

if $f(z)$ is transcendental, $= \alpha^n$ if $f(z)$ is rational of degree n.

We denote the jump points of the central index $\nu(r)$, ordered according to magnitude, by

$$\rho_1, \rho_2, \rho_3, \ldots, \rho_n, \ldots; \quad \rho_1 = \rho_2 = \cdots = \rho_q = 0, \quad \rho_{q+1} > 0, \quad q \geqq 0,$$

where a jump point is represented m times if the magnitude of the jump is m; at the point $r=0$ we regard $\nu(0)$ as the jump [**15**]. This sequence may terminate after finitely many terms.

We denote the zeros of $f(z)$, ordered according to increasing absolute values, and for equal absolute values according to increasing arguments, by

$$w_1,\ w_2,\ w_3, \ldots,\ w_n, \ldots;\quad w_1=w_2=\cdots=w_q=0,\quad w_{q+1}\neq 0,\quad q\geqq 0,$$

where a zero of multiplicity m is represented m times.

Let $|w_n|=r_n$, $n=1, 2, 3, \ldots$, where accordingly

$$r_1\leqq r_2\leqq r_3\leqq\cdots\leqq r_n\leqq\cdots.$$

This sequence may terminate after finitely many terms.

This notation applies to the whole chapter.

25. If $\rho_n<\rho_{n+1}$, then in the half-closed interval $[\rho_n, \rho_{n+1})$

$$\nu(r)=n.$$

26. If $r_n<r_{n+1}$, then in the half-closed interval $[r_n, r_{n+1})$

$$N(r)=n.$$

27. Calculate the numbers ρ_n for the power series

$$1+\frac{z}{1!}+\frac{z^2}{2!}+\cdots+\frac{z^n}{n!}+\cdots,\quad 1+\frac{z}{3!}+\frac{z^2}{5!}+\cdots+\frac{z^n}{(2n+1)!}+\cdots,$$

$$1+\frac{z^2}{2!}+\frac{z^4}{4!}+\cdots+\frac{z^{2n}}{(2n)!}+\cdots.$$

28. Calculate the numbers r_n for the functions

$$e^z+i,\quad \frac{\sin\sqrt{z}}{\sqrt{z}},\quad \cos z.$$

29. If there exist infinitely many ρ_n then

$$\lim_{n\to\infty}\rho_n=R.$$

30. If there exist infinitely many r_n then

$$\lim_{n\to\infty}r_n=R.$$

31. Let $a_0\neq 0$. Then

$$\mu(r)=\frac{|a_0|r^n}{\rho_1\rho_2\ \cdots\ \rho_n},$$

if $\rho_n\leqq r\leqq\rho_{n+1}$.

32. Let $a_0\neq 0$. Then

$$M(r)\geqq\frac{|f(0)|r^n}{r_1r_2\ \cdots\ r_n}.$$

[III **120**.] For which entire functions can the equality sign hold in this inequality?

33. Assuming that $a_0\neq 0$, we have

$$\log\mu(r)-\log|a_0|=\int_0^r\frac{\nu(t)}{t}\,dt.$$

34. Assuming that $a_0 \neq 0$, we have

$$\log M(r) - \log |f(0)| \geqq \int_0^r \frac{N(t)}{t}\, dt.$$

35. For an arbitrary entire function

$$\limsup_{r\to\infty} \frac{\log \nu(r)}{\log r} = \limsup_{r\to\infty} \frac{\log\log \mu(r)}{\log r}.$$

36. For an arbitrary entire function

$$\limsup_{r\to\infty} \frac{\log N(r)}{\log r} \leqq \limsup_{r\to\infty} \frac{\log\log M(r)}{\log r}.$$

37. Consider a power series which is convergent everywhere and let $k>0$. The infinite series $\sum_{n=q+1}^{\infty} \rho_n^{-k}$ and the infinite integral $\int_1^\infty r^{-k-1} \log \mu(r)\, dr$ are either both convergent or both divergent.

38. Consider an entire function and let $k>0$. If the infinite integral $\int_1^\infty r^{-k-1} \log M(r)\, dr$ is convergent, then the infinite series $\sum_{n=q+1}^{\infty} r_n^{-k}$ is also convergent (not vice versa!).

39. Consider a power series which is convergent in the interior of the unit circle, and let $k>0$. The series $\sum_{n=1}^{\infty} (1-\rho_n)^{k+1}$ and the integral $\int_0^1 (1-t)^{k-1} \log \mu(t) dt$ are either both convergent or both divergent.

40. Consider a function which is regular in the interior of the unit circle, and let $k>0$. If the integral $\int_0^1 (1-t)^{k-1} \log M(t)\, dt$ is convergent then the series $\sum_{n=1}^{\infty} (1-r_n)^{k+1}$ is also convergent.

§ 2. Further Results on $\mu(r)$ and $\nu(r)$

41. In a rectangular coordinate system mark the points

$$(0, -\log |a_0|), \quad (1, -\log |a_1|), \quad (2, -\log |a_2|), \ldots, \quad (n, -\log |a_n|), \ldots$$

(omitting meaningless ones with $a_n = 0$) and from each point draw a ray directed vertically upwards. The smallest convex region $\mathfrak{R}$ which includes all these rays extends to infinity. Draw the supporting line with direction coefficient $\log r$ (the straight line on which lies at least one boundary point and no interior point of $\mathfrak{R}$ and which forms an angle with the positive real axis whose tangent is $\log r$). Interpret in the resulting figure $\log \mu(r)$, $\nu(r)$, $\log \rho_n$.

42. If m is a positive integral value assumed by $\nu(r)$, $m > \nu(0)$, then

$$\rho_m = \max\left(\left|\frac{a_0}{a_m}\right|^{\frac{1}{m}}, \left|\frac{a_1}{a_m}\right|^{\frac{1}{m-1}}, \ldots, \left|\frac{a_{m-1}}{a_m}\right| \right).$$

43. Let the power series

$$a_0 + a_1 z + a_2 z^2 + \cdots + a_n z^n + \cdots$$

be such that each term in turn becomes the maximum term, i.e. to every index

$n=0, 1, 2, \ldots$ there belongs at least one real number r, $r>0$, such that $|a_n|r^n$ is not exceeded by any term. For this it is necessary and sufficient that

$$0<\left|\frac{a_0}{a_1}\right|\leqq\left|\frac{a_1}{a_2}\right|\leqq\left|\frac{a_2}{a_3}\right|\leqq\cdots\leqq\left|\frac{a_n}{a_{n+1}}\right|\leqq\cdots.$$

[**31**, I **117**.]

44. Let $0<\alpha<1$. The central index $\nu(r)$ of the power series

$$1+e^{\alpha^{-1}1^\alpha}z+e^{\alpha^{-1}2^\alpha}z^2+\cdots+e^{\alpha^{-1}n^\alpha}z^n+\cdots$$

which is convergent for $|z|<1$ assumes in turn all the values $0, 1, 2, 3 \ldots.$ Its maximum term satisfies

$$\mu(r)\sim\exp\left(\frac{1-\alpha}{\alpha}\left(-\frac{1}{\log r}\right)^{\frac{\alpha}{1-\alpha}}\right),$$

as r tends to 1.

45 (continued). As r tends to 1,

$$\sum_{n=0}^{\infty} e^{\alpha^{-1}n^\alpha}r^n\sim\frac{\sqrt{2\pi\alpha}}{1-\alpha}\left(\frac{\alpha}{1-\alpha}\right)^{\frac{1-\alpha}{\alpha}}\left(\log\mu(r)\right)^{\frac{1}{2}+\frac{1-\alpha}{\alpha}}\mu(r).$$

[Consider $\int_0^\infty e^{\alpha^{-1}x^\alpha}r^x\,dx$; II **208**.]

46. Let $\alpha>0$. Calculate $\rho_1, \rho_2, \rho_3, \ldots$ for the power series

$$1+1^{-\alpha}z+2^{-2\alpha}z^2+\cdots+n^{-n\alpha}z^n+\cdots$$

which is convergent for all finite z, and show that its maximum term satisfies

$$\mu(r)\sim\exp\left(\alpha\, e^{-1}r^{\frac{1}{\alpha}}\right)$$

as r tends to $+\infty$.

47 (continued). For fixed α, $\alpha>0$, as r tends to $+\infty$

$$1+1^{-\alpha}r+2^{-2\alpha}r^2+\cdots+n^{-n\alpha}r^n+\cdots\sim\frac{\sqrt{2\pi}}{\alpha}[\log\mu(r)]^{\frac{1}{2}}\mu(r).$$

[II **209**.]

§ 3. Connection between $\mu(r)$, $\nu(r)$, $M(r)$ and $N(r)$

48. For an entire function $f(z)$ let

$$\limsup_{r\to\infty}\frac{\log M(r)}{r}=l.$$

The convergence of the series

$$f(z)+f'(z)+f''(z)+\cdots=\sum_{n=0}^{\infty}\frac{d^nf(z)}{dz^n}$$

is assured, excluded or undecided according as

$$l<1,\quad l>1\quad\text{or}\quad l=1.$$

49. For the Weierstrass $\sigma(z)$ function, in the case when the period parallelogram is a square with sides of length 1,

$$N(r) \sim \pi r^2, \qquad \log M(r) \sim \frac{\pi}{2} r^2,$$

as r tends to ∞. [Hille, Vol. II pp. 132–133 Hurwitz-Courant, pp. 174–175.]

49.1. There is an entire function for which

$$\rho_n = r_n - 1$$

for $n = 1, 2, 3, \ldots$ [I **50**]. Construct an entire function for which

$$\rho_n = r_n + (-1)^{n-1}$$

for $n = 1, 2, 3, \ldots$.

50. For the entire functions

$$\frac{\sin \sqrt{z}}{\sqrt{z}}, \quad \cos z, \quad \cos^2 z, \quad e^z + i, \quad e^z, \quad \sigma(z)$$

determine the limits of the quotients

$$\frac{r_n}{\rho_n}, \quad \frac{N(r)}{\nu(r)}, \quad \frac{N(r)}{\log M(r)}, \quad \frac{\nu(r)}{\log \mu(r)}, \quad \frac{M(r)}{\mu(r)\sqrt{\log \mu(r)}}, \quad \frac{\log N(r)}{\log r}$$

as r (or n in the first quotient) tends to infinity, and collect the results in a table. [For e^z the first and last quotients are meaningless, for $\sigma(z)$ only the third and the last are to be determined under the assumption that the period parallelogram is a square. The last quotient gives the exponent of convergence of the zeros, II **149**. Use **1–4**, **13**, **27**, **28** and Stirling's formula for the asymptotic value of $\mu(r)$.]

51. For any entire function

$$\limsup_{r \to \infty} \frac{\log \log \mu(r)}{\log r} = \limsup_{r \to \infty} \frac{\log \log M(r)}{\log r}$$

(it is obvious that the inequality $\leqq$ holds). The common value of these two limits is called the *order* of the entire function.

52. The order of an entire function may also be defined as the exponent of convergence [I, Ch. 3, § 2] of the sequence $\rho_1, \rho_2, \ldots, \rho_n, \ldots$.

53. The order of an entire function $f(z) = \sum_{n=0}^{\infty} a_n z^n$ with positive coefficients $a_0, a_1, a_2, \ldots, a_n, \ldots$ may also be defined as the following limit:

$$\limsup_{r \to \infty} \frac{\log [r f'(r) f(r)^{-1}]}{\log r}.$$

54. For entire functions of finite order we have the following result (stronger than **51**):

$$\lim_{r \to \infty} \frac{\log M(r)}{\log \mu(r)} = 1.$$

55. For transcendental entire functions the integral

$$\int_{t_0}^{\infty} \mu(t)[M(t)]^{-l}\,dt, \qquad t_0>0.$$

is convergent for $l>1$.

56. For any entire function, given any arbitrary ε, $\varepsilon>0$, there exist arbitrarily large values of r such that

$$M(r)<\mu(r)\,[\log \mu(r)]^{\frac{1}{2}+\varepsilon}. \qquad [\mathbf{45}, \text{I } \mathbf{122}.]$$

57. For an entire function of finite order λ we have the following result (stronger than **56**): Given any arbitrary ε, $\varepsilon>0$, there exist arbitrarily large values of r such that

$$M(r)<(\lambda+\varepsilon)\sqrt{2\pi \log \mu(r)}\mu(r). \qquad [\mathbf{47}, \text{I } \mathbf{118}.]$$

Let c be a constant, $c\neq 0$, q an integer, $q\geqq 0$, and $w_1, w_2, w_3, \ldots$ an infinite sequence of complex numbers

$$w_1=w_2=\cdots=w_q=0<|w_{q+1}|\leqq|w_{q+2}|\leqq|w_{q+3}|\leqq\cdots,$$

such that

$$\frac{1}{|w_{q+1}|}+\frac{1}{|w_{q+2}|}+\frac{1}{|w_{q+3}|}+\cdots$$

is convergent. An entire function of the form

$$cz^q\left(1-\frac{z}{w_{q+1}}\right)\left(1-\frac{z}{w_{q+2}}\right)\cdots$$

is called a function of *genus zero*, [Titchmarsh p. 253, Hille Vol. II, p. 199]. As Hadamard has shown, every entire function of order <1 is of genus zero. (**36**, **58**, III **332** are important elements of a proof.)

58. The order of an entire function of genus zero is equal to the *exponent of convergence* of its zeros [Vol. I, pp. 25–26, Titchmarsh p. 250, Hille Vol. II, p. 188].

59. For an entire function of finite order λ, $\lambda>0$, whose central index has the jump-points $\rho_1, \rho_2, \rho_3, \ldots, \rho_n, \ldots$ which are regularly spaced in the sense of II, Ch. 4, § 1, we have

$$\lim_{r\to\infty}\frac{\nu(r)}{\log \mu(r)}=\lambda. \qquad [\text{II } \mathbf{159}.]$$

60. For entire functions of finite order λ, $\lambda>0$, without any special conditions of regularity we have

$$\liminf_{r\to\infty}\frac{\nu(r)}{\log\mu(r)}\leqq\lambda\leqq\limsup_{r\to\infty}\frac{\nu(r)}{\log\mu(r)}. \qquad [\text{II } \mathbf{160}.]$$

61. Let λ be the exponent of convergence of the sequence $r_1, r_2, r_3, \ldots, r_n, \ldots$, which is regularly spaced in the sense of II, Ch. 4, § 1, and let $0<\lambda<1$. For the entire function of genus zero

$$f(z)=\left(1+\frac{z}{r_1}\right)\left(1+\frac{z}{r_2}\right)\cdots\left(1+\frac{z}{r_n}\right)\cdots$$

we have

$$\lim_{r\to\infty} \frac{\log f(r\,e^{i\vartheta})}{N(r)} = \frac{e^{i\vartheta\lambda}\pi}{\sin \pi\lambda},$$

if ϑ is a fixed number, $-\pi<\vartheta<\pi$. [II **159**.]

62. For the entire function discussed in **61** we have, even without assuming the regular spacing of the sequence $r_1, r_2, r_3, \ldots, r_n, \ldots$,

$$\limsup_{r\to\infty} \frac{N(r)}{\log M(r)} \geqq \frac{\sin \pi\lambda}{\pi}. \qquad \text{[II 161.]}$$

63. For an entire function whose zeros have a finite exponent of convergence λ

$$\liminf_{r\to\infty} \frac{N(r)}{\log M(r)} \leqq \lambda. \qquad \text{[32.]}$$

64. Let $f(z)$ be an entire function and let $\mathfrak{G}(r)$ and $\mathfrak{g}(r)$ be defined as the geometric means of $|f(z)|$ on the circle $|z|=r$, and in the circular disc $|z|\leqq r$, respectively [III **121**]. If the moduli of the zeros of $f(z)$ are regularly spaced [II, Ch. 4, § 1], then

$$\lim_{r\to\infty} \left(\frac{\mathfrak{g}(r)}{\mathfrak{G}(r)}\right)^{\frac{1}{N(r)}} = e^{-\frac{1}{\lambda+2}},$$

where λ, $0<\lambda<\infty$, denotes the exponent of convergence of the zeros. For polynomials the above limit also exists and equals $e^{-1/2}$.

65 (continued). Without assuming the regular spacing of the zeros we have

$$\liminf_{r\to\infty} \left(\frac{\mathfrak{g}(r)}{\mathfrak{G}(r)}\right)^{\frac{1}{N(r)}} \leqq e^{-\frac{1}{\lambda+2}} \leqq \limsup_{r\to\infty} \left(\frac{\mathfrak{g}(r)}{\mathfrak{G}(r)}\right)^{\frac{1}{N(r)}}.$$

66. Let $f(z)$ be an entire function of finite order λ and $\mathfrak{M}(r)$ and $\mathfrak{m}(r)$ the arithmetic means of $|f(z)|^2$ on the circle $|z|=r$ and in the circular disc $|z|\leqq r$ respectively. Then

$$\liminf_{r\to\infty} \left(\frac{\mathfrak{m}(r)}{\mathfrak{M}(r)}\right)^{\frac{1}{\log r}} = e^{-\lambda}.$$

§ 4. $\mu(r)$ and $M(r)$ under Special Regularity Assumptions

67. Let the entire function

$$f(z) = a_0 + a_1 z + a_2 z^2 + \cdots + a_n z^n + \cdots$$

satisfy the condition

$$\lim_{r\to\infty} r^{-\alpha} \log M(r) = a,$$

where α, a are positive constants. Prove that:

(1) If $b>a$, then for n sufficiently large

$$|a_n| < \left(\frac{\alpha b\, e}{n}\right)^{\frac{n}{\alpha}}.$$

(2) Let k and ε be fixed numbers, k real, $0<\varepsilon<1$. Then there exists a number δ, $\delta>0$, such that for r sufficiently large

$$\left(\sum_{1}^{\alpha a r^{\alpha}(1-\varepsilon)} + \sum_{\alpha a r^{\alpha}(1+\varepsilon)}^{\infty}\right) n^k |a_n| r^n < M(r)\, e^{-\delta r^{\alpha}}.$$

If the summation were extended from 0 to ∞ without omission of the "center" and $k=0$ the sum would be $\geqq M(r)$. Thus the "center" of the series considerably outweighs the two "tails". [(1) serves as a preparation for (2).]

68. If for an entire function any one of the three limiting relations

$$(1)\ \log M(r) \sim a r^{\alpha}, \qquad (2)\ \log \mu(r) \sim a r^{\alpha}, \qquad (3)\ \nu(r) \sim \alpha a r^{\alpha}$$

holds for $r \to \infty$, where a, α are positive constants, then the other two also hold. [Formally, (3) follows from (2) by differentiating both sides of (2), cf. **33**.] Without any conditions of regularity **51**, **54**, **35** and **60** hold.

69. If the coefficients $a_0, a_1, a_2, \ldots$ of the entire function $f(z)$ considered in **67** are positive and satisfy the inequalities

$$a_0 \geqq a_1 \geqq a_2 \geqq \cdots \geqq a_n \geqq \cdots,$$

then we can assert that

$$\log a_n \sim -\frac{n \log n}{\alpha}.$$

If they satisfy the inequalities

$$\frac{a_1}{a_0} \geqq \frac{a_2}{a_1} \geqq \frac{a_3}{a_2} \geqq \cdots \geqq \frac{a_{n+1}}{a_n} \geqq \cdots,$$

then we can deduce the stronger result

$$\sqrt[n]{a_n} \sim \left(\frac{\alpha a\, e}{n}\right)^{\frac{1}{\alpha}}.$$

70. Add to the assumptions of **67** that $a_n \geqq 0$ for $n=0, 1, 2, \ldots$. Then for fixed real k

$$\lim_{r\to\infty} \frac{\sum_{n=1}^{\infty} n^k a_n r^n}{(\alpha a r^{\alpha})^k f(r)} = 1.$$

71. Under the assumptions of **67**, **70** we can differentiate the limiting relation

$$\log f(r) \sim a r^{\alpha},$$

i.e. we can deduce from it

$$\frac{f'(r)}{f(r)} \sim \alpha a r^{\alpha-1}.$$

72. (Generalization of I **94**.) Let

$$b_0 + b_1 z + b_2 z^2 + \cdots + b_n z^n + \cdots = f(z)$$

be an everywhere convergent power series. From the three condition.

$$b_n>0, \quad n=0,1,2,\ldots; \qquad \lim_{r\to+\infty} r^{-\beta}\log f(r)=b; \qquad \lim_{n\to\infty}\frac{a_n}{n^k b}=s$$

(β, b positive, k and s arbitrary real constants) it follows that

$$\lim_{r\to+\infty}\frac{a_0+a_1r+a_2r^2+\cdots+a_nr^n+\cdots}{(\beta b r^\beta)^k f(r)}=s.$$

73. Show that for the *Bessel* function of order 0, $J_0(z)$, we have as $r\to+\infty$

$$J_0(ir)=\sum_{n=0}^{\infty}\frac{1}{n!\,n!}\left(\frac{r}{2}\right)^{2n}\sim\sqrt{\frac{2}{\pi r}}\sum_{n=0}^{\infty}\frac{r^{2n}}{(2n)!}\sim\frac{e^r}{\sqrt{2\pi r}}. \qquad \text{[II 204.]}$$

74. Let $a_1, a_2, \ldots, a_p, b_1, b_2, \ldots, b_q$ be real constants different from $0, -1, -2, -3, \ldots$, where $p<q$,

$$P(z)=(z+a_1-1)(z+a_2-1)\ \ldots\ (z+a_p-1),$$
$$Q(z)=(z+b_1-1)(z+b_2-1)\ \ldots\ (z+b_q-1).$$

As r tends to infinity ($r>0$) the everywhere convergent power series

$$1+\frac{P(1)}{Q(1)}r+\frac{P(1)P(2)}{Q(1)Q(2)}r^2+\cdots+\frac{P(1)P(2)\ \ldots\ P(n)}{Q(1)Q(2)\ \ldots\ Q(n)}r^n+\cdots$$

$$\sim\frac{\Gamma(b_1)\Gamma(b_2)\ \ldots\ \Gamma(b_q)}{\Gamma(a_1)\Gamma(a_2)\ \ldots\ \Gamma(a_p)}(2\pi)^{\frac{1-l}{2}}\,l^{-\frac{1}{2}}\,r^{\frac{2\Delta+l+1}{2l}}\,e^{lr^{\frac{1}{l}}},$$

where we set $l=q-p$, $\Delta=a_1+a_2+\cdots+a_p-b_1-b_2-\cdots-b_q$. [Verify that from this **73** follows as a special case after a change of variables. Use as a "comparison series"

$$1+\frac{l^l}{l!}r+\frac{l^{2l}}{(2l)!}r^2+\cdots+\frac{l^{nl}}{(nl)!}r^n+\cdots\sim\frac{1}{l}e^{lr^{\frac{1}{l}}}.]$$

75. Obtain the asymptotic formula in **74**, without making use of **72**, by means of I **94** and II **207**.

76. Let all the coefficients $a_0, a_1, a_2, \ldots, a_n, \ldots$ of the power series $a_0+a_1z+a_2z^2+\cdots+a_nz^n+\cdots$ be positive and satisfy the condition

$$\lim_{n\to\infty} n\left(\frac{a_n^2}{a_{n-1}a_{n+1}}-1\right)=\frac{1}{\lambda}, \qquad 0<\lambda<\infty.$$

(It is satisfied for example by the power series in **74**.) Prove that

(1) The series represents an entire function of order λ.

(2) As $r\to\infty$, $\nu(r)\sim\lambda\log\mu(r)$.

(3) In a rectangular coordinate system the aggregate of points

$$\left(\frac{l}{\sqrt{\nu(r)}},\ \frac{a_{\nu(r)+l}\,r^{\nu(r)+l}}{\mu(r)}\right), \qquad l=-\nu(r),\ -\nu(r)+1,\ldots,\ -1, 0, 1, 2, \ldots$$

tends to the *Gaussian error curve*, i.e.

$$\lim_{r\to\infty} \frac{a_{\nu(r)+l}\, r^{\nu(r)+l}}{\mu(r)} = e^{-\frac{x^2}{2\lambda}}, \quad \text{if} \quad \lim_{r\to\infty} \frac{l}{\sqrt{\nu(r)}} = x.$$

Draw the rectangles whose base is of length $\nu(r)^{-1/2}$ and lies on the x-axis and whose upper edge is bisected by one of the points described in (3). Compare their total area with the area under the Gaussian error curve.

Chapter 2. Schlicht Mappings

§ 1. Introductory Material

77. Let the polynomial

$$z + a_2 z^2 + a_3 z^3 + \cdots + a_n z^n$$

be schlicht (univalent) in the unit circle $|z| < 1$. Then $n|a_n| \leqq 1$.

78. Let the function $w = f(z)$ effect a schlicht mapping of the unit circle $|z| < 1$ onto a *region* (non-empty connected open set) $\mathfrak{R}$ and let $\varphi(w)$ be schlicht in $\mathfrak{R}$. Then $\varphi[f(z)]$ is schlicht in $|z| < 1$.

79. Let the function $f(z)$ be schlicht in the unit circle $|z| < 1$ and let $f(0) = 0$. Then the function

$$\varphi(z) = \sqrt{f(z^2)} = z\sqrt{\frac{f(z^2)}{z^2}},$$

with a fixed branch of the square root, is also schlicht in $|z| < 1$. A similar result holds for $\sqrt[n]{f(z^n)}$, where n is a positive integer.

80. The function $\varphi(z)$ in **79** is the most general *odd* function that induces a schlicht mapping of the unit circle $|z| < 1$. More precisely: If $\varphi(z)$ is an odd function schlicht for $|z| < 1$, then there exists a function $f(z)$ which is schlicht for $|z| < 1$ from which $\varphi(z)$ is derived as in **79**.

81. Let the open circular disc $\mathfrak{C}$ be mapped one-to-one and conformally onto the region $\mathfrak{R}$. In particular, let the concentric circular disc $\mathfrak{c}$ contained in $\mathfrak{C}$ be mapped onto the sub-region $\mathfrak{r}$ of $\mathfrak{R}$. Denoting the respective areas by $|\mathfrak{C}|$, $|\mathfrak{R}|$, $|\mathfrak{c}|$, $|\mathfrak{r}|$, we have

$$\frac{|\mathfrak{R}|}{|\mathfrak{r}|} \geqq \frac{|\mathfrak{C}|}{|\mathfrak{c}|}.$$

We have equality if and only if the mapping is obtained by means of a linear entire function. [III **124**.]

82. Let the open circular disc $\mathfrak{C}$ be mapped one-to-one and conformally onto the region $\mathfrak{R}$. In particular let the center $\mathfrak{c}$ of $\mathfrak{C}$ be mapped into the point $\mathfrak{r}$ of $\mathfrak{R}$. Denote by a^2 the area enlargement ratio at $\mathfrak{c}$ [Vol. I, p. 117]. Then we have (using the notation of **81**)

$$\frac{|\mathfrak{R}|}{|\mathfrak{C}|} \geqq a^2.$$

We have equality if and only if the mapping is effected by means of a linear entire function (limiting case of **81**).

83. Let the open annular region $\mathfrak{A}$ between two concentric circles be mapped one-to-one and conformally onto the doubly connected region $\mathfrak{S}$. Let the smallest open circular disc which contains $\mathfrak{A}$ be $\mathfrak{C}$. Those points of $\mathfrak{C}$ not belonging to $\mathfrak{A}$ form a closed circular disc $\mathfrak{c}$. Similarly let $\mathfrak{R}$ be the smallest simply connected region containing $\mathfrak{S}$, and let $\mathfrak{r}$ be the set of all points of $\mathfrak{R}$ which do not belong to $\mathfrak{S}$ ($\mathfrak{r}$ is closed). Finally let a circle concentric with $\mathfrak{c}$ and $\mathfrak{C}$ contained in $\mathfrak{A}$ and its image curve in $\mathfrak{S}$ be traversed in the same sense. Then we have the relation between the areas

$$\frac{|\mathfrak{R}|}{|\mathfrak{r}|} \geq \frac{|\mathfrak{C}|}{|\mathfrak{c}|}.$$

We have equality if and only if the mapping is effected by means of a linear entire function. [Generalization of **81**, not contained in it!]

§ 2. Uniqueness Theorems

84. A one-to-one conformal mapping which maps an open circular disc onto itself, with the center a fixed point, is a rotation. [**82**.]

85. Let two open annular regions each bounded by two concentric circles have the same outer boundary and be related by a one-to-one and conformal mapping. Also let the mapping be such that a circle of one region concentric with the boundary corresponds to a curve of the other region traversed in the same sense. Then the two annular regions are identical and the conformal mapping is a rotation. [**83**.]

86. There do not exist two distinct functions $w=f(z)$ which map the unit circle $|z|<1$ in a schlicht manner onto a given region $\mathfrak{R}$, where the origin $z=0$ is mapped onto a given point $w=w_0$ of $\mathfrak{R}$ and $f'(0)>0$.

87. The only functions which map the unit circle $|z|<1$ in a schlicht manner onto itself are the bilinear functions $e^{i\alpha}\dfrac{z-z_0}{1-\bar{z}_0 z}$, α real, $|z_0|<1$.

§ 3. Existence of the Mapping Function

88. In the mapping induced by the formula

$$\frac{a-z}{1-\bar{a}z} = \left(\frac{\sqrt{a}-\eta w}{1-\overline{\sqrt{a}}\eta w}\right)^2,$$

to a point in the z-plane there correspond in general two points in the w-plane. Which z-points form an exception to this? [$a=|a|\,e^{i\alpha}$, $0<|a|<1$, α real, $|\eta|=1$ and η is determined in such a way that $\dfrac{dz}{dw}>0$ for $w=0$.] The mapping is *independent* of the (fixed) sign of $\sqrt{a}$.

89. Let $\mathfrak{R}$ be a simply-connected region in the z-plane, which contains the point $z=0$ and is contained in the circle $|z|<1$. Let a, $|a|<1$, be the boundary point of $\mathfrak{R}$ closest to the point $z=0$. [If there are several such points, choose say the one of smallest argument, $0 \leqq \arg a < 2\pi$.] By means of the mapping **88** there correspond to the region $\mathfrak{R}$ two regions of the w-plane, of which one $\mathfrak{R}^*$ contains the point $w=0$ and the other $\mathfrak{R}^{**}$ does not. $\mathfrak{R}^*$ and $\mathfrak{R}^{**}$ have no common points; however they have at least one common boundary point and both lie in the interior of the unit circle.

The region $\mathfrak{R}^*$ determined in **89** will be termed in the sequel the *Koebe image region* of $\mathfrak{R}$. It is advantageous to interpret $\mathfrak{R}$ and $\mathfrak{R}^*$ as lying in the same plane. $\mathfrak{R}$ and $\mathfrak{R}^*$ correspond to each other in a one-to-one manner, preserving the origin and the directions at the origin. Both $\mathfrak{R}$ and $\mathfrak{R}^*$ form a proper sub-region of the unit circle.

90. Investigate the Koebe image region of the *slit region* which arises from the unit circle $|z|<1$ by deleting ("cutting along") the line segment $a \leqq z \leqq 1$, $0<a<1$. In particular determine the boundary point which is nearest to the origin.

91. Investigate the Koebe image region of the circle $|z|<a$, $0<a<1$ and determine the boundary points which are nearest to and furthest from the origin.

92. Every point of the Koebe image region of $\mathfrak{R}$ is at a greater distance from the origin than the corresponding point of $\mathfrak{R}$.

93. Let $\mathfrak{R}_1$ be the Koebe image region of $\mathfrak{R}$, $\mathfrak{R}_2$ that of $\mathfrak{R}_1$, $\mathfrak{R}_3$ that of $\mathfrak{R}_2, \ldots$. Denote by $a, a_1, a_2, a_3, \ldots$ the nearest point from the origin of $\mathfrak{R}, \mathfrak{R}_1, \mathfrak{R}_2, \mathfrak{R}_3, \ldots$ respectively. Then we have

$$|a| < |a_1| < |a_2| < |a_3| < \cdots.$$

94 (continued). By the Koebe mapping $\mathfrak{R}$ is mapped one-to-one and conformally onto $\mathfrak{R}_1$, $\mathfrak{R}_1$ onto $\mathfrak{R}_2, \ldots, \mathfrak{R}_{n-1}$ onto $\mathfrak{R}_n$. Let the resulting mapping of $\mathfrak{R}$ onto $\mathfrak{R}_n$ be expressed by the analytic function $f_n(z)$ (z runs through the points of $\mathfrak{R}$). Then $f_n(0)=0$, $0<f_n'(0)<\frac{1}{|a|}$. Express $f_n'(0)$ in terms of $a, a_1, a_2, \ldots, a_{n-1}$.

95. Denoting by $a, a_1, a_2, \ldots, a_n, \ldots$ the points introduced in **93**, prove by means of **94** that

$$\lim_{n\to\infty} |a_n| = 1.$$

96. (Continuation of **93, 94**.) The limit $\lim_{n\to\infty} f_n(z) = f(z)$ exists at every point z of $\mathfrak{R}$ and the convergence is uniform in every interior domain of $\mathfrak{R}$. The limit function $f(z)$ maps $\mathfrak{R}$ schlicht onto the interior of the unit circle. [III **258**.]

§ 4. The Inner and the Outer Radius. The Normed Mapping Function

In the sequel [**97–163**] let $\mathfrak{R}$ be an arbitrary, simply-connected region in the z-plane having more than one boundary point, and let a be an arbitrary finite point of $\mathfrak{R}$. The infinite point may be an interior or a boundary point of $\mathfrak{R}$. We map $\mathfrak{R}$ conformally and schlicht onto the interior of a circle in the w-plane, and in such a manner that the point a goes over into the center of the circle and that the *enlargement ratio* [Vol. I, pp. 116–117] at a is equal to 1. The radius of this image circle is uniquely determined by the region $\mathfrak{R}$ and the *central point* a. We term it the *inner radius* of $\mathfrak{R}$ with respect to a and denote its length by r_a. There exists **[96]** a uniquely determined **[86]** function $w=f(a; z)=f(z)$ which effects this mapping, i.e. which maps $\mathfrak{R}$ schlicht onto the interior of the circle $|w|<r_a$ and possesses in the neighbourhood of a a series expansion of the form

$$(*) \qquad w=f(z)=z-a+c_2(z-a)^2+c_3(z-a)^3+\cdots.$$

We term (*) the *normed mapping function* associated with the *central point* a. We define the inner radius of a closed, multiple-point free, continuous curve L (with respect to a point lying in its interior) as the inner radius of the interior domain of L.

Let now $\mathfrak{R}$ be a simply-connected region of the z-plane, containing the point at infinity $z=\infty$. The complement of $\mathfrak{R}$ is then a closed domain $\mathfrak{D}$ lying wholly in the finite part of the z-plane. We define the outer radius of $\mathfrak{D}$ as follows: We map $\mathfrak{R}$ conformally onto the exterior of a circle in the w-plane and in such a manner that the point $z=\infty$ goes over into $w=\infty$ and the absolute value of the derivative of the mapping function for $z=\infty$ (the enlargement ratio at the point at infinity) is equal to 1. The radius of this circle is uniquely determined by the region $\mathfrak{R}$. We term it the *outer radius* of $\mathfrak{D}$ and denote its length by $\bar{r}$. There exists a uniquely determined function $w=f(z)$ which effects this mapping, i.e. that maps $\mathfrak{R}$ schlicht onto the exterior of the circle $|w|>\bar{r}$ and possesses in the neighbourhood of the point at infinity a series expansion of the form

$$(**) \qquad w=f(z)=z+c_0+\frac{c_1}{z}+\frac{c_2}{z^2}+\cdots+\frac{c_n}{z^n}+\cdots.$$

We term (**) the *normed mapping function* associated with the point at infinity as central point. By the outer radius of a continuous curve L, which may consist in part or even wholly of cuts, we mean the outer radius of the closed interior of L. The region previously denoted by $\mathfrak{R}$ in this case is the exterior of L, or the whole plane cut along L if L is simply the arc of a curve.[1]

By the mapping of the interior (or exterior) of a simple closed continuous curve L onto the interior (exterior) of a circle, the *closed* interior (exterior) of L is made to correspond one-to-one and continuously to the closed interior (exterior) of the circle. (Cf. C. Carathéodory, Math. Ann. **73**, pp. 314–320 (1913).)

97. Calculate the inner radius r_a and the outer radius $\bar{r}$ of the circle $|z|=\rho$, $\rho>0$, $|a|<\rho$.

98. Calculate the inner radius r_a of the exterior of the circle $|z|>\rho$, a finite, $|a|>\rho$ and the limit $\lim_{a\to\infty} r_a$.

99. Calculate r_a for the sector $0<\arg z<\vartheta_0,\ 0<\vartheta_0\leqq 2\pi$.

100. Under a similarity transformation $z'=hz+k$ with an enlargement ratio $|h|$ different from zero the inner and the outer radius are magnified in the ratio

[1] More precisely we mean here a continuous curve L with the following property: That simply connected part of the complement of L which contains the point at infinity (i.e. $\mathfrak{R}$) has a boundary which coincides completely with L. Cf. e.g. **101**, **105**.

$1:|h|$. I.e. if the region $\mathfrak{R}$ goes over into $\mathfrak{R}'$ and the finite image point a into $a'=ha+k$ then between the inner radius $r_{a'}$ of $\mathfrak{R}'$ with respect to a' and the inner radius r_a of $\mathfrak{R}$ with respect to a there exists the following relation:

$$r_{a'}=|h|r_a \quad .$$

There is a similar relation for the outer radius.

101. The outer radius of a line segment of length l is

$$\bar{r}=\frac{l}{4} \quad .$$

102. The outer radius of an ellipse whose axes are of length $2a$ and $2b$, respectively, is

$$\bar{r}=\frac{2a+2b}{4} \quad .$$

103. Calculate the inner radius of an open half-plane with respect to a point lying within it whose distance from the boundary is d.

104. Let r_a be the inner radius of a region $\mathfrak{R}$ of the z-plane, and let the schlicht mapping

$$z'-b=\gamma(z-a)+\gamma_2(z-a)^2+\cdots$$

map $\mathfrak{R}$ onto a region $\mathfrak{R}'$ of the z'-plane. Denoting by r_b' the inner radius of the region $\mathfrak{R}'$ with respect to the image point $z'=b$, we have

$$r_b'=|\gamma|r_a.$$

A similar result holds for the outer radius in the case of schlicht mappings of the form

$$z'=\gamma z+\gamma_0+\frac{\gamma_1}{z}+\frac{\gamma_2}{z^2}+\cdots.$$

105. Calculate the outer radius $\bar{r}$ of the curve which is formed by adding to the circle $|z|=1$ the two real line segments $1<z\leqq a_1$, $-a_2\leqq z<-1$, $a_1>1$, $a_2>1$.

106. Calculate the inner radius of an infinite parallel strip of width D with respect to a point lying within it whose least distance from the boundary is d, $2d\leqq D$.

107. Let a be an arbitrary finite point of the region $\mathfrak{R}$ in the z-plane. The rational linear transformation $z'=\dfrac{1}{z-a}$ maps $\mathfrak{R}$ into a region $\mathfrak{R}'$ containing the point at infinity. The inner radius of $\mathfrak{R}$ with respect to a is equal to the reciprocal of the outer radius of the complement of $\mathfrak{R}'$.

108. Consider the region $\mathfrak{R}$ which is obtained from the unit circle $|z|<1$ by removing the two real line segments $b_1\leqq z<1$, $-1<z\leqq -b_2$, $0<b_1<1$, $0<b_2<1$. Calculate the inner radius of $\mathfrak{R}$ with respect to the origin.

109. The inequalities

$$\vartheta_1\leqq \arg\frac{z-z_1}{z-z_2}\leqq\vartheta_2, \qquad 0<\vartheta_1\leqq\vartheta_2<2\pi, \qquad z_1\neq z_2,$$

determine a domain $\mathfrak{C}$ in the z-plane bounded by two circular arcs and with two corners at z_1 and z_2 at which the sides meet at an angle of $\vartheta_2-\vartheta_1$. What is the outer radius of $\mathfrak{C}$? Note the special cases $\vartheta_1+\vartheta_2=2\pi$; $\vartheta_1=\vartheta_2$; $\vartheta_1=\pi/2$, $\vartheta_2=\pi$.

110. Let a and b be points of the region $\mathfrak{R}$ and let $f(z)$ denote the normed mapping function associated with the central point b. Then

$$r_a=\frac{r_b^2-|f(a)|^2}{r_b|f'(a)|}.$$

111. Derive the normed mapping function of the whole plane slit along the real segment $\frac{1}{4}\leqq z<+\infty$ associated with the origin and investigate the variation of the inner radius r_a as a runs through the real values from $-\infty$ to $\frac{1}{4}$.

112. Let L be an analytic curve (i.e. let the normed mapping function associated with an arbitrary interior point of L admit continuation across L and assume different values for distinct points of L). If the central point a in the interior of L converges to a point of L, then the inner radius converges to 0.

113. Let a be a point of the region $\mathfrak{R}$ such that at a we obtain a relative maximum of the inner radius r_a. Then the normed mapping function associated with a

$$f(z)=z-a+c_2(z-a)^2+c_3(z-a)^3+\cdots$$

is such that $c_2=0$.

114. What is the locus of those points a of the upper half-plane $\mathfrak{J}z>0$ with respect to which the inner radius r_a of $\mathfrak{J}z>0$ is constant?

115. Let $\mathfrak{R}$ be a finite region in the z-plane, a an arbitrary point of $\mathfrak{R}$, and n a positive integer. By means of the transformation $z'=\sqrt[n]{z-a}$ there corresponds to $\mathfrak{R}$ a region $\mathfrak{R}'$ "n-fold symmetric" with respect to the origin. A point z' belongs to $\mathfrak{R}'$ if and only if z'^n+a is a point of $\mathfrak{R}$. Between the inner radii r_a of $\mathfrak{R}$ with respect to a and r_0' of $\mathfrak{R}'$ with respect to $z'=0$ we have the relation

$$r_0'=r_a^{\frac{1}{n}}.$$

A similar result holds for the outer radii in the case of the transformation $z'=\sqrt[n]{z}$ if the origin belongs to the complement of $\mathfrak{R}$.

116. Let n be a positive integer. Cut the plane along the n rays

$$\sqrt[n]{\tfrac{1}{4}}\leqq|z|<+\infty,\qquad \arg z=\frac{2\pi\nu}{n},\qquad \nu=1,2,\ldots,n$$

and calculate r_a for the resulting slit region. To what limiting value does r_a tend when a approaches along the line segment $0\leqq|z|<\sqrt[n]{\frac{1}{4}}$, $\arg z=\dfrac{2\pi\nu}{n}$ the end point of the corresponding cut?

117. Calculate the outer radius of the slit region whose boundary consists of the two straight lines $-\alpha\leqq z\leqq\alpha$, $-\beta\leqq iz\leqq\beta$; $\alpha>0$, $\beta>0$.

118. Let $\mathfrak{R}$ be an arbitrary simply-connected region of the z-plane, and a a finite point of $\mathfrak{R}$. The normed mapping function $f(z)$ associated with a and the inner radius r_a have the following minimum property: Of all the functions of the form

$$F(z)=z-a+d_2(z-a)^2+d_3(z-a)^3+\cdots,$$

which are regular in $\mathfrak{R}$, $|f(z)|$ has the least maximum value in $\mathfrak{R}$. This "minimum maximorum" is equal to r_a. More precisely: If M is the supremum of $|F(z)|$ in $\mathfrak{R}$ then

$$M \geqq r_a.$$

The equality sign occurs if and only if $F(z)=f(z)$.

A similar result is true if $\mathfrak{R}$ contains the point at infinity for the mapping function associated with the infinite point as central point and for the outer radius of the complement of $\mathfrak{R}$. The circle mapping is characterized by a "maximum minimorum".

119. Let $\mathfrak{R}$ be a region symmetric with respect to the real axis, a real. Then the power series expansion of the normed mapping function associated with a in powers of $z-a$ has only real coefficients. A similar result holds for the normed mapping function associated with the point at infinity as central point, if $\mathfrak{R}$ contains the point at infinity.

120. Let $\mathfrak{R}$ be a region which is symmetric with respect to the origin. Then the power series expansion of the normed mapping function associated with the origin contains only odd powers of z. A similar result holds for the normed mapping function associated with the point at infinity, if $\mathfrak{R}$ contains the point at infinity.

§ 5. Relations between the Mappings of Different Domains

121. Let $\mathfrak{R}^*$ be a proper sub-region of $\mathfrak{R}$ and r_a and r_a^* the inner radii of $\mathfrak{R}$ and $\mathfrak{R}^*$ respectively, associated with the point a of $\mathfrak{R}^*$. Then

$$r_a^* < r_a.$$

A similar result holds for the outer radius.

122. Let $\mathfrak{R}$ be an arbitrary region, a a finite point of $\mathfrak{R}$, r the radius of the largest open circular region with center a contained in $\mathfrak{R}$ and R the radius of the smallest open circular region with center a containing $\mathfrak{R}$. Then $r>0$, $R \geqq r$, R finite or infinite, and

$$r \leqq r_a \leqq R.$$

The equality sign holds only if $\mathfrak{R}$ is an open circular region with center a.

123. Let the region $\mathfrak{R}$ contain the point at infinity, and let $\mathfrak{D}$ be the complement of $\mathfrak{R}$. We denote by r the radius of the largest closed circular region contained in $\mathfrak{D}$ and by R the radius of the smallest closed circular region containing $\mathfrak{D}$. Then $r \geqq 0$, $R \geqq r$, R finite. Further

$$r \leqq \bar{r} \leqq R.$$

The equality sign holds only if $\mathfrak{D}$ is a closed circular *domain* (closure of a non-empty connected open set).

***124.** Let l denote the length of a closed continuous curve L without double points. Then

$$2\pi r_a \leqq l, \qquad 2\pi \bar{r} \leqq l,$$

where a denotes an arbitrary point in the interior of L, r_a the corresponding inner radius of L and $\bar{r}$ the outer radius of L. In both cases equality is attained only for a circle, and in the first case only for a circle with center a.

124.1 Let l denote the length of the perimeter of an ellipse with unequal semi-axes a and b. Give two different proofs (with and without the use of conformal mapping) for the inequality

$$l > \pi(a+b).$$

(The right-hand side is a convenient approximation to l when a and b are nearly equal.)

125. Let the inner measure of the region $\mathfrak{R}$ be $|\mathfrak{R}|_i$, and let a be an arbitrary finite point of $\mathfrak{R}$. Then

$$|\mathfrak{R}|_i \geqq \pi r_a^2.$$

The equality sign holds only for the open disc $|z-a| < r_a$.

126. Let the complement $\mathfrak{H}$ of the region $\mathfrak{R}$ containing the point at infinity be of outer measure $|\mathfrak{H}|_e$. Then

$$|\mathfrak{H}|_e \leqq \pi \bar{r}^2.$$

The equality sign holds only if $\mathfrak{H}$ is a circular domain.

127. Let the outer radius and the inner radius (with respect to an arbitrary interior point a) of a simple closed continuous curve L be $\bar{r}$ and r_a respectively. Then

$$r_a \leqq \bar{r}.$$

The equality sign holds if and only if L is a circle and a lies at the center of L.

128. Let L be a simple closed continuous curve containing the origin, $\bar{r}$ the outer radius of L and r_0 the inner radius of L associated with the origin. Let $P(z)$ be a polynomial whose lowest order term is $a_k z^k$ and whose highest order term is $a_n z^n$, $P(z) = a_k z^k + a_{k+1} z^{k+1} + \cdots + a_n z^n$, and let M denote the maximum of $|P(z)|$, as z describes the curve L. Then

$$M \geqq |a_k| r_0^k, \qquad M \geqq |a_n| \bar{r}^n.$$

The equality sign holds only if L is a circle with center at the origin and the polynomial in question is a power of z multiplied by a constant.

129. Let the function $f(z)$ be regular and have positive real part in the interior of the simple closed continuous curve L, and be continuous in the closed interior of L. If the real part of $f(z)$ vanishes on an arc L' of L then the imaginary part of $f(z)$ varies always in the same sense, namely monotonically decreasing as z describes the arc L' in the positive sense. [III **233**.]

130. Map the strip $0 < \Im z < D$ onto the circle $|w| < 1$ in such a way that the point $z = i$ corresponds to the center of the circle $w = 0$ $(D > 1)$. How large is the arc of the boundary of the circle $|w| = 1$ which in this mapping corresponds to the real axis $\Im z = 0$? (Obvious for $D = 2$ and $D = \infty$.) How does the arc in question vary as D increases?

131. Let two simple closed continuous curves L_1 and L_2 have a finite number of common arcs and let the interior of L_1 be contained in the interior of L_2. (L_1 consists of an even number of arcs which run alternately in the interior and on the boundary of the region enclosed by L_2.)

Map first the interior of L_1 and then the interior of L_2 onto the interior of one and the same circle, so that in both cases the same point O in the interior of L_1

goes over into the center of the circle. The two mappings associate with the distinct segments of L_1 and L_2 well-defined circular arcs as their images. [Last remark preceding **97**.]

Prove that the length of the image of any *common* arc of L_1 and L_2 will be smaller in the mapping of the smaller region (bounded by L_1) than in the mapping of the larger region (bounded by L_2). Example: **130**. [**129**.]

132. Find an interpretation in electrostatics of **131**.

133. Let two simple closed, continuous curves L_1 and L_2 have only finitely many common points, and let the two regions enclosed by them have a common region $\mathfrak{T}$. Denote by O a point of $\mathfrak{T}$ and by $\mathfrak{T}^*$ that connected part of $\mathfrak{T}$ which contains O. Such arcs of L_1 and L_2 as belong to the boundary of $\mathfrak{T}^*$ are termed "visible" from O, such arcs as do not belong to the boundary of $\mathfrak{T}^*$ are termed "hidden" from O. (The terms "visible" and "hidden" have their usual meaning if L_1 and L_2 are *star-shaped* with respect to O [cf. III **109**].)

Map first the interior of L_1 and then the interior of L_2 onto the interior of the unit circle, in both cases in such a manner that the center of the circle is the image of O. The images of the "visible" parts of L_1 take up a larger percentage of the boundary of the circle than the images of the "hidden" parts of L_2. (To visualize this imagine L_2 as a circle with centre O.) [**131**.]

134. Let $\mathfrak{D}$ be a domain, ζ an interior point of $\mathfrak{D}$, and $\mathfrak{B}$ the set of the boundary points of $\mathfrak{D}$ whose distance from ζ is not greater than ρ. Let those arcs of the circle of radius ρ and with centre ζ that do *not* belong to $\mathfrak{D}$ have total length $\rho\Omega$.

Let the function $f(z)$ be regular and single-valued in the interior of $\mathfrak{D}$ and continuous on the boundary of $\mathfrak{D}$. Further, let $|f(z)|\leqq a$ at points of $\mathfrak{B}$, $|f(z)|<A$ at the remaining boundary points of $\mathfrak{D}$, $a<A$. Then

$$|f(\zeta)|\leqq a^{\frac{\Omega}{2\pi}}A^{1-\frac{\Omega}{2\pi}}.$$

(Sharper than III **276**.)

135. The simply connected region $\mathfrak{R}_n$ lying in the z-plane is subject to the following conditions:

(1) $\mathfrak{R}_n$ is contained in the circle $|z|<a$, $a>1$.

(2) $\mathfrak{R}_n$ contains the circle $|z|<1$.

(3) $\mathfrak{R}_n$ contains the arc determined by the relations

$$|z|=1, \quad -\alpha_n<\arg z<\alpha_n, \qquad 0<\alpha_n<\pi.$$

(4) The arc on the periphery of the unit circle complementary to the arc in (3) for which

$$|z|=1, \quad \alpha_n\leqq \arg z\leqq 2\pi-\alpha_n,$$

belongs to the boundary of $\mathfrak{R}_n$.

Let $\mathfrak{R}_n$ be mapped one-to-one and conformally onto the unit circle by means of $w=f_n(z)$, and in such a manner that $f_n(0)=0$, $f_n'(0)>0$.

If the unit circle is gradually cut off from the remaining part of $\mathfrak{R}_n$ as n increases, i.e. if $\lim_{n\to\infty}\alpha_n=0$, then for $|z|<1$

$$\lim_{n\to\infty} f_n(z)=z,$$

independently of the behaviour for large n of that remaining part of $\mathfrak{R}_n$. [Method III **335**.]

§ 6. Koebe's Distortion Theorem and Related Topics

136. Let the function

$$w=g(z)=z+b_0+\frac{b_1}{z}+\frac{b_2}{z^2}+\cdots$$

be regular for $|z|>1$, and map this region (the exterior of the unit circle) schlicht onto a region $\mathfrak{R}$ containing the point at infinity. Then

$$|b_1|^2+2|b_2|^2+3|b_3|^2+\cdots \leqq 1.$$

In particular $|b_1|\leqq 1$ and in this last inequality the equality sign holds if and only if $\mathfrak{R}$ is the whole plane cut along a line segment of length 4.

137 (continued). For $|z|>1$

$$|g'(z)|\leqq\frac{1}{1-\dfrac{1}{|z|^2}}.$$

The equality sign holds at the point $\frac{\rho}{\varepsilon}$, $|\varepsilon|=1$, $\rho>1$, if and only if $g(z)$ has the form

$$g(z)=z+b_0-\frac{1}{\varepsilon}\left(\rho-\frac{1}{\rho}\right)\frac{1}{\rho\varepsilon z-1}.$$

What is the nature of the image region in this case?

In the case of the mapping of **136** all the curves in the w-plane which correspond to the concentric circles of radius >1 about the origin $z=0$, the so-called level curves (circle images), have the same conformal center of gravity b_0 [III **129**]. We call b_0 the *conformal center of gravity* of the region $\mathfrak{R}$.

138. Let the simply connected region $\mathfrak{R}$ contain the point at infinity and be symmetric with respect to a point P. Then P is the conformal center of gravity of $\mathfrak{R}$.

139. (Continuation of **136**.) If the region $\mathfrak{R}$ does not contain the origin, then the conformal centre of gravity lies inside the circle of radius 2 centered at the origin, that is $|b_0|\leqq 2$. The equality sign holds if and only if $\mathfrak{R}$ is the whole plane cut along a straight line segment of length 4 starting from the origin. [Apply **136** to $\sqrt{g(z^2)}$.]

140 (continued). The distance d of an arbitrary boundary point of $\mathfrak{R}$ from the conformal center of gravity is at most 2. In fact $d<2$ except in the case of the particular mapping specified in **136**.

141 (continued). The maximum distance D between the boundary points of $\mathfrak{R}$ (diameter of the boundary of $\mathfrak{R}$) lies between the limits 2 and 4, i.e.

$$2\leqq D\leqq 4.$$

The equality sign holds for the lower estimate only if $\mathfrak{R}$ is the exterior of a circle of radius 1, and for the upper estimate only if $\mathfrak{R}$ is the slit region of **136**.

142. Of all continuous arcs which connect two fixed points the straight line has the smallest outer radius.

143. Let the region $\mathfrak{R}$ in **136** have the origin as its conformal center of gravity. Then

$$|g(z)| \leqq |z| + \frac{1}{|z|}, \qquad |z| > 1.$$

The equality sign holds only for the mapping $w = z + \frac{e^{i\alpha}}{z}$, α real.

144 (continued). By the mapping in question no point z can be displaced by more than $\frac{3}{|z|}$ from its original position. I.e.

$$|g(z) - z| < \frac{3}{|z|}, \qquad |z| > 1.$$

145. Investigate the displacement [**144**] for the mapping of the exterior of the unit circle onto a slit region of a particular kind, which is bounded by the *horseshoe-shaped* curve that consists of three straight line segments and connects the four points

$$a + i\delta, \qquad -a + i\delta, \qquad -a - i\delta, \qquad a - i\delta$$

in that order; $a > 0$, $\delta > 0$. Note especially the case where a tends to 2, δ tends to 0, and show by means of this example that the constant 3 in **144** cannot be replaced by any smaller constant.

146. Let the function

$$f(z) = z + a_2 z^2 + a_3 z^3 + \cdots$$

be regular and schlicht inside the unit circle $|z| < 1$. Then

$$|a_2| \leqq 2.$$

The equality sign holds only for functions of the form

$$f(z) = \frac{z}{(1 + e^{i\alpha} z)^2}, \qquad \alpha \text{ real. } (\mathbf{111}.)$$

[Apply **139** to $(f(z^{-1}))^{-1}$.]

147. Let the function

$$w = f(z) = z + a_2 z^2 + a_3 z^3 + \cdots$$

be schlicht in the interior of the circle $|z| < R$. If the image region $\mathfrak{R}$ in the w-plane does not contain the point $w = \infty$ then it completely contains the open circular disc $|w| < \frac{R}{4}$. In other words, if d denotes the least distance of the boundary of $\mathfrak{R}$ from the origin $z = 0$, then $d \geqq \frac{R}{4}$. Moreover $d > \frac{R}{4}$ unless $\mathfrak{R}$ is the whole plane slit along the segment $\arg w = \text{const.}$, $\frac{R}{4} \leqq |w| < +\infty$. [Apply **146** to $f(z)/[1 - h^{-1} f(z)]$ where h is a boundary point of $\mathfrak{R}$.]

148. We have the following refinement of **147**: The shortest boundary distance d satisfies the inequality

$$d \geqq \frac{R}{|a_2| + 2}.$$

(Cf. **146.**) Equality sign as in **147**.

149. (Continuation of **147**.) We term the straight line connecting two boundary points of $\mathfrak{R}$ a *principal chord* of $\mathfrak{R}$ if it passes through the origin. Every principal chord of $\mathfrak{R}$ has a length of at least R. The extreme case occurs if and only if $\mathfrak{R}$ is the whole plane slit along the two straight line segments $w=\pm|w|\,e^{i\alpha}$, α real, $\frac{R}{2}\leqq|w|<+\infty$. (**116**.)

150. (Continuation of **146**.) In the interior of the unit circle $|z|<1$ we have the inequality

$$\left|\frac{1-|z|^2}{2}\frac{f''(z)}{f'(z)}-\bar{z}\right|\leqq 2 \quad .$$

The equality sign holds only if the image region is the plane slit along a straight line. [Transform the unit circle into itself in such a manner that an arbitrary fixed point z_0, $|z_0|<1$, goes over into the origin and then apply **146**.]

151. (The Koebe distortion theorem.) Let the function

$$f(z)=z+a_2z^2+a_3z^3+\cdots$$

be regular and schlicht inside the unit circle $|z|<1$. Also let r be a positive number $r<1$. Then in the circular domain $|z|\leqq r$ we have the inequalities

$$\frac{1-r}{(1+r)^3}\leqq|f'(z)|\leqq\frac{1+r}{(1-r)^3}.$$

The equality sign can occur only for the mappings

$$f(z)=\frac{z}{(1+e^{i\alpha}z)^2}, \qquad \alpha \quad \text{real},$$

[**150**.]

152 (continued). In the circular disc $|z|\leqq r$ we have

$$\frac{r}{(1+r)^2}\leqq|f(z)|\leqq\frac{r}{(1-r)^2}.$$

The equality sign as in **151**.

153. We have the following refinement of **152**: For $|z|\leqq r$ we have

$$\frac{r}{1+|a_2|r+r^2}\leqq|f(z)|\leqq\frac{r}{(1-r)^2}.$$

The equality sign as in **151**. (Generalization of **148**.) [Cf. solution **148**, also **143**.]

154. For odd functions $f(z)=-f(-z)$ we can obtain the following refinement of **152**:

$$\frac{r}{1+r^2}\leqq|f(z)|\leqq\frac{r}{1-r^2}.$$

The equality sign holds only for $f(z)=\frac{z}{1-z^2}$.

155 (continued). We have

$$\frac{1}{2\pi}\int_0^{2\pi}|f(r\,e^{i\vartheta})|^2\,d\vartheta\leqq\frac{r^2}{1-r^2}.$$

[The measure of the image of $|z|\leqq\rho<r$ is $\leqq\pi\max|f(z)|^2$, $|z|\leqq\rho$; III **128**.]

156. (Continuation of **152**.) Let n be a positive integer. There exists a function $\omega_n(r)$ of n and r only, $0 \leqq r < 1$, such that for *any* function $f(z)$ of the type mentioned in **151** we have in the circular region $|z| \leqq r$ the inequality

$$|f^{(n)}(z)| \leqq \omega_n(r).$$

157 (continued). Let n be a positive integer, $n \geqq 2$. There exists a constant ω_n depending only on n such that for *any* function $f(z)$ of the type mentioned in **151** we have the inequality

$$|a_n| \leqq \omega_n.$$

Let ω_n be the *smallest* constant of this kind. Then

$$\omega_n \leqq \frac{1}{4} \frac{(n+1)^{n+1}}{(n-1)^{n-1}}.$$ [1]

158 (continued). We have

$$\frac{1}{2\pi} \int_0^{2\pi} |f(r\, e^{i\vartheta})|\, d\vartheta \leqq \frac{r}{1-r}. \qquad \text{[155.]}$$

159 (continued). We have (refinement of **157**)

$$n \leqq \omega_n < en.$$

160. For functions

$$f(z) = z + a_2 z^2 + a_3 z^3 + \cdots + a_n z^n + \cdots,$$

which are schlicht and in absolute value less than M inside the unit circle $|z| < 1$ the estimate of **146** can be refined as follows: $M \geqq 1$ ($M = 1$ only for $f(z) = z$) and

$$|a_2| \leqq 2\left(1 - \frac{1}{M}\right).$$

When does the equality sign occur? [Apply **146** to $\dfrac{f(z)}{[1 + e^{i\alpha} M^{-1} f(z)]^2}$.]

161. Let the function

$$f(z) = z + a_2 z^2 + a_3 z^3 + \cdots + a_n z^n + \cdots$$

be regular inside the unit circle $|z| < 1$ and map the unit circle onto a region which is *star-shaped* with respect to the origin [III **109**]. Then

$$|a_n| \leqq n, \qquad n = 2, 3, 4, \ldots.$$

The equality sign holds if and only if

$$f(z) = \frac{z}{(1 - e^{i\alpha} z)^2}, \qquad \alpha \text{ real}.$$

162. Let the function

$$f(z) = z + a_2 z^2 + a_3 z^3 + \cdots + a_n z^n + \cdots$$

---

[1] This bound is less than $\frac{e^2}{4} n^2$.

be regular inside the unit circle $|z|<1$ and map the unit circle onto a *convex* region [III **108**]. Then

$$|a_n| \leqq 1, \qquad n=2, 3, 4, \ldots.$$

The equality sign holds if and only if

$$f(z)=\frac{z}{1-e^{i\alpha}z};$$

the image region in that case is a half-plane containing the origin, whose boundary line is at a distance $\frac{1}{2}$ from the origin.

163. If the function $f(z)$ is regular and schlicht inside the unit circle $|z|<1$, then the image of the circular region $|z|<r$ is a convex region provided $r \leqq 2-\sqrt{3}=0.26\ldots$. This number (the *curvature bound*) cannot be replaced by any smaller number.

Chapter 3. Miscellaneous Problems

§ 1. Various Propositions

164. If the function $f(z)$ is regular, single valued and *bounded* in the strip $0 \leqq \Re z \leqq \pi$, and vanishes there at the points $z_1, z_2, z_3, \ldots, z_n, \ldots (z_n=x_n+iy_n)$ then either the series of positive terms

$$e^{-|y_1|} \sin x_1+e^{-|y_2|} \sin x_2+e^{-|y_3|} \sin x_3+\cdots+e^{-|y_n|} \sin x_n+\cdots$$

is convergent, or $f(z)$ is identically equal to 0. [III **297**.]

165. Assume that the coefficient $f_1(z)$ of the linear homogeneous differential equation

$$y^{(n)}+f_1(z)y^{(n-1)}+f_2(z)y^{(n-2)}+\cdots+f_n(z)y=0$$

is an entire function. The necessary and sufficient condition that the *general solution* of this differential equation be an entire function is that the remaining coefficients $f_2(z), f_3(z), \ldots, f_n(z)$ are also entire functions.

166. Let the function $w=\varphi(z)$ be regular (possibly many-valued) in the annular region $0<|z|<\rho$, $\rho>0$, and satisfy there identically an equation of the form

$$F_0(z)w^l+F_1(z)w^{l-1}+\cdots+F_{l-1}(z)w+F_l(z)=0,$$

where $F_0(z), F_1(z), \ldots, F_{l-1}(z), F_l(z)$ are regular in a neighbourhood of the point $z=0$. If there exists a power series $c_0+c_1z+c_2z^2+\cdots$ such that the function $(\varphi(z)-c_0-c_1z-c_2z^2-\cdots-c_{n-1}z^{n-1})z^{-n}$ remains bounded in the neighbourhood of $z=0$ for infinitely many values of n, then $\varphi(z)$ is regular in the neighbourhood of $z=0$, and we have $\varphi(z)=c_0+c_1z+c_2z^2+\cdots+c_nz^n+\cdots$.

167. Let the function $f(z)$ be regular and single-valued for $R \leqq |z| < \infty$, $R>0$. Then there exists an integer $p \geqq 0$, an entire function $G(z)$ and a power series

$$\psi(z)=\frac{c_{-1}}{z}+\frac{c_{-2}}{z^2}+\frac{c_{-3}}{z^3}+\cdots,$$

convergent for $|z| \geqq R$, such that for $R \leqq |z| < \infty$ we have

$$f(z) = z^{-p} G(z)\, e^{\psi(z)}.$$

If in the assumption as well as in the conclusion we replaced the inequality $R \leqq |z| < \infty$ by the inequality $R < |z| < \infty$, the theorem obtained would be invalid.

168. Let $f(z) = f(x+iy)$ be a meromorphic periodic function of period 2π that has in the strip $0 \leqq \Re z < 2\pi$ only a finite number of zeros and poles. Denote by $M(y)$ the maximum and by $\mu(y)$ the minimum of $|f(x+iy)|$ for $0 \leqq x \leqq 2\pi$. If it is the case that

$$\limsup_{|y| \to \infty} \frac{\log \log M(y)}{|y|} < 1 \qquad \text{or that} \qquad \liminf_{|y| \to \infty} \frac{\log \log \mu(y)}{|y|} > -1,$$

then $f(z)$ must be a rational function of e^{iz}.

169. Let $f(z)$ be an analytic function, $z = x+iy$, and let the square of the absolute value of $f(z)$

$$|f(z)|^2 = \varphi(x, y)$$

be an algebraic function of the real variables x and y. Then the function $f(z)$ is itself an algebraic function of z.

170. There does not exist a function which is regular analytic along the real axis and which assumes for real values of the variable every value in the interior of a fixed circle. In brief: There does not exist an analytic *Peano-curve.*

171. Find, if possible, $n+1$ analytic functions $f_1(z), f_2(z), \ldots, f_n(z), f(z)$, which differ from each other *not merely by constant factors*, are regular in a domain $\mathfrak{D}$ and for which the relation

$$|f_1(z)| + |f_2(z)| + \cdots + |f_n(z)| = |f(z)|$$

holds in $\mathfrak{D}$.

172. Find, if possible, two non-constant analytic functions $f(z)$ and $g(z)$, which are regular in a domain $\mathfrak{D}$ and for which the relation

$$|g(z)| = \Re f(z)$$

holds in $\mathfrak{D}$. [III **58.**]

172.1 The modular graph [Vol. I, p.130] of an entire function $g(z)$ is everywhere convex from below in just two cases: Either when

$$g(z) = (az+b)^n$$

or when

$$g(z) = e^{az+b},$$

where a and b are arbitrary complex numbers, n a non-negative integer. (See the definition preceding III **130.**)

173. We say that two entire functions $f(z)$ and $g(z)$ have the same a-points if the two functions

$$\frac{f(z)-a}{g(z)-a} \qquad \text{and} \qquad \frac{g(z)-a}{f(z)-a}$$

are both entire. Find two distinct entire functions, which have the same a-, b- and c-points. (Naturally we require $b \neq c$, $c \neq a$, $a \neq b$.)

174. Does there exist an entire function $G(z)$ which satisfies the equations

$$G(0)=a_0, \qquad G'(1)=a_1, \qquad G''(2)=a_2, \qquad \ldots, \qquad G^{(n)}(n)=a_n, \ \ldots$$

if the sequence of numbers $a_0, a_1, a_2, \ldots$ is prescribed arbitrarily? [For the analogous interpolation problem for polynomials cf. VI **75**, VI **76**.]

174.1. Given the numbers

$$z_n; \quad m_n; \quad w_{n,0}, \quad w_{n,1}, \quad \ldots, \quad w_{n,m_n-1}$$

for $n=1, 2, 3, \ldots$ subject to the conditions:

$$z_k \neq z_l \qquad \text{if} \quad k \neq l,$$

$$\lim_{n\to\infty} |z_n| = \infty,$$

m_n is a positive integer, however the $w_{n,k}$ are unrestricted.

Then there exists an entire function such that

$$f^{(k)}(z_n) = k!\, w_{n,k}$$

for $n=0, 1, 2, \ldots, \quad 0 \leqq k < m_n$.

[The condition prescribes the m_n initial terms of the expansion of $f(z)$ about the point z_n

$$f(z) = w_{n,0} + w_{n,1}(z-z_n) + \cdots + w_{n,m_n-1}(z-z_n)^{m_n-1} + \cdots$$

for $n=1, 2, 3, \ldots$.]

174.2. Suppose that $F(z)$ and $G(z)$ are entire functions that have no common zeros. Then there exist two entire functions $f(z)$ and $g(z)$ such that

$$F(z)f(z) + G(z)g(z) = 1.$$

§ 2. A Method of E. Landau

To deduce the theorem, that a function $f(z)$, which is regular and single-valued in a domain $\mathfrak{D}$, assumes its maximum absolute value on the boundary L of the domain $\mathfrak{D}$, directly from the *Cauchy* theorem

$$f(z) = \frac{1}{2\pi i} \oint_L \frac{f(\zeta)}{\zeta - z}\, d\zeta, \qquad \text{for } z \text{ in the interior of } L,$$

one may proceed as follows: Let $|f(\zeta)| \leqq M$ on L, then we have

$$|f(z)| \leqq \frac{M}{2\pi} \int_L \left| \frac{d\zeta}{\zeta - z} \right| = KM,$$

where the constant K depends only on the curve L and on the position of z, and is independent of the specific choice of the function $f(z)$. One may now improve this rough estimate by applying it to the function $[f(z)]^n$, where n is a positive integer, obtaining

$$|f(z)|^n \leqq KM^n, \qquad |f(z)| \leqq K^{\frac{1}{n}} M$$

and then allowing n to tend to infinity. Then it follows that $|f(z)| \leqq M$.

This interesting proof shows that a rough estimate may sometimes be transformed into a sharper estimate by making appropriate use of the generality for which the original estimate is valid. [E. Landau; cf. M. Riesz: Acta Math. Vol. **40**, p. 340, footnote[1] (1916).]

175. Let $f(z)=a_0+a_1z+a_2z^2+\cdots+a_nz^n+\cdots$ be regular for $|z|<R$ and set $\mathfrak{M}(r)=|a_0|+|a_1|r+|a_2|r^2+\cdots+|a_n|r^n+\cdots$. Then we have

$$M(r)\leqq\mathfrak{M}(r)<\frac{r+\delta}{\delta}M(r+\delta),\quad \delta>0,\quad r+\delta<R.$$

176 (continued). If $\mathfrak{M}_n(r)$ is defined for the function $[f(z)]^n$ in the same manner as $\mathfrak{M}(r)$ for $f(z)$, $n=1, 2, 3, \ldots$, then we have

$$\lim_{n\to\infty}[\mathfrak{M}_n(r)]^{\frac{1}{n}}=M(r).$$

177. With the aid of II **123**, obtain a new proof for the *Hadamard* three-circle theorem [III **304**].

178. Prove the following generalization of **160**: If ω_n is defined as in **157**, then under the assumptions of **160** we have the following estimate of the coefficients

$$|a_n|\leqq\omega_n(1-M^{1-n}).$$

179. Assume the *Bernstein* theorem for trigonometric polynomials [VI **82**] in the following imprecise form: There exists an absolute constant K, $K>0$ with the property that, if $\varphi(\vartheta)$ denotes a trigonometric polynomial of nth order whose absolute value does not exceed 1, then we have

$$|\varphi'(\vartheta)|\leqq n+K.$$

Find a method of deducing from this estimate the sharper estimate

$$|\varphi'(\vartheta)|\leqq n.$$

§ 3. Rectilinear Approach to an Essential Singularity

180. There exist entire functions that increase arbitrarily rapidly as $z\to\infty$ along a ray. More precisely: If $\varphi(x)$ is an arbitrary, strictly positive, monotonically increasing function defined for $x\geqq 0$, there exists an entire function $g(z)$, which assumes real values for real z and satisfies for $x\geqq 0$ the inequality $g(x)>\varphi(x)$. (III **290**.)

181. There exist entire functions that tend to 0 along the positive real axis, but tend to ∞ along any other ray emanating from the origin as $z\to\infty$. Can a rational entire function behave in this manner?

182. There exist transcendental entire functions that tend to ∞ along every ray emanating from the origin. — Can this happen uniformly?

183. There exist entire functions that are real-valued along the positive real axis and along it tend to $+\infty$ while tending to 0 along any other ray emanating from the origin. — Can an entire function of finite order behave in this fashion?

184. There exist not identically vanishing entire functions that tend to 0 along every ray emanating from the origin.

185. There exist entire functions that tend to $+1$ along all rays running from the origin into the interior of the upper half-plane, and that tend to -1 along all rays running from the origin into the interior of the lower half-plane. The convergence is in fact uniform in every sector $\varepsilon < \arg z < \pi - \varepsilon$, or $-\pi + \varepsilon < \arg z < -\varepsilon$, where $\varepsilon > 0$.

186. Let the whole plane be divided into n sectors by n rays emanating from the origin. There exists an entire function that tends in these sectors (more precisely, as in **185**) to $a_1, a_2, \ldots, a_n$ respectively, where $a_1, a_2, \ldots a_n$ are arbitrarily assigned complex numbers.

187. Let the rays emanating from the origin be divided in any manner into two classes. Does there exist, corresponding to every such division, an entire function that tends to 0 along the rays of the first class, and that tends to ∞ along the rays of the second class?

§ 4. Asymptotic Values of Entire Functions

If an entire function $g(z)$ which is not a constant tends to a limit a along a *continuous* curve going to infinity, the number a is called an *asymptotic value* of $g(z)$. For example, 0 is an asymptotic value of e^z.

188. The numbers 0 and ∞ are the only asymptotic values of e^z.

189. The numbers $\sqrt{\frac{\pi}{2}}, -\sqrt{\frac{\pi}{2}}$ and ∞ are the only asymptotic values of the function $\int_0^z e^{-\frac{x^2}{2}}\,dx$.

190. Let n be a positive integer. The integral function of order n

$$z - \frac{z^{2n+1}}{3!(2n+1)} + \frac{z^{4n+1}}{5!(4n+1)} - \cdots$$

has exactly $2n$ distinct finite asymptotic values.

191. Let the sequence of positive numbers

$$a_0,\ a_1,\ a_2,\ \ldots,\ a_m,\ \ldots$$

be chosen such that the series

$$g(z) = \sqrt{\frac{2}{\pi}} \sum_{m=0}^{\infty} a_m \int_0^{z^{8^m}} e^{-\frac{x^2}{2}}\,dx$$

converges uniformly in every finite domain of the z-plane and thus represents an entire function $g(z)$. [Set for example $a_m = \exp(-m^{8^m})$.]

Show that the set of asymptotic values of the entire function $g(z)$ thus defined has the power of the continuum. More precisely: All numbers of the form

$$\sum_{m=0}^{\infty} \varepsilon_m a_m, \qquad \varepsilon_m = +1 \quad \text{or} \quad -1,$$

are asymptotic values.

192. If an entire function converges to n distinct finite asymptotic values along n distinct rays emanating from the origin, then its order is not less than $\frac{n}{2}$.

193. For every entire function (that is not a constant) the number ∞ is an asymptotic value.

194. The complex number a is called a *Picard exceptional value* of the transcendental entire function $g(z)$ if the function $g(z)-a$ has only a finite number of zeros. If a Picard exceptional value of the entire function $g(z)$ exists, then it is an asymptotic value of the entire function $g(z)$.

§ 5. Further Applications of the Phragmén-Lindelöf Method

195. Let $f(z)$ be a function that can be analytically continued everywhere in the annular region $0<|z|<1$. If in this process $f(z)$ and all its derivatives $f'(z)$, $f''(z), \ldots$ remain bounded, then $f(z)$ is single-valued in this region and regular at the point $z=0$.

196. If $g(z)$ is an entire function of genus 0 and ε is an arbitrary positive number, then we have

$$|g(z)|<e^{\varepsilon|z|}$$

on all circles with sufficiently large radii, and

$$|g(z)|>e^{-\varepsilon|z|}$$

on certain circles with arbitrarily large radii.

197. Let $M(r)$ be the maximum and $m(r)$ the minimum of the absolute value of an entire function on the circumference of the circle $|z|=r$. If

$$\limsup_{r\to\infty}\frac{\log\log M(r)}{\log r}=\lambda$$

and $\lambda<\frac{1}{2}$, then we have also

$$\limsup_{r\to\infty}\frac{\log\log m(r)}{\log r}=\lambda.$$

[**196**, III **332**.]

198. Let $\lambda_1, \lambda_2, \lambda_3, \ldots, \lambda_n, \ldots$ be distinct positive numbers such that the series

$$\frac{1}{\lambda_1}+\frac{1}{\lambda_2}+\frac{1}{\lambda_3}+\cdots+\frac{1}{\lambda_n}+\cdots$$

is divergent. If a function $h(t)$ which is properly integrable over the interval $0\leqq t\leqq 1$ satisfies the condition

$$\int_0^1 t^{\lambda_n}h(t)\,dt=0, \qquad n=1, 2, 3, \ldots,$$

then $h(t)=0$ at every point of continuity t. (A generalization of II **139**.) [$\int_0^1 t^z h(t)\,dt$ is an analytic function of z, III **298**.]

199. Let $g(z)$ be a transcendental entire function of order $\lambda<\frac{1}{2}$. Let its coefficients be different from 0 and its zeros $w_1, w_2, w_3, \ldots, w_n, \ldots$ be distinct, $w_k\neq w_l$ for $k\neq l$. If a function $h(t)$ which is defined and properly integrable over the interval $0\leqq t\leqq 1$ satisfies the condition

$$\int_0^1 g(w_n t)h(t)\,dt=0, \qquad n=1, 2, 3, \ldots,$$

then $h(t)=0$ at every point of continuity t. [II **139**.]

200. Let $g(z)$ be a transcendental entire function of genus 0 with all its zeros $w_1, w_2, w_3, \ldots, w_n, \ldots$, real and distinct. If a function $h(t)$ defined and properly integrable in the interval $0 \leqq t \leqq 1$ satisfies the condition

$$\int_0^1 g(w_n t)h(t)\,dt = 0, \qquad n = 1, 2, 3, \ldots,$$

then $h(t) = 0$ at every point of continuity t. [We may, for example, set $g(z) = J_0(\sqrt{z})$ or $\cos \pi\sqrt{z}$.]

201. Set

$$a_0 + \frac{a_1}{1!} z + \frac{a_2}{2!} z^2 + \cdots + \frac{a_n}{n!} z^n + \cdots = F(z).$$

Assume that there exist two positive constants ρ and M such that

(1) the sequence $a_0, a_1\rho^{-1}, a_2\rho^{-2}, \ldots, a_n\rho^{-n} \ldots$ is bounded, and

(2) $|F(z)| \leqq M$ for all *real* values of z.

Then we have for all real values z also

$$|F'(z)| \leqq \rho M,$$

and equality holds only if $F(z) = A \cos \rho z + B \sin \rho z$, where A, B are constants. (A generalization of VI **82**.) [III **165**.]

202 (continued). If d is the distance of the point z from the real axis ($d = |\Im z|$), then

$$|F(z)| \leqq M\, e^{\rho d}.$$

(Analogue of III **270**.)

203. Let the entire function $G(z)$ satisfy the same conditions as the function $F(z)$ in **201**, and in addition let $G(z)$ be an odd function, $G(-z) = -G(z)$. Then we have for real z

$$\left|\frac{G(z)}{z}\right| \leqq \rho M.$$

Equality holds only if $G(z) = cM \sin \rho z$, $|c| = 1$ and if $z = 0$. (Analogue of VI **81**.) [III **166**.]

204 (continued). Deduce **201** from **203**.

205. Let $f(z)$ be an entire function of order λ, $\lambda \geqq \frac{1}{2}$, that is bounded on the positive real axis. If ε denotes an arbitrary positive number, then as x tends to infinity through positive values we have

$$\lim_{x \to +\infty} x^{1-\lambda-\varepsilon} f'(x) = 0.$$

§ 6. Supplementary Problems

***206.** We assume that the sequence of complex numbers $z_1, z_2, \ldots, z_k, \ldots$ has the following three properties:

(1) The series

$$\sum_{k=1}^{\infty} z_k, \quad \sum_{k=1}^{\infty} z_k^2, \quad \ldots, \quad \sum_{k=1}^{\infty} z_k^n$$

are convergent.

(2) $|z_1| \geqq |z_2| \geqq \cdots \geqq |z_k| \geqq \cdots$.

(3) There exist two real numbers α and β such that

$$0 < \beta - \alpha < \frac{2\pi n}{n+1}, \qquad \alpha \leqq \arg z_k \leqq \beta, \qquad \text{for} \quad k = 1, 2, 3, \ldots.$$

Then $|z_1|^n + |z_2|^n + \cdots + |z_k|^n + \cdots$ is convergent.

The conclusion holds under a wider assumption: The series listed in (1) need not be convergent; it is sufficient to assume that their partial sums are bounded. Yet the conclusion becomes invalid if we drop assumption (2) or if we require less than the first part of (3), namely that

$$0 < \beta - \alpha \leqq \frac{2\pi n}{n+1}.$$

(Cf. III **36**.)

***207.** (*Unitary transformations of analytic functions.*) The functions $f_1(z), f_2(z), \ldots, f_n(z)$; $g_1(z), g_2(z), \ldots, g_n(z)$ are regular analytic in a region of the z-plane where

$$|f_1(z)|^2 + |f_2(z)|^2 + \cdots + |f_n(z)|^2 = |g_1(z)|^2 + |g_2(z)|^2 + \cdots + |g_n(z)|^2.$$

If $f_1(z), f_2(z), \ldots, f_n(z)$ are linearly independent, then so also are $g_1(z)$, $g_2(z), \ldots, g_n(z)$, and

$$g_\nu(z) = c_{\nu 1} f_1(z) + c_{\nu 2} f_2(z) + \cdots + c_{\nu n} f_n(z)$$

for $\nu = 1, 2, \ldots n$; the constants $c_{\nu\lambda}$ form a unitary matrix.

***208.** (*On the concept of the order of an entire function.*) We consider the approximation of the entire function $g(z)$ by polynomials. We define the integer

$$k = k(r, \alpha)$$

(termed the degree of approximation of the function $g(z)$ on the circle $|z| = r$ to accuracy α) as the degree of the polynomial $P(z)$ of the *lowest possible degree* for which

$$\int_0^{2\pi} |g(r\,e^{i\varphi}) - P(r\,e^{i\varphi})|^2 \, d\varphi < \alpha^{-2} \int_0^{2\pi} |g(r\,e^{i\varphi})|^2 \, d\varphi;$$

$(r > 0,\ \alpha > 1)$. Then

$$\lim_{\alpha \to \infty} \frac{k \log k}{\log \alpha} = \lim_{r \to \infty} \frac{\log k}{\log r} = \rho,$$

where we keep r fixed when computing the first limit and α fixed when computing the second limit; ρ is the order of $g(z)$.

(Cf. III **123**. See **51** for the definition of order.)

***209.** (*On the values assumed by an entire function at the integers.*) Let $g(z)$ be an entire function whose growth does not exceed the minimum type of order 1; that is, $g(z)\,e^{-\varepsilon|z|}$ remains bounded in the whole z-plane provided that ε is a fixed positive number. Assume that the sequence

$$\ldots,\ g(-n),\ \ \ldots,\ g(-2),\ g(-1),\ g(0),\ g(1),\ g(2),\ \ \ldots,\ g(n), \ldots$$

is bounded. Then $g(z)$ reduces to a constant.

(Cf. III **220** (4), III **254**.)

***210.** (*Picard's theorem for matrices.*) We consider square matrices with n rows and n columns, and let capital letters denote such matrices. Let

$$g(z)=a_0+a_1z+a_2z^2+\cdots+a_nz^n+\cdots$$

be an entire function and Z a matrix. Then the series

$$g(Z)=a_0+a_1Z+a_2Z^2+\cdots+a_nZ^n+\cdots$$

converges to a well-defined matrix.

Given the matrix A, when does there exist a matrix Z such that

$$(*) \qquad g(Z)=A\ ?$$

Answer: An entire function $g(z)$ has 0 or 1 or 2 "exceptional values" in the following special sense: A solution of the equation (*) certainly exists if none of the characteristic roots of A coincides with an exceptional value of $g(z)$. On the other hand, the equation (*) has no solution for certain matrices A among whose characteristic roots there are exceptional values of $g(z)$.

For instance, the function e^z-z has no exceptional value, e^z has one, and $\sin z$ has two exceptional values, and the equations

$$\sin\begin{pmatrix} z_{11} & z_{12} \\ z_{21} & z_{22}\end{pmatrix}=\begin{pmatrix} 1 & 1 \\ 0 & 1\end{pmatrix}, \qquad \sin\begin{pmatrix} z_{11} & z_{12} \\ z_{21} & z_{22}\end{pmatrix}=\begin{pmatrix} -1 & -1 \\ 0 & -1\end{pmatrix}$$

admit no solution.

***211.** (*Two power series of the same analytic function converging in complementary regions.*) Let $f(z)$ be a single-valued analytic function which is regular everywhere including the point $z=\infty$, except at the point $z=1$. In a neighborhood of the point $z=0$ it has the expansion

$$f(z)=a_0+a_1z+a_2z^2+\cdots+a_nz^n+\cdots,$$

and in a neighborhood of the point $z=\infty$ the expansion

$$f(z)=b_0+\frac{b_1}{z}+\frac{b_2}{z^2}+\cdots+\frac{b_n}{z^n}+\cdots.$$

If there exists a constant k such that both a_nn^{-k} and b_nn^{-k} are bounded ($n=1, 2, 3, \ldots$), then $f(z)$ is a rational function.

***212.** (*A fraction of the Fourier system of functions.*) Let the numbers a, b, $m_1, m_2, m_3, \ldots$ be real, and satisfy

$$0<m_1<m_2<m_3<\cdots,$$

$$\lim_{n\to\infty}\frac{n}{m_n}>\frac{b-a}{2\pi}>0.$$

Furthermore, let $f(x)$ be continuous in the interval $[a, b]$. Then it follows from

$$\int_a^b f(x)\cos m_nx\,dx=\int_a^b f(x)\sin m_nx\,dx=0$$

for $n=1, 2, 3, \ldots$, that $f(x)$ vanishes identically. (A fraction of the whole system of Fourier functions is closed over the corresponding fraction of the whole interval.) (Cf. **198.**)

Part Five. The Location of Zeros

Chapter 1. Rolle's Theorem and Descartes' Rule of Signs

§ 1. Zeros of Functions, Changes of Sign of Sequences

We investigate in this chapter *real* functions of the real variable x. In particular we assume that the coefficients $a_0, a_1, a_2, \ldots$ of the polynomials $a_0+a_1x+a_2x^2+\cdots+a_nx^n$ and of the power series $a_0+a_1x+a_2x^2+\cdots$ which we shall be considering are real. We assume further, unless the contrary is stated, that all functions are *analytic* in the corresponding intervals. The theorems, however, are changed only slightly or not at all if we introduce more general assumptions, e.g. the existence of derivatives up to some order. The zeros in the following are always to be counted according to their *multiplicity*.

We also investigate sequences of real numbers $a_0, a_1, a_2, \ldots$. These may have finitely or infinitely many terms and the order of the terms is significant. We term the subscript m an *index of change* if either

$$a_{m-1}a_m<0, \qquad m\geqq 1$$

or

$$a_{m-1}=a_{m-2}=\cdots=a_{m-k+1}=0 \quad \text{and} \quad a_{m-k}a_m<0,$$

$m\geqq k\geqq 2$. In the first case a_{m-1} and a_m, and in the second case a_{m-k} and a_m constitute a *change of sign*. The number of changes of sign of a sequence (=the number of indices of change) is unchanged if the zero terms are deleted and the remaining terms are unaltered and their order preserved. The trivial case of the sequence in which all the terms are $=0$ will tacitly be excluded. It has incidentally, according to the definition, no changes of sign.

1. The two sequences

$$a_0,\ a_1,\ a_2,\ \ldots,\ a_n \quad \text{and} \quad a_n,\ a_{n-1},\ a_{n-2},\ \ldots,\ a_0$$

have the same number of indices of change.

2. Omitting terms of a sequence does not increase the number of changes of sign.

3. Inserting zero terms into a sequence does not alter the number of changes of sign. Inserting into a sequence a new term next to an old term with which it agrees in sign again does not alter the number of changes of sign.

4. The sequence

$$a_0,\quad a_0+a_1,\quad a_1+a_2,\quad \ldots,\quad a_{n-1}+a_n,\quad a_n$$

has not more changes of sign than the sequence $a_0, a_1, a_2, \ldots, a_n$.

5. Let the infinite sequence $a_0, a_1, a_2, \ldots, a_n, \ldots$ have only a finite number of changes of sign, and denote this number by C. The sequence

$$a_0, \quad a_0+a_1, \quad a_0+2a_1+a_2, \quad \ldots, \quad a_0+\binom{n}{1}a_1+\binom{n}{2}a_2+\cdots+a_n, \quad \ldots$$

formed from it then has also only finitely many and in fact at most C changes of sign. [**4**.]

In a rectangular coordinate system mark the points $(0, a_0)$, $(1, a_1)$, $(2, a_2), \ldots$, $(n, a_n), \ldots$ and join every two successive points by a straight line (whose horizontal projection thus must be $=1$). In the resulting figure the indices of change of the sequence $a_0, a_1, a_2, \ldots$ are quite evident. The zeros of a real analytic function $f(x)$ are not quite so evident from the graph $y=f(x)$. E.g. a double zero cannot be immediately distinguished from a four-fold zero or a three-fold zero from a five-fold zero even in a very accurately drawn graph.

6. In an interval where $\varphi(x)>0$ the two functions $f(x)$ and $f(x)\varphi(x)$ have the same zeros.

7. If $p_0>0, p_1>0, p_2>0, \ldots$, then the two sequences

$$a_0, \; a_1, \; a_2, \; \ldots \quad \text{and} \quad a_0p_0, \; a_1p_1, \; a_2p_2, \; \ldots$$

have the same indices of change.

8. Let the values $f(a)$ and $f(b)$ be different from 0. The interval $a<x<b$ contains an even or an odd number of zeros of $f(x)$ according as $f(a)$ and $f(b)$ have the same or the opposite sign.

9. Let a_j and a_k be different from 0. The subsequence $a_j, a_{j+1}, \ldots, a_{k-1}, a_k$ has an even or an odd number of changes of sign according as a_j and a_k have the same or the opposite sign.

10. (*Rolle's* Theorem.) Let a and b be successive zeros of $f(x)$ $[f(a)=f(b)=0$, $f(x)\neq 0$ for $a<x<b]$. The derivative $f'(x)$ has an odd number of zeros in the interval $a<x<b$ (and thus at least one zero).

11. Let $j+1$ and $k+1$ be successive indices of change of the sequence $a_0, a_1, a_2, \ldots$. Then the sequence of differences

$$a_{j+1}-a_j, \quad a_{j+2}-a_{j+1}, \quad \ldots, \quad a_k-a_{k-1}, \quad a_{k+1}-a_k$$

has an odd number of changes of sign (and thus at least one change of sign).

12. If $f(x)$ has Z zeros in the interval a, b, then $f'(x)$ has not fewer than $Z-1$ zeros there. This is true whether the interval a, b is open, closed or half open; it may even reduce to a single point.

13. If the sequence

$$a_0, \; a_1, \; a_2, \; \ldots, \; a_n$$

has C indices of change, then the sequence

$$a_1-a_0, \; a_2-a_1, \; \ldots, \; a_n-a_{n-1}$$

has not fewer than $C-1$ indices of change.

14. If $f(x)$ has Z zeros in the finite interval $a<x<b$, and if one of the two conditions

$$\operatorname{sgn} f(a)=\operatorname{sgn} f'(a)\neq 0, \qquad \operatorname{sgn} f(b)=-\operatorname{sgn} f'(b)\neq 0$$

is satisfied, then $f'(x)$ has not fewer than Z zeros in a, b. If both conditions are satisfied then $f'(x)$ has not fewer than $Z+1$ zeros.

15. If the finite sequence

$$a_0,\ a_1,\ a_2,\ \ldots,\ a_n$$

has C changes of sign, then the sequence

$$a_0, \quad a_1-a_0, \quad a_2-a_1, \quad \ldots, \quad a_n-a_{n-1}, \quad -a_n$$

formed from it has not fewer than $C+1$ changes of sign. (With the exception of the obvious trivial case in which all the terms a_ν of the sequence $=0$.)

16. If $\lim_{x\to+\infty} f(x)=0$, then $f'(x)$ has not fewer zeros than $f(x)$ in the interior of the interval a, $+\infty$. (A similar result to that for $+\infty$ of course also holds for $-\infty$.)

17. If $\lim_{n\to\infty} a_n=0$, then the infinite sequence

$$a_0, \quad a_1-a_0, \quad a_2-a_1, \quad \ldots, \quad a_n-a_{n-1}, \quad \ldots$$

has more changes of sign than the sequence

$$a_0,\ a_1,\ a_2,\ \ldots,\ a_n,\ \ldots$$

18. Let α be real, and let $f(x)$ have Z zeros in the interval $0<x<\infty$. Then the function

$$\alpha f(x)+f'(x)$$

has at least $Z-1$ zeros there. Moreover it has at least Z zeros if the condition $\lim_{x\to+\infty} e^{\alpha x}f(x)=0$ is satisfied.

19. Let $\alpha>0$, and let the infinite sequence $a_0, a_1, a_2, \ldots, a_n, \ldots$ have C changes of sign. Then the sequence

$$\alpha a_0, \quad \alpha a_1-a_0, \quad \alpha a_2-a_1, \quad \ldots, \quad \alpha a_n-a_{n-1}, \quad \ldots$$

has at least C changes of sign. Moreover it has at least $C+1$ changes of sign if the condition $\lim_{n\to\infty} a_n\alpha^n=0$ is satisfied.

20. If the function $f(x)$ has Z zeros in the interval $0<x<\infty$, then the function $\int_0^x f(t)\,dt$ has Z or fewer zeros there.

21. If the infinite sequence

$$a_0,\ a_1,\ a_2,\ \ldots,\ a_n,\ \ldots$$

has C changes of sign, then the sequence

$$a_0, \quad a_0+a_1, \quad a_0+a_1+a_2, \quad \ldots, \quad a_0+a_1+a_2+\cdots+a_n, \quad \ldots$$

has C or fewer changes of sign.

§ 2. Reversals of Sign of a Function

We say that the function $f(x)$ is *of constant sign* in an interval in the following two cases: (1) $f(x) \leqq 0$ in the interval, (2) $f(x) \geqq 0$ in the interval. Assume that the interval $a<x<b$ can be divided into $R+1$ sub-intervals such that

(1) $f(x)$ is not identically zero in any sub-interval,
(2) $f(x)$ is of constant sign in each sub-interval,
(3) $f(x)$ is of opposite sign in adjacent sub-intervals.

Under these three conditions we say that $f(x)$ has R *reversals of sign* in the interval $a<x<b$. A zero of an analytic function does or does not induce a reversal of sign (is or is not a "crossing point") according as the multiplicity of the zero is odd or even. However the concept "reversal of sign" can usefully be applied also to certain non-analytic functions.

22. Let the function $f(x)$ be different from 0 and of constant sign both in a certain neighbourhood of the point a and in a certain neighbourhood of the point b. The interval $a<x<b$ contains an even or an odd number of reversals of sign of the function $f(x)$, according as $f(a)$ and $f(b)$ have the same sign or opposite signs.

23. If the function $f(x)$ has R reversals of sign in the interval $0<x<\infty$, then the function $\int_0^x f(t)\,dt$ has in this interval R or fewer reversals of sign.

24. If R is the number of reversals of sign and Z the number of zeros of $f(x)$ in the same open interval, then $Z-R$ is a non-negative even number.

25. If $f(a)=f(b)=0$ and $f(x)\neq 0$ for $a<x<b$, then in the interval $a<x<b$ there are an odd number of reversals of sign of the derivative $f'(x)$.

26. Let $A_1, A_2, \ldots, A_n$ be non-zero real numbers and let $a_1<a_2<a_3<\cdots<a_n$. One can prove in various cases that the rational function

$$f(x)=\frac{A_1}{x-a_1}+\frac{A_2}{x-a_2}+\cdots+\frac{A_n}{x-a_n}$$

has only real zeros, in particular in the following cases:

(1) $A_1>0, \quad A_2>0, \quad \ldots, \quad A_{n-1}>0;$
(2) $A_1>0, \quad A_2>0, \quad \ldots, \quad A_{k-1}>0, \quad A_{k+1}>0, \quad \ldots, \quad A_n>0,$
$A_1+A_2+\cdots+A_n<0, \quad 1<k<n.$

27. The trigonometric polynomial

$$f(x)=a_0+a_1\cos x+a_2\cos 2x+\cdots+a_n\cos nx$$

with real coefficients $a_0, a_1, a_2, \ldots, a_n$ has only real zeros if

$$|a_0|+|a_1|+|a_2|+\cdots+|a_{n-1}|<a_n.$$

§ 3. First proof of Descartes' Rule of Signs

By the *changes of sign* and the *indices of change* of the polynomial

$$a_0+a_1x+a_2x^2+\cdots+a_nx^n$$

or the power series

$$a_0+a_1x+a_2x^2+a_3x^3+\cdots$$

we mean the changes of sign and indices of change of the finite or the infinite *sequence of coefficients*

$$a_0,\ a_1,\ a_2,\ \ldots,\ a_n \qquad \text{or} \qquad a_0,\ a_1,\ a_2,\ a_3,\ \ldots$$

respectively.

28. The polynomials $P(x)$ and $P(\alpha x)$ have the same number of changes of sign if α is positive.

29. Denote the number of changes of sign of the polynomials

$$P(x)=a_0+a_1x+a_2x^2+\cdots+a_nx^n,$$
$$P(-x)=a_0-a_1x+a_2x^2-\cdots+(-1)^n a_nx^n$$

by C^+ and C^- respectively. We then have

$$C^+ + C^- \leqq n.$$

30. Let $\alpha>0$. Passing from the polynomial

$$a_0+a_1x+a_2x^2+\cdots+a_nx^n$$

to the polynomial

$$(\alpha-x)(a_0+a_1x+a_2x^2+\cdots+a_nx^n)$$
$$=\alpha a_0+(\alpha a_1-a_0)x+(\alpha a_2-a_1)x^2+\cdots-a_nx^{n+1}$$

the number of changes of sign increases; in fact it increases by an odd number. [For the case $\alpha=1$ cf. **15**.]

31. Let $\alpha>0$. Passing from the power series

$$a_0+a_1x+a_2x^2+a_3x^3+\cdots$$

to the power series

$$(\alpha-x)(a_0+a_1x+a_2x^2+\cdots)=\alpha a_0+(\alpha a_1-a_0)x+(\alpha a_2-a_1)x^2+\cdots$$

the number of changes of sign does not decrease. In fact it increases if the original series $a_0+a_1x+a_2x^2+\cdots$ converges for $x=\alpha$.

32. Let $\alpha>0$. Passing from the power series

$$a_0+a_1x+a_2x^2+a_3x^3+\cdots$$

to the power series

$$(\alpha+x)(a_0+a_1x+a_2x^2+\cdots)=\alpha a_0+(\alpha a_1+a_0)x+(\alpha a_2+a_1)x^2+\cdots$$

the number of changes of sign does not increase.

33. Passing from the power series

$$a_0+a_1x+a_2x^2+a_3x^3+\cdots$$

to the power series

$$\frac{a_0+a_1x+a_2x^2+\cdots}{1-x}=a_0+(a_0+a_1)x+(a_0+a_1+a_2)x^2+\cdots$$

the number of changes of sign does not increase.

34. Let $\alpha>0$. Passing from the power series

$$a_0+\frac{a_1}{1!}x+\frac{a_2}{2!}x^2+\frac{a_3}{3!}x^3+\cdots$$

to the power series

$$e^{\alpha x}\left(a_0+\frac{a_1}{1!}x+\frac{a_2}{2!}x^2+\cdots\right)=\sum_{n=0}^{\infty}\frac{a_0\alpha^n+\binom{n}{1}a_1\alpha^{n-1}+\cdots+a_n}{n!}x^n$$

the number of changes of sign does not increase.

35. Let $p_1, p_2, \ldots, p_n$ be positive numbers. Set

$$a_0+\frac{a_1x}{p_1-x}+\frac{a_2x^2}{(p_1-x)(p_2-x)}+\cdots+\frac{a_nx^n}{(p_1-x)(p_2-x)\cdots(p_n-x)}$$
$$=A_0+A_1x+A_2x^2+\cdots$$

(the power series expansion is convergent for sufficiently small values of x). The number of changes of sign of the finite sequence $a_0, a_1, a_2, \ldots, a_n$ is not less than the number of changes of sign of the infinite sequence $A_0, A_1, A_2, \ldots$. [Mathematical induction, **31**.]

36. (*Descartes' rule of signs.*) Let Z be the number of positive zeros of the polynomial $a_0+a_1x+a_2x^2+\cdots+a_nx^n$ and C the number of changes of sign of the sequence of its coefficients. We then have

$$C-Z\geqq 0 \qquad \textbf{[30.]}$$

37 (continued). $C-Z$ is an even number.

38. Let the radius of convergence of the power series $a_0+a_1x+a_2x^2+\cdots$ be ρ, let the number of its zeros in the interval $0<x<\rho$ be Z and let the number of changes of sign in the sequence of its coefficients be C. Then

$$Z\leqq C.$$

Hence, in particular: If C is finite then Z is also finite. [Besides **31**, the theory of functions of a complex variable must be used.]

39. The power series

$$2-\frac{x}{1\cdot 2}-\frac{x^2}{2\cdot 3}-\frac{x^3}{3\cdot 4}-\cdots$$

has no zeros in its circle of convergence. (The sequence of its coefficients has one change of sign—**37** cannot be extended to power series without some essential additional condition.)

40. (Continuation of **38**). If $\rho=\infty$ or if ρ is finite and $a_0+a_1\rho+a_2\rho^2+\cdots$ is divergent, then, provided C is finite, $C-Z$ is a non-negative even number.

41. Let the minimum of the real numbers $\xi_1, \xi_2, \ldots, \xi_n$ be denoted by ξ_α and their maximum by ξ_ω. The number of zeros of the polynomial

$$a_0+a_1(x-\xi_1)+a_2(x-\xi_1)(x-\xi_2)+\cdots+a_n(x-\xi_1)(x-\xi_2)\ldots(x-\xi_n)$$

in the interval $\xi_\omega<x<\infty$ is equal to, or less by an even number than, the number of changes of sign of the sequence

$$a_0,\ a_1,\ a_2,\ \ldots,\ a_n,$$

and the number of zeros in the interval $-\infty < x < \xi_\alpha$ is equal to, or less by an even number than, the number of changes of sign of the sequence

$$a_0, \; -a_1, \; a_2, \; -a_3, \; \ldots, \; (-1)^n a_n.$$

(For $\xi_1 = \xi_2 = \cdots = \xi_n$ equivalent to **36, 37**.) [**35**.]

§ 4. Applications of Descartes' Rule of Signs

42. Let λ be a positive proper fraction, n a positive integer. The transcendental equation

$$1 + \frac{x}{1!} + \frac{x^2}{2!} + \cdots + \frac{x^n}{n!} = \lambda \, e^x$$

has a single positive root. This increases monotonically to infinity as n tends to infinity.

43. The function $x^{-5}\left(e^{\frac{1}{x}} - 1\right)^{-1}$ of the positive variable x tends to 0 for $x=0$ and for $x=+\infty$ and between these values has a maximum and no minimum.

43.1. Let $R(f)$ denote the number of reversals of sign of the function $f(x)$ in $(0, 1)$ and $Z(K_n)$ the number of zeros of the polynomial $K_n(x)$ (see introduction to II **144**) in $(0, 1)$. Prove that

$$Z(K_n) \leqq R(f).$$

44. Let the radius of convergence of the power series $a_0 + a_1 x + a_2 x^2 + \cdots$ be $\geqq 1$. The number of its zeros in the interval $0 < x < 1$ does not exceed the number of changes of sign of the sequence

$$a_0, \quad a_0 + a_1, \quad a_0 + a_1 + a_2, \quad \ldots, \quad a_0 + a_1 + \cdots + a_n, \quad \ldots.$$

44.1. Let C be the number of changes of sign of the finite sequence

$$a_0, \quad a_0 + a_1, \quad a_0 + a_1 + a_2, \quad \ldots, \quad a_0 + a_1 + \cdots + a_n,$$

set

$$P(x) = a_0 + a_1 x + a_2 x^2 + \cdots + a_n x^n,$$

and assume that $P(1)=0$. Then $P(x)$ has at most $C+1$ positive zeros. Show also that the bound $C+1$ can be attained.

45. Let $\left(\frac{n}{p}\right)$ denote the *Legendre* (*Jacobi*) symbol, defined for p an odd prime and n any number not divisible by p as $\left(\frac{n}{p}\right) = 1$ if n is a *quadratic residue* of p and $\left(\frac{n}{p}\right) = -1$ if n is a *quadratic non-residue* of p (Hardy-Wright, p. 68). The equation

$$\left(\frac{1}{19}\right) x + \left(\frac{2}{19}\right) x^2 + \left(\frac{3}{19}\right) x^3 + \cdots + \left(\frac{18}{19}\right) x^{18} = 0$$

has only one positive root, namely $x=1$. [The equation is reciprocal so that it is sufficient to investigate the zeros in the interval $0 < x < 1$ with the aid of **44, 33**.]

46. The equation of degree 162

$$\left(\frac{1}{163}\right) x + \left(\frac{2}{163}\right) x^2 + \left(\frac{3}{163}\right) x^3 + \cdots + \left(\frac{162}{163}\right) x^{162} = 0$$

has exactly 5 positive roots which are all simple. [Investigate the point $x=0.7$.]

47. (Supplement to **36**.) If the polynomial $a_0+a_1x+\cdots+a_nx^n$ has only real zeros, then $Z=C$.

48. Let $\nu_1, \nu_2, \ldots, \nu_n$ be integers, $0\leqq\nu_1<\nu_2<\cdots<\nu_n$, and $0<\alpha_1<\alpha_2<\cdots<\alpha_n$. Then the determinant (a generalization of the *Vandermonde* determinant, which corresponds to the case $\nu_1=0$, $\nu_2=1, \ldots, \nu_n=n-1$)

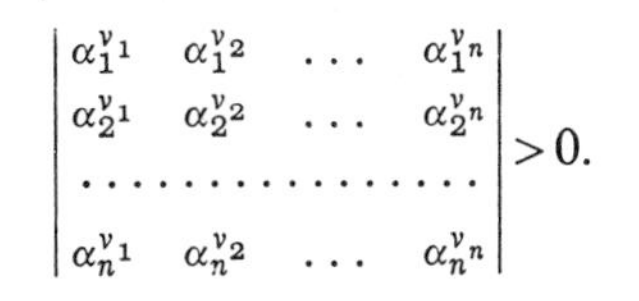

$$\begin{vmatrix} \alpha_1^{\nu_1} & \alpha_1^{\nu_2} & \ldots & \alpha_1^{\nu_n} \\ \alpha_2^{\nu_1} & \alpha_2^{\nu_2} & \ldots & \alpha_2^{\nu_n} \\ \cdots & \cdots & \cdots & \cdots \\ \alpha_n^{\nu_1} & \alpha_n^{\nu_2} & \ldots & \alpha_n^{\nu_n} \end{vmatrix} > 0.$$

[First prove that it is $\neq 0$, **36**.]

49. Let $a_0\neq 0$, $a_n\neq 0$, and assume that $2m$ *consecutive* coefficients of the polynomial $a_0+a_1x+\cdots+a_nx^n$ vanish, where m is an integer, $m\geqq 1$. Then the polynomial has at least $2m$ non-real zeros.

50. Let the polynomial $P(x)$ have only real zeros, and let $P(0)=1$ and $P(x)$ not a constant. Set

$$\frac{1}{P(x)}=1+b_1x+b_2x^2+\cdots+b_nx^n+\cdots.$$

Then the polynomial $1+b_1x+\cdots+b_{2m}x^{2m}$ has only non-real zeros. [**49**.]

51. Let m be an integer, $m\geqq 1$, and

$$S(x_1, x_2, \ldots, x_n)=\sum x_1^{l_1}x_2^{l_2}\ldots x_n^{l_n},$$

where the sum $\sum$ is taken over all sets of non-negative integers $l_1, l_2, \ldots, l_n$ for which

$$l_1+l_2+\cdots+l_n=2m.$$

The homogeneous symmetric function $S(x_1, x_2, \ldots, x_n)$ of the n real variables $x_1, x_2, \ldots, x_n$ is positive definite, i.e. >0 for all sets of values of $x_1, x_2, \ldots, x_n$ with the exception of the single set $x_1=0, x_2=0, \ldots, x_n=0$.

52. Let the polynomial $P(x)=x^n+\cdots$ have only positive zeros. In the power series expansion

$$\frac{1}{P(x)}=\frac{1}{x^n}+\frac{B_n}{x^{n+1}}+\frac{B_{n+1}}{x^{n+2}}+\cdots$$

all the coefficients $B_n, B_{n+1}, \ldots$ are positive.

§ 5. Applications of Rolle's Theorem

53. The derivative of a polynomial has not more non-real zeros than the polynomial itself.

54. Multiple zeros of the derivative of a polynomial with only real zeros are also multiple zeros of the polynomial itself.

55. If a polynomial has only real and simple zeros, then its higher derivatives have the same property, and moreover every zero of the $(\nu+1)$th derivative lies between two consecutive zeros of the νth.

56. The higher derivatives of the function $(1+x^2)^{-1/2}$ have only real and simple zeros; moreover every zero of the νth derivative lies between two consecutive zeros of the $(\nu+1)$th.

57. The higher derivatives of the function $x(1+x^2)^{-1}$ also have the properties described in **56**.

58. One may define the *Legendre*, the *Laguerre* and the *Hermite* polynomials by the formulae

$$P_n(x)=\frac{1}{2^n n!}\frac{d^n}{dx^n}(x^2-1)^n, \qquad e^{-x}L_n(x)=\frac{1}{n!}\frac{d^n}{dx^n}e^{-x}x^n,$$

$$e^{-\frac{x^2}{2}}H_n(x)=\frac{1}{n!}\frac{d^n}{dx^n}e^{-\frac{x^2}{2}}$$

respectively [VI **84**, VI **99**, VI **100**]. Deduce from this definition that these polynomials have only real and simple zeros, which lie in the interior of the intervals

$$(-1, +1), \qquad (0, +\infty), \qquad (-\infty, +\infty)$$

respectively. (VI **97**, VI **99**, VI **100**.)

59. Let q be an integer, $q \geqq 2$. Let

$$\frac{d^{n-1}}{dx^{n-1}}\left(\frac{x^{q-1}}{1+x^q}\right)=\frac{Q_n(x)}{(1+x^q)^n}, \qquad \frac{d^n}{dx^n}e^{-x^q}=e^{-x^q}R_n(x).$$

In the complex plane draw q rays from the origin which divide the plane into equal sectors. Let one of these rays be the positive real axis. Show that the zeros of the polynomials $Q_n(x)$, $R_n(x)$ lie on these q rays and are arranged in the same way on each ray and finally that those which are $\neq 0$ are simple. (The special case $q=2$ is already covered by **57, 58.**)

60. Let μ, ν be integers satisfying $0 \leqq \mu < \mu+\nu \leqq n$. None of the polynomials

$$a_\mu+\binom{\nu}{1}a_{\mu+1}x+\binom{\nu}{2}a_{\mu+2}x^2+\cdots+\binom{\nu}{\nu-1}a_{\mu+\nu-1}x^{\nu-1}+a_{\mu+\nu}x^\nu$$

has more nonreal zeros than the polynomial

$$a_0+\binom{n}{1}a_1x+\binom{n}{2}a_2x^2+\cdots+\binom{n}{n-1}a_{n-1}x^{n-1}+a_nx^n.$$

(Generalization of **53**.)

61. Let $a_1, a_2, \ldots, a_n$ be positive numbers which are not all equal. Let

$$(x+a_1)(x+a_2)\ldots(x+a_n)=x^n+\binom{n}{1}m_1x^{n-1}+\binom{n}{2}m_2^2x^{n-2}+\cdots+m_n^n.$$

The numbers $m_1, m_2, \ldots, m_n$ are to be determined by taking the appropriate roots and are to be chosen positive. m_1 is the arithmetic mean and m_n the geometric mean of $a_1, a_2, \ldots a_n$. Show that

$$m_1>m_2>m_3>\cdots>m_{n-1}>m_n.$$

62. Let α be real and $P(x)$ a polynomial. The polynomial $\alpha P(x)+P'(x)$ has not more non-real zeros than the polynomial $P(x)$ itself. (Generalization of **53**.)

62.1. The polynomials

$$s_1^nx+s_2^nx^2+s_3^nx^3+\cdots+s_n^nx^n, \qquad S_1^nx+S_2^nx^2+S_3^nx^3+\cdots+S_n^nx^n$$

have only real, non-positive, simple zeros. [See definitions in the introduction to I **197** and I **186**, respectively.]

63. Assume that the equation

$$a_0+a_1x+a_2x^2+\cdots+a_nx^n=0$$

has only real roots. Then the polynomial

$$a_0P(x)+a_1P'(x)+\cdots+a_nP^{(n)}(x)$$

has not more non-real zeros than the polynomial $P(x)$ itself.

64.

$$P(x)-\frac{P''(x)}{1!}+\frac{P^{(IV)}(x)}{2!}-\frac{P^{(VI)}(x)}{3!}+\cdots$$

has not more non-real zeros than the polynomial $P(x)$ itself.

65. If the equation

$$a_0+a_1x+a_2x^2+\cdots+a_nx^n=0$$

has only real roots, then the following equation

$$a_0+\frac{a_1}{1!}x+\frac{a_2}{2!}x^2+\cdots+\frac{a_n}{n!}x^n=0$$

has also only real roots.

65.1. (Continuation of **63**.) Assume further that $a_0\neq 0$ and consider the power series

$$\frac{1}{a_0+a_1x+a_2x^2+\cdots+a_nx^n}=b_0+b_1x+b_2x^2+\cdots.$$

Then the polynomial

$$b_0P(x)+b_1P'(x)+b_2P''(x)+\cdots+b_kP^{(k)}(x)+\cdots$$

has not less non-real zeros than the polynomial $P(x)$ itself.

66. Let $P(x)$ be a polynomial of nth degree, and let the real number α lie outside the interval $[-n, 0]$. Then $\alpha P(x)+xP'(x)$ has not more non-real zeros than $P(x)$. (A generalization of **53** different from **60** and **62**.)

67. Let $Q(x)$ be a polynomial whose zeros are all real and lie outside the interval $[0, n]$. Then the equation

$$a_0Q(0)+a_1Q(1)x+a_2Q(2)x^2+\cdots+a_nQ(n)x^n=0$$

has not more non-real roots than the equation

$$a_0+a_1x+a_2x^2+\cdots+a_nx^n=0.$$

68. Let $0<q<1$. The equation

$$a_0+a_1qx+a_2q^4x^2+\cdots+a_nq^{n^2}x^n=0$$

has not more non-real roots than the equation

$$a_0+a_1x+a_2x^2+\cdots+a_nx^n=0.$$

69. Let $q>0$. The equation

$$2a_0+(q+q^{-1})a_1x+(q^{\sqrt{2}}+q^{-\sqrt{2}})a_2x^2+\cdots+(q^{\sqrt{n}}+q^{-\sqrt{n}})a_nx^n=0$$

has not more non-real roots than the equation

$$a_0+a_1x+a_2x^2+\cdots+a_nx^n=0.$$

70. If the curve $y=f(x)$ meets a straight line in three distinct points, then between the outermost intersections there lies at least one point of inflection.

71. If a function coincides at $n+1$ points with a polynomial of degree $n-1$, then its nth derivative vanishes at an intermediate point.

72. Let α be a real constant, different from $0, 1, 2, \ldots, n-1$. The difference

$$(1+x)^\alpha-\left(1+\frac{\alpha}{1}x+\frac{\alpha(\alpha-1)}{1.2}x^2+\cdots+\frac{\alpha(\alpha-1)\ldots(\alpha-n+2)}{1.2\ldots(n-1)}x^{n-1}\right)$$

vanishes for $x=0$, but at no other point in the interval $(-1,\infty)$.

73. The remainder of the exponential series

$$\frac{x^n}{n!}+\frac{x^{n+1}}{(n+1)!}+\frac{x^{n+2}}{(n+2)!}+\cdots$$

vanishes for $x=0$, but for no other real value of x.

74. The nth partial sum of the exponential series

$$1+\frac{x}{1!}+\frac{x^2}{2!}+\cdots+\frac{x^n}{n!}$$

has no real zero or one real zero according as n is even or odd.

75. Let the polynomials $P_1(x), P_2(x), \ldots, P_l(x)$ be $\not\equiv 0$ and of degree m_1-1, $m_2-1, \ldots, m_l-1$ respectively, and let the real constants $a_1, a_2, \ldots, a_l$ be distinct. The function

$$g(x)=P_1(x)\,e^{a_1x}+P_2(x)\,e^{a_2x}+\cdots+P_l(x)\,e^{a_lx}$$

has at most $m_1+m_2+\cdots+m_l-1$ real zeros.

76. Let $\alpha_1<\alpha_2<\cdots<\alpha_n$, $\beta_1<\beta_2<\cdots<\beta_n$. Then the determinant

$$\begin{vmatrix} e^{\alpha_1\beta_1} & e^{\alpha_1\beta_2} & \ldots & e^{\alpha_1\beta_n} \\ e^{\alpha_2\beta_1} & e^{\alpha_2\beta_2} & \ldots & e^{\alpha_2\beta_n} \\ \ldots & \ldots & \ldots & \ldots \\ e^{\alpha_n\beta_1} & e^{\alpha_n\beta_2} & \ldots & e^{\alpha_n\beta_n} \end{vmatrix}>0.$$

(Generalization of **48**.)

§ 6. Laguerre's Proof of Descartes' Rule of Signs

77. Let $a_1, a_2, \ldots, a_n, \lambda_1, \lambda_2, \ldots, \lambda_n$ be real constants, $\lambda_1<\lambda_2<\lambda_3<\cdots<\lambda_n$. Denote by Z the number of real zeros of the entire function

$$F(x)=a_1\,e^{\lambda_1x}+a_2\,e^{\lambda_2x}+\cdots+a_n\,e^{\lambda_nx}$$

and by C the number of changes of sign in the sequence of numbers $a_1, a_2, \ldots, a_n$.

Then $C-Z$ is a non-negative even integer. (A generalization of **36**, **37** different from **41**: Replace e^x by x.)

Proof. We may assume without loss of generality that $a_1, a_2, \ldots a_n$ are all different from 0. That $C-Z$ is even is evident from the fact that as $x \to -\infty$ the term $a_1\, e^{\lambda_1 x}$ is the dominant term, and as $x \to +\infty$ the term $a_n\, e^{\lambda_n x}$ is the dominant term. [**8**, **9**, **37**.] That $C-Z \geqq 0$ may be proved by mathematical induction; we deduce the relation for C from that for $C-1$, with the aid of Rolle's theorem [solution **75**]. Indeed, if there are *no* changes of sign, evidently Z is also $=0$ and the statement holds. Assume that it holds in the case that there are $C-1$ changes of sign. $F(x)$ has C changes of sign, $C \geqq 1$; e.g. let $\alpha+1$ be an index of change, $1 \leqq \alpha < n$, $a_\alpha a_{\alpha+1} < 0$. Choose a number λ in the interval $\lambda_\alpha < \lambda < \lambda_{\alpha+1}$ and consider the function

$$F^*(x) = e^{\lambda x} \frac{d[e^{-\lambda x} F(x)]}{dx}$$
$$= a_1(\lambda_1 - \lambda)\, e^{\lambda_1 x} + \cdots + a_\alpha(\lambda_\alpha - \lambda)\, e^{\lambda_\alpha x}$$
$$+ a_{\alpha+1}(\lambda_{\alpha+1} - \lambda)\, e^{\lambda_{\alpha+1} x} + \cdots + a_n(\lambda_n - \lambda)\, e^{\lambda_n x}.$$

Let the number of zeros of $F^*(x)$ be Z^*. We have [**6**, **12**]

$$Z^* \geqq Z-1.$$

Let the number of changes of sign of the sequence of coefficients

$$-a_1(\lambda-\lambda_1), \quad -a_2(\lambda-\lambda_2), \quad \ldots, \quad -a_\alpha(\lambda-\lambda_\alpha),$$
$$a_{\alpha+1}(\lambda_{\alpha+1}-\lambda), \quad \ldots, \quad a_n(\lambda_n-\lambda)$$

be C^*. Clearly

$$C^* = C-1.$$

By the assumption of the mathematical induction argument we have

$$C^* \geqq Z^*.$$

From the three relations above it follows that

$$C \geqq Z,$$

q.e.d.—Was it necessary to assume from the outset that $a_1, a_2, \ldots, a_n$ are all $\neq 0$? Could we not also have chosen $\lambda = \lambda_\alpha$ or $\lambda = \lambda_{\alpha+1}$?

77.1. The entire function

$$F(x) = \sum_{k=1}^{n} (-1)^{n-k} \binom{n}{k} k^x$$

has simple zeros at the points

$$x = 1,\ 2,\ 3,\ \ldots,\ n-1$$

and no other real zeros. (Interpret k^x as $e^{x \log k}$, with the principal value of $\log k$.)

78. Let $\lambda_1 < \lambda_2 < \lambda_3 < \cdots$, $\lim_{n\to\infty} \lambda_n = \infty$. In the interior of its region of convergence (a half-plane bounded on the left), the *Dirichlet* series

$$a_1\, e^{-\lambda_1 x} + a_2\, e^{-\lambda_2 x} + \cdots + a_n\, e^{-\lambda_n x} + \cdots$$

has not more real zeros than the number of changes of sign of the sequence of coefficients $a_1, a_2, a_3, \ldots, a_n, \ldots$. (Generalization of **38.**)

79. Let m and n be integers and $a_1, a_2, \ldots, a_n, \lambda_1, \lambda_2, \ldots, \lambda_n$ real numbers that satisfy the conditions

$$m \geqq 1; \qquad n \geqq 2; \qquad a_\nu \neq 0, \quad \nu = 1, 2, \ldots, n; \qquad \lambda_1 < \lambda_2 < \cdots < \lambda_n.$$

Assume that the polynomial

$$P(x) = a_1(x-\lambda_1)^m + a_2(x-\lambda_2)^m + \cdots + a_n(x-\lambda_n)^m$$

does not vanish identically. Show that the number Z of real zeros of $P(x)$ does not exceed the number C of changes of sign of the sequence

$$a_1, \ a_2, \ a_3, \ \ldots, \ a_{n-1}, \ a_n, \ (-1)^m a_1$$

[**77**, **14**].

80. Denote the number of reversals of sign of the function $\varphi(\lambda)$ in the interval $0 < \lambda < \infty$ by R, and the number of real zeros of the integral

$$F(x) = \int_0^\infty \varphi(\lambda)\, e^{-\lambda x}\, d\lambda$$

by Z. Then we have $Z \leqq R$.

In the number of zeros Z we naturally include only the zeros lying in the interior of the region of convergence (a half-plane bounded on the left). [Proof not by application of a limiting process to **77** but by appropriate translation of the proof of **77** to the present case!]

81. The integrals

$$\int_0^\infty f(x) x^n \, dx = M_n, \qquad n = 0, 1, 2, 3, \ldots,$$

assumed convergent, are called the *moments* of the function $f(x)$. Consider the case that not all the moments vanish [II **139**, III **153**]. Assume e.g. $M_\mu \neq 0$ and set

(1) $a_n = M_n$ if $M_n \neq 0$,

(2) $a_n = -\operatorname{sgn} a_{n+1}$ if $M_n = 0$ and $n < \mu$,

(3) $a_n = -\operatorname{sgn} a_{n-1}$ if $M_n = 0$ and $n > \mu$.

(The a_n are defined by recursion; first of all the sign of a_μ must be determined.) The number of reversals of sign of $f(x)$ is not less than the number of changes of sign of the sequence $a_0, a_1, a_2, a_3, \ldots$. (II **140** is a special case.)

82. (Continuation of **80.**) The number of positive zeros lying in the interior of the region of convergence does not exceed the number of reversals of sign of the function $\Phi(\lambda) = \int_0^\lambda \varphi(x)\, dx$. (Analogue of **44.**)

83. (Continuation of **78.**) The number of positive zeros lying in the interior of the region of convergence does not exceed the number of changes of sign of the sequence

$$a_1, \ a_1 + a_2, \ a_1 + a_2 + a_3, \ \ldots.$$

(Generalization of **44.**) [**80.**]

83.1. Let C be the number of changes of sign of the finite sequence

$$a_1, \quad a_1+a_2, \quad a_1+a_2+a_3, \quad \ldots, \quad a_1+a_2+\cdots+a_n,$$

$\lambda_i (i=1, 2, \ldots, n)$ real numbers satisfying

$$\lambda_1 < \lambda_2 < \lambda_3 < \cdots < \lambda_n,$$

and

$$D(x) = a_1\, e^{-\lambda_1 x} + a_2\, e^{-\lambda_2 x} + \cdots + a_n\, e^{-\lambda_n x},$$

where we assume that $D(0)=0$. Then $D(x)$ has at most $C+1$ real zeros.

(Both the hypothesis and the conclusion are stronger than in **83**. Generalization of **44.1**.)

84. The number of positive zeros lying within the region of convergence of the factorial series

$$a_0 + \frac{1!a_1}{x} + \frac{2!a_2}{x(x+1)} + \frac{3!a_3}{x(x+1)(x+2)} + \cdots$$

does not exceed the number of changes of sign of the sequence

$$a_0, \quad a_0+a_1, \quad a_0+a_1+a_2, \quad \ldots .$$

[It does not even exceed the number of zeros lying within the interval $0<x<1$ of the power series

$$f(x) = a_0 + a_1 x + a_2 x^2 + \cdots; \qquad \textbf{80, 44.}]$$

85. Let the coefficients $p_0, p_1, p_2, \ldots$ of the non-terminating infinite series

$$F(x) = p_0 + p_1 x + p_2 x^2 + \cdots$$

be non-negative, and let its radius of convergence be ρ. Let $a_1, a_2, \ldots, a_n, \alpha_1, \alpha_2, \ldots, \alpha_n$ be real and

$$0 < \alpha_1 < \alpha_2 < \alpha_3 < \cdots < \alpha_n \leqq 1.$$

The number of zeros of the function

$$a_1 F(\alpha_1 x) + a_2 F(\alpha_2 x) + \cdots + a_n F(\alpha_n x)$$

in the interval $0<x<\rho$ is not larger than the number of changes of sign of the sequence

$$a_n, \quad a_n+a_{n-1}, \quad a_n+a_{n-1}+a_{n-2}, \quad \ldots, \quad a_n+a_{n-1}+\cdots+a_2+a_1 \qquad \textbf{[38, 83]}.$$

86 (continued). If $0<\beta_1<\beta_2<\cdots<\beta_n$ and $\alpha_n\beta_n$ lies in the interior of the circle of convergence of $F(x)$, then the determinant

$$\begin{vmatrix} F(\alpha_1\beta_1) & F(\alpha_1\beta_2) & \ldots & F(\alpha_1\beta_n) \\ F(\alpha_2\beta_1) & F(\alpha_2\beta_2) & \ldots & F(\alpha_2\beta_n) \\ \ldots & \ldots & \ldots & \ldots \\ F(\alpha_n\beta_1) & F(\alpha_n\beta_2) & \ldots & F(\alpha_n\beta_n) \end{vmatrix} \neq 0.$$

(**76** readily follows from this.)

§ 7. What is the Basis of Descartes' Rule of Signs?

We see from **36, 41, 77, 84, 85** that the sequences of functions

$$\begin{array}{lllll} 1, & x, & x^2, & x^3, & \ldots, \\ 1, & x-\xi_1, & (x-\xi_1)(x-\xi_2), & \ldots, & \\ e^{\lambda_1 x}, & e^{\lambda_2 x}, & e^{\lambda_3 x}, & \ldots, & \\ 1, & \dfrac{1}{x}, & \dfrac{1}{x(x+1)}, & \dfrac{1}{x(x+1)(x+2)}, & \ldots, \\ F(\alpha_1 x), & F(\alpha_2 x), & F(\alpha_3 x), & \ldots & \end{array}$$

considered there have a common property: The number of zeros lying in a certain interval of their linear combinations with constant coefficients never exceeds the number of changes of sign of these coefficients. What is the basis for this frequent validity of Descartes' rule of signs?

87. Let the sequence of functions

$$h_1(x),\ h_2(x),\ h_3(x),\ \ldots,\ h_n(x)$$

obey Descartes' rule of signs in the open inverval $a<x<b$. More precisely: If $a_1, a_2, \ldots, a_n$ denote any real numbers which are not all zero, then the number of zeros lying in $a<x<b$ of the linear combination

$$a_1h_1(x)+a_2h_2(x)+\cdots+a_nh_n(x)$$

never exceeds the number of changes of sign of the sequence

$$a_1,\ a_2,\ \ldots,\ a_n.$$

For this to hold, the following property of the sequence $h_1(x), h_2(x), \ldots, h_n(x)$ is a necessary condition: If $\nu_1, \nu_2, \ldots, \nu_l$ denote integers with $1\leqq\nu_1<\nu_2<\nu_3<\cdots<\nu_l\leqq n$, then the Wronskian determinants [VII, §5]

$$W[h_{\nu_1}(x),\ h_{\nu_2}(x),\ h_{\nu_3}(x),\ \ldots,\ h_{\nu_l}(x)]$$

do not vanish in the interval (a, b) and further any two Wronskian determinants with the same number l of rows have the same sign, where $l=1, 2, 3, \ldots, n-1$. [Look at multiple zeros!]

88 (continued). In particular for the validity of Descartes' rule of signs it is necessary that in the interval $a<x<b$ the quotients

$$\frac{h_2(x)}{h_1(x)},\ \frac{h_3(x)}{h_2(x)},\ \ldots,\ \frac{h_n(x)}{h_{n-1}(x)}$$

are all positive and are either all monotonically decreasing or all monotonically increasing.

89 (continued). Let $1\leqq\alpha\leqq n$. If $h_1(x), h_2(x), \ldots, h_n(x)$ satisfy the determinantal conditions stated in **87**, then so do the $n-1$ functions

$$H_1=-\frac{d}{dx}\frac{h_1}{h_\alpha},\qquad H_2=-\frac{d}{dx}\frac{h_2}{h_\alpha},\ldots,\qquad H_{\alpha-1}=-\frac{d}{dx}\frac{h_{\alpha-1}}{h_\alpha},$$

$$H_\alpha=\frac{d}{dx}\frac{h_{\alpha+1}}{h_\alpha},\ldots,\qquad H_{n-2}=\frac{d}{dx}\frac{h_{n-1}}{h_\alpha},\qquad H_{n-1}=\frac{d}{dx}\frac{h_n}{h_\alpha}.$$

[VII **58**.]

90 (continued). The criterion given in **87** as a necessary condition for the validity of Descartes' rule of signs is also a sufficient condition [**77**].

91. Verify that the criterion **87** is satisfied by the functions $e^{\lambda_1 x}, e^{\lambda_2 x}, \ldots, e^{\lambda_n x}$, considered in **77**.

§ 8. Generalizations of Rolle's Theorem

92. Let $0<a<b$. If $f(x)$ vanishes at $n+1$ points of the interval $[a, b]$ and if all the zeros of the polynomial with real coefficients $a_0+a_1x+a_2x^2+\cdots+a_nx^n$ are real, then at an interior point ξ of $[a, b]$

$$a_0 f(\xi)+a_1 f'(\xi)+a_2 f''(\xi)+\cdots+a_n f^{(n)}(\xi)=0. \qquad [\mathbf{63}.]$$

93. (Generalization of Rolle's theorem to homogeneous linear differential expressions.) Let the differential equation of nth order

$$(*) \qquad y^{(n)}+\varphi_1(x)y^{(n-1)}+\varphi_2(x)y^{(n-2)}+\cdots+\varphi_n(x)y=0$$

have $n-1$ solutions $h_1(x), h_2(x), \ldots, h_{n-1}(x)$ such that, for $a<x<b$, we have

$$(**) \qquad h_1(x)>0, \qquad \begin{vmatrix} h_1(x) & h_1'(x) \\ h_2(x) & h_2'(x) \end{vmatrix}>0, \quad \ldots, \quad \begin{vmatrix} h_1(x) & h_1'(x) & \ldots & h_1^{(n-2)}(x) \\ h_2(x) & h_2'(x) & \ldots & h_2^{(n-2)}(x) \\ \ldots & \ldots & \ldots & \ldots \\ h_{n-1}(x) & h_{n-1}'(x) & \ldots & h_{n-1}^{(n-2)}(x) \end{vmatrix}>0.$$

If in the interval $a<x<b$ there are $n+1$ zeros of the function $f(x)$, then there is also a point ξ in the interval such that

$$f^{(n)}(\xi)+\varphi_1(\xi)f^{(n-1)}(\xi)+\varphi_2(\xi)f^{(n-2)}(\xi)+\cdots+\varphi_n(\xi)f(\xi)=0. \qquad [\text{VII } \mathbf{62}.]$$

94 (continued). The conclusion also holds if the assumption is replaced by the weaker assumption that $f(x)$ has the same values at $n+1$ points as a solution of the differential equation (*). (A generalization of **71** where we are dealing with the equation $y^{(n)}=0$.)

94.1. (An analogue of Rolle's theorem for a partial differential operator.) The closed curve $\mathfrak{C}$ without double points is the boundary of the region $\mathfrak{R}$, a part of the x, y-plane. The function $f(x, y)$ is continuous in the closure of $\mathfrak{R}$ (the union of $\mathfrak{R}$ and $\mathfrak{C}$) and has continuous partial derivatives of the second order in $\mathfrak{R}$. If $f(x, y)$ vanishes along $\mathfrak{C}$ and also at some interior point of $\mathfrak{R}$, then there exists a point ξ, η in $\mathfrak{R}$ at which

$$\frac{\partial^2 f}{\partial x^2}+\frac{\partial^2 f}{\partial y^2}=0.$$

95. (Generalization of the mean-value theorem of differential calculus to a system of functions.) Let

$$x_1<x_2<x_3<\cdots<x_n.$$

The ratio of the two determinants

$$\begin{vmatrix} f_1(x_1) & f_1(x_2) & \ldots & f_1(x_n) \\ f_2(x_1) & f_2(x_2) & \ldots & f_2(x_n) \\ \ldots & \ldots & \ldots & \ldots \\ f_n(x_1) & f_n(x_2) & \ldots & f_n(x_n) \end{vmatrix} : \begin{vmatrix} \varphi_1(x_1) & \varphi_1(x_2) & \ldots & \varphi_1(x_n) \\ \varphi_2(x_1) & \varphi_2(x_2) & \ldots & \varphi_2(x_n) \\ \ldots & \ldots & \ldots & \ldots \\ \varphi_n(x_1) & \varphi_n(x_2) & \ldots & \varphi_n(x_n) \end{vmatrix}$$

is equal to the ratio of the two determinants

$$\begin{vmatrix} f_1(\xi_1) & f_1'(\xi_2) & \dots & f_1^{(n-1)}(\xi_n) \\ f_2(\xi_1) & f_2'(\xi_2) & \dots & f_2^{(n-1)}(\xi_n) \\ \dots & \dots & \dots & \dots \\ f_n(\xi_1) & f_n'(\xi_2) & \dots & f_n^{(n-1)}(\xi_n) \end{vmatrix} : \begin{vmatrix} \varphi_1(\xi_1) & \varphi_1'(\xi_2) & \dots & \varphi_1^{(n-1)}(\xi_n) \\ \varphi_2(\xi_1) & \varphi_2'(\xi_2) & \dots & \varphi_2^{(n-1)}(\xi_n) \\ \dots & \dots & \dots & \dots \\ \varphi_n(\xi_1) & \varphi_n'(\xi_2) & \dots & \varphi_n^{(n-1)}(\xi_n) \end{vmatrix},$$

where $\xi_1, \xi_2, \dots, \xi_n$ are suitably chosen intermediate points; in fact

$$\xi_1 = x_1, \qquad \xi_1 < \xi_2 < x_2, \qquad \xi_2 < \xi_3 < x_3, \qquad \dots, \qquad \xi_{n-1} < \xi_n < x_n.$$

96. Assuming $x_1 < x_2 < x_3 < \dots < x_n$, we have

$$\begin{vmatrix} f_1(x_1) & f_1(x_2) & \dots & f_1(x_n) \\ f_2(x_1) & f_2(x_2) & \dots & f_2(x_n) \\ \dots & \dots & \dots & \dots \\ f_n(x_1) & f_n(x_2) & \dots & f_n(x_n) \end{vmatrix}$$

$$= \begin{vmatrix} 1 & 1 & \dots & 1 \\ \binom{x_1}{1} & \binom{x_2}{1} & \dots & \binom{x_n}{1} \\ \dots & \dots & \dots & \dots \\ \binom{x_1}{n-1} & \binom{x_2}{n-1} & \dots & \binom{x_n}{n-1} \end{vmatrix} \cdot \begin{vmatrix} f_1(\xi_1) & f_1'(\xi_2) & \dots & f_1^{(n-1)}(\xi_n) \\ f_2(\xi_1) & f_2'(\xi_2) & \dots & f_2^{(n-1)}(\xi_n) \\ \dots & \dots & \dots & \dots \\ f_n(\xi_1) & f_n'(\xi_2) & \dots & f_n^{(n-1)}(\xi_n) \end{vmatrix},$$

where the intermediate points $\xi_1, \xi_2, \dots, \xi_n$ satisfy the inequalities given in **95**.

97.

$$\sum_{\nu=1}^{n} \frac{f(x_\nu)}{(x_\nu - x_1)\dots(x_\nu - x_{\nu-1})(x_\nu - x_{\nu+1})\dots(x_\nu - x_n)} = \frac{f^{(n-1)}(\xi)}{(n-1)!},$$

where ξ is a suitable interior point of the smallest interval that contains the n distinct points $x_1, x_2, \dots, x_n$.

98.

$$f(x+nh) - \binom{n}{1} f(x+(n-1)h) + \binom{n}{2} f(x+(n-2)h) - \cdots + (-1)^n f(x) = h^n f^{(n)}(\xi),$$

where either $x < \xi < x+nh$ or $x+nh < \xi < x$. The special case $n=1$ is the ordinary mean value theorem.

99. (A generalization of the mean-value theorem to systems of functions different from **95**.) Let the functions $h_1(x), h_2(x), \dots, h_{n-1}(x)$ satisfy the inequalities (**) of **93** and let the function $f(x)$ be arbitrary. If $x_1 < x_2 < x_3 < \dots < x_n$, then there exists a number ξ, $x_1 < \xi < x_n$, such that

$$\operatorname{sgn} \begin{vmatrix} h_1(x_1) & h_1(x_2) & \dots & h_1(x_n) \\ h_2(x_1) & h_2(x_2) & \dots & h_2(x_n) \\ \dots & \dots & \dots & \dots \\ h_{n-1}(x_1) & h_{n-1}(x_2) & \dots & h_{n-1}(x_n) \\ f(x_1) & f(x_2) & \dots & f(x_n) \end{vmatrix} = \operatorname{sgn} \begin{vmatrix} h_1(\xi) & h_1'(\xi) & \dots & h_1^{(n-1)}(\xi) \\ h_2(\xi) & h_2'(\xi) & \dots & h_2^{(n-1)}(\xi) \\ \dots & \dots & \dots & \dots \\ h_{n-1}(\xi) & h_{n-1}'(\xi) & \dots & h_{n-1}^{(n-1)}(\xi) \\ f(\xi) & f'(\xi) & \dots & f^{(n-1)}(\xi) \end{vmatrix}$$

[Mathematical induction; use **89**.]

100. Deduce **76** from **91**.

Chapter 2. The Geometry of the Complex Plane and the Zeros of Polynomials

§ 1. Center of Gravity of a System of Points with respect to a Point

101. Let $z_1, z_2, \ldots, z_n$ be arbitrary points lying in the finite part of the complex plane and let $m_1, m_2, \ldots, m_n$ be non-negative masses whose sum is 1 which are placed at the points $z_1, z_2, \ldots, z_n$ respectively. In any linear transformation of the complex plane that carries the point at infinity into itself, the center of gravity ζ of such a mass-distribution is mapped by the *same transformation*, i.e. if we again place the point mass m_ν at the image-point z'_ν of z_ν $(1 \leqq \nu \leqq n)$, then the center of gravity ζ' of this new mass distribution is identical with the image of the point ζ.

In the sequel (**102–156**) we mean by a "point" of the complex plane an arbitrary finite point or the point at infinity, by a "circle" the circumference of a circle or a straight line—the latter passes through the point at infinity—and by a "circular domain" a domain whose boundary is a circle. A circular domain is either the closed interior or the closed exterior of a circle or a closed half-plane, according as it does not contain the point at infinity or contains it in its interior or on its boundary.

102. Let the points $z_1, z_2, \ldots, z_n, z$ and the masses $m_1, m_2, \ldots, m_n$ be given. Let z be different from all the points $z_1, z_2, \ldots, z_n$ while of the latter some may coincide. Let $m_1 \geqq 0$, $m_2 \geqq 0$, $\ldots$, $m_n \geqq 0$, $m_1 + m_2 + \cdots + m_n = 1$.

Find a point ζ situated in such a manner that, in a linear transformation of the plane that maps the $n+2$ points

$$z_1, \; z_2, \; z_3, \; \ldots, \; z_n; \quad z; \quad \zeta$$

into the points

$$z'_1, \; z'_2, \; z'_3, \; \ldots, \; z'_n; \quad \infty; \quad \zeta'$$

respectively, ζ' becomes the center of gravity of the mass-distribution determined by the masses m_ν placed at the points z'_ν $(\nu = 1, 2, 3, \ldots, n)$.

There are infinitely many linear transformations that map z into ∞; the point ζ however is *independent* of the choice of the underlying transformation.

The point $\zeta = \zeta_z$, defined in **102**, which is determined uniquely by the system of points $z_1, z_2, \ldots, z_n$, by the masses $m_1, m_2, \ldots, m_n$, $m_1 + m_2 + \cdots + m_n = 1$, placed there, and by the "point of reference" z which must be distinct from all the points z_ν, is called the *center of gravity* of the mass distribution $m_1, m_2, \ldots, m_n$ *with respect to* z. If z is the point at infinity, $z = \infty$, then $\zeta_z = \zeta_\infty$ is the ordinary center of gravity.

103. Consider all possible mass distributions with total mass 1 over the fixed points $z_1, z_2, \ldots, z_n$ and let z be a point distinct from $z_1, z_2, \ldots, z_n$. The centers of gravity ζ_z of all mass distributions of this kind with respect to z cover a "circular-arc polygon" $\mathfrak{C}_z$.

104. Let $\mathfrak{C}_z$ be the circular-arc polygon defined in **103** and w_1, w_2 two points of $\mathfrak{C}_z$. Consider the circle through w_1, w_2 and z. The arc of this circle with end-points w_1, w_2, that does not contain z, is contained in $\mathfrak{C}_z$, in other words $\mathfrak{C}_z$ is *circularly-convex with respect to z*. (We call $\mathfrak{C}_z$ the *smallest circularly-convex domain with respect to z* that contains the points $z_1, z_2, \ldots, z_n$.)

105. If the points $z_1, z_2, \ldots, z_n$ all belong to a circular domain C and if the point z lies outside C, then the circular-arc polygon $\mathfrak{C}_z$ [**103**] lies completely inside C.

Let $z_1, z_2, \ldots, z_n$ be arbitrary points and z a point distinct from these. Let the *same* mass $1/n$ be placed at each of the points z_ν. In the sequel we consider only the center of gravity $\zeta = \zeta_z$ of this special mass distribution and call it for brevity *the* center of gravity of the points $z_1, z_2, \ldots, z_n$ with respect to z.

106. Let ζ_z be the center of gravity of $z_1, z_2, \ldots, z_n$ with respect to z. Every circle through z and ζ_z *separates* the points $z_1, z_2, \ldots, z_n$, i.e. either both circular domains bounded by the circle contain some of the points $z_1, z_2, \ldots, z_n$ in their interior or all the points $z_1, z_2, \ldots, z_n$ lie on the circle.

107 (continued). If all the points $z_1, z_2, \ldots, z_n$ belong to a circular domain C, then the points z and ζ_z cannot *both* lie outside C. If one of them, e.g. z, lies outside C, then the other, ζ_z, lies in the *interior* of C, except in the case where all the points $z_1, z_2, \ldots, z_n$ coincide at a boundary point of C and ζ_z then also lies there.

108. Let the fixed points $w_1, w_2, \ldots, w_k, z$ be pair-wise distinct. Let the complex numbers $z_1, z_2, \ldots, z_n$ take only the k different values $w_1, w_2, \ldots, w_k$. The centers of gravity with respect to z of all systems $z_1, z_2, \ldots, z_n$ of this kind lie everywhere dense in the smallest domain, circularly convex with respect to z, that contains the points $w_1, w_2, \ldots, w_k$.

109. The points $z_1, z_2, \ldots, z_n$ are fixed and the variable point z converges to z_1. To what limiting position does the center of gravity ζ_z of the points $z_1\ z_2, \ldots, z_n$ with respect to z tend?

110. The center of gravity ζ_z of the finite fixed points $z_1, z_2, \ldots, z_n$ with respect to the variable point z may for sufficiently large z be expanded in decreasing powers of z. Calculate the first two terms of the expansion.

§ 2. Center of Gravity of a Polynomial with respect to a Point. A Theorem of Laguerre

In the sequel (**111–156**) we shall mean by a rational integral function (polynomial) of degree n a function $f(z)$ not vanishing identically of the form

$$f(z) = a_0 + \binom{n}{1} a_1 z + \binom{n}{2} a_2 z^2 + \cdots + \binom{n}{n-1} a_{n-1} z^{n-1} + a_n z^n.$$

Here a_n is not necessarily different from 0. In the special case when $a_n \neq 0$, we say that n is the *precise degree* of $f(z)$. If $a_n = a_{n-1} = \cdots = a_{n-k+1} = 0$, $a_{n-k} \neq 0$, we shall say that $f(z)$ has the k-fold zero $z = \infty$.[1] With this interpretation every rational

[1] By introducing improper elements and using homogeneous coordinates, as would of course be in harmony with the projective character of these questions, the special discussions about the point at infinity and the distinction of cases connected with it could be avoided.

integral function of degree n has a total of n zeros, some of which may possibly lie at infinity. Thus there is determined by $f(z)$ a system of points $z_1, z_2, \ldots, z_n$, in the complex plane, the zeros of $f(z)$. Conversely, to every system of n points there belongs a polynomial, whose coefficients are uniquely determined up to a constant of proportion. We shall often disregard this unimportant constant factor and speak of *the* polynomial, whose zeros are $z_1, z_2, \ldots, z_n$.

111. By the center of gravity ζ of a polynomial we mean the center of gravity of its zeros. Express the center of gravity of $f(z)$ with respect to a variable point z: (1) in terms of $f(z)$ and $f'(z)$, (2) in terms of the coefficients of $f(z)$. Consider the special cases $z=\infty$ and $f(z)=z^n$.

112. Let ζ be the center of gravity of $f(z)$ with respect to z. A necessary and sufficient condition that $f(z)$ has only real zeros is that the imaginary parts of z and ζ have opposite signs if they do not both vanish however z may vary. ($z=\infty$ counts as a real zero.)

113. With respect to which points z of the plane does the center of gravity of $f(z)$ lie at infinity? What do the statements in **106** and **107** say about these points?

114. Let $f(z)$ be a polynomial of precise degree n, C a circular domain that contains the zeros of $f(z)$ and $c\neq 0$. The zeros of the derivative of the transcendental function $e^{-\frac{z}{c}} f(z)$ lie either in C or in $C+nc$, i.e. in the circular domain that is obtained from C by parallel displacement by the vector nc. [Cf. III **33**; consider the center of gravity of $f(z)$ with respect to one of the zeros in question.]

115. Let z_1 be a zero of the polynomial $f(z)$ of degree n, x finite and $x\neq z_1$, and $f'(x)=0$. Then $f(z)$ has at least one zero in every circle that passes through both the points

$$x \quad \text{and} \quad x-(n-1)(z_1-x).$$

116. Let $f(z)$ be a polynomial of degree n, whose zeros are all of absolute value $\geqq 1$. Let α_1 and α_2 be arbitrary positive numbers, such that $n\alpha_1/\alpha_2\neq 1$. Then all the zeros of $\alpha_1 z f'(z)-\alpha_2 f(z)$ are in absolute value

$$\geqq \min\left(1, \left|1-n\frac{\alpha_1}{\alpha_2}\right|^{-1}\right).$$

117. Let $f(z)$ be a polynomial of degree n, all of whose zeros lie in the annulus $r\leqq |z|\leqq R$. Let α_1 and α_2 be arbitrarily positive numbers such that $n\alpha_1/\alpha_2\neq 1$. Then all the zeros of $\alpha_1 z f'(z)-\alpha_2 f(z)$ lie in the annulus

$$r \min\left(1, \left|1-n\frac{\alpha_1}{\alpha_2}\right|^{-1}\right) \leqq |z| \leqq R \max\left(1, \left|1-n\frac{\alpha_1}{\alpha_2}\right|^{-1}\right).$$

118. If z_1 is a finite zero of the polynomial $f(z)$ of degree n, then the center of gravity with respect to z_1 of the remaining $n-1$ zeros of $f(z)$ is

$$=z_1-\frac{2(n-1)f'(z_1)}{f''(z_1)}.$$

How is this formula to be modified if z_1 is at infinity?

119. The nth *Hermite* polynomial satisfies the differential equation

$$f''(z)-zf'(z)+nf(z)=0.$$

[VI **100**, solution (g).] Show by means of **118** that the Hermite polynomials have only real zeros [VI **100**, solution (i)]. [Consider the assumed complex zero with largest imaginary part.]

120. The polynomial of degree n that satisfies the differential equation

$$(1-z^2)f''(z)-2zf'(z)+n(n+1)f(z)=0$$

(the nth *Legendre* polynomial, cf. VI **90**), has only real zeros. (VI **97**.)

121. *Gauss's* theorem [III **31**] is equivalent to the following: If all the zeros of a polynomial $f(z)$ lie in a circular disc C, then the zeros of its derivative $f'(z)$ also lie in C [**113**]. Decide whether the following theorem is true or false: If all the zeros of a polynomial $f(z)$ lie in *two* circular discs C_1, C_2, then the zeros of its derivative $f'(z)$ also lie in C_1 or C_2.

Let C_1 and C_2 be two circular discs or, more generally, two arbitrary circular domains. Denoting by n_1 and n_2 fixed positive numbers, we call the set of all points

$$z=\frac{n_1z_2+n_2z_1}{n_1+n_2},$$

where z_1 and z_2 independently run through C_1 and C_2 respectively, a *mean domain* of C_1 and C_2, and use for it the following notation

$$C=\frac{n_1C_2+n_2C_1}{n_1+n_2}.$$

C is completely determined by C_1, C_2 and n_1, n_2. [Cf. H. Minkowski: Werke, Vol. 2, p. 176, Leipzig: B. G. Teubner 1911.]

122. Let C_1 and C_2 be two circular discs whose centers are $z_1^{(0)}$, $z_2^{(0)}$ and radii r_1, r_2 respectively. Show that then

$$C=\frac{n_1C_2+n_2C_1}{n_1+n_2}$$

is again a circular disc whose center $z^{(0)}$ and radius r are given by the following equations:

$$z^{(0)}=\frac{n_1z_2^{(0)}+n_2z_1^{(0)}}{n_1+n_2}, \qquad r=\frac{n_1r_2+n_2r_1}{n_1+n_2}.$$

C is homothetic to C_1 and C_2 with respect to the same center of similitude. [H. S. M. Coxeter: Introduction to Geometry. New York: John Wiley & Sons 1961, pp. 68–70.]

123. Let C_1 and C_2 be two half-planes with parallel boundary lines, one of which contains the other. The mean domain $C=n_1C_2+n_2C_1/(n_1+n_2)$ is then again a half-plane, whose boundary line runs parallel to those of C_1 and C_2 and divides the distance between them in the ratio $n_1:n_2$.

124. Let $f(z)=f_1(z)f_2(z)$, and $f_1(z)$ be a polynomial of degree n_1, $f_2(z)$ a polynomial of degree n_2. Let all the zeros of $f_1(z)$ lie in the circular domain C_1 and all the zeros of $f_2(z)$ in the circular domain C_2. Then the zeros of the derivative $f'(z)$ lie either in C_1 or in C_2 or in the mean domain

$$C=\frac{n_1C_2+n_2C_1}{n_1+n_2}.$$

(By degree we mean the precise degree.)

125. Let $f(z)$ be a rational (non-integral) function, whose numerator and denominator have no common factors. Let it have n_1 zeros (poles) which lie in the circular domain C_1 and n_2 poles (zeros) which lie in the circular domain C_2. (Only finite zeros and poles are considered.) Let $n_1 \gtrless n_2$. Then the zeros of $f'(z)$ lie either in C_1 or in C_2 or in the domain C defined by the numbers

$$z=\frac{n_1z_2-n_2z_1}{n_1-n_2},$$

where z_1 and z_2 independently run through C_1 and C_2, for which we use the notation

$$C=\frac{n_1C_2-n_2C_1}{n_1-n_2}.$$

If $n_1=n_2$ and the circular domains C_1 and C_2 have no common points, then all the zeros of $f'(z)$ lie either in C_1 or in C_2.

126. Let $f(z)$ be a polynomial, and let the equation

$$f(z)=a$$

have all its roots in a convex domain (oval) $\mathfrak{O}_1$. Then all the numbers a for which this is the case also lie in a convex domain $\mathfrak{O}_2$.

127. Let $f(z)$ be a polynomial and let its a-points (the zeros of $f(z)-a$) lie in a circular domain C_1 and its b-points in a circular domain C_2. If n_1, n_2 are positive numbers and we set $c=(n_1b+n_2a)/(n_1+n_2)$, those c-points of $f(z)$ that lie neither in C_1 nor in C_2, lie in the mean domain

$$C=\frac{n_1C_2+n_2C_1}{n_1+n_2}$$

[We may take n_1, n_2 to be integers.]

§ 3. Derivative of a Polynomial with respect to a Point. A Theorem of Grace

By the *derivative* of a polynomial $f(z)$ of degree n *with respect to* the point ζ, denoted by $A_\zeta f(z)$, we mean the ordinary derivative if $\zeta=\infty$, and

$$A_\zeta f(z)=(\zeta-z)f'(z)+nf(z),$$

if ζ is finite. $A_\zeta f(z)$ is a polynomial of degree $n-1$.[1]

---

[1] When using homogeneous coordinates this expression is called the *first polar*, or, by older authors, *émanant*; cf. Laguerre, Oeuvres, Vol. 1, p. 48. Paris: Gauthier-Villars, 1898.

128. Let the $n+1$ coefficients of $f(z)$ (in the notation of §2 preceding **111**) be $a_0, a_1, a_2, \ldots, a_n$. What are the n coefficients of $[A_\zeta f(z)]/n$?

129. Let ζ be arbitrary. Show that the operation $A_\zeta f(z)$ is distributive: If we denote by $f_1(z)$ and $f_2(z)$ two polynomials of degree n, and by c_1 and c_2 arbitrary constants, then we have

$$A_\zeta[c_1 f_1(z)+c_2 f_2(z)]=c_1 A_\zeta f_1(z)+c_2 A_\zeta f_2(z).$$

130. Let ζ be arbitrary and $f(z)$ a polynomial of degree n, $f(z)=g(z)h(z)$ a factorization of $f(z)$ into the two factors $g(z)$ and $h(z)$ of degree k and l respectively, $k+l=n$. Show that

$$A_\zeta f(z)=g(z)A_\zeta h(z)+h(z)A_\zeta g(z).$$

131. Let $f(z)$ be a polynomial of degree n, ζ_1 and ζ_2 two arbitrary constants. The operations $A_{\zeta_1}f(z)$ and $A_{\zeta_2}f(z)$ are *commutative*, that is to say

$$A_{\zeta_1}[A_{\zeta_2}f(z)]=A_{\zeta_2}[A_{\zeta_1}f(z)]=A_{\zeta_1}A_{\zeta_2}f(z).$$

132. Show that $A_\zeta f(z)$ vanishes identically if and only if all the zeros of $f(z)$ coincide with ζ.

By the *derived system* of the n points $z_1, z_2, \ldots, z_n$ *with respect to* a $(n+1)$th point ζ we mean the $n-1$ zeros $z_1', z_2', \ldots, z_{n-1}'$ of $A_\zeta f(z)$, where $f(z)$ is the polynomial of degree n whose zeros are $z_1, z_2, \ldots, z_n$, all the zeros being correctly counted (§ 2, preceding **111**). The derived system is completely determined, except in the single exceptional case where the $n+1$ points $z_1, z_2, \ldots, z_n, \zeta$ all coincide (cf. **132**).

133. If ζ is finite, then to the derived system with respect to ζ in general there belong four different kinds of points:

(1) The finite points distinct from $z_1, z_2, \ldots, z_n, \zeta$, with respect to which the center of gravity of $z_1, z_2, \ldots, z_n$ is precisely ζ.

(2) The multiple zeros of $f(z)$, distinct from ζ, each with multiplicity reduced by 1.

(3) ζ itself only if it is a zero of $f(z)$ and in that case with the same multiplicity.

(4) The point ∞ if it is at least a double zero of $f(z)$; moreover also if it is not a zero of $f(z)$ but if ζ is the ordinary center of gravity of $f(z)$, i.e. the one constructed with respect to ∞.

134. A circular domain, whose boundary contains ζ as well as a point of the derived system with respect to ζ, itself contains points of the original system.

135. If a circular domain contains the points $z_1, z_2, \ldots, z_n$ and does not contain the point ζ, then it contains also the derived system of $z_1, z_2, \ldots, z_n$ with respect to ζ.

136. The derived system of the points $z_1, z_2, \ldots, z_n$ lies in the smallest circular-arc polygon, circularly convex with respect to ζ, that contains $z_1, z_2, \ldots, z_n$. (Generalization of III **31**.)

137. If $f(z)$ is of degree n, then $(A_\zeta)^{n-1}f(z)$ is of the first degree. Calculate the only zero of $(A_\zeta)^{n-1}f(z)$!

138. Let

$$f(z)=a_0+\binom{n}{1}a_1z+\binom{n}{2}a_2z^2+\cdots+\binom{n}{n-1}a_{n-1}z^{n-1}+a_nz^n$$

be an arbitrary polynomial of degree n and let $\zeta_1, \zeta_2, \ldots, \zeta_n$ be arbitrary points in the complex plane. The expression

$$A(\zeta_1, \zeta_2, \ldots, \zeta_n)f(z)=\frac{1}{n!}A_{\zeta_1}A_{\zeta_2}\ldots A_{\zeta_n}f(z)$$

is a symmetric multilinear function of the variables $\zeta_1, \zeta_2, \ldots, \zeta_n$, which we can write as follows:

$$A(\zeta_1, \zeta_2, \ldots, \zeta_n)f(z)=a_0\Sigma_0+a_1\Sigma_1+a_2\Sigma_2+\cdots+a_{n-1}\Sigma_{n-1}+a_n\Sigma_n.$$

Here $\Sigma_0, \Sigma_1, \Sigma_2, \ldots, \Sigma_{n-1}, \Sigma_n$ are the elementary symmetric functions of $\zeta_1, \zeta_2, \ldots, \zeta_n$:

$$\Sigma_0=1, \qquad \Sigma_1=\zeta_1+\zeta_2+\cdots+\zeta_n,$$

$$\Sigma_2=\zeta_1\zeta_2+\zeta_1\zeta_3+\cdots+\zeta_{n-1}\zeta_n, \qquad \ldots, \qquad \Sigma_n=\zeta_1\zeta_2\ldots\zeta_n.$$

If the last k of the numbers $\zeta_1, \zeta_2, \ldots, \zeta_n$ (and only these) lie at infinity, then we set $\Sigma_0=\Sigma_1=\cdots=\Sigma_{k-1}=0$, $\Sigma_k=1$, $\Sigma_{k+1}=\zeta_1+\zeta_2+\cdots+\zeta_{n-k}$, $\Sigma_{k+2}=\zeta_1\zeta_2+\zeta_1\zeta_3+\cdots+\zeta_{n-k-1}\zeta_{n-k}, \ldots, \Sigma_n=\zeta_1\zeta_2\ldots\zeta_{n-k}$.

139. Let

$$f(z)=a_0+\binom{n}{1}a_1z+\binom{n}{2}a_2z^2+\cdots+\binom{n}{n-1}a_{n-1}z^{n-1}+a_nz^n$$

be an arbitrary polynomial of degree n with zeros $z_1, z_2, \ldots, z_n$ and

$$g(z)=b_0+\binom{n}{1}b_1z+\binom{n}{2}b_2z^2+\cdots+\binom{n}{n-1}b_{n-1}z^{n-1}+b_nz^n$$

an arbitrary polynomial of degree n with zeros $\zeta_1, \zeta_2, \ldots, \zeta_n$. Calculate $A(\zeta_1, \zeta_2, \ldots, \zeta_n)\,f(z)$ and $A(z_1, z_2, \ldots, z_n)\,g(z)$.

If the two polynomials in **139** are related in such a manner that

$$A(\zeta_1, \zeta_2, \ldots, \zeta_n)f(z)=0$$

or, equivalently [solution **139**], $A(z_1, z_2, \ldots, z_n)g(z)=0$, then the two polynomials $f(z)$ and $g(z)$ are termed *apolar*. The term suggests the vanishing of the nth polar $A_{\zeta_1}A_{\zeta_2}\ldots A_{\zeta_n}f(z)$. The condition of apolarity is

$$a_0b_n-\binom{n}{1}a_1b_{n-1}+\binom{n}{2}a_2b_{n-2}-\cdots+(-1)^{n-1}\binom{n}{n-1}a_{n-1}b_1+(-1)^na_nb_0=0.$$

[**139**.] The two systems $z_1, z_2, \ldots, z_n$ and $\zeta_1, \zeta_2, \ldots, \zeta_n$ are also termed mutually apolar.

140. What is the geometrical significance of the apolarity of z_1 and ζ_1, and that of z_1, z_2 and ζ_1, ζ_2?

141. Which are the systems $\zeta_1, \zeta_2, \ldots, \zeta_n$, that are apolar to the system consisting of the roots of the equation $z^n - 1 = 0$?

142. Let $z_1, z_2, \ldots, z_n$ be an arbitrary system of points. There are n systems $\zeta, \zeta, \ldots, \zeta$ of coincident points that are apolar to the given system. Which are they?

143. The polynomial of degree n

$$f(z) = 1 - z + cz^n$$

is apolar to any system of numbers whose sum is n and product is 0.

144. Let $f(z)$ be an arbitrary polynomial of degree n, all of whose zeros lie in the circular domain C. If $\zeta_1, \zeta_2, \ldots, \zeta_k$, $k < n$, are arbitrary points lying outside C, then all the zeros of the polynomial of degree $n-k$, $A_{\zeta_1} A_{\zeta_2} \ldots A_{\zeta_k} f(z)$ lie in C.

145. Let the two polynomials $f(z)$ and $g(z)$ of degree n be apolar to each other. Every circular domain that contains all the zeros of one polynomial also contains at least one zero of the other.

146. Let the two polynomials $f(z)$ and $g(z)$ of degree n be apolar to each other. The two smallest convex polygons that contain all the zeros of $f(z)$ and $g(z)$ respectively have at least one common point. More generally, the two smallest circular-arc polygons that are circularly convex with respect to the same point and that contain all the zeros of $f(z)$ and $g(z)$ respectively, also have at least one common point.

147. The polynomial

$$1 - z + cz^n$$

always has one zero in the circle $|z| \leqq 2$.

148. The polynomial

$$1 - z + cz^n$$

always has one zero in the circle $|z - 1| \leqq 1$.

149. The polynomial of $k+1$ terms

$$1 - z + c_2 z^{\nu_2} + c_3 z^{\nu_3} + \cdots + c_k z^{\nu_k}, \qquad 1 = \nu_1 < \nu_2 < \nu_3 < \cdots < \nu_k$$

always has one zero in the circle

$$|z| \leqq \left[\left(1 - \frac{1}{\nu_2}\right)\left(1 - \frac{1}{\nu_3}\right) \cdots \left(1 - \frac{1}{\nu_k}\right)\right]^{-1},$$

and hence in the circle

$$|z| \leqq k.$$

(Generalization of **147**.)

150. If a polynomial of degree n assumes the same value at two points a and b, $a \neq b$, then its derivative has at least one zero in the circular disc with center at the midpoint of the line segment ab and radius $(|a-b|/2) \cot(\pi/n)$. (Analogue of Rolle's theorem in the complex domain.)

151. Let

$$f(z) = a_0 + \binom{n}{1} a_1 z + \binom{n}{2} a_2 z^2 + \cdots + \binom{n}{n-1} a_{n-1} z^{n-1} + a_n z^n$$

be a polynomial of degree n all of whose zeros lie in a circular domain C. Also let

$$g(z)=b_0+\binom{n}{1}b_1z+\binom{n}{2}b_2z^2+\cdots+\binom{n}{n-1}b_{n-1}z^{n-1}+b_nz^n$$

be a polynomial of degree n with zeros $\beta_1, \beta_2, \ldots, \beta_n$. Every zero γ of the polynomial

$$h(z)=a_0b_0+\binom{n}{1}a_1b_1z+\binom{n}{2}a_2b_2z^2+\cdots+\binom{n}{n-1}a_{n-1}b_{n-1}z^{n-1}+a_nb_nz^n$$

formed by "composition" of $f(z)$ and $g(z)$ then has the form

$$\gamma=-\beta_\nu k,$$

where ν denotes a suitably chosen index and k a suitably chosen point in C. (Here we must set $\infty\cdot\infty=\infty$ and $0\cdot\infty=$ indeterminate.)

152. Let the zeros of the polynomials $f(z)$ and $g(z)$ of degree n all lie in the unit circle $|z|\leqq 1$, and the zeros of at least one of the two polynomials in fact in $|z|<1$. Then also all the zeros of the polynomial of degree n formed from $f(z)$ and $g(z)$ by composition [**151**] lie in the interior of the unit circle $|z|<1$.

153. Let the polynomial $f(z)$ of degree n have all its zeros in a convex domain $\mathfrak{C}$ that contains the origin. Let the polynomial $g(z)$ of the same degree have only real zeros which lie in the interval $-1, 0$. The polynomial $h(z)$ of degree n derived from the two polynomials by composition [**151**] then also has all its zeros in $\mathfrak{C}$.

154. If all the zeros of the polynomial

$$f(z)=a_0+\binom{n}{1}a_1z+\binom{n}{2}a_2z^2+\cdots+\binom{n}{n-1}a_{n-1}z^{n-1}+a_nz$$

of degree n lie in the real interval $-a$, a, and those of the polynomial

$$g(z)=b_0+\binom{n}{1}b_1z+\binom{n}{2}b_2z^2+\cdots+\binom{n}{n-1}b_{n-1}z^{n-1}+b_nz^n$$

of degree n in the interval $-b$, 0 (or 0, b) then the zeros of the polynomial

$$h(z)=a_0b_0+\binom{n}{1}a_1b_1z+\binom{n}{2}a_2b_2z^2+\cdots+\binom{n}{n-1}a_{n-1}b_{n-1}z^{n-1}+a_nb_nz^n$$

of degree n lie in the interval $-ab$, ab. (a, b are positive numbers.)

155. Let all the zeros of the polynomial

$$a_0+a_1z+a_2z^2+\cdots+a_nz^n$$

of degree n be real and those of the polynomial

$$b_0+b_1z+b_2z^2+\cdots+b_nz^n$$

of degree n be real and of the same sign. Then all the zeros of the polynomial

$$a_0b_0+a_1b_1z+a_2b_2z^2+\cdots+a_nb_nz^n$$

are also real. ($z=\infty$ counts as a real zero.)

156. Under the assumptions of **155**, all the zeros of the polynomial

$$a_0b_0+1!a_1b_1z+2!a_2b_2z^2+\cdots+n!a_nb_nz^n$$

are real.

Chapter 3. Miscellaneous Problems

§ 1. Approximation of the Zeros of Transcendental Functions by the Zeros of Rational Functions

157. Deduce from

$$\lim_{n\to\infty}\left(1+\frac{iz}{n}\right)^n=\cos z+i\sin z,$$

that $\cos z$ and $\sin z$ have no non-real zeros.

158. Deduce from

$$\lim_{n\to\infty}\left(1-\frac{z^2}{n}\right)^n=e^{-z^2},$$

that no derivative of e^{-z^2} has non-real zeros.

159. The Bessel function of order 0

$$J_0(z)=1-\frac{1}{1!1!}\left(\frac{z}{2}\right)^2+\frac{1}{2!2!}\left(\frac{z}{2}\right)^4-\frac{1}{3!3!}\left(\frac{z}{2}\right)^6+\cdots$$

has no non-real zeros. One can obtain from the four different representations

(a) $J_0(z)=\lim_{n\to\infty}(-1)^n\left(1+\frac{z^2}{4n^2}\right)^nP_n\left(\frac{z^2-4n^2}{z^2+4n^2}\right)$ [VI **85**],

(b) $J_0(z)=\lim_{n\to\infty}L_n\left(\frac{z^2}{4n}\right)$ [VI **99**, solution (b)],

(c) $J_0(z)=\frac{1}{2\pi}\int_{-\pi}^{\pi}e^{iz\sin\vartheta}\,d\vartheta$ [III **148, 56**]

(d) $J_0(z)=\frac{2}{\pi}\int_0^1\frac{\cos zt}{\sqrt{1-t^2}}\,dt$ [III **205**]

four different proofs of this result.

160. Let q be an integer, $q\geqq 2$. The entire function

$$F(z)=1+\frac{z}{q!}+\frac{z^2}{(2q)!}+\frac{z^3}{(3q)!}+\cdots$$

has no non-real zeros. One can obtain from **59** two different proofs of this result.

161. Let $\alpha\geqq 0$, $0<\alpha_1\leqq\alpha_2\leqq\alpha_3\leqq\cdots$, and $1/\alpha_1+1/\alpha_2+1/\alpha_3+\cdots$ be convergent. The transcendental entire function

$$g(z)=e^{-\alpha z}\left(1-\frac{z}{\alpha_1}\right)\left(1-\frac{z}{\alpha_2}\right)\left(1-\frac{z}{\alpha_3}\right)\cdots$$

may be represented as the limit of a sequence of polynomials with only real positive zeros.

162 (continued). If

$$g(z)=1+\frac{a_1}{1!}z+\frac{a_2}{2!}z^2+\frac{a_3}{3!}z^3+\cdots,$$

then the polynomials

$$1+\binom{n}{1}a_1z+\binom{n}{2}a_2z^2+\cdots+\binom{n}{n-1}a_{n-1}z^{n-1}+a_nz^n, \qquad n=1,\ 2,\ 3,\ \ldots$$

have only real, positive zeros [**63**].

163 (continued). In the interval $0<x\leqq\alpha_1$

$$g(x)<1$$

and furthermore we have there more generally

$$1+\frac{a_1}{1!}x+\frac{a_2}{2!}x^2+\cdots+\frac{a_{2m-1}}{(2m-1)!}x^{2m-1}<g(x)$$

$$<1+\frac{a_1}{1!}x+\frac{a_2}{2!}x^2+\cdots+\frac{a_{2m}}{(2m)!}x^{2m}, \qquad m=1,\ 2,\ 3,\ \ldots.$$

[Thus $g(x)$ is enveloped by its Maclaurin series for $0<x\leqq\alpha_1$; **55**, **72**.]

164 (continued). Assume $\alpha<1$. The entire function of z defined by the integral

$$\int_0^\infty e^{-t^2}g(-t^2)\cos zt\,dt$$

has only real zeros [**63**].

165. Let $\alpha, \beta, \beta_1, \beta_2, \beta_3, \ldots$ be real, $\alpha\geqq 0$, $\beta_\nu\neq 0$, and the series $1/\beta_1^2+1/\beta_2^2+1/\beta_3^2+\cdots$ convergent. The transcendental entire function

$$G(z)=e^{-\alpha z^2+\beta z}\left(1-\frac{z}{\beta_1}\right)e^{\frac{z}{\beta_1}}\left(1-\frac{z}{\beta_2}\right)e^{\frac{z}{\beta_2}}\ldots$$

may be represented as the limit of a sequence of polynomials with only real zeros.

166 (continued). If

$$G(z)=1+\frac{b_1}{1!}z+\frac{b_2}{2!}z^2+\frac{b_3}{3!}z^3+\cdots,$$

then

$$b_m^2+b_{m+1}^2>0, \qquad m=1,\ 2,\ 3,\ \ldots. \qquad [\mathbf{49}.]$$

167 (continued). If $G(z)$ has no positive zeros and $a_0, a_1, a_2, \ldots, a_n$ are real numbers, then the polynomial

$$a_0G(0)+a_1G(1)z+a_2G(2)z^2+\cdots+a_nG(n)z^n$$

can have not more non-real zeros than the polynomial

$$a_0+a_1z+a_2z^2+\cdots+a_nz^n.$$

[**68**, **69** are special cases.]

168. Prove from **167** that the Bessel function $J_0(z)$ has no non-real zeros [II **31**].

169. Prove from **167** that the entire functions considered in **160** have no non-real zeros.

170. Let α be an even integer, $\alpha \geqq 2$. Then the integral

$$\int_0^\infty e^{-t^\alpha} \cos zt\, dt = F_\alpha(z)$$

represents an entire function that has no non-real zeros. [**167**.]

171 (continued). If we have only $\alpha > 1$, then the integral still represents an entire function. If α is *not* an even integer, this function has only finitely many real zeros.

171.1. We make the following assumptions:

The polynomial

$$a_0 + a_1 z + a_2 z^2 + \cdots + a_m z^m$$

of degree m has only positive zeros.

The entire function $G(z)$ is either defined as in **165** or it is equal to the function defined in **165** multiplied by z. $G(z)$ has precisely s zeros in the interval $[0, m]$; these zeros are simple and the distance between any two of them is not less than 1; $s \leqq m$.

Then the polynomial

$$a_0 G(0) + a_1 G(1) z + a_2 G(2) z^2 + \cdots + a_m G(m) z^m$$

has precisely $m - s$ positive zeros. [**67**, **167**.]

171.2. We make the following assumptions:

The entire function

$$g(z) = a_0 + \frac{a_1}{1!} z + \frac{a_2}{2!} z^2 + \frac{a_3}{3!} z^3 + \cdots$$

is a limit of polynomials with only positive zeros, but not itself a polynomial [**161**].

The entire function $G(z)$ is either defined as in **165** or it is equal to the function defined in **165** multiplied by z; it has precisely s non-negative zeros which are simple and the distance between any two of them is not less than 1. (We obtain the present $G(z)$ from the $G(z)$ of **171.1** by making m arbitrarily large.)

Then the entire function

$$a_0 G(0) + \frac{a_1 G(1)}{1!} z + \frac{a_2 G(2)}{2!} z^2 + \cdots$$

has precisely s non-positive zeros.

171.3. If $\nu \geqq 0$ the entire function

$$z^{\frac{\nu}{2}} J_{-\nu}(2z^{1/2}) = \sum_{n=0}^{\infty} \frac{(-z)^n}{n!} \frac{1}{\Gamma(n+1-\nu)}$$

has precisely $[\nu]$ non-positive zeros. (This contains the result of **159**.)

171.4. Let k, l, m and q be non-negative integers,

$$0 \leqq k < l, \qquad k+ql \leqq m < k+(q+1)l.$$

If all the zeros of the polynomial

$$b_0 + b_1 z + b_2 z^2 + \cdots + b_m z^m$$

are negative, then the zeros of

$$b_k + b_{k+l} z + b_{k+2l} z^2 + \cdots + b_{k+ql} z^q$$

are also all negative. (This implies, for $n \to \infty$, the result of **160**.)

171.5. If all the zeros of the polynomial

$$b_0 + \frac{b_1}{1!} z + \frac{b_2}{2!} z^2 + \cdots + \frac{b_m}{m!} z^m$$

are negative, then the zeros of

$$b_k + \frac{b_{k+l}}{1!} z + \frac{b_{k+2l}}{2!} z^2 + \cdots + \frac{b_{k+ql}}{q!} z^q$$

are also all negative; k, l, m and q are subject to the same conditions as in **171.4**.

172. The equation

$$\tan z - z = 0$$

has only real roots. [**26**.]

173. Let $f(t)$ be twice continuously differentiable, $f(t) > 0$, $f'(t) < 0$, $f''(t) < 0$ for $0 \leqq t \leqq 1$. Then the even entire function

$$F(z) = \int_0^1 f(t) \cos zt \, dt$$

has infinitely many, and in fact only real, zeros. (Related to but different from III **205**.) [**26**, III **165**.]

174. Let $\varphi(t)$ be properly integrable for $0 \leqq t \leqq 1$. If we have

$$\int_0^1 |\varphi(t)| \, dt \leqq 1,$$

then the entire function

$$F(z) = \sin z - \int_0^1 \varphi(t) \sin zt \, dt$$

has only real zeros. [**27**.]

175. Let $f(t)$ be real and continuously differentiable for $0 \leqq t \leqq 1$. If we have

$$|f(1)| \geqq \int_0^1 |f'(t)| \, dt,$$

then the entire function

$$F(z) = \int_0^1 f(t) \cos zt \, dt$$

has only real zeros. (The condition is satisfied in particular if $f(0) \geqq 0$, $f'(t) \geqq 0$ for $0 \leqq t \leqq 1$; cf. III **205**.)

176. Let $a \geqq 2$. The entire function

$$F(z)=1+\frac{z}{a}+\frac{z^2}{a^4}+\frac{z^3}{a^9}+\cdots+\frac{z^n}{a^{n^2}}+\cdots$$

as well as all the partial sums of its power series, has only real negative simple zeros. [III **200**.]

177. Let the function $f(t)$ be continuously differentiable and positive for $0<t<1$, and also let $\int_0^1 f(t)\,dt$ exist. The entire function defined by the integral

$$\int_0^1 f(t)\,e^{zt}\,dt=F(z)$$

has no zeros

$$\begin{aligned} &\text{in the half-plane} \quad \Re z \geqq 0, \quad \text{if} \quad f'(t)>0,\\ &\text{in the half-plane} \quad \Re z \leqq 0, \quad \text{if} \quad f'(t)<0. \end{aligned}$$

[Not analogously to III **205** through a limiting process from III **22**, but by appropriate translation of the proof of III **22**.]

178. Deduce III **189** from **177**.

179. The remainder of the exponential series

$$\frac{z^{n+1}}{(n+1)!}+\frac{z^{n+2}}{(n+2)!}+\frac{z^{n+3}}{(n+3)!}+\cdots$$

vanishes for $z=0$, but otherwise at no point of the half-plane $\Re z \leqq n$, if $n=1, 2, 3, \ldots$; $n=0$ is an exception: Here the zeros all lie on the boundary of the half-plane $\Re z \leqq 0$. (This contains **73**.) [**177**.]

180. If $\sum_{n=0}^{\infty} \left|\frac{a_{n+1}}{a_n}\right|^2$ is convergent, then

$$a_0+a_1z+a_2z^2+\cdots+a_nz^n+\cdots=F(z)$$

is an entire function. If we denote the zeros of $F(z)$ by $z_1, z_2, \ldots, z_n, \ldots$, then

$$\frac{1}{|z_1|^2}+\frac{1}{|z_2|^2}+\cdots+\frac{1}{|z_n|^2}+\cdots$$

is convergent.

181. If the zeros of a polynomial with real coefficients are all real and simple, then in the interval between two successive zeros lies only one zero of the derivative. Is this theorem also true for transcendental entire functions?

182. Let the degree of the polynomial $H(x)$ be $\geqq 3$. If

$$F(x)=e^{H(x)},$$

then at least one of the two functions dF/dx and d^2F/dx^2 has not only real zeros.

§ 2. Precise Determination of the Number of Zeros by Descartes' Rule of Signs

Along with the polynomial

$$f(z)=a_0+a_1z+a_2z^2+\cdots+a_mz^m$$

of degree m with real coefficients $a_0, a_1, \ldots a_m$, $a_m \gtrless 0$, we consider the following three *associated polynomials*:

$$P(z, \omega) = a_0 + a_1 z + a_2 z(z-\omega) + \cdots + a_m z(z-\omega)\ldots(z-\overline{m-1}\omega),$$

$$\begin{aligned} Q(z, \omega) = a_0 \quad & (1+z-\overline{m-1}\omega)(1+z-\overline{m-2}\omega)\ldots(1+z-\omega)(1+z) \\ +a_1 \quad & (1+z-\overline{m-1}\omega)(1+z-\overline{m-2}\omega)\ldots(1+z-\omega)z \\ +a_2 \quad & (1+z-\overline{m-1}\omega)(1+z-\overline{m-2}\omega)\ldots(z-\omega)z \\ & \ldots\ldots\ldots\ldots\ldots\ldots\ldots\ldots\ldots\ldots \\ +a_{m-1} & (1+z-\overline{m-1}\omega)(z-\overline{m-2}\omega) \quad \ldots(z-\omega)z \\ +a_m \quad & (z-\overline{m-1}\omega) \quad (z-\overline{m-2}\omega) \quad \ldots(z-\omega)z, \end{aligned}$$

$$\begin{aligned} R(z, \omega) = a_0 \quad & (1-z+\overline{m-1}\omega)(1-z+\overline{m-2}\omega)\ldots(1-z+\omega)(1-z) \\ +a_1 \quad & (1-z+\overline{m-1}\omega)(1-z+\overline{m-2}\omega)\ldots(1-z+\omega)z \\ +a_2 \quad & (1-z+\overline{m-1}\omega)(1-z+\overline{m-2}\omega)\ldots(z-\omega)z \\ & \ldots\ldots\ldots\ldots\ldots\ldots\ldots\ldots\ldots\ldots \\ +a_{m-1} & (1-z+\overline{m-1}\omega)(z-\overline{m-2}\omega) \quad \ldots(z-\omega)z \\ +a_m \quad & (z-\overline{m-1}\omega) \quad (z-\overline{m-2}\omega) \quad \ldots(z-\omega)z. \end{aligned}$$

The three associated polynomials depend on a parameter ω. If we simultaneously replace

z by $-z$, ω by $-\omega$, a_ν by $(-1)^\nu a_\nu$, $\nu = 0, 1, 2, \ldots, m$, then

$Q(z, \omega)$ is transformed into $R(z, \omega)$.

183. If we set

$$f(z)\, e^{kz} = \sum_{n=0}^{\infty} \frac{k^n}{n!} A_n^{(k)} z^n, \qquad k > 0,$$

$$f(z)(1-z)^{-k-1} = \sum_{n=0}^{\infty} \frac{\Gamma(k+n+1-m)}{\Gamma(k+1)} \frac{k^m}{n!} B_n^{(k)} z^n, \qquad k \text{ integral}, \quad k > m-1,$$

$$f(z)(1+z)^{k-1} = \sum_{n=0}^{k+m-1} \frac{(k-1)!\,k^m}{n!\,(k-n+m-1)!} C_n^{(k)} z^n, \qquad k \text{ integral}, \quad k \geqq 1,$$

then the coefficients $A_n^{(k)}$, $B_n^{(k)}$ and $C_n^{(k)}$ may be expressed in terms of the associated polynomials of $f(z)$, $P(z, \omega)$, $Q(z, \omega)$, and $R(z, \omega)$.

184. Prove the formula, valid for $\omega > 0$, $-1+(m-1)\omega < \Re z < 0$,

$$\frac{Q(z, \omega)}{\omega^m} = \frac{\Gamma\left(\frac{1}{\omega}+1\right)}{\Gamma\left(\frac{1+z}{\omega}-m+1\right)\Gamma\left(-\frac{z}{\omega}\right)} \int_0^1 (1-t)^{\frac{1+z}{\omega}} t^{-\frac{z+\omega}{\omega}} f\left(\frac{t}{t-1}\right) dt$$

and find analogous formulae for $P(z, \omega)$ and $R(z, \omega)$.

185. Denote by

$$\mathfrak{F}_a^b, \quad \mathfrak{P}_a^b, \quad \mathfrak{Q}_a^b, \quad \mathfrak{R}_a^b$$

the number of zeros of the polynomials

$$f(z), \quad P(z, \omega), \quad Q(z, \omega), \quad R(z, \omega),$$

respectively, in the open interval $a<z<b$. We have the following inequalities:

$$\mathfrak{F}_0^\infty \leqq \mathfrak{P}_0^\infty, \qquad \mathfrak{F}_{-\infty}^0 \geqq \mathfrak{P}_{-\infty}^0,$$
$$\mathfrak{F}_0^1 \leqq \mathfrak{Q}_0^\infty, \qquad \mathfrak{F}_{-\infty}^0 \geqq \mathfrak{Q}_{-1+(m-1)\omega}^0,$$
$$\mathfrak{F}_0^\infty \leqq \mathfrak{R}_0^{1+(m-1)\omega}, \qquad \mathfrak{F}_{-1}^0 \geqq \mathfrak{R}_{-\infty}^0, \qquad \mathfrak{F}_{-\infty}^{-1} \geqq \mathfrak{R}_{1+(m-1)\omega}^\infty.$$

Here we have assumed throughout that $\omega>0$; in the inequalities in the second line we require moreover that $(m-1)\omega<1$, and in the last inequality of the first column we require that ω^{-1} be an integer. [**38**, **80**.]

186 (continued). Assuming that $f(z)$ does not vanish at the end-points of the corresponding interval (that is at $z=0$, $z=1$, $z=-1$ respectively) we can add to the above inequalities the statement that the difference between the two sides is an even number (possibly 0).—Why must no additional assumptions be made concerning the point $z=+\infty$?

187. The three power series considered in **183** have the following properties:

(1) They have not fewer changes of sign than the polynomial $f(z)$ has zeros in the interior of the intervals

$$(0, +\infty), \quad (0, 1), \quad (0, +\infty),$$

respectively.

(2) The number of changes of sign decreases or remains unaltered as k increases.

(3) The number of changes of sign *attains* the number of zeros of $f(z)$ in the corresponding interval, if k is sufficiently large.

188. Along with the real polynomial $f(z)$ of degree m, consider the fourth associated polynomial

$$J(z, \omega)=f(m\omega)-\binom{m}{1}f(\overline{m-1}\omega)z+\binom{m}{2}f(\overline{m-2}\omega)z^2-\cdots+(-1)^m f(0)z^m,$$

$\omega>0$. Analogously to the notation of **185**, let $\mathfrak{J}_a^b$ denote the number of zeros of $J(z, \omega)$ in the open interval $a<z<b$. Then we have the inequalities

$$\mathfrak{F}_{m\omega}^\infty \leqq \mathfrak{J}_0^1, \qquad \mathfrak{F}_{-\infty}^0 \leqq \mathfrak{J}_1^\infty, \qquad \mathfrak{F}_0^{m\omega} \geqq \mathfrak{J}_{-\infty}^0.$$

If we assume that $f(0)\gtrless 0$, $f(m\omega)\gtrless 0$, then in addition we can assert that the difference between the two sides is an even number.

189. The number of zeros of the real polynomial $f(z)$ of degree m in the interval $m\omega<z<\infty$ is not greater than the number of zeros of the polynomial

$$\Delta^m f(0)+\binom{m}{1}\Delta^{m-1}f(0)z+\binom{m}{2}\Delta^{m-2}f(0)z^2+\cdots+f(0)z^m$$

in the interval $0<z<1$; here we use the notation

$$\Delta^\nu f(0)=f(\nu\omega)-\binom{\nu}{1}f(\overline{\nu-1}\omega)+\binom{\nu}{2}f(\overline{\nu-2}\omega)-\cdots+(-1)^\nu f(0),$$
$$\nu=0, 1, 2, \ldots, m.$$

189.1. The polynomial

$$f(x)=a_0+a_1x+a_2x^2+a_3x^3+\cdots+a_mx^m$$

with real coefficients $a_0, a_1, \ldots, a_m$ has not more zeros in the interval $0<x<1$ than the polynomial

$$S(x)=a_0+a_1x+a_2x^4+a_3x^9+\cdots+a_mx^{m^2}$$

has in the interval $0<x<1$, even if in the case of the latter only distinct zeros with odd multiplicity are taken into account.

189.2 (continued). The entire function

$$D(x)=a_1 1^{-x}+a_2 2^{-x}+a_3 3^{-x}+\cdots+a_m m^{-x}$$

has not more zeros in the interval $0<x<\infty$ than the polynomial $f(x)-a_0$ has in the interval $0<x<1$, even if in the case of the latter only distinct zeros with odd multiplicity are taken into account.

189.3. Let $F(x_1, x_2, \ldots, x_m)$ denote a homogeneous polynomial with real coefficients in the variables $x_1, x_2, \ldots, x_m$ and assume that

$$F(x_1, x_2, \ldots, x_m)>0$$

where

$$x_1\geqq 0, \quad x_2\geqq 0, \quad \ldots, \quad x_m\geqq 0, \quad x_1+x_2+\cdots+x_m>0.$$

Then the coefficients of the polynomial

$$(x_1+x_2+\cdots+x_m)^k F(x_1, x_2, \ldots, x_m)$$

are non-negative for all sufficiently large positive integers k. (The particular case $m=2$ coincides with a particular case of **187** (3) concerning the third expression considered in **183** when $\mathfrak{F}_0^\infty=0$.)

§ 3. Additional Problems on the Zeros of Polynomials

190.[1] We can represent any given polynomial as a quotient of two polynomials such that the denominator has no changes of sign and the number of changes of sign of the numerator is equal to the number of positive zeros of the given polynomial.

191. Let the polynomial $f(x)$ of degree m be such that, if $P(x)$ is an arbitrary polynomial, $P(x)f(x)$ has at least m more changes of sign than $P(x)$. For this it is necessary and sufficient that $f(x)$ has only real positive zeros.

192. Let the polynomial $f(x)$ be such that, if $P(x)$ is an arbitrary polynomial, $P(x)f(x)+P'(x)$ has not more non-real zeros than $P(x)$.

For this it is necessary and sufficient, that $f(x)=a-bx$, where a is arbitrary, $b\geqq 0$.

193. If the polynomial $f(x)$ has the property that for all positive p the equation $f(x)+p=0$ has only real roots, then $f(x)$ is at most of the second degree. If the roots of the same equation are all real and all of the same sign, then $f(x)$ is of the first degree.

This is evident geometrically; we require a proof based upon other principles.

---

[1] In **190–195** we are concerned with polynomials with real coefficients.

194. The expression

$$a_0^2b_2^2-a_0a_1b_1b_2+a_0a_2b_1^2-2a_0a_2b_0b_2+a_1^2b_0b_2-a_1a_2b_0b_1+a_2^2b_0^2$$

formed from the real constants a_0, a_1, a_2, b_0, b_1, b_2, is negative if and only if the two polynomials

$$f(x)=a_0x^2+a_1x+a_2, \qquad g(x)=b_0x^2+b_1x+b_2$$

have real and simple zeros that separate each other, i.e. that lie in such a manner that the two zeros of one polynomial contain between them one and only one zero of the other polynomial.

195. Let $P(x)$ be a polynomial with only real zeros. Then in general there exists no primitive function of $P(x)$ that has only real zeros. More precisely: If n denotes the precise degree of $P(x)$, determine a real number a such that the polynomial of degree $n+1$

$$Q(x)=\int_a^x P(x)\,dx$$

has a maximum number of real zeros. Then we can assert that this maximum number is

$$\begin{aligned} &=2 \quad \text{for} \quad n=1\\ &\geqq 3 \quad \text{for} \quad n=2,\ 4,\ 6,\ 8,\ \ldots,\\ &\geqq 4 \quad \text{for} \quad n=3,\ 5,\ 7,\ 9,\ \ldots \end{aligned}$$

and we cannot assert more than this, for the equality sign can actually hold for given n for a suitable $P(x)$.

196. Let L be the maximum of the absolute values of the coefficients of the polynomial

$$f(z)=z^n+a_1z^{n-1}+a_2z^{n-2}+\cdots+a_n,$$

and let $z_1, z_2, \ldots, z_n$ be the zeros of $f(z)$. Then we have

$$(1+|z_1|)(1+|z_2|)\ldots(1+|z_n|)\leqq 2^n\sqrt{n+1}L. \qquad \text{[II } \mathbf{52}.]$$

196.1. Consider the polynomials $P(z)$, $Q(z)$, $R(z)$ and

$$S(z)=a_0z^n+a_1z^{n-1}+\cdots+a_\nu z^{n-\nu}+\cdots+a_n.$$

Assume that

$P(z)$ has real coefficients,
$Q(z)$ has only negative zeros,
$R(z)$ has only real zeros,
$S(z)$ has only zeros whose absolute value equals 1.

Then

$$a_0P(z+ni)+\cdots+a_\nu P(z+(n-2\nu)i)+\cdots+a_nP(z-ni)$$

has no more non-real zeros than $P(z)$,

$$a_0Q(niz)+\cdots+a_\nu Q((n-2\nu)iz)+\cdots+a_nQ(-niz)$$

has only real zeros, and

$$a_0R(ni)z^n+\cdots+a_\nu R((n-2\nu)i)z^{n-\nu}+\cdots+a_nR(-ni)$$

has only zeros whose absolute value equals 1.

Part Six. Polynomials and Trigonometric Polynomials

§ 1. Tchebychev Polynomials

Setting $\cos\vartheta = x$, the expressions

$$T_n(x) = \cos n\vartheta, \qquad U_n(x) = \frac{1}{n+1}\, T'_{n+1}(x) = \frac{\sin(n+1)\vartheta}{\sin\vartheta}, \qquad n = 0,\ 1,\ 2,\ \ldots$$

are polynomials in x of degree n (the Tchebychev polynomials); the leading coefficient of $T_n(x)$ is equal to 2^{n-1}, and that of $U_n(x)$ is equal to 2^n, $n = 1, 2, 3, \ldots$. We list the first five polynomials of both kinds:

	$T_n(x)$	$U_n(x)$
$n=1$	x	$2x$
$n=2$	$2x^2-1$	$4x^2-1$
$n=3$	$4x^3-3x$	$8x^3-4x$
$n=4$	$8x^4-8x^2+1$	$16x^4-12x^2+1$
$n=5$	$16x^5-20x^3+5x$	$32x^5-32x^3+6x$

1. The zeros of $T_n(x)$ and $U_n(x)$ are all real and distinct and lie in the interior of the interval $-1, 1$. Determine these zeros.

2. Prove that the following relations hold:

$$T_{n+1}(x) = xT_n(x) - (1-x^2)U_{n-1}(x),$$
$$U_n(x) = xU_{n-1}(x) + T_n(x), \qquad n = 1,\ 2,\ 3, \ldots$$

3. The polynomials $T_n(x)$ and $U_n(x)$ satisfy the following differential equations:

$$(1-x^2)T_n''(x) - xT_n'(x) + n^2T_n(x) = 0,$$
$$(1-x^2)U_n''(x) - 3xU_n'(x) + n(n+2)U_n(x) = 0.$$

4. $T_n(x)$ and $U_n(x)$ satisfy the following orthogonality relations:

$$\left.\begin{aligned} &\int_{-1}^{1} \frac{1}{\sqrt{1-x^2}}\, T_m(x)T_n(x)\, dx = 0, \\ &\int_{-1}^{1} \sqrt{1-x^2}\; U_m(x)U_n(x)\, dx = 0, \end{aligned}\right\} \qquad m,\ n = 0,\ 1,\ 2,\ \ldots;\quad m \gtrless n.$$

Evaluate these integrals for $m = n$.

5. $T_n(x)$ and $U_n(x)$ satisfy the identities

$$\frac{T_n(x)}{\sqrt{1-x^2}}=\frac{(-1)^n}{1\cdot 3\cdot 5\ldots(2n-1)}\frac{d^n}{dx^n}(1-x^2)^{n-\frac{1}{2}},$$

$$\sqrt{1-x^2}\,U_n(x)=\frac{(-1)^n(n+1)}{1\cdot 3\cdot 5\ldots(2n+1)}\frac{d^n}{dx^n}(1-x^2)^{n+\frac{1}{2}}.$$

6. Let $f(x)$ be a function defined in the interval $-1\leqq x\leqq 1$ with a continuous nth derivative in this interval. Then we have

$$\int_0^\pi f(\cos\vartheta)\cos n\vartheta\,d\vartheta=\frac{1}{1\cdot 3\cdot 5\ldots(2n-1)}\int_0^\pi f^{(n)}(\cos\vartheta)\sin^{2n}\vartheta\,d\vartheta.$$

7. For $n=1, 2, 3, \ldots, -1\leqq x\leqq 1$, we have the inequalities

$$|T_n(x)|\leqq 1, \qquad |U_n(x)|\leqq n+1.$$

In the first inequality we have equality for precisely $n+1$ values of x, namely at the $n-1$ zeros of $U_{n-1}(x)$ and, in addition, at $x=-1$ and $x=1$. In the second inequality we have equality only for $x=-1$ and $x=1$.

§ 2. General Problems on Trigonometric Polynomials

An expression of the form

$$g(\vartheta)=\lambda_0+\lambda_1\cos\vartheta+\mu_1\sin\vartheta+\lambda_2\cos 2\vartheta+\mu_2\sin 2\vartheta+\cdots$$
$$+\lambda_n\cos n\vartheta+\mu_n\sin n\vartheta.$$

is termed a *trigonometric polynomial* of order n. If all the μ_ν are equal to zero, then $g(\vartheta)$ is termed a *cosine polynomial* of order n. If all the λ_ν are equal to zero, then $g(\vartheta)$ is termed a *sine polynomial* of order n.

8. A cosine polynomial of order n may always be written in the form $P(\cos\vartheta)$, where $P(x)$ is a polynomial of degree n. The converse is also true.

9. A sine polynomial of order n may always be written in the form $\sin\vartheta\, P(\cos\vartheta)$, where $P(x)$ is a polynomial of degree $n-1$. The converse is also true.

10. The product of two trigonometric polynomials of order m and n respectively is a trigonometric polynomial of order $m+n$.

11. A trigonometric polynomial of order n with real coefficients λ_r, μ_s, $r=0, 1, 2, \ldots, s=1, 2, \ldots,$

$$g(\vartheta)=\lambda_0+\lambda_1\cos\vartheta+\mu_1\sin\vartheta+\lambda_2\cos 2\vartheta+\mu_2\sin 2\vartheta+\cdots$$
$$+\lambda_n\cos n\vartheta+\mu_n\sin n\vartheta \qquad (\vartheta \text{ real})$$

may be written in the form

$$g(\vartheta)=e^{-in\vartheta}G(e^{i\vartheta}),$$

where $G(z)=u_0+u_1z+u_2z^2+\cdots+u_{2n}z^{2n}$ denotes a polynomial of degree $2n$ which remains unchanged if one forms the reciprocal polynomial and simultaneously replaces the coefficients by their complex conjugates:

$$G(z)=\bar{u}_{2n}+\bar{u}_{2n-1}z+\bar{u}_{2n-2}z^2+\cdots+\bar{u}_0z^{2n}=z^{2n}\bar{G}(z^{-1}).$$

Calculate the coefficients $u_0, u_1, u_2, \ldots, u_{2n}$.

12. Denote by $G(z)$ a polynomial of degree $2n$ satisfying the identity

$$z^{2n}\bar{G}(z^{-1})=G(z).$$

Then for ϑ real

$$e^{-in\vartheta}G(e^{i\vartheta})=g(\vartheta)$$

is a trigonometric polynomial in ϑ of order n with real coefficients.

13. Let $G(z)$ be a polynomial of degree $2n$ and

$$z^{2n}\bar{G}(z^{-1})=G(z).$$

How are the zeros of $G(z)$ distributed in the complex plane?

14. A trigonometric polynomial of order n with real coefficients

$$g(\vartheta)=\lambda_0+\lambda_1\cos\vartheta+\mu_1\sin\vartheta+\lambda_2\cos 2\vartheta+\mu_2\sin 2\vartheta+\cdots$$
$$+\lambda_n\cos n\vartheta+\mu_n\sin n\vartheta,$$

in which λ_n and μ_n are not both zero, has exactly $2n$ zeros, if both real and complex values are admitted and multiple zeros are counted according to their multiplicity. [ϑ and $\vartheta+2\pi$ are regarded as not distinct.]

15. Determine all trigonometric polynomials of order n

$$g(\vartheta)=\lambda_0+\lambda_1\cos\vartheta+\mu_1\sin\vartheta+\lambda_2\cos 2\vartheta+\mu_2\sin 2\vartheta+\cdots$$
$$+\lambda_n\cos n\vartheta+\mu_n\sin n\vartheta,$$

which have real coefficients and satisfy identically in α and β the relation

$$\sum_{\nu=0}^{n} g\left(\alpha-\frac{\nu\pi}{n+1}\right)g\left(\frac{\nu\pi}{n+1}-\beta\right)=g(\alpha-\beta).$$

§ 3. Some Special Trigonometric Polynomials

16.

$$\tfrac{1}{2}+\cos\vartheta+\cos 2\vartheta+\cdots+\cos n\vartheta=\frac{\sin\frac{2n+1}{2}\vartheta}{2\sin\frac{\vartheta}{2}},$$

$$\cos\vartheta+\cos 2\vartheta+\cos 3\vartheta+\cdots+\cos n\vartheta=\frac{\sin\frac{n}{2}\vartheta\cos\frac{n+1}{2}\vartheta}{\sin\frac{\vartheta}{2}},$$

$$\cos\vartheta+\cos 3\vartheta+\cos 5\vartheta+\cdots+\cos(2n-1)\vartheta=\frac{\sin 2n\vartheta}{2\sin\vartheta},$$

$$\sin\vartheta+\sin 2\vartheta+\sin 3\vartheta+\cdots+\sin n\vartheta=\frac{\sin\frac{n}{2}\vartheta\sin\frac{n+1}{2}\vartheta}{\sin\frac{\vartheta}{2}}.$$

17.

$$\frac{\sin\vartheta}{\sin\vartheta}+\frac{\sin 3\vartheta}{\sin\vartheta}+\frac{\sin 5\vartheta}{\sin\vartheta}+\cdots+\frac{\sin(2n-1)\vartheta}{\sin\vartheta}=\left(\frac{\sin n\vartheta}{\sin\vartheta}\right)^2.$$

What may be deduced from this for $\vartheta=0$?

18.

$$\frac{n+1}{2}+n\cos\vartheta+(n-1)\cos 2\vartheta+\cdots+\cos n\vartheta=\frac{1}{2}\left(\frac{\sin(n+1)\frac{\vartheta}{2}}{\sin\frac{\vartheta}{2}}\right)^2.$$

19. Where do the zeros of the following trigonometric polynomials lie:

$$\tfrac{1}{2}+\cos\vartheta+\cos 2\vartheta+\cdots+\cos n\vartheta,\qquad \cos\vartheta+\cos 2\vartheta+\cdots+\cos n\vartheta,$$

$$\cos\vartheta+\cos 3\vartheta+\cos 5\vartheta+\cdots+\cos(2n-1)\vartheta,$$

$$\sin\vartheta+\sin 2\vartheta+\cdots+\sin n\vartheta,\qquad \sin\vartheta+\sin 3\vartheta+\cdots+\sin(2n-1)\vartheta,$$

$$\frac{n+1}{2}+n\cos\vartheta+(n-1)\cos 2\vartheta+\cdots+\cos n\vartheta\quad ?$$

20. Derive the identity:

$$\cos\vartheta+\cos 2\vartheta+\cdots+\cos n\vartheta=\frac{1}{2}\sin(n+1)\vartheta\cot\frac{\vartheta}{2}-\cos^2\frac{n+1}{2}\vartheta.$$

21. Show that

$$\sin\vartheta+\sin 2\vartheta+\cdots+\sin n\vartheta+\frac{\sin(n+1)\vartheta}{2}$$

is non-negative for $0\leqq\vartheta\leqq\pi$.

22. The arithmetic means of the partial sums of the series

$$\tfrac{1}{2}+\cos\vartheta+\cos 2\vartheta+\cos 3\vartheta+\cdots+\cos n\vartheta+\cdots$$

are non-negative for all ϑ; they converge as n tends to infinity uniformly to 0 if $\varepsilon\leqq\vartheta\leqq 2\pi-\varepsilon$, $\varepsilon>0$.

23. The trigonometric polynomial

$$A(n,\vartheta)=\sin\vartheta+\frac{\sin 2\vartheta}{2}+\frac{\sin 3\vartheta}{3}+\cdots+\frac{\sin n\vartheta}{n}$$

has in the interval $0\leqq\vartheta\leqq\pi$ at each of the points

$$\frac{\pi}{n+1},\quad 3\frac{\pi}{n+1},\quad 5\frac{\pi}{n+1},\quad \ldots,\quad (2q-1)\frac{\pi}{n+1}$$

(and only at these points) a relative maximum, and at each of the points

$$\frac{2\pi}{n},\quad 2\frac{2\pi}{n},\quad 3\frac{2\pi}{n},\quad \ldots,\quad (q-1)\frac{2\pi}{n}$$

(and only at these points) a relative minimum, where $q=[(n+1)/2]$.

24 (continued). The maxima of $A(n,\vartheta)$ in the interval $0\leqq\vartheta\leqq\pi$ decrease monotonically, so that the absolute maximum of $A(n,\vartheta)$ in the whole interval $0\leqq\vartheta\leqq\pi$ is equal to $A(n,\pi/(n+1))$. [**20**.]

25 (continued). The maxima $A(n,\pi/(n+1))$ increase monotonically with n, and we have

$$\lim_{n\to\infty} A\left(n,\frac{\pi}{n+1}\right)=\int_0^{\pi}\frac{\sin\vartheta}{\vartheta}\,d\vartheta=1.8519\ldots.$$

26. The trigonometric polynomial

$$B(n, \vartheta)=\cos \vartheta+\frac{\cos 2\vartheta}{2}+\frac{\cos 3\vartheta}{3}+\cdots+\frac{\cos n\vartheta}{n}$$

has in the interval $0 \leqq \vartheta \leqq \pi$ at each of the points

$$0, \quad \frac{2\pi}{n}, \quad 2\frac{2\pi}{n}, \quad \ldots, \quad p\frac{2\pi}{n}$$

(and only at these points) a relative maximum, and at each of the points

$$\frac{2\pi}{n+1}, \quad 2\frac{2\pi}{n+1}, \quad 3\frac{2\pi}{n+1}, \quad \ldots, \quad q\frac{2\pi}{n+1}$$

(and only at these points) a relative minimum, where $p=[n/2]$, $q=[(n+1)/2]$.

27 (continued). The trigonometric polynomial $B(n, \vartheta)$ assumes its least value for $\vartheta=[(n+1)/2]\, 2\pi/(n+1)$. [**21**.]

28. For $0 \leqq \vartheta \leqq 2\pi$ and $n=1, 2, 3, \ldots$

$$B(n, \vartheta) = \cos \vartheta+\frac{\cos 2\vartheta}{2}+\frac{\cos 3\vartheta}{3}+\cdots+\frac{\cos n\vartheta}{n} \geqq -1.$$

§ 4. Some Problems on Fourier Series

Let $f(\vartheta)$ be a periodic function of period 2π, properly integrable in the interval $0 \leqq \vartheta \leqq 2\pi$. The constants

$$a_n=\frac{1}{2\pi}\int_0^{2\pi} f(\vartheta) \cos n\vartheta \, d\vartheta, \qquad b_n=\frac{1}{2\pi}\int_0^{2\pi} f(\vartheta) \sin n\vartheta \, d\vartheta, \qquad n=0,\ 1,\ 2,\ \ldots; \quad b_0=0,$$

are termed the *Fourier coefficients*, and the formally constructed series

$$a_0+2a_1 \cos \vartheta+2b_1 \sin \vartheta+2a_2 \cos 2\vartheta+2b_2 \sin 2\vartheta+\cdots$$
$$+2a_n \cos n\vartheta+2b_n \sin n\vartheta+\cdots$$

the *Fourier series* of $f(\vartheta)$. If $f(\vartheta)$ is of bounded variation, then this series is convergent to the sum

$$\frac{f(\vartheta+0)+f(\vartheta-0)}{2}.$$

29. Let $f(\vartheta)$ (ϑ real) be a real-valued periodic function, $f(\vartheta+2\pi)=f(\vartheta)$. Each of the following equations (or pairs of equations respectively) characterizes a special symmetry property of the graph of the curve $y=f(x)$:

(1) $f(-\vartheta)=f(\vartheta)$,
(2) $f(-\vartheta)=-f(\vartheta)$,
(3) $f(\vartheta+\pi)=-f(\vartheta)$,
(4a) $f(-\vartheta)=f(\vartheta), \quad f(\vartheta+\pi)=-f(\vartheta)$,
(4b) $f(-\vartheta)=-f(\vartheta), \quad f(\vartheta+\pi)=-f(\vartheta)$,
(5) $f(\vartheta+\pi)=f(\vartheta)$.

Show that the existence of such a symmetry property is equivalent to the vanishing of certain (infinitely many) Fourier coefficients.

30. What is the Fourier series of a trigonometric polynomial $f(\vartheta)$ of order n?

31. Let n and ν be positive integers, $0 \leqq \nu \leqq n$. Then

$$\frac{1}{2\pi} \int_{-\pi}^{\pi} \left(2 \cos \frac{\vartheta}{2}\right)^n \cos \left(\frac{n}{2} - \nu\right)\vartheta \, d\vartheta = \binom{n}{\nu}.$$

32. Let the series of numbers

$$a_0, \; a_1, \; b_1, \; a_2, \; b_2, \; \ldots, \; a_n, \; b_n, \; \ldots$$

be such that the "trigonometric series"

$$a_0 + 2a_1 \cos \vartheta + 2b_1 \sin \vartheta + 2a_2 \cos 2\vartheta + 2b_2 \sin 2\vartheta + \cdots$$
$$+ 2a_n \cos n\vartheta + 2b_n \sin n\vartheta + \cdots$$

is uniformly convergent for all values of ϑ and therefore represents a periodic continuous function $f(\vartheta)$ of period 2π. What is the Fourier series of $f(\vartheta)$?

33.
$$\frac{1}{2} - \frac{1}{\pi} \sum_{n=1}^{\infty} \frac{\sin 2\pi n x}{n} = \begin{cases} x - [x], & \text{if } x \text{ is not an integer,} \\ \frac{1}{2}, & \text{if } x \text{ is an integer.} \end{cases}$$

34. Show that

$$|\sin \vartheta| = \frac{2}{\pi} - \frac{4}{\pi} \sum_{n=1}^{\infty} \frac{\cos 2n\vartheta}{4n^2 - 1} = \frac{8}{\pi} \sum_{n=1}^{\infty} \frac{\sin^2 n\vartheta}{4n^2 - 1}.$$

35. The constants

$$\rho_m = \frac{2}{\pi} \int_0^{\frac{\pi}{2}} \frac{|\sin m\vartheta|}{\sin \vartheta} \, d\vartheta, \qquad m = 1, 2, 3, \ldots$$

(which are closely related to the so-called Lebesgue constants of the Fourier series) increase steadily with m. They may be represented in the following form:

$$\rho_m = \frac{16}{\pi^2} \sum_{n=1}^{\infty} \frac{\frac{1}{1} + \frac{1}{3} + \frac{1}{5} + \cdots + \frac{1}{2nm - 1}}{4n^2 - 1}.$$

[Substitute for $|\sin m\vartheta|$ the series of **34** and apply **17**.]

36. Let $f(\vartheta)$ (ϑ real) be a real-valued periodic function of period π, and in addition an "even function", i.e. $f(-\vartheta) = f(\vartheta)$. If $f(\vartheta)$ is concave in the interval $0 \leqq \vartheta \leqq \pi$ [Vol. I p. 65], then its Fourier series has the form

$$c_0 - c_1 \cos 2\vartheta - c_2 \cos 4\vartheta - \cdots - c_n \cos 2n\vartheta - \cdots,$$

where all the coefficients $c_1, c_2, \ldots, c_n, \ldots$ are non-negative.

37. Let the function $f(\vartheta)$ of **36** satisfy in addition the condition $f(0) = 0$. Further let $\vartheta^{-p} f(\vartheta)$ be bounded, $p > 0$. Then the sequence

$$\rho_m = \int_0^{\frac{\pi}{2}} \frac{f(m\vartheta)}{\sin \vartheta} \, d\vartheta, \qquad m = 1, 2, 3, \ldots,$$

increases monotonically with m. [A generalization of **35**.]

38. Let M_n denote the maximum of

$$\Gamma(n, \vartheta)=\frac{|\sin \vartheta|}{1}+\frac{|\sin 2\vartheta|}{2}+\frac{|\sin 3\vartheta|}{3}+\cdots+\frac{|\sin n\vartheta|}{n}$$

for all values of ϑ. Then we have the inequalities

$$\frac{2}{\pi}\sum_{\nu=1}^{n}\frac{1}{\nu}<M_n<\frac{2}{\pi}\sum_{\nu=1}^{n}\frac{1}{\nu}+\frac{2}{\pi}. \qquad \textbf{[34, 28.]}$$

§ 5. Real Non-negative Trigonometric Polynomials

39. Let $x_0, x_1, x_2, \ldots, x_n$ be arbitrary complex numbers, $x_0 \neq 0$, $x_n \neq 0$. The expression

$$|x_0+x_1z+x_2z^2+\cdots+x_nz^n|^2, \qquad z=e^{i\vartheta}, \qquad (\vartheta \text{ real}),$$

represents a real non-negative trigonometric polynomial of order precisely n. Calculate the coefficients of this trigonometric polynomial.

40. If the real trigonometric polynomial of nth order

$$g(\vartheta)=\lambda_0+\lambda_1 \cos \vartheta+\mu_1 \sin \vartheta+\lambda_2 \cos 2\vartheta+\mu_2 \sin 2\vartheta+\cdots$$
$$+\lambda_n \cos n\vartheta+\mu_n \sin n\vartheta, \qquad (\lambda_i, \mu_j \text{ real}, 0 \leq i \leq n, 1 \leq j \leq n)$$

can assume only non-negative values for all real ϑ, then it may be represented in the following form:

$$g(\vartheta)=|h(z)|^2, \qquad z=e^{i\vartheta}, \qquad (\vartheta \text{ real}),$$

where $h(z)=x_0+x_1z+x_2z^2+\cdots+x_nz^n$ is a polynomial of nth degree. [Factorize the polynomial $G(z)$ considered in **11** into linear factors.]

41. If $g(\vartheta)$ (ϑ real) denotes a real non-negative cosine polynomial, then one can always find a polynomial $h(z)=x_0+x_1z+x_2z^2+\cdots+x_nz^n$ with only *real* coefficients such that we have

$$g(\vartheta)=|h(z)|^2, \qquad z=e^{i\vartheta}.$$

42. Show that the representation of a real non-negative trigonometric polynomial as given in **40** for real ϑ is in general possible in several ways.

43. Show that in **40** there always exists a representation such that $h(z)$ satisfies the following conditions:

(1) $h(z)$ is different from 0 for $|z|<1$;
(2) $h(0)$ is real and positive.

Here it is assumed that $g(\vartheta)$ is not identically zero.

§ 6. Real Non-negative Polynomials

44. If a polynomial is real and non-negative for all real x then it may be written in the form $[A(x)]^2+[B(x)]^2$, where $A(x)$ and $B(x)$ are suitable polynomials with only real coefficients.

45. Every polynomial which is real and non-negative for non-negative real values of x, may be written in the form

$$[A(x)]^2+[B(x)]^2+x\{[C(x)]^2+[D(x)]^2\},$$

where $A(x)$, $B(x)$, $C(x)$, $D(x)$ denote polynomials with only real coefficients.

46. Every polynomial of degree n which can assume only real and non-negative values for $-1\leqq x\leqq 1$ may be represented in the form

$$[A(x)]^2+(1-x^2)[B(x)]^2,$$

where $A(x)$ and $B(x)$ are polynomials of nth and $(n-1)$th degrees, respectively, with only real coefficients. [**41**.]

47. Every polynomial $P(x)$ of nth degree which can assume only real and non-negative values for $-1\leqq x\leqq 1$ may be written in the form

$$[A(x)]^2+(1-x)[B(x)]^2+(1+x)[C(x)]^2+(1-x^2)[D(x)]^2$$

and in fact in such a manner that all the four terms are at most of nth degree.

If $P(x)$ is of degree $2m$, there exists a representation for which $B(x)=C(x)=0$, $A(x)$ is of degree m and $D(x)$ of degree $m-1$.

48. Can every polynomial $P(x)$ of degree n which is real and positive for $-1<x<1$ be represented in the form

$$P(x)=\sum A(1-x)^{\alpha}(1+x)^{\beta},$$

where $A\geqq 0$, $\alpha+\beta\leqq n$, and α and β are non-negative integers?

49. Every polynomial $P(x)$ which is real and positive in the interval $-1<x<1$, can be represented in the form

$$P(x)=\sum A(1-x)^{\alpha}(1+x)^{\beta},$$

where $A\geqq 0$, and α and β are non-negative integers.

§ 7. Maximum–Minimum Problems on Trigonometric Polynomials

50. Let

$$g(\vartheta)=\lambda_0+\lambda_1\cos\vartheta+\mu_1\sin\vartheta+\lambda_2\cos 2\vartheta+\mu_2\sin 2\vartheta+\cdots$$
$$+\lambda_n\cos n\vartheta+\mu_n\sin n\vartheta$$

(ϑ real) be a real non-negative trigonometric polynomial of nth order, and let the constant term in $g(\vartheta)$ be equal to 1, i.e.

$$\lambda_0=\frac{1}{2\pi}\int_0^{2\pi} g(\vartheta)\,d\vartheta=1.$$

Then

$$g(\vartheta)\leqq n+1.$$

The equality sign here holds only if

$$g(\vartheta)=\frac{1}{n+1}|1+z+z^2+\cdots+z^n|^2$$
$$=1+2\frac{n}{n+1}\cos(\vartheta-\vartheta_0)+2\frac{n-1}{n+1}\cos 2(\vartheta-\vartheta_0)+$$
$$\cdots+2\frac{1}{n+1}\cos n(\vartheta-\vartheta_0),\qquad z=e^{i(\vartheta-\vartheta_0)}$$

and $\vartheta=\vartheta_0$ (or $\vartheta=\vartheta_0+2k\pi$, $k=0,\ \pm 1,\ \pm 2,\ldots$).

51 (continued). We have

$$\lambda_n^2+\mu_n^2\leqq 1.$$

The equality sign here holds only for

$$g(\vartheta)=1+\cos n(\vartheta-\vartheta_0).$$

52 (continued). We have

$$\lambda_1^2+\mu_1^2\leqq 4\cos^2\frac{\pi}{n+2}.$$

53. The geometric mean [II **48**] of a real and non-negative trigonometric polynomial $g(\vartheta)$ (ϑ real) which does not vanish identically is

$$e^{\frac{1}{2\pi}\int_0^{2\pi}\log g(\vartheta)\,d\vartheta}=|h(0)|^2=[h(0)]^2,$$

where $g(\vartheta)=|h(e^{i\vartheta})|^2$ is the "standard representation" of $g(\vartheta)$ defined in **43**.

54. Let the real non-negative trigonometric polynomial of order n which does not vanish identically

$$g(\vartheta)=\lambda_0+\lambda_1\cos\vartheta+\mu_1\sin\vartheta+\lambda_2\cos 2\vartheta+\mu_2\sin 2\vartheta+\cdots$$
$$+\lambda_n\cos n\vartheta+\mu_n\sin n\vartheta$$

(ϑ real) have the geometric mean 1. Then

$$g(\vartheta)\leqq 4^n.$$

The equality sign here holds only for

$$g(\vartheta)=\left(2\cos\frac{\vartheta-\vartheta_0}{2}\right)^{2n}$$

and for $\vartheta=\vartheta_0$ (or $\vartheta=\vartheta_0+2k\pi$, $k=0,\ \pm 1,\ \pm 2,\ldots$).

55 (continued). The arithmetic mean of $g(\vartheta)$ is

$$\lambda_0=\frac{1}{2\pi}\int_0^{2\pi}g(\vartheta)\,d\vartheta\leqq\binom{2n}{n}.$$

When does the equality sign hold?

56 (continued).

$$\sqrt{\lambda_\nu^2+\mu_\nu^2}\leqq 2\binom{2n}{n+\nu},\qquad \nu=1,\ 2,\ 3,\ \ldots,\ n.$$

When does the equality sign hold here?

57. A real trigonometric polynomial without a constant term

$$g(\vartheta)=\lambda_1\cos\vartheta+\mu_1\sin\vartheta+\lambda_2\cos 2\vartheta+\mu_2\sin 2\vartheta+\cdots$$
$$+\lambda_n\cos n\vartheta+\mu_n\sin n\vartheta$$

cannot have the same sign for all real ϑ unless it is identically zero. This theorem is to be proved *without* the use of integral calculus.

58. Let $-m$ and M be the minimum and maximum, respectively, of a real trigonometric polynomial of nth order of the form

$$g(\vartheta)=\lambda_1\cos\vartheta+\mu_1\sin\vartheta+\lambda_2\cos 2\vartheta+\mu_2\sin 2\vartheta+\cdots+\lambda_n\cos n\vartheta+\mu_n\sin n\vartheta,$$

(ϑ real) $m\geqq 0$, $M\geqq 0$. [**57**.] Then we have

$$M\leqq nm, \qquad m\leqq nM.$$

59. The first mean-value theorem of integral calculus may be strengthened for real trigonometric polynomials as follows: Let $g(\vartheta)$ be a real trigonometric polynomial of nth order and let m be the minimum and M the maximum of $g(\vartheta)$ (ϑ real). Then we have

$$m+\frac{M-m}{n+1}\leqq\frac{1}{2\pi}\int_0^{2\pi} g(\vartheta)\,d\vartheta\leqq M-\frac{M-m}{n+1}.$$

60. Let $-m$ be the minimum and M the maximum of a real trigonometric polynomial of nth order

$$g(\vartheta)=\lambda_0+\lambda_1\cos\vartheta+\mu_1\sin\vartheta+\lambda_2\cos 2\vartheta+\mu_2\sin 2\vartheta+\cdots$$
$$+\lambda_n\cos n\vartheta+\mu_n\sin n\vartheta,$$

(ϑ real). Then either m or M is greater than $\sqrt{\lambda_n^2+\mu_n^2}$. We have $m=M=\sqrt{\lambda_n^2+\mu_n^2}$ only if $g(\vartheta)=c\cos n(\vartheta-\vartheta_0)$.

61. Compound n simple harmonic motions with periods in the ratio

$$1:\frac{1}{2}:\frac{1}{3}:\ldots:\frac{1}{n}$$

and with arbitrary phases. The greatest displacement of the resulting motion is then equal to at least the arithmetic mean of the amplitudes of the separate simple harmonic motions.

With the notation of **58**, we are dealing with the inequality:

$$M\geqq\frac{\sqrt{\lambda_1^2+\mu_1^2}+\sqrt{\lambda_2^2+\mu_2^2}+\cdots+\sqrt{\lambda_n^2+\mu_n^2}}{n}.$$

(Strengthening of **50**.)

§ 8. Maximum–Minimum Problems on Polynomials

62. Let $P(x)$ be a polynomial of degree n with leading coefficient 1. Then the maximum of $|P(x)|$ in the interval $-1\leqq x\leqq 1$ is equal to at least $1/2^{n-1}$. If $|P(x)|\leqq 1/2^{n-1}$ for $-1\leqq x\leqq 1$, then $P(x)$ differs only by a constant factor from the polynomial $T_n(x)$ defined at the beginning of Part VI. [**60**.]

Consider the set of all polynomials of degree n whose leading coefficient is equal to 1, i.e. the polynomials

$$P(x)=x^n+a_1x^{n-1}+a_2x^{n-2}+\cdots+a_n$$

with arbitrary complex coefficients $a_1, a_2, \ldots, a_n$. Denote by $\mu_n(\alpha, \beta)$ the minimum of the maxima of all $|P(x)|$ in $\alpha\leqq x\leqq\beta$. The theorem of **62** may then be formulated as follows: For $\alpha=-1, \beta=1$ this "minimum maximorum" is equal to

$$\mu_n(-1, 1)=\frac{1}{2^{n-1}}.$$

63. We have

$$\mu_n(\alpha, \beta)=2\left(\frac{\beta-\alpha}{4}\right)^n=2\left(\frac{\mathfrak{l}}{4}\right)^n.$$

The necessary and sufficient condition in order that there exist polynomials with leading coefficient 1 which assume arbitrarily small values in a prescribed interval is therefore that the length $\mathfrak{l}$ of this interval is less than 4.

64. Let the independent variable x range through the two intervals of equal length which are formed from the interval $\alpha\leqq x\leqq\beta$ by removing from it the interval of length d with center $(\alpha+\beta)/2$; $d<\beta-\alpha=\mathfrak{l}$. Let μ_n be the greatest lower bound of the maxima of all polynomials of degree n with leading coefficient 1. Then we have

$$\mu_n=2\left(\frac{\mathfrak{l}^2-d^2}{16}\right)^{\frac{n}{2}},$$

if n is even, and

$$d\left(\frac{\mathfrak{l}^2-d^2}{16}\right)^{\frac{n-1}{2}}\leqq\mu_n\leqq\mathfrak{l}\left(\frac{\mathfrak{l}^2-d^2}{16}\right)^{\frac{n-1}{2}}$$

if n is odd. [**63**.]

65. Consider the set of all polynomials of degree n of the form

$$Q(z)=z^n+b_1z^{n-1}+b_2z^{n-2}+\cdots+b_n$$

with arbitrary complex coefficients $b_1, b_2, \ldots, b_n$. The maximum of $|Q(z)|$ on the unit circle $|z|=1$ is greater than or equal to 1 and is equal to 1 only if $Q(z)=z^n$.

66. One can give the following geometric formulation to **63** and **65**: Take n fixed points $P_1, P_2, \ldots, P_n$ in a plane, and let P be a variable point in that plane. The function of the point P given by

$$\overline{PP_1}\cdot\overline{PP_2}\ \cdots\ \overline{PP_n}$$

(where $\overline{PP_\nu}$ is the distance between the points P and P_ν) assumes on every straight line segment of length $\mathfrak{l}$ a maximum value greater than or equal to $2(\mathfrak{l}/4)^n$ and on every circle of radius r a maximum value greater than or equal to r^n. The only possible extreme cases are those in which the points $P_1, P_2, \ldots, P_n$ are distributed on the given line segment in the same way as the zeros of the Tchebychev polynomial $T_n(x)$ [**1**] in the interval $-1, 1$ or in which the n points all coincide with the centre of the circle.

Prove that the same assertion holds if the points $P_1, P_2, \ldots, P_n$ are arbitrarily distributed in space.

§ 9. The Lagrange Interpolation Formula

Let $x_1, x_2, \ldots, x_n$ be arbitrary distinct real or complex numbers. Set

$$f(x)=a_0(x-x_1)(x-x_2)\ldots(x-x_n), \qquad a_0\neq 0,$$

$$f_\nu(x)=\frac{1}{f'(x_\nu)}\frac{f(x)}{x-x_\nu}$$

$$=\frac{(x-x_1)\ldots(x-x_{\nu-1})(x-x_{\nu+1})\ldots(x-x_n)}{(x_\nu-x_1)\ldots(x_\nu-x_{\nu-1})(x_\nu-x_{\nu+1})\ldots(x_\nu-x_n)}, \qquad \nu=1,\ 2,\ \ldots,\ n.$$

Every polynomial of degree $n-1$ may be represented by means of its values at the points $x_1, x_2, \ldots, x_n$ as follows:

$$(*) \qquad P(x)=P(x_1)f_1(x)+P(x_2)f_2(x)+\cdots+P(x_n)f_n(x)$$

(the *Lagrange interpolation formula*). The polynomials $f_\nu(x)$ are called the *basis polynomials* of the interpolation.

67. Let the polynomial of nth degree

$$f(x)=a_0x^n+a_1x^{n-1}+\cdots+a_{n-1}x+a_n$$

have only distinct zeros $x_1, x_2, \ldots, x_n$. Then we have

$$\sigma_k=\sum_{\nu=1}^{n}\frac{x_\nu^k}{f'(x_\nu)}=\begin{cases}0, & \text{if } 0\leqq k\leqq n-2,\\ a_0^{-1}, & \text{if } k=n-1.\end{cases}$$

Furthermore σ_k is independent of a_n for $k=n, n+1, \ldots, 2n-2$ and is linear in a_n with coefficient $-a_0^{-2}$ for $k=2n-1$.

68 (continued). We have

$$\sum_{\nu=1}^{n}\frac{kx_\nu^{k-1}f'(x_\nu)-x_\nu^kf''(x_\nu)}{[f'(x_\nu)]^3}=\begin{cases}0, & \text{if } 0\leqq k\leqq 2n-2,\\ a_0^{-2}, & \text{if } k=2n-1.\end{cases}$$

69 (continued). Let $x_1, x_2, \ldots, x_n$ be different from 0 and -1. Prove that

$$\sum_{\nu=1}^{n}\frac{x_\nu^nf(x_\nu^{-1})}{f'(x_\nu)(1+x_\nu)}=(-1)^{n-1}(1-x_1x_2\ldots x_n).$$

70. Let $x_0, x_1, x_2, \ldots, x_n$ be arbitrary integers, $x_0<x_1<x_2\ldots<x_n$. Every polynomial of nth degree of the form

$$x^n+a_1x^{n-1}+a_2x^{n-2}+\cdots+a_n$$

assumes at the points $x_0, x_1, x_2, \ldots, x_n$ values, at least one of which is of absolute value $\geqq n!/2^n$.

71. If one chooses the zeros of $T_n(x)$ [**1**], i.e. the numbers $x_\nu=\cos(2\nu-1)\pi/(2n)$, $\nu=1, 2, \ldots, n$, as interpolation points, then the basis polynomials are

$$f_\nu(x)=\frac{(-1)^{\nu-1}\sqrt{1-x_\nu^2}}{n}\frac{T_n(x)}{x-x_\nu}, \qquad \nu=1,\ 2,\ \ldots,\ n.$$

Hence every polynomial $P(x)$ of degree $n-1$ may be represented as follows:

$$P(x)=\frac{1}{n}\sum_{\nu=1}^{n}(-1)^{\nu-1}\sqrt{1-x_\nu^2}\,P(x_\nu)\frac{T_n(x)}{x-x_\nu}.$$

The square roots are to be taken everywhere as positive.

72. If one chooses the zeros of $U_n(x)$ [**1**], that is the numbers

$$x_\nu=\cos\nu\frac{\pi}{n+1},\qquad \nu=1,\,2,\,\ldots,\,n,$$

as interpolation points, then the basis polynomials are

$$f_\nu(x)=(-1)^{\nu-1}\frac{1-x_\nu^2}{n+1}\frac{U_n(x)}{x-x_\nu},\qquad \nu=1,\,2,\,\ldots,\,n.$$

Hence every polynomial of degree $n-1$ may be represented as follows:

$$P(x)=\frac{1}{n+1}\sum_{\nu=1}^{n}(-1)^{\nu-1}(1-x_\nu^2)P(x_\nu)\frac{U_n(x)}{x-x_\nu}.$$

73. If one chooses the zeros of $U_{n-1}(x)\,(x^2-1)$, that is the numbers

$$x_\nu=\cos\nu\frac{\pi}{n},\qquad \nu=1,\,2,\,\ldots,\,n-1,$$

and in addition $x_0=1$, $x_n=-1$ as interpolation points, then the basis polynomials $f_0(x), f_1(x), f_2(x), \ldots, f_n(x)$ are the following:

$$f_\nu(x)=\frac{(-1)^\nu}{n}\frac{U_{n-1}(x)(x^2-1)}{x-x_\nu},\qquad \nu=1,\,2,\,\ldots,\,n-1,$$

$$f_0(x)=\frac{1}{2n}U_{n-1}(x)(x+1),\qquad f_n(x)=\frac{(-1)^n}{2n}U_{n-1}(x)(x-1).$$

Hence every polynomial $P(x)$ of degree n may be represented in the following form:

$$P(x)=\frac{1}{2n}U_{n-1}(x)[P(1)(x+1)+(-1)^nP(-1)(x-1)]$$

$$+\frac{1}{n}\sum_{\nu=1}^{n-1}(-1)^\nu P(x_\nu)\frac{U_{n-1}(x)(x^2-1)}{x-x_\nu}.$$

74. If one chooses the nth roots of unity $\varepsilon_\nu=e^{\frac{2\pi i\nu}{n}}$, $\nu=1, 2, \ldots, n$, as interpolation points, then the basis polynomials are

$$f_\nu(x)=\frac{\varepsilon_\nu}{n}\frac{x^n-1}{x-\varepsilon_\nu},\qquad \nu=1,\,2,\,\ldots,\,n.$$

Hence for an arbitrary polynomial $P(x)$ of degree $n-1$ we have the representation

$$P(x)=\frac{1}{n}\sum_{\nu=1}^{n}\varepsilon_\nu P(\varepsilon_\nu)\frac{x^n-1}{x-\varepsilon_\nu}.$$

75. Let the polynomials $P(x)$ and $Q(x)$ be both of degree n, and let x_0, x_1,

$x_2, \ldots, x_n$ be any $n+1$ distinct or in part coinciding real or complex numbers. From the $n+1$ equations

$$P(x_0)=Q(x_0), \quad P'(x_1)=Q'(x_1), \quad P''(x_2)=Q''(x_2), \quad \ldots, \quad P^{(n)}(x_n)=Q^{(n)}(x_n)$$

it follows that we have identically

$$P(x)=Q(x).$$

76. It is possible, and in fact in a unique manner, to determine, corresponding to $n+1$ arbitrary given numbers $c_0, c_1, c_2, \ldots, c_n$, a polynomial $P(x)$ of degree $\leqq n$ such that

$$P(0)=c_0, \quad P'(1)=c_1, \quad P''(2)=c_2, \ldots, \quad P^{(n)}(n)=c_n$$

[**75**]. This polynomial $P(x)$ may be represented in the form

$$P(x)=P(0)A_0(x)+P'(1)A_1(x)+P''(2)A_2(x)+\cdots+P^{(n)}(n)A_n(x),$$

where $A_0(x), A_1(x), \ldots, A_n(x)$ are arithmetically determined polynomials, independent of the arbitrarily assigned numbers $c_0=P(0), c_1=P'(1), \ldots, c_n=P^{(n)}(n)$. (They may be thought of as the "basis polynomials" of this interpolation, which is essentially different from the Lagrange interpolation.)

§ 10. The Theorems of S. Bernstein and A. Markov

77. For the set of all polynomials of degree $n-1$, $P(x)=a_0x^{n-1}+a_1x^{n-2}+\cdots+a_{n-1}$, which satisfy in the interval $-1\leqq x\leqq 1$ the inequality

$$\sqrt{1-x^2}\,|P(x)|\leqq 1,$$

the maximum value that $|a_0|$ can assume is equal to 2^{n-1}, i.e.

$$|a_0|\leqq 2^{n-1}.$$

The equality sign holds if and only if $P(x)=\gamma U_{n-1}(x)$, $|\gamma|=1$. [**71**]

78. Deduce from **73** a new proof of **62**.

79. Deduce from **74** a new proof of **65**.

80. Consider the set of all polynomials $P(x)$ of degree $n-1$ which satisfy in the interval $-1\leqq x\leqq 1$ the inequality

$$\sqrt{1-x^2}\,|P(x)|\leqq 1.$$

Then we have

$$|P(x)|\leqq n, \qquad -1\leqq x\leqq 1.$$

The equality sign holds only if $P(x)=\gamma U_{n-1}(x)$ with $|\gamma|=1$ and $x=\pm 1$.

81. Prove the following generalization of the second inequality in **7**: Let

$$S(\vartheta)=\mu_1 \sin \vartheta+\mu_2 \sin 2\vartheta+\mu_3 \sin 3\vartheta+\cdots+\mu_n \sin n\vartheta$$

be an arbitrary sine polynomial of nth order with real coefficients for which we have $|S(\vartheta)|\leqq 1$. Then

$$\left|\frac{S(\vartheta)}{\sin \vartheta}\right|\leqq n.$$

The equality sign holds only for $S(\vartheta)=\pm \sin n\vartheta$.

82. Let the trigonometric polynomial of nth order with real coefficients

$$g(\vartheta)=\lambda_0+\lambda_1\cos\vartheta+\mu_1\sin\vartheta+\lambda_2\cos 2\vartheta+\mu_2\sin 2\vartheta+\cdots$$
$$+\lambda_n\cos n\vartheta+\mu_n\sin n\vartheta$$

satisfy for all real ϑ the inequality

$$|g(\vartheta)|\leqq 1.$$

Then we have

$$|g'(\vartheta)|\leqq n.$$

[Consider $S(\vartheta)=[g(\vartheta_0+\vartheta)-g(\vartheta_0-\vartheta)]/2$, $\vartheta=0$.]

83. Let the polynomial $P(x)$ of nth degree satisfy in the interval $-1\leqq x\leqq 1$ the inequality $|P(x)|\leqq 1$. Then we have

$$|P'(x)|\leqq n^2.$$

The equality sign holds only if $P(x)=\gamma T_n(x)$, $|\gamma|=1$, $x=\pm 1$. [Consider $P(\cos\vartheta)$.]

§ 11. Legendre Polynomials and Related Topics

We define the *Legendre polynomials*

$$P_0(x),\quad P_1(x),\quad P_2(x),\ldots,\quad P_n(x),$$

by the following conditions:

(1) $P_n(x)$ is of nth degree, has real coefficients,[1] and satisfies

$$\int_{-1}^{1} P_n(x)x^\nu\,dx=0,\qquad \nu=0,\ 1,\ 2,\ \ldots,\ n-1;\quad n\geqq 1;$$

(2) $P_n(x)$ satisfies

$$\int_{-1}^{1}[P_n(x)]^2\,dx=\frac{2}{2n+1},\qquad n=0,\ 1,\ 2,\ \ldots;$$

(3) the coefficient of x^n in $P_n(x)$ is positive, $n=0, 1, 2, \ldots$.

The first condition implies that, if $K(x)$ denotes an arbitrary polynomial of degree $n-1$, then the equation

$$\int_{-1}^{1} P_n(x)K(x)\,dx=0$$

holds. From this it follows that $P_n(x)$ is uniquely determined by (1), apart from a constant factor. For if $P_n^*(x)$ were another polynomial with the same property, then $aP_n(x)-bP_n^*(x)$, where a, b are constants, $a\neq 0$, $b\neq 0$, would also have this property. Choosing a and b such that $aP_n(x)-bP_n^*(x)$ is of degree $n-1$, it would then follow from

$$\int_{-1}^{1}[aP_n(x)-bP_n^*(x)]^2\,dx$$
$$=a\int_{-1}^{1}P_n(x)[aP_n(x)-bP_n^*(x)]\,dx-b\int_{-1}^{1}P_n^*(x)[aP_n(x)-bP_n^*(x)]\,dx=0,$$

[1] The requirement that the coefficients be real can be omitted if certain details are appropriately amended. Cf. also **98, 99, 100.**

that $aP_n(x)-bP_n^*(x)=0$. From (2) it then follows that $|a|=|b|$ and from (3) that $a=b$.

The integral conditions (1), (2) may be combined in the following form:

$$\int_{-1}^{1} P_m(x)P_n(x)\,dx=\begin{cases}0, & \text{if } m\gtrless n,\\ \dfrac{2}{2n+1}, & \text{if } m=n;\quad m,\ n=0,\ 1,\ 2,\ \ldots\end{cases}$$

(the *orthogonality condition*).

84. $P_n(x)$ is, apart from a constant factor, equal to the nth derivative of $(x^2-1)^n$:

$$P_n(x)=\frac{1}{2^n n!}\frac{d^n}{dx^n}(x^2-1)^n \qquad (\textit{Rodrigues' formula}).$$

85. We have

$$(1-t)^nP_n\left(\frac{1+t}{1-t}\right)=1+\binom{n}{1}^2 t+\binom{n}{2}^2 t^2+\binom{n}{3}^2 t^3+\cdots+\binom{n}{n-1}^2 t^{n-1}+t^n.$$

86. $P_n(x)$ may be represented in the following integral form:

$$P_n(x)=\frac{1}{\pi}\int_0^{\pi}(x+\sqrt{x^2-1}\cos\varphi)^n\,d\varphi \qquad (\textit{Laplace's formula}).$$

87. Between three successive Legendre polynomials there holds the following recurrence relation:

$$P_n(x)=\frac{2n-1}{n}xP_{n-1}(x)-\frac{n-1}{n}P_{n-2}(x), \qquad n=2,\ 3,\ 4,\ \ldots.$$

88. There exists a unique polynomial $S_n(x)$ of nth degree with real coefficients such that the equation

$$\int_{-1}^{1} S_n(x)K(x)\,dx=K(1)$$

is satisfied for all polynomials $K(x)$ of nth degree. Express $S_n(x)$ and $(1-x)S_n(x)$ as a linear combination of Legendre polynomials. [Express $K(x)$ similarly.]

89 (continued). The polynomials

$$S_0(x),\quad S_1(x),\quad S_2(x),\ldots,\quad S_n(x),\ldots$$

satisfy the orthogonality condition

$$\int_{-1}^{1}(1-x)S_m(x)S_n(x)\,dx=\begin{cases}0, & \text{if } m\gtrless n,\\ \dfrac{n+1}{2}, & \text{if } m=n;\quad m,\ n=0,\ 1,\ 2,\ \ldots.\end{cases}$$

90. $P_n(x)$ satisfies a homogeneous linear differential equation of second order

$$(1-x^2)P_n''(x)-2xP_n'(x)+n(n+1)P_n(x)=0.$$

91. For sufficiently small w, we have identically in w and x

$$\frac{1}{\sqrt{1-2xw+w^2}}=P_0(x)+P_1(x)w+P_2(x)w^2+\cdots+P_n(x)w^n+\cdots.$$

(*Generating function* for the Legendre polynomials.)

92. Deduce **84**, **86**, **87**, **90** directly from **91**. (**91** may also be regarded as offering a definition of the Legendre polynomials, and in fact a "centrally located" definition: From here there are particularly convenient paths leading to most of the properties of the Legendre polynomials.)

93. $P_n(\cos\vartheta)$ is a cosine polynomial with non-negative coefficients. Determine these coefficients.

Deduce from this the following inequality:

$$|P_n(x)|\leqq 1, \qquad -1\leqq x\leqq 1.$$

For $n>0$ the equality sign holds only if $x=1$ or $x=-1$.

94. Let $x>1$. The sequence

$$P_0(x), \quad P_1(x), \quad P_2(x),\ldots, \quad P_n(x),\ldots$$

increases monotonically.

95. The sum of the first n Legendre polynomials is non-negative in the interval $-1\leqq x\leqq 1$:

$$P_0(x)+P_1(x)+P_2(x)+\cdots+P_n(x)\geqq 0, \qquad -1\leqq x\leqq 1.$$

The equality sign holds only if n is odd and $x=-1$. [III **157**.]

96. The sum of the first n Legendre polynomials

$$P_0(x)+P_1(x)+P_2(x)+\cdots+P_n(x)$$

is positive for every value of x if n is even; it has its only change of sign at the point $x=-1$ if n is odd. [**94**.]

97. The nth Legendre polynomial $P_n(x)$ has only real, simple zeros all of which lie in the interior of the interval -1, 1. [II **140**.]

98. Generalize the theorems in **84–91**, **97** for the *Jacobi* (*hypergeometric*) *polynomials*

$$P_0^{(\alpha,\beta)}(x), \quad P_1^{(\alpha,\beta)}(x), \quad P_2^{(\alpha,\beta)}(x),\ldots, \quad P_n^{(\alpha,\beta)}(x),\ldots, \quad \alpha,\beta>-1,$$

defined by the following conditions:

(1), (2) $P_n^{(\alpha,\beta)}(x)$ is of degree n, has real coefficients and orthogonality property

$$\int_{-1}^{1}(1-x)^\alpha(1+x)^\beta P_m^{(\alpha,\beta)}(x)P_n^{(\alpha,\beta)}(x)\,dx$$

$$=\begin{cases}0, & \text{if } m\gtrless n,\\ \dfrac{2^{\alpha+\beta+1}}{2n+\alpha+\beta+1}\,\dfrac{\Gamma(n+\alpha+1)\Gamma(n+\beta+1)}{\Gamma(n+1)\Gamma(n+\alpha+\beta+1)}, & \text{if } m=n; \quad m,\ n=0,\ 1,\ 2,\ \ldots\end{cases}$$

(for $n=0$ the above expression is to be taken as:

$$2^{\alpha+\beta+1}\frac{\Gamma(\alpha+1)\Gamma(\beta+1)}{\Gamma(\alpha+\beta+2)} \qquad);$$

(3) the coefficient of x^n in $P_n^{(\alpha,\beta)}(x)$ is positive.

Some cases already encountered, corresponding to special values of α and β, are collected in the following table:

$$\begin{array}{lllll} \alpha= & 0, & 1, & -\frac{1}{2}, & \frac{1}{2}, \\ \beta= & 0, & 0, & -\frac{1}{2}, & \frac{1}{2}, \\ P_n^{(\alpha,\beta)}(x)= & P_n(x), & \frac{2}{n+1} S_n(x), & \frac{1\cdot 3\ldots(2n-1)}{2\cdot 4\ldots 2n} T_n(x), & 2\frac{1\cdot 3\ldots(2n+1)}{2\cdot 4\ldots(2n+2)} U_n(x). \end{array}$$

Cf. p. 85, **89**, **4**, **4**.

(The coefficient of $T_n(x)$ must be replaced by 1 for $n=0$).

99. Prove the analogues of **84**, **85**, **87–91**, **97** for the *generalized Laguerre polynomials*

$$L_0^{(\alpha)}(x),\quad L_1^{(\alpha)}(x),\quad L_2^{(\alpha)}(x),\ldots,\quad L_n^{(\alpha)}(x),\ldots,\qquad \alpha>-1,$$

defined by the following conditions:

(1), (2) $L_n^{(\alpha)}(x)$ is of degree n, has real coefficients and orthogonality property

$$\int_0^\infty e^{-x}x^\alpha L_m^{(\alpha)}(x)L_n^{(\alpha)}(x)\,dx$$

$$=\begin{cases} 0, & \text{for } m\gtrless n,\\ \Gamma(\alpha+1)\binom{n+\alpha}{n}, & \text{for } m=n;\quad m,\ n=0,\ 1,\ 2,\ \ldots; \end{cases}$$

(3) the coefficient of x^n in $L_n^{(\alpha)}(x)$ has the sign $(-1)^n$.

100. Prove the analogues of **84**, **87**, **88**, **90**, **91**, **97** for the *Hermite polynomials*

$$H_0(x),\quad H_1(x),\quad H_2(x),\ldots,\quad H_n(x),\ldots$$

defined by the following conditions:

(1), (2) $H_n(x)$ is of degree n, has real coefficients and orthogonality property

$$\int_{-\infty}^\infty e^{-\frac{x^2}{2}} H_m(x)H_n(x)\,dx=\begin{cases} 0, & \text{for } m\gtrless n,\\ \frac{\sqrt{2\pi}}{n!} & \text{for } m=n;\quad m,\ n=0,\ 1,\ 2,\ \ldots; \end{cases}$$

(3) the coefficient of x^n in $H_n(x)$ has the sign $(-1)^n$.

101. If $P_n^{(\alpha,\beta)}(x)$ and $L_n^{(\alpha)}(x)$ are the functions defined in **98** and **99**, respectively, then we have

$$\lim_{\beta\to+\infty} P_n^{(\alpha,\beta)}(1-\varepsilon)=L_n^{(\alpha)}(x),$$

if ε converges to 0 as β increases in such a manner that

$$\lim_{\beta\to+\infty} \varepsilon\beta=2x.$$

102. If $L_n^{(\alpha)}(x)$ and $H_n(x)$ are the functions defined in **99** and **100**, respectively, then we have

$$H_{2q}(x)=\frac{(-1)^q}{1\cdot 3\cdot 5\ldots(2q-1)} L_q^{(-\frac{1}{2})}\left(\frac{x^2}{2}\right),$$

$$H_{2q+1}(x)=\frac{(-1)^{q+1}}{1\cdot 3\cdot 5\ldots(2q+1)} xL_q^{(\frac{1}{2})}\left(\frac{x^2}{2}\right),\qquad q=0,\ 1,\ 2,\ \ldots.$$

§ 12. Further Maximum–Minimum Problems on Polynomials

103. Let $P(x)$ be an arbitrary polynomial of nth degree with real coefficients satisfying

$$\int_{-1}^{1} [P(x)]^2\,dx=1.$$

Then for $-1\leqq x\leqq 1$ we have

$$|P(x)|\leqq\frac{n+1}{\sqrt{2}}.$$

The equality sign here holds if and only if either

$$P(x)=\pm\frac{\sqrt{2}}{n+1}\,S_n(x)\ \textbf{[88]}\quad\text{and}\quad x=1$$

or

$$P(x)=\pm\frac{\sqrt{2}}{n+1}\,S_n(-x)\qquad\text{and}\quad x=-1$$

($n>0$). [The integral of $[P(x)]^2$ is a quadratic form in $n+1$ variables. Choose these variables in such a manner that the quadratic form becomes a sum of $n+1$ squares.]

104. Let $P(x)$ be an arbitrary polynomial of nth degree with real coefficients satisfying

$$\int_{-1}^{1} (1-x)[P(x)]^2\,dx=1.$$

Then we have

$$|P(1)|\leqq\frac{1}{\sqrt{2}}\binom{n+2}{2},\qquad |P(-1)|\leqq\frac{1}{\sqrt{2}}\sqrt{\binom{n+2}{2}}.$$

These bounds cannot be replaced by any smaller bounds.

105. Let α and β be constants, $\alpha>-1, \beta>-1$, and $P(x)$ an arbitrary polynomial of nth degree with real coefficients satisfying

$$\int_{-1}^{1} (1-x)^{\alpha}(1+x)^{\beta}[P(x)]^2\,dx=1.$$

Determine the maximum of $|P(1)|$ and $|P(-1)|$, as $P(x)$ ranges through the set of all polynomials of the specified kind. How do these maximum values behave for large values of n?

106. Let α be a constant, $\alpha>-1$. Determine the maximum of $|P(0)|$ for all polynomials $P(x)$ of nth degree which satisfy the condition

$$\int_{0}^{\infty} e^{-x}x^{\alpha}[P(x)]^2\,dx=1.$$

How does this maximum behave for large values of n?

107. Determine the maximum of $|P(0)|$ for all polynomials $P(x)$ of nth degree, which satisfy the condition

$$\int_{-\infty}^{\infty} e^{-\frac{x^2}{2}}[P(x)]^2\,dx=1.$$

How does this maximum behave for large values of n?

108. Let $P(x)$ be a polynomial of nth degree which assumes only non-negative values in the interval $-1\leqq x\leqq 1$, and which satisfies

$$\int_{-1}^{1} P(x)\,dx=1.$$

Then we have

$$P(1)\leqq\begin{cases}\dfrac{q(q+1)}{2} & \text{for odd } n;\quad n=2q-1,\\[2ex] \dfrac{(q+1)^2}{2} & \text{for even } n,\quad n=2q.\end{cases}$$

The same estimate holds for $P(-1)$. These bounds cannot be replaced by any smaller bounds.

109. The first mean-value theorem of integral calculus may be strengthened for polynomials of nth degree as follows: Let $P(x)$ be a polynomial of nth degree and m the minimum, M the maximum of $P(x)$ in the interval $a\leqq x\leqq b$. Then we have

$$m+\frac{M-m}{\alpha_n}\leqq\frac{1}{b-a}\int_a^b P(x)\,dx\leqq M-\frac{M-m}{\alpha_n},$$

where $\alpha_n=q(q+1)$ for odd n, $n=2q-1$, and $\alpha_n=(q+1)^2$ for even n, $n=2q$.

110. Let α and β be constants, $\alpha>-1$, $\beta>-1$, and $P(x)$ a polynomial of nth degree which assumes only non-negative values in the interval $-1\leqq x\leqq 1$ and which satisfies

$$\int_{-1}^{1} (1-x)^\alpha(1-x)^\beta P(x)\,dx=1.$$

Then we have

$$P(1)\leqq\begin{cases}\dfrac{1}{2^{\alpha+\beta+1}}\dfrac{\Gamma(q+\alpha+1)\Gamma(q+\alpha+\beta+2)}{\Gamma(\alpha+1)\Gamma(\alpha+2)\Gamma(q)\Gamma(q+\beta+1)} & \text{for odd } n,\quad n=2q-1,\\[2ex] \dfrac{1}{2^{\alpha+\beta+1}}\dfrac{\Gamma(q+\alpha+2)\Gamma(q+\alpha+\beta+2)}{\Gamma(\alpha+1)\Gamma(\alpha+2)\Gamma(q+1)\Gamma(q+\beta+1)} & \text{for even } n,\quad n=2q.\end{cases}$$

These bounds cannot be replaced by any smaller bounds.

The corresponding bounds for $P(-1)$ are obtained by interchanging α and β.

111. Let α be a constant, $\alpha>-1$, and $P(x)$ a polynomial of nth degree which assumes only non-negative values for non-negative values of x, and which satisfies

$$\int_0^{\infty} e^{-x}x^\alpha P(x)\,dx=1.$$

Then we have

$$P(0) \leqq \frac{\Gamma(p+\alpha+2)}{\Gamma(\alpha+1)\Gamma(\alpha+2)\Gamma(p+1)}, \qquad p = \left[\frac{n}{2}\right].$$

This bound cannot be replaced by any smaller bound.

112. Let $P(x)$ be a polynomial of nth degree, which is non-negative for non-negative values of x, and which satisfies

$$\int_0^\infty e^{-x} P(x)\, dx = 1.$$

Then we have

$$P(0) \leqq \left[\frac{n}{2}\right] + 1.$$

113 (continued). Let ξ be an arbitrary non-negative number. Then we have

$$e^{-\xi} P(\xi) \leqq \left[\frac{n}{2}\right] + 1.$$

Part Seven. Determinants and Quadratic Forms

§ 1. Evaluation of Determinants. Solution of Linear Equations

1. Let the n vertices of a polyhedron be numbered in a definite order. Define a determinant of nth order as follows:

If the λth and the μth vertices are the two end-points of an edge of the polyhedron, let $a_{\lambda\mu}=a_{\mu\lambda}=1$.

If the line joining the λth and the μth vertices is not an edge of the polyhedron, let $a_{\lambda\mu}=0$. In particular $a_{\lambda\lambda}=0$, $\lambda=1, 2, \ldots, n$.

Show that the value of this determinant is independent of the manner of numbering the vertices. Form and calculate the determinants for the regular tetrahedron, hexahedron and octahedron.

1.1 (continued). If exactly p edges end in each vertex the determinant is divisible by p.

2. Calculate

$$\begin{vmatrix} 1 & 1 & 1 & \ldots & 1 \\ b_1 & a_1 & a_1 & \ldots & a_1 \\ b_1 & b_2 & a_2 & \ldots & a_2 \\ \ldots & \ldots & \ldots & \ldots & \ldots \\ b_1 & b_2 & b_3 & \ldots & a_n \end{vmatrix}$$

3. Prove the identity:

$$\left|\frac{1}{a_\lambda+b_\mu}\right|_1^n=\frac{\prod_{j>k}^{1,2,\ldots,n}(a_j-a_k)(b_j-b_k)}{\prod_{\lambda,\mu}^{1,2,\ldots,n}(a_\lambda+b_\mu)}.$$

4. Denoting the determinant of the quadratic form

$$\sum_{\lambda=1}^{n}\sum_{\mu=1}^{n}\frac{x_\lambda x_\mu}{\lambda+\mu}$$

by D_n, we have

$$D_n=\frac{[1!2!\ldots(n-1)!]^3 n!}{(n+1)!(n+2)!\ldots(2n)!}.$$

Also calculate the determinant $D_n(\alpha)$ of the quadratic form

$$\sum_{\lambda=1}^{n}\sum_{\mu=1}^{n}\frac{x_\lambda x_\mu}{\lambda+\mu+\alpha}, \qquad \alpha>-2.$$

5.

$$|(a_\lambda-b_\mu)^{n-1}|_1^n=\prod_{\nu=1}^{n-1}(n-\nu)^{n-2\nu}\prod_{j>k}^{1,2,\ldots,n}(a_j-a_k)(b_j-b_k).$$

6. Derive Theorem V **86** from Theorem V **48** by means of a transformation of the determinant $|F(\alpha_\lambda\beta_\mu)|$.

7. Setting $f(x)=(r_1-x)(r_2-x)\dots(r_n-x)$, we have

$$\begin{vmatrix} r_1 & a & a & \dots & a \\ b & r_2 & a & \dots & a \\ b & b & r_3 & \dots & a \\ \dots & \dots & \dots & \dots & \dots \\ b & b & b & \dots & r_n \end{vmatrix} = \frac{af(b)-bf(a)}{a-b}.$$

[Add to all the n^2 elements the variable x. The determinant formed in this way is a *linear* function of x and as such can be determined from two particular values.]

8. Set $\Delta=ad-bc$. Then the functional determinant

$$\frac{\partial(a\Delta, b\Delta, c\Delta, d\Delta)}{\partial(a, b, c, d)}=3\Delta^4.$$

9. Prove that $\rho-2$ is a divisor of the expression

$$\begin{vmatrix} \rho & \frac{l}{m}+\frac{m}{l} & \frac{n}{l}+\frac{l}{n} \\ \frac{l}{m}+\frac{m}{l} & \rho & \frac{m}{n}+\frac{n}{m} \\ \frac{n}{l}+\frac{l}{n} & \frac{m}{n}+\frac{n}{m} & \rho \end{vmatrix}, \qquad l\neq 0, \quad m\neq 0, \quad n\neq 0.$$

Determine the remaining factors.

10. The determinant

$$|1, x_\nu, x_\nu^2, \dots, x_\nu^{n-q-1}, x_\nu^{n-q+1}, \dots, x_\nu^{n-1}, x_\nu^n|, \qquad \nu=1, 2, \dots, n$$

is an alternating rational entire function of the n numbers $x_1, x_2, \dots, x_n$ and thus divisible by the product of differences

$$\prod_{j>k}^{1, 2, \dots, n} (x_j-x_k).$$

Show that the quotient S_q is equal to the qth elementary symmetric function of the n numbers $x_1, x_2, \dots, x_n$.

11. Let the numbers $a_0, a_1, a_2, \dots, a_n$ be different from 0. Then we have identically in z

$$a_0\begin{vmatrix} z+\frac{a_1}{a_0} & \frac{a_2}{a_1} & \frac{a_3}{a_2} & \dots & \frac{a_{n-1}}{a_{n-2}} & \frac{a_n}{a_{n-1}} \\ -\frac{a_1}{a_0} & z & 0 & \dots & 0 & 0 \\ 0 & -\frac{a_2}{a_1} & z & \dots & 0 & 0 \\ \dots & \dots & \dots & \dots & \dots & \dots \\ 0 & 0 & 0 & \dots & -\frac{a_{n-1}}{a_{n-2}} & z \end{vmatrix} = a_0z^n+a_1z^{n-1}+\dots+a_n.$$

11.1. Let $a_0, a_1, a_2, \ldots, b_0, b_1, b_2, \ldots$ denote constants, z a variable. Suppose that $a_0 \neq 0$ and that we have identically

$$(a_0+a_1z+a_2z^2+\cdots)(b_0+b_1z+b_2z^2+\cdots)=1.$$

Then

$$a_0^{n+1}b_n=(-1)^n\begin{vmatrix} a_1 & a_0 & 0 & 0 & \ldots & 0 \\ a_2 & a_1 & a_0 & 0 & \ldots & 0 \\ a_3 & a_2 & a_1 & a_0 & \ldots & 0 \\ \cdots & \cdots & \cdots & \cdots & \cdots & \cdots \\ a_n & a_{n-1} & a_{n-2} & a_{n-3} & \ldots & a_1 \end{vmatrix}.$$

(This serves as a method of computing the reciprocal of a given power series. The supposed identity may be merely formal; no considerations of convergence are involved.)

11.2. Show that

$$\begin{vmatrix} a-x & 1 & 0 & 0 & \ldots & 0 \\ \binom{a}{2} & a-x & 1 & 0 & \ldots & 0 \\ \binom{a}{3} & \binom{a}{2} & a-x & 1 & \ldots & 0 \\ \cdots & \cdots & \cdots & \cdots & \cdots & \cdots \\ \binom{a}{n} & \binom{a}{n-1} & \binom{a}{n-2} & \binom{a}{n-3} & \ldots & a-x \end{vmatrix}$$

$$=\binom{a+n-1}{n}-\binom{2a+n-2}{n-1}x+\binom{3a+n-3}{n-2}x^2-\cdots+(-1)^nx^n.$$

11.3. Show that

$$\begin{vmatrix} 1-x & 1 & 0 & 0 & \ldots & 0 \\ \frac{1}{2!} & 1-x & 1 & 0 & \ldots & 0 \\ \frac{1}{3!} & \frac{1}{2!} & 1-x & 1 & \ldots & 0 \\ \cdots & \cdots & \cdots & \cdots & \cdots & \cdots \\ \frac{1}{n!} & \frac{1}{(n-1)!} & \frac{1}{(n-2)!} & \frac{1}{(n-3)!} & \ldots & 1-x \end{vmatrix}$$

$$=\frac{1}{n!}-\frac{2^{n-1}}{(n-1)!}x+\frac{3^{n-2}}{(n-2)!}x^2-\frac{4^{n-3}}{(n-3)!}x^3+\cdots+(-1)^nx^n.$$

12. Let $a_1, a_2, a_3, b_1, b_2, b_3, c$ be real numbers. The system

$$\begin{aligned} -a_3x_2+a_2x_3&=b_1, \\ a_3x_1 \qquad\quad -a_1x_3&=b_2, \\ -a_2x_1+a_1x_2 \qquad\quad &=b_3, \\ a_1x_1+a_2x_2+a_3x_3&=c \end{aligned}$$

can be consistent in two different ways. In one case the unknowns x_1, x_2, x_3 are completely undetermined, in the other case they are completely determined.

13. Let the entire function

$$1+c_1z+c_2z^2+c_3z^3+\cdots$$

have all distinct zeros $a_1, a_2, \ldots, a_n, \ldots$

$$0<|a_1|<|a_2|<\cdots<|a_n|<\cdots.$$

Consider the system of n equations

$$\begin{aligned}
1+a_1u_1^{(n)}+a_1^2u_2^{(n)}+\cdots+a_1^nu_n^{(n)}&=0,\\
1+a_2u_1^{(n)}+a_2^2u_2^{(n)}+\cdots+a_2^nu_n^{(n)}&=0,\\
\cdots\cdots\cdots\cdots\cdots\cdots\cdots\cdots&\\
1+a_nu_1^{(n)}+a_n^2u_2^{(n)}+\cdots+a_n^nu_n^{(n)}&=0
\end{aligned}$$

that completely determine the $u_k^{(n)}$. If

$$\frac{1}{|a_1|}+\frac{1}{|a_2|}+\cdots+\frac{1}{|a_n|}+\cdots$$

is convergent, then $\lim_{n\to\infty} u_k^{(n)}$ exists, but it is not necessarily true that

$$\lim_{n\to\infty} u_k^{(n)}=c_k.$$

(*One* solution of the *infinite* system

$$1+a_\nu u_1+a_\nu^2u_2+\cdots=0, \qquad \nu=1,\ 2,\ 3,\ \ldots$$

is: $u_1=c_1,\ u_2=c_2,\ u_3=c_3, \ldots$.)

14. In the equations

$$\begin{aligned}
c_{11}z_1+c_{12}z_2+\cdots+c_{1n}z_n&=0,\\
c_{21}z_1+c_{22}z_2+\cdots+c_{2n}z_n&=0,\\
\cdots\cdots\cdots\cdots\cdots\cdots\cdots&\\
c_{n1}z_1+c_{n2}z_2+\cdots+c_{nn}z_n&=0
\end{aligned}$$

let the coefficients and the unknowns be complex:

$$c_{\lambda\mu}=a_{\lambda\mu}+ib_{\lambda\mu}, \qquad z_\mu=x_\mu+iy_\mu,$$

$a_{\lambda\mu}, b_{\lambda\mu}, x_\mu, y_\mu$ real. For the equations to admit not only the identically zero solution

$$z_1=z_2=\cdots=z_n=0, \qquad \text{i.e.}\quad x_1=x_2=\cdots=x_n=y_1=y_2=\cdots=y_n=0,$$

it is necessary and sufficient that the determinant $|c_{\lambda\mu}|_1^n$ vanishes. This yields *two equations* in the $2n^2$ real numbers $a_{\lambda\mu}, b_{\lambda\mu}$. On the other hand the given equations may be written as $2n$ linear homogeneous equations for $2n$ real unknowns. The necessary and sufficient condition for the existence of a not identically vanishing solution now consists in the vanishing of a real determinant, i.e. in *one equation* in the $a_{\lambda\mu}, b_{\lambda\mu}$. How can this be correct?

15. The six terms in the expansion of a third-order determinant cannot all be positive.

16. The rule for the expansion of a determinant consists of two parts. The first part specifies what products are to be formed from the elements, while the second part determines the sign of these products.

The second part may be simplified in the case of second-order determinants in the following manner:

Associate with the elements

$$\begin{matrix} a_{11} & a_{12} \\ a_{21} & a_{22} \end{matrix} \quad \text{the signs} \quad \begin{matrix} + & + \\ - & + \end{matrix} \quad \text{respectively.}$$

Then form the products prescribed in the first part from the elements, each with its fixed associated sign.

Now prove that a corresponding simplification is impossible in the case of a determinant of order higher than two, i.e. it is impossible to associate with the n^2 elements n^2 fixed signs in such a way that, if we form the products prescribed in the first part of the expansion rule from the elements each with its associated sign, we automatically obtain the correct sign in all the products.

§ 2. Power Series Expansion of Rational Functions

In **17–34** we consider the *Hankel* (or *recurrent*) determinants

$$\begin{vmatrix} a_n & a_{n+1} & a_{n+2} & \cdots & a_{n+r-1} \\ a_{n+1} & a_{n+2} & a_{n+3} & \cdots & a_{n+r} \\ a_{n+2} & a_{n+3} & a_{n+4} & \cdots & a_{n+r+1} \\ \cdots & \cdots & \cdots & \cdots & \cdots \\ a_{n+r-1} & a_{n+r} & a_{n+r+1} & \cdots & a_{n+2r-2} \end{vmatrix} = A_n^{(r)},$$

formed from the coefficients of the power series

$$a_0 + a_1 z + a_2 z^2 + \cdots + a_n z^n + \cdots;$$

a_n is the leading entry, r is the number of rows.

17. Let the power series $a_0 + a_1 z + a_2 z^2 + \cdots$ represent a rational function whose denominator is of precise degree q and whose numerator is of precise degree $p-1$. Assume that the numerator and the denominator have no common divisor. Setting $d = \max(0, p-q)$ we have

$$A_d^{(q+1)} = A_{d+1}^{(q+1)} = A_{d+2}^{(q+1)} = \cdots = 0.$$

18. Let d, q be non-negative integers. If we have

$$\begin{matrix} A_d^{(q)} \neq 0, & A_{d+1}^{(q)} \neq 0, & A_{d+2}^{(q)} \neq 0, & A_{d+3}^{(q)} \neq 0, & \ldots \\ & A_d^{(q+1)} = 0, & A_{d+1}^{(q+1)} = 0, & A_{d+2}^{(q+1)} = 0, & \ldots \end{matrix}$$

then the power series may be represented as the quotient of two polynomials, where the denominator is of degree q and the numerator is of degree $\leqq q+d-1$. [Investigate the dependence of the linear forms

$$L_n(x) = a_n x_0 + a_{n+1} x_1 + a_{n+2} x_2 + \cdots + a_{n+q} x_q.]$$

19.

$$A_n^{(r)} A_{n+2}^{(r)} - A_n^{(r+1)} A_{n+2}^{(r-1)} = (A_{n+1}^{(r)})^2.$$

20. If we have

$$A_m^{(q+1)} = A_{m+1}^{(q+1)} = A_{m+2}^{(q+1)} = A_{m+3}^{(q+1)} = \cdots = A_{m+t-1}^{(q+1)} = 0,$$

then the t determinants $A_{m+1}^{(q)}, \quad A_{m+2}^{(q)}, \quad A_{m+3}^{(q)}, \quad \ldots, \quad A_{m+t}^{(q)}$ are either all $=0$, or all $\neq 0$.

21. The triangular array

$$\begin{array}{ccccccccc}
A_0^{(1)} & A_1^{(1)} & A_2^{(1)} & A_3^{(1)} & A_4^{(1)} & * & * & * & * \\
& A_0^{(2)} & A_1^{(2)} & A_2^{(2)} & * & * & * & * & * \\
& & A_0^{(3)} & * & * & * & A_{n+2}^{(r-1)} & * & * \\
& & & * & * & A_n^{(r)} & A_{n+1}^{(r)} & A_{n+2}^{(r)} & * \\
& & & & * & * & A_n^{(r+1)} & * & *
\end{array}$$

is arranged so that a right angle opening upwards and bisected by the vertical contains only minors of the determinant standing at the vertex of the angle.

22. In the array **21** $A_n^{(r)}$ and $A_{n+1}^{(r)}$ constitute a *horizontal pair*, $A_{n+1}^{(r)}$ and $A_n^{(r+1)}$ constitute a *vertical pair*, $A_n^{(r)}$ and $A_n^{(r+1)}$ constitute a *diagonal pair*, $A_{n+1}^{(r)}$ and $A_{n-1}^{(r+1)}$ also form a diagonal pair, *cross-wise* to $A_n^{(r)}$ and $A_n^{(r+1)}$.

(1) If a diagonal pair vanishes, then the diagonal pair cross-wise to it also vanishes.

(2) If a horizontal pair vanishes, then either the horizontal pair lying above it or the one lying below it also vanishes.

(3) If a vertical pair vanishes, then either the neighbouring vertical pair on the left or the one on the right also vanishes.

Theorems (1), (2), (3) also hold on the slanting boundary line of the array **21**.

23. If of the infinitely many determinants

$$A_0^{(k+1)}, \quad A_1^{(k+1)}, \quad A_2^{(k+1)}, \quad \ldots, \quad A_n^{(k+1)}, \quad \ldots$$

only finitely many are different from 0, then the power series $a_0+a_1z+a_2z^2+\cdots+a_nz^n+\cdots$ represents a rational function whose denominator is of degree not greater than k. [**20**, **18**.]

24. If of the infinitely many determinants

$$A_0^{(1)}, \quad A_0^{(2)}, \quad A_0^{(3)}, \quad \ldots, \quad A_0^{(n)}, \quad \ldots$$

only finitely many are different from 0, then the power series $a_0+a_1z+a_2z^2+\cdots+a_nz^n+\cdots$ represents a rational function.

25. If of the infinitely many determinants

$$\begin{array}{ccccccc}
A_0^{(1)}, & A_1^{(2)}, & A_2^{(3)}, & A_3^{(4)}, & \ldots, & A_n^{(n+1)}, & \ldots \\
& A_1^{(1)}, & A_2^{(2)}, & A_3^{(3)}, & \ldots, & A_n^{(n)}, & \ldots
\end{array}$$

only finitely many are different from 0, then the power series $a_0+a_1z+a_2z^2+\cdots+a_nz^n+\cdots$ represents a rational function.

26. Show that the sufficient criteria given in **23, 24, 25** for the power series $a_0+a_1z+a_2z^2+\cdots+a_nz^n+\cdots$ to represent a rational function are also necessary.

An infinite matrix

$$\begin{matrix} a_{00} & a_{01} & a_{02} & \cdots \\ a_{10} & a_{11} & a_{12} & \cdots \\ \cdots & \cdots & \cdots & \cdots \end{matrix}$$

is said to be *of finite rank r* if all determinants of order $r+1$ contained in it vanish but not all those of order r.

In the infinite *Hankel matrix* $\mathfrak{H}$

$$(\mathfrak{H}) \qquad \begin{matrix} a_0 & a_1 & a_2 & \cdots \\ a_1 & a_2 & a_3 & \cdots \\ a_2 & a_3 & a_4 & \cdots \\ \cdots & \cdots & \cdots & \cdots \end{matrix}$$

$(a_{\lambda\mu}=a_{\lambda+\mu})$ we term the rank just defined also the *gross rank* of $\mathfrak{H}$. By striking out the first k columns (or rows) we obtain from $\mathfrak{H}$ the matrix $\mathfrak{H}_k$

$$(\mathfrak{H}_k) \qquad \begin{matrix} a_k & a_{k+1} & a_{k+2} & \cdots \\ a_{k+1} & a_{k+2} & a_{k+3} & \cdots \\ a_{k+2} & a_{k+3} & a_{k+4} & \cdots \\ \cdots & \cdots & \cdots & \cdots \end{matrix}$$

whose rank we shall denote by r_k. We have $\mathfrak{H}_0=\mathfrak{H}$, $r_0=r$, $r_0\geqq r_1\geqq r_2\geqq\cdots$. The minimum of the sequence of numbers $r_0, r_1, r_2, \ldots$, which is necessarily attained after finitely many steps, we term the *net rank* of $\mathfrak{H}$.

27. The rank of the matrix $\mathfrak{H}$ is finite if and only if the power series $a_0+a_1z+a_2z^2+\cdots+a_nz^n+\cdots$ represents a rational function.

28. The gross rank of $\mathfrak{H}$ is equal to the order of the *last* non-vanishing determinant in the infinite sequence

$$A_0^{(1)},\quad A_0^{(2)},\quad A_0^{(3)},\quad A_0^{(4)},\quad \ldots.$$

If this is $A_0^{(p)}$ then the net rank is equal to the order of the *first* non-vanishing determinant in the finite sequence

$$A_1^{(p)},\quad A_2^{(p-1)},\quad A_3^{(p-2)},\ldots,\quad A_p^{(1)}.$$

(The net rank $=0$ if these p determinants all vanish.)

29. The net rank of $\mathfrak{H}$ (assumed finite) is equal to the degree of the denominator of the rational function represented by $a_0+a_1z+a_2z^2+\cdots+a_nz^n+\cdots$. The rational function is a proper fraction if and only if the gross rank is equal to the net rank. If the gross rank is greater than the net rank, then the gross rank exceeds the degree of the numerator by one. (The numerator and the denominator of the rational function are assumed to have no common divisors; the degree here is the precise degree.)

30. The power series

$$a_0+\frac{a_1z}{1!}+\frac{a_2z^2}{2!}+\cdots+\frac{a_nz^n}{n!}+\cdots$$

satisfies a linear homogeneous differential equation with constant coefficients if and only if the determinants

$$A_0^{(1)},\quad A_0^{(2)},\quad A_0^{(3)},\ldots,\quad A_0^{(n)},\ldots,$$

apart from a finite number, all vanish.

31. Let $Q_n(z)$ be a polynomial of degree n, $Q_n(0)=1, n=0, 1, 2, \ldots$. For brevity set

$$a_0+a_1z+a_2z^2+\cdots=f(z),\quad Q_k(z)f(z)=\Box_k a_0+\Box_k a_1z+\Box_k a_2z^2+\cdots,$$
$$Q_k(z)Q_l(z)f(z)=\Box_k\,\Box_l a_0+\Box_k\,\Box_l a_1z+\Box_k\,\Box_l a_2z^2+\cdots$$

($\Box_k\,\Box_l a_n$ is a homogeneous linear expression in $a_n, a_{n-1}, a_{n-2}, \ldots, a_{n-k-l}$).

We have

$$A_0^{(n+1)}=\begin{vmatrix} a_0 & \Box_1\, a_1 & \cdots & \Box_n\, a_n \\ \Box_1\, a_1 & \Box_1\,\Box_1\, a_2 & \cdots & \Box_1\,\Box_n\, a_{n+1} \\ \cdots & \cdots & \cdots & \cdots \\ \Box_n\, a_n & \Box_n\,\Box_1\, a_{n+1} & \cdots & \Box_n\,\Box_n\, a_{2n} \end{vmatrix}.$$

32. Let $a_0/z+a_1/z^2+a_2/z^3+\cdots$ be the power series expansion of a rational function whose denominator (with no common divisors with the numerator) is $z^q-c_1z^{q-1}-c_2z^{q-2}-\cdots-c_q$. Between the matrices

$$\mathfrak{A}_m=\begin{bmatrix} a_m & a_{m+1} & \cdots & a_{m+q-1} \\ a_{m+1} & a_{m+2} & \cdots & a_{m+q} \\ \cdots & \cdots & \cdots & \cdots \\ a_{m+q-1} & a_{m+q} & \cdots & a_{m+2q-2} \end{bmatrix},\qquad \mathfrak{C}=\begin{bmatrix} 0 & 0 & 0 & 0 & \cdots & 0 & c_q \\ 1 & 0 & 0 & 0 & \cdots & 0 & c_{q-1} \\ 0 & 1 & 0 & 0 & \cdots & 0 & c_{q-2} \\ 0 & 0 & 1 & 0 & \cdots & 0 & c_{q-3} \\ \cdots & \cdots & \cdots & \cdots & \cdots & \cdots & \cdots \\ 0 & 0 & 0 & 0 & \cdots & 1 & c_1 \end{bmatrix}$$

there exists the relation

$$\mathfrak{A}_m=\mathfrak{A}_0\mathfrak{C}^m,\qquad m=0,\ 1,\ 2,\ \ldots.$$

Let the rank of the infinite matrix $\mathfrak{M}$ with finitely many $(=m)$ rows

$$(\mathfrak{M})\qquad \begin{matrix} a_{10} & a_{11} & a_{12} & \cdots & a_{1n} & \cdots \\ a_{20} & a_{21} & a_{22} & \cdots & a_{2n} & \cdots \\ \cdots & \cdots & \cdots & \cdots & \cdots & \cdots \\ a_{m0} & a_{m1} & a_{m2} & \cdots & a_{mn} & \cdots \end{matrix}$$

be denoted by r_0 and let the rank of the matrix that is formed from $\mathfrak{M}$ by deleting the first n columns (in whose left upper corner we have the element a_{1n}) be r_n. We have $r_0\geqq r_1\geqq r_2\geqq\cdots$, and we term $\lim_{n\to\infty} r_n$ the *net rank* of $\mathfrak{M}$.

The matrix $\mathfrak{M}$ corresponds to the system of m power series

$$\begin{aligned}
f_1(z) &= a_{10} + a_{11}z + a_{12}z^2 + \cdots + a_{1n}z^n + \cdots \\
f_2(z) &= a_{20} + a_{21}z + a_{22}z^2 + \cdots + a_{2n}z^n + \cdots \\
&\cdots\cdots\cdots\cdots\cdots\cdots\cdots\cdots\cdots \\
f_m(z) &= a_{m0} + a_{m1}z + a_{m2}z^2 + \cdots + a_{mn}z^n + \cdots.
\end{aligned}$$

We say that these are *linearly dependent* if there exist constants $c_1, c_2, \ldots, c_m$ satisfying $|c_1|+|c_2|+\cdots+|c_m|>0$, such that we have

$$c_1 f_1(z) + c_2 f_2(z) + \cdots + c_m f_m(z) = 0$$

identically in z. They are said to be *quasilinearly dependent* if there exist constants $c_1, c_2, \ldots, c_m$ satisfying $|c_1|+|c_2|+\cdots+|c_m|>0$, such that we have

$$c_1 f_1(z) + c_2 f_2(z) + \cdots + c_m f_m(z) = P(z)$$

identically in z, where $P(z)$ is a polynomial.

"The power series $f(z), zf(z), z^2 f(z), \ldots, z^m f(z)$ are quasilinearly dependent for sufficiently large m" and "$f(z)$ represents a rational function": these two statements are equivalent. We say that $f(z)$ represents an *algebraic function* if the $(m+1)^2$ power series $z^\mu [f(z)]^\nu$ $(\mu, \nu = 0, 1, 2, \ldots, m)$ are linearly dependent for sufficiently large m. We say that $f(z)$ satisfies an *algebraic differential equation* of the rth order, if the $(m+1)^{r+2}$ power series

$$z^\mu [f(z)]^\nu [f'(z)]^{\nu_1} [f''(z)]^{\nu_2} \ldots [f^{(r)}(z)]^{\nu_r} \qquad (\mu, \nu, \nu_1, \nu_2, \ldots, \nu_r = 0, 1, 2, \ldots, m)$$

are linearly dependent for sufficiently large m.

33. In a system of finitely many power series there are r linearly independent series if the rank of the corresponding matrix is r, and any power series of the system is linearly dependent on those r power series.

34. In a system of finitely many power series there are r quasilinearly independent series if the net rank of the corresponding matrix is r, and any power series of the system is quasilinearly dependent on those r power series.

§ 3. Generation of Positive Quadratic Forms

35. Consider the two quadratic forms

$$\sum_{\lambda=1}^{n} \sum_{\mu=1}^{n} a_{\lambda\mu} x_\lambda x_\mu \quad \text{and} \quad \sum_{\lambda=1}^{n} \sum_{\mu=1}^{n} b_{\lambda\mu} x_\lambda x_\mu.$$

If they are positive, then the quadratic form

$$\sum_{\lambda=1}^{n} \sum_{\mu=1}^{n} a_{\lambda\mu} b_{\lambda\mu} x_\lambda x_\mu$$

is also positive. If furthermore one of the given forms is definite and if in the matrix of the other the entries on the principal diagonal are different from 0, then the third form is also definite.

36. Consider the two symmetric matrices

$$\begin{pmatrix} a_{11} & a_{12} & \dots & a_{1n} \\ a_{21} & a_{22} & \dots & a_{2n} \\ \dots & \dots & \dots & \dots \\ a_{n1} & a_{n2} & \dots & a_{nn} \end{pmatrix}, \qquad \begin{pmatrix} e^{a_{11}} & e^{a_{12}} & \dots & e^{a_{1n}} \\ e^{a_{21}} & e^{a_{22}} & \dots & e^{a_{2n}} \\ \dots & \dots & \dots & \dots \\ e^{a_{n1}} & e^{a_{n2}} & \dots & e^{a_{nn}} \end{pmatrix}.$$

If the quadratic form corresponding to the matrix $(a_{\lambda\mu})$ is positive, then the quadratic form corresponding to $(e^{a_{\lambda\mu}})$ is also positive. If moreover among the rows of $(a_{\lambda\mu})$ there are no two identical, then the form corresponding to $(e^{a_{\lambda\mu}})$ is in fact definite. [V **76.**]

37. Let the power series

$$p_0+p_1x+p_2x^2+\cdots=F(x)$$

have no negative coefficients and converge for $x=a_{11}, a_{12}, \ldots, a_{nn}$. If the quadratic form corresponding to the symmetric matrix $(a_{\lambda\mu})$ of order n is positive, then the quadratic form corresponding to $(F(a_{\lambda\mu}))$ is also positive. If moreover there are among the coefficients $p_0, p_2, p_4, \ldots$ at least n different from 0 and if of the rows of $(a_{\lambda\mu}^2)$ no two are identical, then the form corresponding to $(F(a_{\lambda\mu}))$ is in fact definite.

38. Let the real numbers $a_0, a_1, a_2, \ldots, a_{2n}$ have the following property: If $f(x)$ denotes an arbitrary polynomial, with real coefficients of degree at most $2n$, that does not vanish identically and does not take a negative value for any real value of x, then also

$$a_0 f(x)+\frac{a_1}{1!}f'(x)+\frac{a_2}{2!}f''(x)+\cdots+\frac{a_{2n}}{(2n)!}f^{(2n)}(x)\geqq 0 \qquad (\text{or } >0)$$

for all real values of x.

Show that for this to be true it is necessary and sufficient that the quadratic form

$$\sum_{\lambda=0}^{n}\sum_{\mu=0}^{n} a_{\lambda+\mu}x_\lambda x_\mu$$

is positive (or positive definite, respectively).

39. Let the numbers $a_0, a_1, \ldots, a_n$, $n\geqq 1$ have the following property: If $f(x)$ is an arbitrary polynomial with real coefficients of degree at most n, that does not vanish identically and is non-negative for non-negative values of x, then also

$$a_0 f(x)+\frac{a_1}{1!}f'(x)+\frac{a_2}{2!}f''(x)+\cdots+\frac{a_n}{n!}f^{(n)}(x)\geqq 0 \qquad (\text{or } >0)$$

for all non-negative values of x (x real).

Show that for this to be true it is necessary and sufficient that the two quadratic forms

$$\sum_{\lambda=0}^{\left[\frac{n}{2}\right]}\sum_{\mu=0}^{\left[\frac{n}{2}\right]} a_{\lambda+\mu}x_\lambda x_\mu, \qquad \sum_{\lambda=0}^{\left[\frac{n-1}{2}\right]}\sum_{\mu=0}^{\left[\frac{n-1}{2}\right]} a_{\lambda+\mu+1}x_\lambda x_\mu$$

are positive (or positive definite, respectively).

40. Let the numbers $a_0, a_1, a_2, \ldots, a_n$, $n \geqq 1$ have the following property: If $f(x)$ denotes a polynomial with real coefficients of degree $\leqq n$ that does not vanish identically and is non-negative in the interval $-1 \leqq x \leqq 1$, then also

$$a_0 f(x) + \frac{a_1}{1!} f'(x) + \frac{a_2}{2!} f''(x) + \cdots + \frac{a_n}{n!} f^{(n)}(x) \geqq 0$$

for $-1 \leqq x \leqq 1$. Show that the only set of numbers $a_0, a_1, a_2, \ldots, a_n$ of this kind is determined by the conditions

$$a_0 \geqq 0, \qquad a_1 = a_2 = \cdots = a_n = 0.$$

41. Let the two quadratic forms

$$\sum_{\lambda=0}^{n} \sum_{\mu=0}^{n} a_{\lambda+\mu} x_\lambda x_\mu, \qquad \sum_{\lambda=0}^{n} \sum_{\mu=0}^{n} b_{\lambda+\mu} x_\lambda x_\mu$$

be positive. If we set

$$c_\nu = a_0 b_\nu + \binom{\nu}{1} a_1 b_{\nu-1} + \binom{\nu}{2} a_2 b_{\nu-2} + \cdots + a_\nu b_0,$$

then the quadratic form

$$\sum_{\lambda=0}^{n} \sum_{\mu=0}^{n} c_{\lambda+\mu} x_\lambda x_\mu$$

is also positive, and it is in fact definite if at least one of the two preceding forms is definite and the other not identically vanishing.

42. Let

$$c_\nu = a_0 b_\nu + \binom{\nu}{1} a_1 b_{\nu-1} + \binom{\nu}{2} a_2 b_{\nu-2} + \cdots + a_\nu b_0.$$

From the positiveness of the four quadratic forms

$$\sum_{\lambda=0}^{m} \sum_{\mu=0}^{m} a_{\lambda+\mu} x_\lambda x_\mu, \qquad \sum_{\lambda=0}^{m-1} \sum_{\mu=0}^{m-1} a_{\lambda+\mu+1} x_\lambda x_\mu,$$

$$\sum_{\lambda=0}^{m} \sum_{\mu=0}^{m} b_{\lambda+\mu} x_\lambda x_\mu, \qquad \sum_{\lambda=0}^{m-1} \sum_{\mu=0}^{m-1} b_{\lambda+\mu+1} x_\lambda x_\mu$$

follows the positiveness of the following two forms:

$$\sum_{\lambda=0}^{m} \sum_{\mu=0}^{m} c_{\lambda+\mu} x_\lambda x_\mu, \qquad \sum_{\lambda=0}^{m-1} \sum_{\mu=0}^{m-1} c_{\lambda+\mu+1} x_\lambda x_\mu.$$

Furthermore, from the positiveness of the four quadratic forms

$$\sum_{\lambda=0}^{m} \sum_{\mu=0}^{m} a_{\lambda+\mu} x_\lambda x_\mu, \qquad \sum_{\lambda=0}^{m} \sum_{\mu=0}^{m} a_{\lambda+\mu+1} x_\lambda x_\mu,$$

$$\sum_{\lambda=0}^{m} \sum_{\mu=0}^{m} b_{\lambda+\mu} x_\lambda x_\mu, \qquad \sum_{\lambda=0}^{m} \sum_{\mu=0}^{m} b_{\lambda+\mu+1} x_\lambda x_\mu$$

follows the positiveness of the following two forms:

$$\sum_{\lambda=0}^{m}\sum_{\mu=0}^{m} c_{\lambda+\mu}x_\lambda x_\mu, \qquad \sum_{\lambda=0}^{m}\sum_{\mu=0}^{m} c_{\lambda+\mu+1}x_\lambda x_\mu.$$

43. Let ϑ be real and let the complex numbers

$$c_{-n},\ c_{-n+1},\dots,\ c_{-1},\ c_0,\ c_1,\dots,\ c_{n-1},\ c_n, \qquad c_{-\nu}=\bar{c}_\nu, \qquad \nu=0,1,2,\dots,n$$

have the following property: If

$$\begin{aligned} g(\vartheta) &= \alpha_0+2(\alpha_1\cos\vartheta+\beta_1\sin\vartheta+\alpha_2\cos 2\vartheta+\beta_2\sin 2\vartheta+\cdots \\ &\quad +\alpha_n\cos n\vartheta+\beta_n\sin n\vartheta) \\ &= \sum_{\nu=-n}^{n}\gamma_\nu e^{-i\nu\vartheta}, \\ \gamma_\nu &= \bar{\gamma}_{-\nu}=\alpha_\nu+i\beta_\nu, \qquad \nu=0,1,2,\dots,n; \qquad \beta_0=0, \end{aligned}$$

denotes an arbitrary trigonometric polynomial of degree at most n, that does not vanish identically and does not have a negative value for any value of ϑ, let

$$\sum_{\nu=-n}^{n} c_\nu\gamma_\nu e^{-i\nu\vartheta}\geqq 0 \qquad (\text{or } >0).$$

Show that for this to be true, it is necessary and sufficient that the Hermitian form

$$\sum_{\lambda=0}^{n}\sum_{\mu=0}^{n} c_{\mu-\lambda}x_\lambda\bar{x}_\mu$$

is positive (or positive definite, respectively).

43.1. Let $G(z)$ be an entire function that takes real values for real z, with $G(0)=1$, and

$$-\frac{G'(z)}{G(z)}=s_1+s_2z+s_3z^2+\cdots+s_mz^{m-1}+\cdots.$$

If $G(z)$ has infinitely many zeros, all real, and is of the form V **165**, then, for $n=1,2,3,\dots,$

$$\begin{vmatrix} s_2 & s_3 & s_4 & \cdots & s_{n+1} \\ s_3 & s_4 & s_5 & \cdots & s_{n+2} \\ s_4 & s_5 & s_6 & \cdots & s_{n+3} \\ \cdots & \cdots & \cdots & \cdots & \cdots \\ s_{n+1} & s_{n+2} & s_{n+3} & \cdots & s_{2n} \end{vmatrix} >0.$$

(If $G(z)$ of V **165** has only a finite number of zeros, then only a finite number of these recurrent determinants can be $\neq 0$ [**27**].)

43.2. Let $G(z)$ be an entire function of the form V **165**, but not of the form $e^{\beta z}$, and

$$\frac{1}{G(z)}=c_0+\frac{c_1}{1!}z+\frac{c_2}{2!}z^2+\frac{c_3}{3!}z^3+\cdots.$$

Then, for $n=0,1,2,\dots,$

$$\begin{vmatrix} c_0 & c_1 & c_2 & \cdots & c_n \\ c_1 & c_2 & c_3 & \cdots & c_{n+1} \\ c_2 & c_3 & c_4 & \cdots & c_{n+2} \\ \cdots & \cdots & \cdots & \cdots & \cdots \\ c_n & c_{n+1} & c_{n+2} & \cdots & c_{2n} \end{vmatrix} >0.$$

[**38**, V **65.1.**]

§ 4. Miscellaneous Problems

44. Let the n^2 elements $a_{\lambda\mu}$ of a determinant of nth order be independent variables. Show that, of the $n!$ terms $\pm a_{1k_1}a_{2k_2}\dots a_{nk_n}$ in the expansion of the determinant, only $N=n^2-2n+2$ are independent, and determine N terms by means of which all the remaining terms may be expressed rationally.

45. In the expansion of a symmetric nth order determinant (with arbitrary elements) let s_n' be the number of distinct positive terms and s_n'' the number of distinct negative terms. For $s_n=s_n'+s_n''$ we have already in the works of Cayley the recurrence relation

$$s_{n+1}=(n+1)s_n-\binom{n}{2}s_{n-2}.$$

Show that for $d_n=s_n'-s_n''$ we have the recurrence relation

$$d_{n+1}=-(n-1)d_n-\binom{n}{2}d_{n-2},$$

and that

$$\lim_{n\to\infty}\frac{n^{1/2}s_n}{n!}=\frac{e^{3/4}}{\sqrt{\pi}},\qquad \lim_{n\to\infty}\frac{(-1)^{n-1}n^{3/2}d_n}{n!}=\frac{e^{-3/4}}{2\sqrt{\pi}}.$$

46. In the expansion of a symmetric nth order determinant $|a_{\lambda\mu}|$, in which the n elements $a_{\lambda\lambda}$ in the leading diagonal are equal to zero, let σ_n' be the number of distinct positive terms and σ_n'' the number of distinct negative terms. Setting

$$\sigma_n=\sigma_n'+\sigma_n'',\qquad \delta_n=\sigma_n'-\sigma_n'',$$

we have

$$\sigma_{n+1}=n\sigma_n+n\sigma_{n-1}-\binom{n}{2}\sigma_{n-2},\qquad \delta_{n+1}=-n\delta_n-n\delta_{n-1}-\binom{n}{2}\delta_{n-2}$$

and

$$\lim_{n\to\infty}\frac{n^{1/2}\sigma_n}{n!}=\frac{e^{-1/4}}{\sqrt{\pi}},\qquad \lim_{n\to\infty}\frac{(-1)^{n-1}n^{3/2}\,\delta_n}{n!}=\frac{e^{1/4}}{2\sqrt{\pi}}.$$

46.1. Let the capital letters $A, B, \dots I, \dots Z$ denote $n\times n$ matrices. In particular, let I denote the identity matrix with entries 1 along the main diagonal and 0 elsewhere, and Z denote the matrix in which all entries are 1. Show that

$$Z\cdot Z=n.Z.$$

46.2 (continued). If all the entries of A are positive numbers,

$$A\cdot B=I,$$

and $n>1$, then there are among the entries of B at least n positive numbers and at least n negative numbers.

Prove this, and show by examples that the two bounds for the number of positive entries, n and n^2-n, can be attained.

47. Denote the difference product $\prod_{j<k}^{1,2,\ldots,n}(x_j-x_k)$ by Δ. It is well known that an expression of the form

$$\Phi=\frac{1}{\Delta}\sum\pm\varphi(x_{\lambda_1}, x_{\lambda_2}, \ldots, x_{\lambda_n})$$

is a symmetric function of $x_1, x_2, \ldots, x_n$. Here the sum is to be taken over all the $n!$ permutations of $1, 2, \ldots, n$ and for the even permutations the positive sign is to be chosen and for the odd permutations the negative sign. Prove that, if

$$\varphi=\prod_{\nu=1}^{n}\frac{1}{1-x_1x_2\ldots x_\nu},$$

then

$$\Phi=\frac{1}{\prod_{\nu=1}^{n}(1-x_\nu)\prod_{j>k}^{1,2,\ldots,n}(1-x_jx_k)}.$$

48. Let the characteristic equation of the matrix $\|a_{\lambda\mu}\|$, $\lambda, \mu=1, 2, \ldots, n$,

$$\chi(z)=\begin{vmatrix} a_{11}-z & a_{12} & \ldots & a_{1n} \\ a_{21} & a_{22}-z & \ldots & a_{2n} \\ \cdots & \cdots & \cdots & \cdots \\ a_{n1} & a_{n2} & \ldots & a_{nn}-z \end{vmatrix}=0$$

have the roots $\alpha_1, \alpha_2, \ldots, \alpha_n$, that need not necessarily be all distinct. Denote the minors of order $n-1$ of $\chi(z)$ by $\chi_{\lambda\mu}(z)$. Prove that the characteristic equation of the matrix $\chi_{\lambda\mu}(z)$ has the roots

$$\zeta_\rho=\frac{\chi(z)}{\alpha_\rho-z}, \qquad \rho=1, 2, \ldots, n.$$

49. The linear transformations

$$S_\alpha: \qquad y_p=\frac{\sin\alpha\pi}{n+1}\sum_{q=0}^{n}\frac{x_q}{\sin\frac{\pi}{n+1}(p-q+\alpha)}, \qquad p=0, 1, \ldots, n,$$

α arbitrary, form a group in the following sense: We have

$$S_\alpha S_\beta=S_{\alpha+\beta}.$$

50. Let $g(\vartheta)$ be a trigonometric polynomial of order at most n, with purely real coefficients. Determine the conditions that $g(\vartheta)$ must satisfy, in order that the linear transformations

$$S_\alpha: \qquad y_p=\sum_{q=0}^{n}x_q g\left(\frac{\pi}{n+1}(p-q+\alpha)\right), \qquad p=0, 1, \ldots, n,$$

α arbitrary, form a group in the following sense: We have

$$S_\alpha S_\beta=S_{\alpha+\beta}.$$

51 (continued). The determinant of the linear transformation S_α vanishes for all values of α. The single exceptional case, when it is identically $=1$, is

$$g(\vartheta)=\frac{1}{n+1}\frac{\sin(n+1)\vartheta}{\sin\vartheta}.$$

52. (Continuation of **49.**) The transformation S_α is orthogonal, i.e. we have identically in $x_0, x_1, x_2, \ldots, x_n$ that

$$y_0^2+y_1^2+y_2^2+\cdots+y_n^2=x_0^2+x_1^2+x_2^2+\cdots+x_n^2.$$

We say that the linear transformation

$$\begin{aligned}
y_1&=a_{11}x_1+a_{12}x_2+\cdots+a_{1n}x_n+\cdots\\
y_2&=a_{21}x_1+a_{22}x_2+\cdots+a_{2n}x_n+\cdots\\
&\cdots\cdots\cdots\cdots\cdots\cdots\cdots\cdots\cdots\\
y_m&=a_{m1}x_1+a_{m2}x_2+\cdots+a_{mn}x_n+\cdots\\
&\cdots\cdots\cdots\cdots\cdots\cdots\cdots\cdots\cdots
\end{aligned}$$

or that the corresponding matrix $\|a_{mn}\|$ is *orthogonal*, if the relations

$$\begin{aligned}
&a_{m1}^2+a_{m2}^2+a_{m3}^2+\cdots+a_{mn}^2+\cdots=1, && m=1,2,3,\ldots,\\
&a_{\lambda1}a_{\mu1}+a_{\lambda2}a_{\mu2}+a_{\lambda3}a_{\mu3}+\cdots+a_{\lambda n}a_{\mu n}+\cdots=0, && \lambda\lessgtr\mu,\quad \lambda,\mu=1,2,3,\ldots
\end{aligned}$$

and also the further relations

$$\begin{aligned}
&a_{1n}^2+a_{2n}^2+a_{3n}^2+\cdots+a_{mn}^2+\cdots=1, && n=1,2,3,\ldots,\\
&a_{1\lambda}a_{1\mu}+a_{2\lambda}a_{2\mu}+a_{3\lambda}a_{3\mu}+\cdots+a_{m\lambda}a_{m\mu}+\cdots=0, && \lambda\lessgtr\mu,\quad \lambda,\mu=1,2,3,\ldots
\end{aligned}$$

are satisfied. We define similarly the orthogonality of a matrix that goes to infinity in four directions:

$$\begin{pmatrix}
\cdots\cdots\cdots\cdots\cdots\cdots\cdots\cdots\cdots\cdots\cdots\cdots\\
\ldots, a_{-m,-n},\ldots, a_{-m,-1}, a_{-m,0}, a_{-m,1},\ldots, a_{-m,n},\ldots\\
\cdots\cdots\cdots\cdots\cdots\cdots\cdots\cdots\cdots\cdots\cdots\cdots\\
\ldots, a_{m,-n},\ldots, a_{m,-1}, a_{m,0}, a_{m,1},\ldots, a_{m,n},\ldots\\
\cdots\cdots\cdots\cdots\cdots\cdots\cdots\cdots\cdots\cdots\cdots\cdots
\end{pmatrix}.$$

53. The *Fibonacci numbers* 0, 1, 1, 2, 3, 5, 8, 13, 21, 34, ... are defined as follows: Set $u_0=0$, $u_1=1$, $u_n+u_{n+1}=u_{n+2}$ for $n=0,1,2,\ldots$. The linear transformation

$$y_n=\frac{u_1x_1+u_2x_2+\cdots+u_nx_n-u_nx_{n+1}}{\sqrt{u_nu_{n+2}}},\qquad n=1,2,3,\ldots$$

is orthogonal.

54. Let α be real and not integral. The linear transformation

$$y_m=\frac{\sin\alpha\pi}{\pi}\sum_{n=-\infty}^{\infty}\frac{x_n}{m+n-\alpha},\qquad m=\ldots,-2,-1,0,1,2,\ldots$$

is orthogonal.

In the sequel, **54.1–54.4**, we consider infinite triangular matrices

$$\begin{pmatrix} a_{11} & & & \\ a_{21} & a_{22} & & \\ a_{31} & a_{32} & a_{33} & \\ \cdots & \cdots & \cdots & \cdots \end{pmatrix}$$

It is understood that all elements above the principal diagonal are 0. An equation between such matrices is regarded as valid if, and only if, it is valid between the corresponding submatrices with n^2 elements obtained by deleting all rows and columns except the first n, for $n=1, 2, 3, \ldots$.

54.1. Prove the relation, involving binomial coefficients,

$$\begin{pmatrix} 1 & & & \\ 1 & 1 & & \\ 1 & 2 & 1 & \\ 1 & 3 & 3 & 1 \\ \cdots & \cdots & \cdots & \cdots \end{pmatrix} \begin{pmatrix} 1 & & & \\ -1 & 1 & & \\ 1 & -2 & 1 & \\ -1 & 3 & -3 & 1 \\ \cdots & \cdots & \cdots & \cdots \end{pmatrix} = \begin{pmatrix} 1 & & & \\ 0 & 1 & & \\ 0 & 0 & 1 & \\ 0 & 0 & 0 & 1 \\ \cdots & \cdots & \cdots & \cdots \end{pmatrix}.$$

54.2. Prove the relation, involving Stirling numbers s_k^n and S_k^n of the first and second kind respectively (see the introductions to I **197** and I **186**),

$$\begin{pmatrix} 1 & & & \\ 1 & 1 & & \\ 1 & 3 & 1 & \\ 1 & 7 & 6 & 1 \\ \cdots & \cdots & \cdots & \cdots \end{pmatrix} \begin{pmatrix} 1 & & & \\ -1 & 1 & & \\ 2 & -3 & 1 & \\ -6 & 11 & -6 & 1 \\ \cdots & \cdots & \cdots & \cdots \end{pmatrix} = \begin{pmatrix} 1 & & & \\ 0 & 1 & & \\ 0 & 0 & 1 & \\ 0 & 0 & 0 & 1 \\ \cdots & \cdots & \cdots & \cdots \end{pmatrix}.$$

In other words, the matrices (S_k^n) and $((-1)^{n-k}s_k^n)$ are inverse to each other.

54.3. Prove

$$\begin{pmatrix} 1 & & & & \\ 1 & -1 & & & \\ 1 & -2 & 1 & & \\ 1 & -3 & 3 & -1 & \\ 1 & -4 & 6 & -4 & 1 \\ \cdots & \cdots & \cdots & \cdots & \cdots \end{pmatrix}^2 = \begin{pmatrix} 1 & & & & \\ 0 & 1 & & & \\ 0 & 0 & 1 & & \\ 0 & 0 & 0 & 1 & \\ 0 & 0 & 0 & 0 & 1 \\ \cdots & \cdots & \cdots & \cdots & \cdots \end{pmatrix}.$$

[I **34.1**, **54.1**.]

54.4. Generalizing **54.3**, find all infinite sequences of numbers $a_0, a_1, a_2, \ldots$ such that the system of relations

$$y_n = a_n x_0 - \binom{n}{1} a_{n-1} x_1 + \binom{n}{2} a_{n-2} x_2 - \cdots + (-1)^n a_0 x_n,$$

$n=0, 1, 2, 3, \ldots$, should imply the same system with the roles of x and y reversed, that is,

$$x_n = a_n y_0 - \binom{n}{1} a_{n-1} y_1 + \binom{n}{2} a_{n-2} y_2 - \cdots + (-1)^n a_0 y_n$$

for $n=0, 1, 2, 3, \ldots$. [I **34.1**.]

§ 5. Determinants of Systems of Functions

We term the determinant

$$\begin{vmatrix} f_1(x) & f_1'(x) & f_1''(x) & \dots & f_1^{(n-1)}(x) \\ f_2(x) & f_2'(x) & f_2''(x) & \dots & f_2^{(n-1)}(x) \\ f_3(x) & f_3'(x) & f_3''(x) & \dots & f_3^{(n-1)}(x) \\ \dots & \dots & \dots & \dots & \dots \\ f_n(x) & f_n'(x) & f_n''(x) & \dots & f_n^{(n-1)}(x) \end{vmatrix} = W[f_1(x), f_2(x), \dots, f_n(x)]$$

the *Wronskian* determinant of the system of functions $f_1(x), f_2(x), \dots, f_n(x)$.

55. If $c_{\lambda\mu}$ are constants, then we have

$$W(c_{11}f_1+c_{12}f_2+\cdots+c_{1n}f_n,\ c_{21}f_1+c_{22}f_2+\cdots+c_{2n}f_n,\ \dots,\\ c_{n1}f_1+c_{n2}f_2+\cdots+c_{nn}f_n)=|c_{\lambda\mu}|_1^n\cdot W(f_1, f_2, \dots, f_n).$$

56.

$$W[f_1(\varphi(x)), f_2(\varphi(x)), \dots, f_n(\varphi(x))]=\varphi'(x)^{\frac{n(n-1)}{2}}W[f_1(y), f_2(y), \dots, f_n(y)],$$

setting $y=\varphi(x)$ on the right-hand side.

57. $$W(\varphi f_1, \varphi f_2, \dots, \varphi f_n)=\varphi^n W(f_1, f_2, \dots, f_n).$$

58. $$\frac{1}{f_1^n}W(f_1, f_2, \dots, f_n)=W\left[\left(\frac{f_2}{f_1}\right)', \left(\frac{f_3}{f_1}\right)', \dots, \left(\frac{f_n}{f_1}\right)'\right].$$

59. $$\frac{d}{dx}\frac{W(f_1, \dots, f_{n-2}, f_n)}{W(f_1, \dots, f_{n-2}, f_{n-1})}=\frac{W(f_1, \dots, f_{n-2})W(f_1, \dots, f_{n-2}, f_{n-1}, f_n)}{[W(f_1, \dots, f_{n-2}, f_{n-1})]^2}.$$

60. If $W(f_1, f_2, \dots, f_{n-1}, f_n)$ vanishes at every point and $W(f_1, f_2, \dots, f_{n-1})$ at no point of the interval a, b, then there exist $n-1$ constants $c_1, c_2, \dots, c_{n-1}$ such that in the whole interval a, b

$$f_n(x)=c_1f_1(x)+c_2f_2(x)+\cdots+c_{n-1}f_{n-1}(x).$$

61. If $f_1(x), f_2(x), \dots, f_n(x)$ are n linearly independent solutions of the homogeneous linear nth order differential equation

$$y^{(n)}+\varphi_1(x)y^{(n-1)}+\varphi_2(x)y^{(n-2)}+\cdots+\varphi_n(x)y=0,$$

then we have for an arbitrary function y

$$y^{(n)}+\varphi_1(x)y^{(n-1)}+\varphi_2(x)y^{(n-2)}+\cdots+\varphi_n(x)y\equiv\frac{W(f_1, f_2, \dots, f_n, y)}{W(f_1, f_2, \dots, f_n)}.$$

62 (continued). Set

$$1=W_0, \qquad f_1=W_1, \qquad W(f_1, f_2)=W_2, \qquad \dots, \qquad W(f_1, f_2, \dots, f_n)=W_n.$$

Then we have for an arbitrary function y

$$y^{(n)}+\varphi_1(x)y^{(n-1)}+\varphi_2(x)y^{(n-2)}+\cdots+\varphi_n(x)y\\ \equiv\frac{W_n}{W_{n-1}}\cdot\frac{d}{dx}\frac{W_{n-1}^2}{W_{n-2}W_n}\cdots\frac{d}{dx}\frac{W_2^2}{W_1W_3}\cdot\frac{d}{dx}\frac{W_1^2}{W_0W_2}\cdot\frac{d}{dx}\frac{y}{W_1}.$$

This representation of the linear differential expression, in terms of n independent solutions of the homogeneous differential equation, is analogous to the factorization of a polynomial of degree n in terms of its n zeros.

63. Let $f_1(x), f_2(x), \ldots, f_n(x)$ be continuous real functions, defined in the interval a, b. The determinant

$$\left| \int_a^b f_\lambda(x) f_\mu(x)\, dx \right|_{\lambda, \mu = 1, 2, \ldots, n}$$

is never negative. It vanishes if and only if between the functions $f_1(x), f_2(x), \ldots, f_n(x)$ there exists a homogeneous linear relation with constant coefficients, not all 0.

64 (continued). The determinant

$$\begin{vmatrix} \int_a^b (f_2^2+f_3^2+\cdots+f_n^2)\, dx & -\int_a^b f_1 f_2\, dx & \cdots & -\int_a^b f_1 f_n\, dx \\ -\int_a^b f_2 f_1\, dx & \int_a^b (f_1^2+f_3^2+\cdots+f_n^2)\, dx & \cdots & -\int_a^b f_2 f_n\, dx \\ \cdots & \cdots & \cdots & \cdots \\ -\int_a^b f_n f_1\, dx & -\int_a^b f_n f_2\, dx & \cdots & \int_a^b (f_1^2+f_2^2+\cdots+f_{n-1}^2)\, dx \end{vmatrix}$$

is never negative. It vanishes if and only if the functions $f_1(x), f_2(x), \ldots, f_n(x)$ differ from one another only by constant factors, or, in other words, if they are all multiples of one of them.

65 (continued). The determinant

$$\left| e^{\int_a^b f_\lambda(x) f_\mu(x)\, dx} \right|_{\lambda, \mu = 1, 2, \ldots, n}$$

is never negative. It vanishes if and only if among the functions $f_1(x), f_2(x), \ldots, f_n(x)$ there are two identically equal.

66. A theorem mentioned in passing by Gauss in his Theoria combinationis observationum [Werke, Vol. 4, p. 12, cf. the last two lines of Art. 11] may be generalized as follows:

Let $f(x)$ be defined for $x > 0$, non-increasing there, $f(x) \geqq 0$, but not identically $= 0$. Moreover let the real numbers $a_1, a_2, \ldots, a_n$ be distinct. Assuming the existence of all the integrals that occur, the determinant

$$\left| (a_\lambda + a_\mu + 1) \int_0^\infty x^{a_\lambda + a_\mu} f(x)\, dx \right|_{\lambda, \mu = 1, 2, \ldots, n}$$

is never negative. It is $= 0$ if and only if $f(x)$ is a piecewise constant function with not more than $n-1$ jump discontinuities.

Let the function $f(\vartheta)$ be properly integrable in the interval $0 \leqq \vartheta < 2\pi$. Let its Fourier series [VI § 4] be

$$f(\vartheta) \sim a_0 + 2 \sum_{n=1}^{\infty} (a_n \cos n\vartheta + b_n \sin n\vartheta).$$

The *Hermitian* forms

$$T_n(f) = \sum_{\lambda=0}^{n} \sum_{\mu=0}^{n} c_{\mu-\lambda} x_\lambda \bar{x}_\mu$$

formed from the constants $c_0 = a_0$, $c_n = a_n + ib_n$, $c_{-n} = \bar{c}_n$, $n = 1, 2, 3, \ldots$, are called the *Toeplitz* forms associated with $f(\vartheta)$. (Cf. **43**).

67. Form the Toeplitz forms that belong to the following functions:

$$f(\vartheta) = c = \text{const.}, \qquad f(\vartheta) = a_0 + 2(a_1 \cos\vartheta + b_1 \sin\vartheta),$$

$$f(\vartheta) = \frac{1 - r^2}{1 - 2r\cos\vartheta + r^2}, \qquad 0 < r < 1.$$

68. If the function $f(\vartheta)$ is positive in the interval $0 \leqq \vartheta < 2\pi$, then all its associated Toeplitz forms are positive definite. More precisely: If $f(\vartheta)$ lies between the bounds $m \leqq f(\vartheta) \leqq M$ in the interval $0 \leqq \vartheta < 2\pi$, then we have

$$m \leqq T_n(f) \leqq M,$$

provided that the variables $x_0, x_1, \ldots, x_n$ satisfy the condition $T_n(1) = 1$. The equality sign can hold here only if $f(\vartheta) = \text{const.}$ at every point of continuity of $f(\vartheta)$.

69. Calculate the determinant $D_n(f)$ of the form $T_n(f)$ for the functions in **67** and show that, if $f(\vartheta)$ is everywhere positive,

$$\lim_{n\to\infty} \sqrt[n+1]{D_n(f)} = e^{\frac{1}{2\pi}\int_0^{2\pi} \log f(\vartheta)\,d\vartheta} = \mathfrak{G}(f).$$

70. Let $f(\vartheta)$ be a trigonometric polynomial of first order and let h be a parameter. The determinant $D_n(f-h)$ of the Toeplitz form $T_n(f-h)$ associated with $f(\vartheta) - h$ is a polynomial of degree $n+1$ in h with only real zeros. Calculate these zeros.

71 (continued). The zeros $h_{0n}, h_{1n}, \ldots, h_{nn}$ of $D_n(f-h)$ all lie between the minimum and the maximum of $f(\vartheta)$, that is $m \leqq h_{\nu n} \leqq M$, if $m \leqq f(\vartheta) \leqq M$ in the interval $0, 2\pi$. Moreover if $F(h)$ denotes an arbitrary properly integrable function in the interval in $m \leqq h \leqq M$, then we have

$$\lim_{n\to\infty} \frac{F(h_{0n}) + F(h_{1n}) + \cdots + F(h_{nn})}{n+1} = \frac{1}{2\pi} \int_0^{2\pi} F[f(\vartheta)]\,d\vartheta.$$

72. Let the Hermitian form

$$\sum_{\lambda=0}^{n} \sum_{\mu=0}^{n} c_{\lambda-\mu} x_\lambda \bar{x}_\mu, \qquad c_{-\nu} = \bar{c}_\nu,$$

be positive definite. Then all the zeros of the polynomial

$$\begin{vmatrix} c_0 & c_1 & c_2 & \ldots & c_n \\ c_{-1} & c_0 & c_1 & \ldots & c_{n-1} \\ c_{-2} & c_{-1} & c_0 & \ldots & c_{n-2} \\ \ldots & \ldots & \ldots & \ldots & \ldots \\ c_{-n+1} & c_{-n+2} & c_{-n+3} & \ldots & c_1 \\ 1 & z & z^2 & \ldots & z^n \end{vmatrix}$$

lie in the interior of the unit circle.

Part Eight. Number Theory*

Chapter 1. Arithmetical Functions

§ 1. Problems on the Integral Parts of Numbers

Let x be a real number. Denote by $[x]$ the *integral part* of x, i.e. the integer that satisfies the inequalities

$$[x] \leqq x < [x]+1.$$

We have for example

$$[\pi]=3, \qquad [2]=2, \qquad [-0.73]=-1.$$

1. Let n be an integer and x arbitrary. We then have

$$[x+n]=[x]+n.$$

2. In the expansion of a determinant of nth order, the product of the elements in the secondary diagonal (from north-east to south-west) has the sign $(-1)^{\left[\frac{n}{2}\right]}$.

3. We have

$$[2x]-2[x]=0 \quad \text{or} \quad 1,$$

according as

$$x-[x]<\tfrac{1}{2} \quad \text{or} \quad \geqq \tfrac{1}{2}.$$

4. If $0<\alpha<1$, then we have

$$[x]-[x-\alpha]=0 \quad \text{or} \quad 1$$

according as

$$x-[x] \geqq \alpha \quad \text{or} \quad <\alpha.$$

5. Let x be a number that does not lie at the mid-point between two consecutive integers. Express the integer nearest to x in terms of the symbol [ ].

6. One may term $[x]$ the integer *left-adjacent* to x. Express the integer *right-adjacent* to x in terms of the symbol [ ]. (Postal rates and other such rates are often based on the right-adjacent integer.)

---

* Number theory is concerned with the integers

$$\ldots, \ -3, \ -2, \ -1, \ 0, \ 1, \ 2, \ 3, \ \ldots$$

(mainly, although not exclusively, see Chap. 4). Throughout the present Part VIII we shall sometimes omit to mention that a number under consideration is an integer (or, more specifically, a positive or a non-negative integer) if this is sufficiently clear from the context or from the notation; thus n usually denotes an integer.

7. Prove that

$$[\alpha]+[\beta] \text{ either } =[\alpha+\beta], \text{ or } =[\alpha+\beta]-1,$$
$$[\alpha]-[\beta] \text{ either } =[\alpha-\beta], \text{ or } =[\alpha-\beta]+1.$$

8. We have

$$[2\alpha]+[2\beta] \geqq [\alpha]+[\alpha+\beta]+[\beta].$$

9. Let n be a positive integer, x arbitrary; then we have

$$[x]+\left[x+\frac{1}{n}\right]+\left[x+\frac{2}{n}\right]+\cdots+\left[x+\frac{n-1}{n}\right]=[nx].$$

10. Let n be a positive integer, x arbitrary. We have

$$\left[\frac{[nx]}{n}\right]=[x].$$

11. Let m be a positive integer. The highest power of 2 that is a divisor of

$$[(1+\sqrt{3})^{2m+1}]$$

is 2^{m+1}.

§ 2. Counting Lattice Points

12. Let a and n be positive integers. The number of those of the numbers $1, 2, 3, \ldots, n$ that are divisible by a is $\left[\frac{n}{a}\right]$.

13. How many zeros does the function $\sin x$ have in the interval $a<x\leqq b$? How many does it have in the interval $a\leqq x<b$?

14. Let $0\leqq\alpha\leqq\pi$. Denote by $V_n(\alpha)$ the number of changes of sign in the sequence

$$1, \quad \cos\alpha, \quad \cos 2\alpha, \ldots, \quad \cos(n-1)\alpha, \quad \cos n\alpha.$$

Then we have

$$\lim_{n\to\infty}\frac{V_n(\alpha)}{n}=\frac{\alpha}{\pi}.$$

15. Let ϑ be an irrational number, $0<\vartheta<1$ and $g_n=0$ or 1, according as $[n\vartheta]$ and $[(n-1)\vartheta]$ are equal or different. Show that

$$\lim_{n\to\infty}\frac{g_1+g_2+\cdots+g_n}{n}=\vartheta.$$

16. What is the number $N(r, a, \alpha)$ of zeros of the entire function $e^z-ae^{i\alpha}$ in the circular disc $|z|\leqq r$? (r, a, α real constants, $r>0$, $a>0$.)

17. Let a and b be integers, $f(x)$ a function defined in $a\leqq x\leqq b$, $f(x)>0$. Using the symbol [ ] find an expression for the number of *lattice points* (points with integral coordinates, cf. introduction to I **60.9**) that are situated in the part of the plane bounded by the inequalities

$$a\leqq x\leqq b, \qquad 0<y\leqq f(x).$$

18. Let p and q be positive integers with no common divisor different from 1, i.e. p, q relatively prime. By counting lattice points prove the formula

$$\left[\frac{q}{p}\right]+\left[\frac{2q}{p}\right]+\left[\frac{3q}{p}\right]+\cdots+\left[\frac{(p-1)q}{p}\right]=\frac{(p-1)(q-1)}{2}.$$

19. If p and q are odd positive integers with no common divisor, then with

$$\frac{p-1}{2}=p', \qquad \frac{q-1}{2}=q',$$

$$\left(\left[\frac{q}{p}\right]+\left[\frac{2q}{p}\right]+\cdots+\left[\frac{p'q}{p}\right]\right)+\left(\left[\frac{p}{q}\right]+\left[\frac{2p}{q}\right]+\cdots+\left[\frac{q'p}{q}\right]\right)=p'q'.$$

20. Let p be a prime number of the form $4n+1$. We then have

$$[\sqrt{p}]+[\sqrt{2p}]+[\sqrt{3p}]+\cdots+\left[\sqrt{\frac{p-1}{4}p}\right]=\frac{p^2-1}{12}.$$

§ 3. The Principle of Inclusion and Exclusion

21. Consider a set of N objects. Let N_α be the number of those objects that have a certain property α, N_β the number of those that have the property $\beta, \ldots,$ N_κ and N_λ the number of those that have the property κ and λ respectively. Similarly let $N_{\alpha\beta}, N_{\alpha\gamma}, \ldots, N_{\alpha\beta\gamma}, \ldots, N_{\alpha\beta\gamma\ldots\kappa\lambda}$ denote the number of those objects that have *simultaneously* the properties α and β, α and $\gamma, \ldots,$ α, β and $\gamma, \ldots,$ α, β, $\gamma, \ldots,$ κ and λ, respectively. Then the number N_0 of those objects that have *none* of the properties $\alpha, \beta, \gamma, \ldots, \kappa, \lambda$, is given by

$$\begin{aligned}
&N-N_\alpha-N_\beta-N_\gamma-\cdots-N_\kappa-N_\lambda\\
&\quad+N_{\alpha\beta}+N_{\alpha\gamma}+\cdots \qquad +N_{\kappa\lambda}\\
&\quad-N_{\alpha\beta\gamma}-\cdots\\
&\quad\ \cdots\\
&\quad\pm N_{\alpha\beta\gamma\ldots\kappa\lambda}.
\end{aligned}$$

The principle stated by **21** deserves another proof which, although needing some introduction, exhibits more clearly the underlying basic set-theoretical (logical) ideas. We consider a finite or infinite set U (the "universe of discourse"). The *characteristic function* $A(x)$ of a subset A of U is a function of the variable element x of U:

$$A(x)=1 \quad \text{or} \quad 0$$

according as x does, or does not, belong to A. The characteristic function of U has the constant value 1 (U thus stands also for unity) and the characteristic function of the empty set (the null set) has the constant value 0.

The characteristic functions of the subsets $A, B, C, \ldots$ will de denoted by $A(x), B(x), C(x), \ldots,$ respectively. If there is no danger of misunderstanding, we may simplify the notation by writing A for $A(x)$, etc., that is, by using the same symbol for a subset and its characteristic function.

A is a subset of B if, and only if,

$$A(x) \leqq B(x)$$

for all values of x, or, more simply, $A \leqq B$.

If U is a finite set, the number of elements of the subset A is $\Sigma A(x)$, where the summation is extended over all the elements of U.

21.1. What is the characteristic function of

(1) the complement of A,
(2) the intersection of A and B,
(3) the union of A and B?

21.2. Give another proof for **21**.

22. Assume that we have n objects, $n > 1$. Let these all have property α with the exception of the first object, all property β with the exception of the second object, . . . , and all property λ with the exception of the last nth object. What do we obtain from **21** applied to this case?

22.1. Prove I **189** by a combinatorial argument [**21**, I **192**].

22.2. Prove I **208** by a combinatorial argument.

22.3. Prove that

$$\tilde{S}_k^n = S_k^n - \binom{n}{1} S_{k-1}^{n-1} + \binom{n}{2} S_{k-2}^{n-2} - \cdots + (-1)^n S_{k-n}^0.$$

[See definitions in Part I, Chap. 4, § 3; if the condition $1 \leqq k \leqq n$ is not satisfied, S_k^n must be interpreted as having the value 0].

23. How many of the $n!$ terms in the expansion of a determinant of the nth order remain, if all the elements of the principal diagonal are set equal to 0? [Observe **21**: Let those terms have property α that contain a_{11} as a factor, etc.]

24. Let $a, b, c, \ldots, k, l$ be positive integers any two of which are relatively prime. How many numbers are there among the numbers $1, 2, 3, \ldots, n$, that are not divisible by any of the numbers $a, b, c, \ldots, k, l$? [**12.**]

25. Let the different prime factors of the number n be denoted by $p, q, r, \ldots$. The number of numbers less than n that have no common divisor different from 1 with n is equal to

$$n\left(1-\frac{1}{p}\right)\left(1-\frac{1}{q}\right)\left(1-\frac{1}{r}\right)\cdots.$$

This number is generally denoted by $\varphi(n)$ (the *Euler* function). We set $\varphi(1)=1$; $\varphi(n)$ is thus the number of those numbers $\leqq n$ that are relatively prime to n. This is the case for $n \geqq 1$, since 1 has no common divisor different from 1 with itself.

26. Assume that we have any N objects, that, as in **21**, can have the properties $\alpha, \beta, \ldots, \kappa, \lambda$. Associate with each individual object a *value*. Denote by W_α the total value (the sum of the values) of those objects that have property α, W_β the total value of those with property β, etc. Similarly let $W_{\alpha\beta}, W_{\alpha\gamma}, \ldots, W_{\alpha\beta\gamma}, \ldots, W_{\alpha\beta\gamma\ldots\kappa\lambda}$ denote the total value of those objects that have *simultaneously* the properties α and β, α and $\gamma, \ldots, \alpha, \beta$ and $\gamma, \ldots, \alpha, \beta, \gamma, \ldots, \kappa$ and λ, respectively. If

W is the total value of all the objects, then the total value of those objects that have *none* of the properties $\alpha, \beta, \ldots, \kappa, \lambda$ is equal to

$$\begin{aligned}
W &- W_\alpha - W_\beta - W_\gamma - \cdots - W_\kappa - W_\lambda \\
&+ W_{\alpha\beta} + W_{\alpha\gamma} + \cdots \qquad + W_{\kappa\lambda} \\
&- W_{\alpha\beta\gamma} - \cdots \\
&\cdots \\
&\pm W_{\alpha\beta\gamma\ldots\kappa\lambda}.
\end{aligned}$$

27. Let $n > 1$; and $r_1, r_2, \ldots, r_{\varphi(n)}$ be the numbers less than n that have no common divisors different from 1 with n. We then have

$$r_1^2 + r_2^2 + \cdots + r_{\varphi(n)}^2 = \frac{\varphi(n)}{3}\left(n^2 + \frac{(-1)^\nu}{2} pqr \ldots\right),$$

where $p, q, r, \ldots$ are the different prime factors of n, and ν is their number.

27.1. The number N_0 sought in **21** is

$$\begin{aligned}
&\leqq N \\
&\geqq N - N_\alpha - N_\beta - N_\gamma - \cdots - N_\kappa - N_\lambda \\
&\leqq N - N_\alpha - N_\beta - N_\gamma - \cdots - N_\kappa - N_\lambda + N_{\alpha\beta} + N_{\alpha\gamma} + \cdots + N_{\kappa\lambda}. \\
&\cdots\cdots\cdots\cdots\cdots\cdots\cdots\cdots\cdots\cdots\cdots\cdots
\end{aligned}$$

Briefly, in a somewhat extended meaning of the term (see the introduction to I **140** and I **144**), the expression given in **21** "envelops" N_0.

27.2. Let l stand for the number of properties considered in **21**.

Define

$$\begin{aligned}
S_0 &= N, \\
S_1 &= N_\alpha + N_\beta + N_\gamma + \cdots + N_\kappa + N_\lambda, \\
S_2 &= N_{\alpha\beta} + N_{\alpha\gamma} + \cdots + N_{\kappa\lambda}, \\
&\cdots\cdots\cdots\cdots\cdots\cdots \\
S_l &= N_{\alpha\beta\gamma\ldots\kappa\lambda},
\end{aligned}$$

and let s_k denote the number of those objects that possess precisely k of the l properties considered, $0 \leqq k \leqq l$.

It is easily seen (recognize the relation to binomial coefficients) that

$$\begin{aligned}
S_0 &= s_0 + s_1 + s_2 + s_3 + \cdots + s_l, \\
S_1 &= \qquad s_1 + 2s_2 + 3s_3 + \cdots + ls_l, \\
S_2 &= \qquad\qquad s_2 + 3s_3 + \cdots + \binom{l}{2} s_l, \\
S_3 &= \qquad\qquad\qquad s_3 + \cdots + \binom{l}{3} s_l, \\
&\cdots\cdots\cdots\cdots\cdots\cdots\cdots\cdots \\
S_l &= \qquad\qquad\qquad\qquad\qquad s_l.
\end{aligned}$$

Hence prove that

$$\begin{aligned}
s_0 &= S_0 - S_1 + S_2 - S_3 + \cdots + (-1)^l S_l.\\
s_1 &= \quad S_1 - 2S_2 + 3S_3 - \cdots + (-1)^{l-1} l S_l,\\
s_2 &= \qquad S_2 - 3S_3 + \cdots + (-1)^{l-2}\binom{l}{2} S_l,\\
s_3 &= \qquad\quad S_3 + \cdots + (-1)^{l-3}\binom{l}{3} S_l,\\
&\cdots\cdots\cdots\cdots\cdots\cdots\cdots\cdots\cdots\cdots\\
s_l &= \qquad\qquad\qquad S_l.
\end{aligned}$$

The first line just restates the result of **21**. [VII **54.1**.]

§ 4. Parts and Divisors

Let n be a non-negative integer. By a *part* of n we mean any non-negative integer that is $\leqq n$. The numbers 0 and n are termed "improper parts" of n. The number 0 contains only itself as a part and in fact as an improper one.

Let n be a positive integer. By a *divisor* of n we mean any positive integer that divides n without remainder. The numbers 1 and n are termed "improper divisors" of n. The number 1 contains only itself as a divisor and in fact as an improper one.

Let m and n be non-negative integers. By the *greatest common part* of m and n we mean that number whose parts are identical with the common parts of m and n. It is the smaller of the two numbers m and n or equal to both if they are equal. It is denoted by $\min(m, n)$.

Let m and n be positive integers. By the *greatest common divisor* of m and n we mean that number whose divisors are identical with the common divisors of m and n. It is denoted by (m, n).

By $\min(l, m, n, \ldots)$ we denote in general the smallest of the numbers $l, m, n, \ldots$, and by $(l, m, n, \ldots)$ we denote the greatest common divisor of the numbers $l, m, n, \ldots$.

There exists a smallest number, among whose parts are contained those of m as well as those of n. It is the larger of the two numbers m and n or equal to both if they are equal. It is denoted by $\max(m, n)$.

There exists a smallest number among whose divisors are contained those of m as well as those of n. It is termed the *least common multiple* of m and n.

Let $a_0, a_1, a_2, \ldots$ be any numbers. By $\sum_{t\leqq n} a_t$ we mean the sum taken over all the parts of n (including the improper parts 0 and n); $\sum_{t\leqq n} a_t = \sum_{t=0}^{n} a_t$.

Let $a_1, a_2, a_3, \ldots$ be any numbers. By $\sum_{t/n} a_t$ we mean the sum taken over all the divisors of n (including the improper divisors 1 and n). E.g. $\sum_{t/6} a_t = a_1 + a_2 + a_3 + a_6$.

28. Let $a, b, c, \ldots, k, l$ be any non-negative integers. We have

$$\begin{aligned}
\max(a, b, c, \ldots, k, l) = a + b + c + \cdots + k + l&\\
-\min(a, b) - \min(a, c) - \cdots - \min(k, l)&\\
+\min(a, b, c) + \cdots&\\
\cdots\cdots\cdots\cdots\cdots\cdots\cdots\cdots&\\
\pm\min(a, b, c, \ldots, k, l).&
\end{aligned}$$

29. The least common multiple M of the positive integers $a, b, c, \ldots, k, l$ may be represented as follows:

$$M = abc\ldots kl(a, b)^{-1}(a, c)^{-1}\ldots(k, l)^{-1}(a, b, c)\ldots(a, b, c, \ldots, k, l)^{\pm 1}.$$

30. We have

$$1 = \begin{vmatrix} 1 & 0 & 0 & 0 & \ldots & 0 \\ 1 & 1 & 0 & 0 & \ldots & 0 \\ 1 & 1 & 1 & 0 & \ldots & 0 \\ 1 & 1 & 1 & 1 & \ldots & 0 \\ \multicolumn{6}{c}{\dotfill} \\ 1 & 1 & 1 & 1 & \ldots & 1 \end{vmatrix}^2 = \begin{vmatrix} 1 & 1 & 1 & 1 & \ldots & 1 \\ 1 & 2 & 2 & 2 & \ldots & 2 \\ 1 & 2 & 3 & 3 & \ldots & 3 \\ 1 & 2 & 3 & 4 & \ldots & 4 \\ \multicolumn{6}{c}{\dotfill} \\ 1 & 2 & 3 & 4 & \ldots & n+1 \end{vmatrix}.$$

The general element of the first determinant $\eta_{\lambda\mu} = 1$, if μ is a (proper or improper) part of λ, otherwise it $= 0$ ($\lambda, \mu = 0, 1, \ldots, n$); in the second determinant the general element $c_{\lambda\mu}$ is equal to the number of common parts of λ and μ, i.e. the smaller of the two numbers $\lambda + 1$ and $\mu + 1$ ($\lambda, \mu = 0, 1, \ldots, n$).

31. Prove that $|c_{\lambda\mu}|$ is equal to 1, where $|c_{\lambda\mu}|$ denotes the determinant whose general element $c_{\lambda\mu}$ is equal to the number of common divisors of λ and μ, in other words to the number of divisors of the greatest common divisor of λ and μ, where $\lambda, \mu = 1, 2, \ldots, n$.

32. Let $a_0, a_1, \ldots, a_n$ be arbitrary, $A_\nu = \sum_{t \leq \nu} a_t$, $\nu = 0, 1, \ldots, n$. We have

$$\begin{vmatrix} A_0 & A_0 & A_0 & A_0 & \ldots & A_0 \\ A_0 & A_1 & A_1 & A_1 & \ldots & A_1 \\ A_0 & A_1 & A_2 & A_2 & \ldots & A_2 \\ A_0 & A_1 & A_2 & A_3 & \ldots & A_3 \\ \multicolumn{6}{c}{\dotfill} \\ A_0 & A_1 & A_2 & A_3 & \ldots & A_n \end{vmatrix} = a_0 a_1 a_2 \ldots a_n.$$

The general element $c_{\lambda\mu}$ of the determinant is $= A_r$ where $r = \min(\lambda, \mu)$, $\lambda, \mu = 0, 1, \ldots, n$. (Generalization of **30**.)

33. Let $a_1, a_2, \ldots, a_n$ be arbitrary, $A_\nu = \sum_{t/\nu} a_t$, $\nu = 1, 2, \ldots, n$. We have

$$\begin{vmatrix} A_1 & A_1 & A_1 & A_1 & \ldots & A_1 \\ A_1 & A_2 & A_1 & A_2 & \ldots & A_{(2, n)} \\ A_1 & A_1 & A_3 & A_1 & \ldots & A_{(3, n)} \\ A_1 & A_2 & A_1 & A_4 & \ldots & A_{(4, n)} \\ \multicolumn{6}{c}{\dotfill} \\ A_1 & A_{(n, 2)} & A_{(n, 3)} & A_{(n, 4)} & \ldots & A_n \end{vmatrix} = a_1 a_2 a_3 \ldots a_n.$$

The general element $c_{\lambda\mu}$ of the determinant is $= A_r$ where $r = (\lambda, \mu)$, $\lambda, \mu = 1, 2, \ldots, n$. (Generalization of **31**.)

34. If $a_0, a_1, a_2, \ldots$ are arbitrary and $A_n = \sum_{t \leq n} a_t$, $n = 0, 1, 2, \ldots$, then we clearly have

$$a_0 = A_0, \qquad a_1 = A_1 - A_0, \qquad a_2 = A_2 - A_1, \ldots, \qquad a_n = A_n - A_{n-1}, \qquad \ldots.$$

If $a_1, a_2, a_3, \ldots$ are arbitrary and $A_n = \sum_{t/n} a_t$, $n = 1, 2, 3, \ldots$, then we have

$$a_1 = A_1, \qquad a_2 = A_2 - A_1, \qquad a_3 = A_3 - A_1, \qquad a_4 = A_4 - A_2,$$
$$a_5 = A_5 - A_1, \qquad a_6 = A_6 - A_3 - A_2 + A_1, \qquad \ldots$$

and in general

$$a_n = \sum_{t/n} \mu(t) A_{\frac{n}{t}}, \qquad n = 1, 2, 3, \ldots,$$

where $\mu(n)$ is the Möbius function (see definition in § 5 preceding **38**).

35. Denote by $\psi(y)$ an arbitrary function defined for $0 \leqq y \leqq 1$. Let

$$g(n) = \sum_{\nu=1}^{n} \psi\left(\frac{\nu}{n}\right), \qquad f(n) = \sum_{(r,\, n)=1} \psi\left(\frac{r}{n}\right),$$

where the last sum is extended over the numbers r that are $\leqq n$ and are relatively prime to n. We then have

$$f(n) = \sum_{t/n} \mu(t) g\left(\frac{n}{t}\right) = \sum_{t/n} \mu\left(\frac{n}{t}\right) g(t).$$

36. As is well known, we have

$$\prod_{\nu=1}^{n} \left(x - e^{\frac{2\pi i \nu}{n}}\right) = x^n - 1.$$

Set

$$\prod_{(r,\, n)=1} \left(x - e^{\frac{2\pi i r}{n}}\right) = K_n(x),$$

where the product is extended over those numbers r that are $\leqq n$ and relatively prime to n. ($K_n(x)$ is termed the nth *cyclotomic polynomial.*) The zeros of $x^n - 1$ are the nth roots of unity, the zeros of $K_n(x)$ are the *primitive* nth roots of unity. Derive the formula

$$K_n(x) = \prod_{t/n} \left(x^{\frac{n}{t}} - 1\right)^{\mu(t)}.$$

37. If $\mu(n)$ is the Möbius function, then we have

$$\sum_{(r,\, n)=1} e^{\frac{2\pi i r}{n}} = \mu(n).$$

§ 5. Arithmetical Functions, Power Series, Dirichlet Series

By an *arithmetical function* $f(n)$ we mean a function defined for $n = 1, 2, 3, \ldots$. In this general sense defining an "arithmetical function" is equivalent to defining an arbitrary infinite sequence of numbers. Some functions of this kind that are of special significance in number theory are the following:

$\varphi(n)$ (the *Euler function*), the number of numbers less than n that are relatively prime to n; $\varphi(1) = 1$ [**25**];

$\tau(n) = \sum_{t/n} 1$, the number of divisors of n;

$\sigma(n) = \sum_{t/n} t$, the sum of the divisors of n;

$\sigma_\alpha(n) = \sum_{t/n} t^\alpha$, the sum of the αth powers of the divisors of n; $\sigma_1(n) = \sigma(n)$, $\sigma_0(n) = \tau(n)$;

$\nu(n)$, the number of distinct prime factors of n;

$\mu(n)$ (the *Möbius function*), $\mu(1) = 1$, $\mu(n) = 0$, if n is divisible by a square (apart from 1) and $\mu(n) = (-1)^{\nu(n)}$ in all other cases;

$\lambda(n)$ (the *Liouville function*), $\lambda(1)=1$, $\lambda(n)=(-1)^k$ where k is the number of prime factors of n (counting multiple factors according to multiplicity);

$\Lambda(n)$ (the *Mangoldt function*), $\Lambda(n)=\log p$ if $n=p^m$ is a power of a prime number, and otherwise $\Lambda(n)=0$.

38. Construct a table of these functions from $n=1$ to $n=10$. (In the case of $\sigma_\alpha(n)$ for $\alpha=0, 1, 2$.)

38.1. Study the table in **38** and prove that:

(1) If $n>2$, then $\varphi(n)$ is even.

(2) $\varphi(n)=1$ only for $n=1$ and 2.

(3) $\varphi(n)=2$ only for $n=3$, 4 and 6.

(4) $\tau(n)$ is odd or even according as n is, or is not, a perfect square.

(5) $\sigma(n)$ is odd if n is a perfect square multiplied by 1 or 2. Is it odd in any other case?

Continue to explore the table in **38**, extend it successively, try to guess further results and attempt to prove or disprove your guesses.

Consider power series of the form

$$a_0+a_1z+a_2z^2+\cdots+a_nz^n+\cdots=\sum_{n=0}^{\infty} a_nz^n$$

together with the corresponding *Dirichlet series* of the form

$$a_1 1^{-s}+a_2 2^{-s}+\cdots+a_nn^{-s}+\cdots=\sum_{n=1}^{\infty} a_nn^{-s}.$$

Power series are the appropriate tool for *additive* number theory (cf. I, Chap. 1), while Dirichlet series are the appropriate tool for *multiplicative* number theory[1].

The *Cauchy product* of two power series

$$\sum_{k=0}^{\infty} a_kz^k \sum_{l=0}^{\infty} b_lz^l=\sum_{n=0}^{\infty} c_nz^n$$

is defined by

$$c_n=\sum_{k+l=n} a_kb_l=\sum_{t\leqq n} a_tb_{n-t}$$

[I **34**]. Here t runs through all the parts of n, including the improper parts 0 and n. The *Dirichlet product* of two Dirichlet series

$$\sum_{k=1}^{\infty} a_kk^{-s} \sum_{l=1}^{\infty} b_ll^{-s}=\sum_{n=1}^{\infty} c_nn^{-s}$$

is defined by

$$c_n=\sum_{kl=n} a_kb_l=\sum_{t/n} a_tb_{\frac{n}{t}}.$$

Here t runs through all the divisors of n, including the improper divisors 1 and n.

---

[1] In this chapter we do not consider questions of convergence. In the case of absolute convergence all the calculations are valid.

If all the coefficients of a power series are set equal to 1, then we obtain the geometric series

$$1+z+z^2+\cdots+z^n+\cdots=\frac{1}{1-z}.$$

If all the coefficients of a Dirichlet series are set equal to 1, then we obtain the *Riemann zeta function*

$$1^{-s}+2^{-s}+3^{-s}+\cdots+n^{-s}+\cdots=\zeta(s).$$

39. The number of parts of n is $\sum_{t\leqq n} 1=n+1$. The number of divisors of n is $\sum_{t/n} 1=\tau(n)$. We have

$$\sum_{n=0}^{\infty}(n+1)z^n=\frac{1}{(1-z)^2},\qquad \sum_{n=1}^{\infty}\tau(n)n^{-s}=\zeta(s)^2.$$

40. The nth coefficients in the expansion of the products

$$\frac{1}{1-z}\sum_{n=0}^{\infty}a_nz^n \quad\text{and}\quad \zeta(s)\sum_{n=1}^{\infty}a_nn^{-s}$$

are

$$\sum_{t\leqq n}a_t \quad\text{and}\quad \sum_{t/n}a_t$$

respectively.

41. Show that

$$1-z=\frac{1}{1+z+z^2+\cdots+z^n+\cdots},$$

$$\begin{aligned}&\mu(1)1^{-s}+\mu(2)2^{-s}+\mu(3)3^{-s}+\cdots+\mu(n)n^{-s}+\cdots\\&=\frac{1}{1^{-s}+2^{-s}+3^{-s}+\cdots+n^{-s}+\cdots}=\frac{1}{\zeta(s)}.\end{aligned}$$

42. Let us set as in **32**: $\sum_{t\leqq n} a_t=A_n$, $n=0, 1, 2, \ldots$. Then we have

$$\begin{aligned}&(A_0+A_1z+A_2z^2+\cdots+A_nz^n+\cdots)(1-z)\\&=a_0+a_1z+a_2z^2+\cdots+a_nz^n+\cdots.\end{aligned}$$

Now set as in **33**: $\sum_{t/n} a_t=A_n$, $n=1, 2, 3, \ldots$. Then we have

$$\begin{aligned}&(A_11^{-s}+A_22^{-s}+A_33^{-s}+\cdots+A_nn^{-s}+\cdots)\\&\times(\mu(1)1^{-s}+\mu(2)2^{-s}+\mu(3)3^{-s}+\cdots+\mu(n)n^{-s}+\cdots)\\&=a_11^{-s}+a_22^{-s}+a_33^{-s}+\cdots+a_nn^{-s}+\cdots.\end{aligned}$$

§ 6. Multiplicative Arithmetical Functions

By a *multiplicative arithmetical function* $f(n)$ we mean an arithmetical function such that $f(1)=1$, and that satisfies for relatively prime numbers m and n the equation

$$f(m)f(n)=f(mn).$$

43. Show that

$$n^{\alpha},\quad \sigma_{\alpha}(n),\quad 2^{\nu(n)},\quad \mu(n),\quad \lambda(n),\quad \varphi(n)$$

are multiplicative arithmetical functions. [For $\varphi(n)$ observe **25**.]

44. Let $n=p_1^{k_1}p_2^{k_2}\dots p_\nu^{k_\nu}$, where $p_1, p_2, \dots, p_\nu$ are distinct prime numbers. Then we have

$$\sigma_\alpha(n)=\frac{1-p_1^{\alpha(k_1+1)}}{1-p_1^\alpha}\cdot\frac{1-p_2^{\alpha(k_2+1)}}{1-p_2^\alpha}\cdots\frac{1-p_\nu^{\alpha(k_\nu+1)}}{1-p_\nu^\alpha};$$

for example

$$\sigma(n)=\frac{1-p_1^{k_1+1}}{1-p_1}\cdot\frac{1-p_2^{k_2+1}}{1-p_2}\cdots\frac{1-p_\nu^{k_\nu+1}}{1-p_\nu},\qquad \tau(n)=(k_1+1)(k_2+1)\dots(k_\nu+1).$$

45. Show that, for $n>30$, we have

$$\varphi(n)>\tau(n).$$

46. Let $a, b, c, d, \dots, k, l$ be positive integers, M their least common multiple, let $(a, b), (a, c), \dots, (a, b, c), \dots$ denote as usual the greatest common divisors of a and b, a and $c, \dots, a, b$ and $c, \dots$ respectively. If $f(n)$ is a multiplicative function, then we have

$$f(M)f((a,b))f((a,c))\dots f((k,l))f((a,b,c,d))\cdots=f(a)f(b)\dots f(l)f((a,b,c))\dots.$$

(On the right-hand side we have those $f(n)$ for which n is the greatest common divisor of an odd number of the numbers $a, b, c, \dots, k, l$.) [**29** is the special case $f(n)=n$.]

47. Let $f(n)$ be a multiplicative arithmetical function. Then we have

$$\sum_{n=1}^{\infty} f(n)n^{-s}=\prod_p (1^{-s}+f(p)p^{-s}+f(p^2)p^{-2s}+f(p^3)p^{-3s}+\cdots),$$

where the infinite product is extended over all prime numbers p and formed by taking products only of finitely many factors different from 1^{-s}.

48.

$$\zeta(s)=\prod_p\frac{1}{1-p^{-s}}.$$

49. Show that

$$\sum_{n=1}^{\infty}\sigma_\alpha(n)n^{-s}=\zeta(s)\zeta(s-\alpha),\qquad \sum_{n=1}^{\infty}2^{\nu(n)}n^{-s}=\frac{\zeta(s)^2}{\zeta(2s)},$$

$$\sum_{n=1}^{\infty}\lambda(n)n^{-s}=\frac{\zeta(2s)}{\zeta(s)},\qquad \sum_{n=1}^{\infty}\varphi(n)n^{-s}=\frac{\zeta(s-1)}{\zeta(s)}.$$

[**43**, **44**, **25**.]

50. Let $a(n)$ be the greatest odd divisor of n. Then we have

$$a(1)1^{-s}+a(2)2^{-s}+a(3)3^{-s}+\cdots+a(n)n^{-s}+\cdots=\frac{1-2^{1-s}}{1-2^{-s}}\zeta(s-1).$$

51. Prove that

$$\sum_{n=1}^{\infty}\Lambda(n)n^{-s}=-\frac{\zeta'(s)}{\zeta(s)}.$$

52.

$$\sum_{t/n}\mu(t)=\begin{cases}1 & \text{for } n=1,\\ 0 & \text{for } n>1.\end{cases}$$

53.

$$\sum_{t/n} \lambda(t) = \begin{cases} 1, & \text{if } n \text{ is a square,} \\ 0, & \text{if } n \text{ is not a square.} \end{cases}$$

54.

$$\sum_{t/n} \varphi(t) = n.$$

55.

$$\sum_{t/n} \frac{\mu(t)}{t} = \frac{\varphi(n)}{n}.$$

56.

$$\sum_{t/n} \Lambda(t) = \log n.$$

57.

$$\begin{vmatrix} (1,1) & (1,2) & \dots & (1,n) \\ (2,1) & (2,2) & \dots & (2,n) \\ \dots & \dots & \dots & \dots \\ (n,1) & (n,2) & \dots & (n,n) \end{vmatrix} = \varphi(1)\varphi(2)\dots\varphi(n).$$

58. Consider all possible factorizations of an even integer of the form $n = \alpha\beta$, such that α is odd (possibly 1) and β is even. Then

$$\sum \beta - \sum \alpha$$

is equal to the sum of the divisors of $n/2$.

58.1. Let $P(n, k)$ denote the number of decompositions of n into a product of k factors; only positive integers greater than 1 are admissible as factors, and two decompositions are regarded as equal if, and only if, they involve the same factors in the same order. For instance, $P(12, 2) = 4$ and $P(12, 3) = 3$ since

$$\begin{aligned} 12 &= 2 \cdot 6 = 6 \cdot 2 = 3 \cdot 4 = 4 \cdot 3 \\ &= 3 \cdot 2 \cdot 2 = 2 \cdot 3 \cdot 2 = 2 \cdot 2 \cdot 3. \end{aligned}$$

Show that, for $n \geqq 2$,

$$\mu(n) = -P(n, 1) + P(n, 2) - P(n, 3) + \cdots,$$

the difference between the number of "even" and "odd" decompositions, that is, decompositions into an even and an odd number of factors, respectively. For instance

$$\mu(12) = -1 + 4 - 3 = 0.$$

58.2 (continued). Show that

$$P(n, 1) - \tfrac{1}{2}P(n, 2) + \tfrac{1}{3}P(n, 3) - \cdots$$

equals $1/m$ if $n = p^m$, the mth power of a prime number p, but equals 0 if n is divisible by more than one prime. In the case $n = 12$ we have

$$1 - \tfrac{4}{2} + \tfrac{3}{3} = 0.$$

58.3 (continued). If n is the product of m different prime factors, then

$$P(n, k) = k!\,S_k^m$$

[introduction to I **186**].

58.4. If p is a prime and m a positive integer, then

$$P(p^m, k)=\binom{m-1}{k-1}.$$

(For this particular case, verify **58.1** by using the binomial theorem; also cf. **58.2** with I **38**.)

58.5. For $n>1$ let $Q(n)$ denote the number of different decompositions of the integer n into a product of integers greater than 1; two decompositions are not regarded as different if they involve the same factors, and so the order of factors does not matter (in contrast to **58.1**); define $Q(1)=1$. For instance $Q(12)=4$. If n is the product of m different prime factors, then

$$Q(n)=T_m$$

[introduction to I **186**].

58.6 (continued). Show that

$$\sum_{n=1}^{\infty} Q(n)n^{-s}=\exp\left(\sum_{n=1}^{\infty} (\zeta(ns)-1)/n\right).$$

58.7. Let n be a positive integer.

$$n=2^j p_1^{k_1} p_2^{k_2} \ldots p_\mu^{k_\mu} q_1^{l_1} q_2^{l_2} \ldots q_\nu^{l_\nu},$$

where $2, p_1, \ldots, p_\mu, q_1, \ldots, q_\nu$ are distinct primes,

$$p_1 \equiv \cdots \equiv p_\mu \equiv 1, \qquad q_1 \equiv \cdots \equiv q_\nu \equiv 3 \qquad (\text{mod } 4).$$

Set

$$\delta(n)=(k_1+1)(k_2+1) \ \ldots \ (k_\mu+1)$$

if $l_1, l_2, \ldots, l_\nu$ are all even, and

$$\delta(n)=0$$

otherwise. Show that $\delta(n)$ is, in fact, the difference between the numbers of two different kinds of divisors of n:

$$\delta(n)=\sum_{t'/n} (-1)^{(t'-1)/2},$$

where the summation is extended over the odd divisors t' of n.

($4\delta(n)$ is the number of lattice points on the circle

$$x^2+y^2=n.$$

This fact was discovered by Gauss.)

58.8. Which of the functions $P(n, 2), P(n, 3), \ldots, Q(n)$ and $\delta(n)$ are multiplicative?

59. Let $f(n)$ and $g(n)$ be multiplicative arithmetical functions. Then the arithmetical function

$$h(n)=\sum_{t/n} f(t)g\left(\frac{n}{t}\right)$$

is also multiplicative.

60. The number of different kinds of regular n-gons, one convex, the others self-intersecting, is equal to $\frac{1}{2}\varphi(n)$.

61. The sum of all positive proper fractions that, when reduced to their lowest terms, have denominator n, is equal to $\frac{1}{2}\varphi(n)$, $n \geqq 2$.

62. Let a, b, c be positive integers. Among the c fractions

$$\frac{a}{c}, \quad \frac{a+b}{c}, \quad \frac{a+2b}{c}, \quad \ldots, \quad \frac{a+(c-1)b}{c}$$

there are $\varphi(bc)/\varphi(b)$ irreducible fractions, if $(a, b, c)=1$. If $(a, b, c)>1$, then they are of course all reducible.

63. How many irreducible fractions are there among the following n^2 fractions:

$$\begin{array}{cccccc}
\frac{1}{1}, & \frac{1}{2}, & \frac{1}{3}, & \frac{1}{4}, & \ldots, & \frac{1}{n}, \\
\frac{2}{1}, & \frac{2}{2}, & \frac{2}{3}, & \frac{2}{4}, & \ldots, & \frac{2}{n}, \\
\frac{3}{1}, & \frac{3}{2}, & \frac{3}{3}, & \frac{3}{4}, & \ldots, & \frac{3}{n}, \\
\cdots & \cdots & \cdots & \cdots & \cdots & \cdots \\
\frac{n}{1}, & \frac{n}{2}, & \frac{n}{3}, & \frac{n}{4}, & \ldots, & \frac{n}{n}?
\end{array}$$

64. Let $\Phi(n)$ be the number of irreducible fractions among the following n^2 fractions:

$$\begin{array}{cccc}
\frac{1+i}{n}, & \frac{1+2i}{n}, & \ldots, & \frac{1+ni}{n}, \\
\frac{2+i}{n}, & \frac{2+2i}{n}, & \ldots, & \frac{2+ni}{n}, \\
\cdots & \cdots & \cdots & \cdots \\
\frac{n+i}{n}, & \frac{n+2i}{n}, & \ldots, & \frac{n+ni}{n}.
\end{array}$$

($i=\sqrt{-1}$; the fraction $(a+ib)/n$ is called reducible if $(a, b, n)>1$, irreducible if $(a, b, n)=1$.) The function $\Phi(n)$ has the following properties:

(1) $$\Phi(m)\Phi(n)=\Phi(mn) \quad \text{for} \quad (m, n)=1, \quad \Phi(1)=1,$$
$$\Phi(p^k)=p^{2k}-p^{2k-2}, \quad p \text{ a prime number}, \quad k=1, 2, 3, \ldots;$$

(2) $$\Phi(n)=n^2\left(1-\frac{1}{p^2}\right)\left(1-\frac{1}{q^2}\right)\left(1-\frac{1}{r^2}\right)\cdots,$$

where $p, q, r, \ldots$ are the distinct prime factors of n;

(3) $$\sum_{t/n} \Phi(t)=n^2;$$

$$(4) \qquad \sum_{n=1}^{\infty} \Phi(n) n^{-s} = \frac{\zeta(s-2)}{\zeta(s)},$$

Prove (1), (2), (3) independently of each other directly from the definition. Show further that:

$$(1) \to (4), \quad (2) \to (4), \quad (3) \to (4),$$
$$(4) \to (1), \quad (4) \to (2), \quad (4) \to (3).$$[1]

§7. Lambert Series and Related Topics

65. From the identity

$$\zeta(s) \sum_{n=1}^{\infty} a_n n^{-s} = \sum_{n=1}^{\infty} A_n n^{-s}$$

it follows that

$$\sum_{n=1}^{\infty} \frac{a_n x^n}{1-x^n} = \sum_{n=1}^{\infty} A_n x^n$$

and conversely, where $a_1, a_2, a_3, \ldots, A_1, A_2, A_3, \ldots$ are constants. (The series occurring on the left-hand side of the second equation is called a *Lambert series.*)

66. The two identities

$$\zeta(s)(1-2^{1-s}) \sum_{n=1}^{\infty} a_n n^{-s} = \sum_{n=1}^{\infty} B_n n^{-s}, \qquad \sum_{n=1}^{\infty} \frac{a_n x^n}{1+x^n} = \sum_{n=1}^{\infty} B_n x^n$$

are equivalent.

67. Assume that between the numbers $a_1, a_2, a_3, \ldots,$ and $A_1, A_2, A_3, \ldots$ the same relation subsists as in **65**. Then

$$\prod_{n=1}^{\infty} \left(\frac{n}{x} \sin \frac{x}{n}\right)^{a_n} = \prod_{n=1}^{\infty} \left(1 - \frac{x^2}{n^2 \pi^2}\right)^{A_n}.$$

68. Assume that between the numbers $a_1, a_2, a_3, \ldots$ and $B_1, B_2, B_3, \ldots$ the same relation subsists as in **66**. Then

$$\prod_{n=1}^{\infty} \left(\frac{x}{2n} \cot \frac{x}{2n}\right)^{a_n} = \prod_{n=1}^{\infty} \left(1 - \frac{x^2}{n^2 \pi^2}\right)^{B_n}.$$

68.1. Define

$$f(x) + f(2x) + \cdots + f(nx) + \cdots = F(x).$$

Then, with the notation of **65**,

$$\sum_{n=1}^{\infty} a_n F(nx) = \sum_{n=1}^{\infty} A_n f(nx).$$

(How is this result related to **65** and **67**?)

---

[1] I.e. from (1) follows (4), from (2) follows (4), and so on.

68.2. Define

$$f(x)-f(2x)+\cdots+(-1)^{n-1}f(nx)+\cdots=G(x).$$

Then, with the notation of **66**,

$$\sum_{n=1}^{\infty} a_n G(nx)=\sum_{n=1}^{\infty} B_n f(nx).$$

(How is this result related to **66** and **68**?)

68.3. Each of the following two equations implies the other:

$$F(x)=\sum_{n=1}^{\infty} f(nx),$$

$$f(x)=\sum_{n=1}^{\infty} \mu(n)F(nx).$$

69. Show that

$$\sum_{n=1}^{\infty} \frac{\mu(n)x^n}{1-x^n} \quad \text{and} \quad \sum_{n=1}^{\infty} \frac{\varphi(n)x^n}{1-x^n}$$

are rational functions of x. What functions?

70.

$$\sum_{n=1}^{\infty} \frac{\lambda(n)x^n}{1-x^n}=x+x^4+x^9+x^{16}+x^{25}+\cdots.$$

71. Show that

$$\sum_{n=1}^{\infty} \frac{\mu(n)x^n}{1+x^n}=x-2x^2, \qquad \sum_{n=1}^{\infty} \frac{\phi(n)x^n}{1+x^n}=x\,\frac{1+x^2}{(1-x^2)^2}.$$

72.

$$\sum_{n=1}^{\infty} \lambda(n)\,\frac{x^n}{1+x^n}=x-2x^2+x^4-2x^8+x^9+x^{16}-2x^{18}+x^{25}-\cdots$$

$$=\sum_{n=1}^{\infty} b_n x^n,$$

where $b_n=1$ if n is a square, $b_n=-2$ if n is twice a square, and in all other cases $b_n=0$.

72.1.

$$\prod_{1}^{\infty} (1-x^n)^{\frac{\mu(n)}{n}}=e^{-x}, \qquad \prod_{1}^{\infty} (1-x^n)^{\frac{\varphi(n)}{n}}=e^{-\frac{x}{1-x}}.$$

73.

$$\sum_{n=1}^{\infty} \Phi(n)\,\frac{x^n}{1-x^n}=x\,\frac{1+x}{(1-x)^3},$$

$$\sum_{n=1}^{\infty} \Phi(n)\,\frac{x^n}{1+x^n}=x\,\frac{1+2x+6x^2+2x^3+x^4}{(1-x^2)^3}. \qquad \textbf{[64]}.$$

74.

$$\sum_{n=1}^{\infty} \tau(n)x^n=\sum_{n=1}^{\infty} \frac{x^n}{1-x^n}=x\,\frac{1+x}{1-x}+x^4\,\frac{1+x^2}{1-x^2}+x^9\,\frac{1+x^3}{1-x^3}+\cdots.$$

74.1.

$$\sum_{n=1}^{\infty} \delta(n)x^n = \frac{x}{1-x} - \frac{x^3}{1-x^3} + \frac{x^5}{1-x^5} - \frac{x^7}{1-x^7} + \cdots \qquad \textbf{[58.7]}.$$

***75.**

$$\sum_{n=1}^{\infty} \sigma(n)x^n = \sum_{n=1}^{\infty} n \frac{x^n}{1-x^n}$$

$$= \frac{x}{(1-x)^2} + \frac{x^2}{(1-x^2)^2} + \frac{x^3}{(1-x^3)^2} + \cdots.$$

75.1.

$$\sum_{n=1}^{\infty} \sigma(n)x^n = \frac{x + 2x^2 - 5x^5 - 7x^7 + + - - \cdots}{1 - x - x^2 + x^5 + x^7 - - + + \cdots}. \qquad [\text{I } \textbf{54}].$$

75.2.

$$3 \sum_{n=1}^{\infty} \sigma(n)x^n = \frac{3x - 15x^3 + 42x^6 - 90x^{10} + \cdots}{1 - 3x + 5x^3 - 7x^6 + 9x^{10} - \cdots} \qquad [\text{A I } \textbf{54.1}]$$

75.3.

$$\sigma(n) = \sigma(n-1) + \sigma(n-2) - \sigma(n-5) - \sigma(n-7) + + - - \cdots.$$

The terms on the right-hand side are of the form

$$(-1)^{k-1}\sigma\left(n - \frac{3k^2 \pm k}{2}\right), \qquad \text{where} \quad 1 \leqq \frac{3k^2 \pm k}{2} \leqq n.$$

Where the symbol $\sigma(0)$, till now undefined, occurs, substitute for it n.

75.4.

$$\sigma(n) = 3\sigma(n-1) - 5\sigma(n-3) + 7\sigma(n-6) - 9\sigma(n-10) + - \cdots.$$

The terms on the right-hand side are of the form

$$(-1)^{k-1}(2k+1)\sigma\left(n - \frac{k(k+1)}{2}\right), \qquad \text{where} \quad 1 \leqq \frac{k(k+1)}{2} \leqq n.$$

Where the symbol $\sigma(0)$ occurs, substitute for it $n/3$.

75.5. Using the definition of $p(n)$, the number of partitions of n, in A I **20.1**, prove that

$$\sigma(n) + \sigma(n-1) + 2\sigma(n-2) + \cdots + p(n-1)\sigma(1) = \sum_{k=0}^{n-1} p(k)\sigma(n-k) = np(n).$$

76. The expression

$$\frac{1}{1-q} + \frac{p}{1-qx} + \frac{p^2}{1-qx^2} + \frac{p^3}{1-qx^3} + \cdots$$

is unaltered, if p and q are interchanged.

77.

$$\frac{x}{1+x^2} + \frac{x^2}{1+x^4} + \frac{x^3}{1+x^6} + \frac{x^4}{1+x^8} + \cdots$$

$$= \frac{x}{1-x} - \frac{x^3}{1-x^3} + \frac{x^5}{1-x^5} - \frac{x^7}{1-x^7} + \cdots.$$

78.

$$\frac{x}{1-x}=\frac{x}{1-x^2}+\frac{x^2}{1-x^4}+\frac{x^4}{1-x^8}+\frac{x^8}{1-x^{16}}+\cdots$$

$$=\frac{x}{1+x}+\frac{2x^2}{1+x^2}+\frac{4x^4}{1+x^4}+\frac{8x^8}{1+x^8}+\cdots.$$

§ 8. Further Problems on Counting Lattice Points

79. We have

$$\tau(1)+\tau(2)+\tau(3)+\cdots+\tau(n)=\left[\frac{n}{1}\right]+\left[\frac{n}{2}\right]+\left[\frac{n}{3}\right]+\cdots+\left[\frac{n}{n}\right].$$

[Both sides represent the number of lattice points in the region satisfying $x>0$, $y>0$, $xy\leqq n$.][1]

80. If $\nu=[\sqrt{n}]$, we have

$$\tau(1)+\tau(2)+\tau(3)+\cdots+\tau(n)=2\left[\frac{n}{1}\right]+2\left[\frac{n}{2}\right]+\cdots+2\left[\frac{n}{\nu}\right]-\nu^2.$$

80.1. Let $U(n)$ denote the least common multiple of the numbers $1, 2, 3, \ldots, n$. Show that

$$\Lambda(1)+\Lambda(2)+\Lambda(3)+\cdots+\Lambda(n)=\log U(n).$$

80.2. Define $\pi(x)$ to be the number of primes which do not exceed x. Thus

$$\pi(1)=0, \qquad \pi(10)=4, \qquad \pi(100)=25.$$

Show that

$$\Lambda(1)+\Lambda(2)+\cdots+\Lambda(n)\leqq\pi(n)\log n.$$

81. Let

$$\sum_{k=1}^{\infty} a_k k^{-s} \sum_{l=1}^{\infty} b_l l^{-s}=\sum_{n=1}^{\infty} c_n n^{-s}.$$

Then between the sums of the coefficients

$$a_1+a_2+\cdots+a_n=A_n, \qquad b_1+b_2+\cdots+b_n=B_n,$$
$$c_1+c_2+\cdots+c_n=\Gamma_n,$$

we have the following relation:

$$\Gamma_n=\sum_{r=1}^{n} a_r B_{\left[\frac{n}{r}\right]}=\sum_{s=1}^{n} b_s A_{\left[\frac{n}{s}\right]}.$$

82.

$$\log n!=\sum_{p\leqq n}\log p\left(\left[\frac{n}{p}\right]+\left[\frac{n}{p^2}\right]+\left[\frac{n}{p^3}\right]+\cdots\right),$$

[1] The analogy (part, divisor) presented in **28–42** could here be pursued somewhat further.

where the summation is to be extended over all prime numbers p that do not exceed n. [Apply **81** to the Dirichlet product

$$-\zeta'(s)=\sum_{k=1}^{\infty} \Lambda(k)k^{-s} \sum_{l=1}^{\infty} l^{-s}.]$$

83. With the notation of **81** we have also

$$\Gamma_n=\sum_{r=1}^{\nu} a_r B_{\left[\frac{n}{r}\right]}+\sum_{s=1}^{\nu} b_s A_{\left[\frac{n}{s}\right]}-A_\nu B_\nu,$$

where we have set $\nu=[\sqrt{n}]$.

Chapter 2. Polynomials with Integral Coefficients and Integral-Valued Functions

§ 1. Integral Coefficients and Integral-Valued Polynomials

A polynomial

$$P(x)=a_0x^m+a_1x^{m-1}+\cdots+a_{m-1}x+a_m$$

is termed *a polynomial with integral coefficients* if all its coefficients $a_0, a_1, a_2, \ldots, a_{m-1}, a_m$ are integers. The polynomial $P(x)$ is called *integral-valued* if its values at the integers $P(0), P(1), P(2), \ldots, P(n), \ldots$ are integers. If a polynomial has integral coefficients, then it also is integral-valued.

84. The polynomial

$$\frac{x(x-1)(x-2)\ldots(x-m+1)}{1\cdot 2\cdot 3\ldots m}=\binom{x}{m}$$

is integral-valued, but it does not have integral coefficients if $m\geqq 2$.

85. Every polynomial $P(x)$ of degree m may be expressed in the form

$$P(x)=b_0\binom{x}{m}+b_1\binom{x}{m-1}+\cdots+b_{m-1}\binom{x}{1}+b_m.$$

$P(x)$ is integral-valued if and only if the numbers $b_0, b_1, \ldots, b_{m-1}, b_m$ are integers.

86. If the polynomial $P(x)$ of degree m is integral-valued, then $m!P(x)$ has integral coefficients.

87. If a polynomial of degree m assumes integral values for $m+1$ consecutive integral values of the variable, then it assumes integral values for all integral values of the variable.

88. Every *odd* polynomial $P(x)$ of degree $2m-1$ may be written in the form

$$P(x)=c_1\binom{x}{1}+c_2\binom{x+1}{3}+c_3\binom{x+2}{5}+\cdots+c_m\binom{x+m-1}{2m-1}.$$

$P(x)$ is integral-valued if and only if the numbers $c_1, c_2, \ldots, c_{m-1}, c_m$ are integers.

89. Every *even* polynomial $P(x)$ of degree $2m$ may be written in the form

$$P(x)=d_0+d_1\frac{x}{1}\binom{x}{1}+d_2\frac{x}{2}\binom{x+1}{3}+\cdots+d_m\frac{x}{m}\binom{x+m-1}{2m-1}.$$

$P(x)$ is integral-valued if and only if the numbers $d_0, d_1, d_2, \ldots, d_{m-1}, d_m$ are integers.

90. There exist polynomials with integral coefficients of degree m, whose absolute value is equal to 1 at $m+1$ integral points if $m \leqq 3$, but there exist no such polynomials if $m \geqq 4$. [VI **70**.]

91. If a rational integral function (a polynomial) of degree m assumes rational values at $m+1$ integral points, then its coefficients are rational numbers.

92. If a rational function that is not a polynomial assumes rational values at all positive integral points, then it is the quotient of two relatively prime polynomials with integral coefficients.

93. If a rational function has integral values for infinitely many integral values of the variable, then it is a rational *integral* function (a polynomial).

§ 2. Integral-Valued Functions and their Prime Divisors

94. In the sequence

$$2^1+1, \quad 2^2+1, \quad 2^4+1, \quad 2^8+1, \quad \ldots, \quad 2^{2^n}+1, \quad \ldots$$

the terms are pairwise relatively prime. (We can deduce from this that there exist infinitely many primes.)

95. In the sequence of numbers

$$a, \quad a+d, \quad a+2d, \quad \ldots, \quad a+nd, \quad \ldots \quad (a, d \text{ integers}, d \gtrless 0),$$

there exist several, in fact infinitely many, terms with the same prime factors.

96. In the sequence

$$5, \quad 11, \quad 17, \quad 23, \quad 29, \quad 35, \quad \ldots, \quad 6n-1, \quad \ldots$$

every term has a prime factor that is $\equiv -1 \pmod 6$.

97. Let $P(x)$ be an integral-valued polynomial. Is it possible that all the terms of the sequence of numbers

$$P(1), \quad P(2), \quad P(3), \quad \ldots, \quad P(n), \quad \ldots$$

are prime numbers? (With the particular choice of $P(x)=x^2-x+41$ the first 40 terms are prime numbers, as Euler noted.)

If the function $f(x)$ is such that the values

$$f(0), \quad f(1), \quad f(2), \quad \ldots, \quad f(n), \quad \ldots$$

are all integral, then $f(x)$ is termed *integral-valued.* E.g. 2^x is integral-valued. A rational function that is not a polynomial cannot be integral-valued [**93**]. A prime number p is termed a *prime divisor* of the integral-valued function $f(x)$ if there exists an integer n, $n \geqq 0$, such that $f(n) \gtrless 0$ and $f(n) \equiv 0 \pmod p$. 2^x has only the single prime divisor 2. By **94** the function 2^x+1 has infinitely many prime divisors.

98. The prime divisors of the polynomial x^2+1 are 2, 5, 13, 17, ..., i.e. 2 and the odd prime numbers of the form $4n+1$; the prime numbers of the form $4n+3$ are not prime divisors of x^2+1.

99. Determine the prime divisors of x^2+15.

100. There are infinitely many prime numbers that are not prime divisors of a given irreducible polynomial of second degree with integral coefficients. [To be proved with the aid of deeper fundamental results. Cf. footnote to **110**.]

101. If an integral-valued polynomial that does not vanish identically has a rational zero, then every prime number is a prime divisor of the polynomial with the possible exception of finitely many primes.

102. Every prime number is a prime divisor of the polynomial

$$x^6-11x^4+36x^2-36,$$

which has no rational zeros.

103. An odd prime divisor of $K_m(x)$, the mth cyclotomic polynomial [**36**], is either a divisor of m or $\equiv 1 \pmod{m}$. [**104**.]

104. If p is a prime number, $p>2$, p not a divisor of m and $K_m(a)\equiv 0 \pmod{p}$, then a is relatively prime to p and m is the order of $a \pmod{p}$, i.e. m is the smallest integer such that $a^m\equiv 1 \pmod{p}$.

105. There are infinitely many prime numbers of the form $6n-1$. [Consider the prime factors of $6P-1$, where $P=5\cdot 11\cdot 17\cdot 23\cdot 29\cdot 41\ldots$ is the product of all primes of the form $6n-1$ that are already known.]

106. There are infinitely many prime numbers of the form $4n-1$.

107. Let a, b, c be integers, $a\gtrless 0$, $b\geqq 2$, $c\gtrless 0$. The integral-valued function ab^x+c has infinitely many prime divisors. [ab^x+c is periodic modulo n for every n that is relatively prime to b; the absolute value of the function ab^x+c tends to infinity.]

108. Let $P(x)$ be an integral-valued polynomial that is not a constant. Then $P(x)$ has infinitely many prime divisors.

109. The fact stated in **108** may be expressed as follows: If $P(x)$ is an integral-valued polynomial that is not a constant, then not all the terms of the sequence of numbers

$$P(0), \quad P(1), \quad P(2), \quad \ldots, \quad P(n), \quad \ldots$$

can be constructed from finitely many prime numbers. Is it possible that *infinitely many* terms of the sequence can be constructed from finitely many prime numbers?

110. Let m be a positive integer. In the arithmetical progression

$$1, \quad 1+m, \quad 1+2m, \quad 1+3m, \quad \ldots$$

there are infinitely many prime numbers.[1]

111. Let the two integral-valued non-constant polynomials $P(x)$ and $Q(x)$ be relatively prime. Then there exist arbitrarily large integers n such that the least common multiple of the numbers $P(n)$ and $Q(n)$ contains prime factors that are not contained in their greatest common divisor.

112. Let $P(x)$ be an *irreducible* integral-valued polynomial that is not a constant.

---

[1] **105**, **106**, **110** are special cases of the important theorem proved by Dirichlet: In every arithmetic progression whose first term and common difference have no common divisors there occur infinitely many prime numbers.

(See § 3, introduction to problem **116**.) Then there exist arbitrarily large integers n, such that in $P(n)$ there occurs at least one simple prime factor p; i.e. $P(n) \equiv 0 \pmod p$, $P(n) \not\equiv 0 \pmod{p^2}$.

113. Let $P(x)$ be an integral-valued polynomial. Let the zero of $P(x)$ of *smallest multiplicity* have multiplicity m. (Then all the prime factors of $P(n)$, with the possible exception of finitely many, must when raised to the power m also be factors of $P(n)$, $n=0, 1, 2, \ldots$.) There exist arbitrarily large integers n such that in $P(n)$ at least one prime factor occurs *not more than m times.*

114. If the polynomial $P(x)$ assumes for every positive integer x a value that is the square of an integer, then $P(x)$ is the square of a polynomial. (Analogous results are true for higher powers.)

115. Let $b_1, b_2, \ldots, b_k$ be distinct positive integers, $0 < b_1 < b_2 < \cdots < b_k$ and $P_1(x), P_2(x), \ldots, P_k(x)$ polynomials with integral coefficients. Then the function

$$P_1(x)b_1^x + P_2(x)b_2^x + \cdots + P_k(x)b_k^x$$

has infinitely many prime divisors.

§ 3. Irreducibility of Polynomials

A polynomial of degree n with rational coefficients is termed *reducible*, if it is the product of two polynomials, each of which is of degree $<n$ and has rational coefficients. Every polynomial is either reducible or *irreducible*; in the former case it is the product of irreducible polynomials.

116. If a polynomial with integral coefficients is reducible, then it is the product of polynomials *with integral coefficients* of lower degree.

117. A polynomial with integral coefficients $f(x)$ cannot vanish at an integral value of x, if $f(0)$ and $f(1)$ are both odd.

118. If the polynomial $P(x)$ of degree n with integral coefficients assumes at n distinct integral values of x values that are all different from 0 and in absolute value less than

$$\frac{\left(n - \left[\frac{n}{2}\right]\right)!}{2^{n-\left[\frac{n}{2}\right]}},$$

then $P(x)$ is irreducible. [VI **70**.]

119. The bound given in **118** may be replaced by

$$\left(\frac{d}{2}\right)^{n-\left[\frac{n}{2}\right]} \left(n - \left[\frac{n}{2}\right]\right)!,$$

where d is the minimum distance between any two of the n integral values of x where $P(x)$ assumes the integral values considered.

120. In order that the value of the polynomial $P(x)$ with integral coefficients should be equal to a prime number for infinitely many integral values of x, $P(x)$ must be irreducible and the greatest common divisor of the coefficients of $P(x)$ must be 1. The converse of this obvious statement is not true. The polynomial

$$x(x-(n!+1))(x-2(n!+1))\ldots(x-(n-1)(n!+1))+n! = x^n + \cdots$$

is irreducible but it does not represent a prime number for any integral value of x, if $n \geqq 3$.

121. Let $a_1, a_2, \ldots, a_n$ be distinct integers. Prove that the function

$$(x-a_1)(x-a_2)\ldots(x-a_n)-1$$

is always irreducible.

122 (continued). Investigate the cases in which

$$(x-a_1)(x-a_2)\ldots(x-a_n)+1$$

is reducible.

123 (continued). The function

$$(x-a_1)^2(x-a_2)^2\ldots(x-a_n)^2+1$$

is irreducible.

124 (continued). The function

$$(x-a_1)^4(x-a_2)^4\ldots(x-a_n)^4+1$$

is also irreducible.

125 (continued). If $F(z)=z^4+Az^3+Bz^2+Az+1$ is a positive definite irreducible polynomial with integral coefficients, then

$$F((x-a_1)(x-a_2)\ldots(x-a_n))$$

is reducible only if $F(z)=z^4-z^2+1$ (the 12th cyclotomic polynomial) and $(x-a_1)(x-a_2)\ldots(x-a_n)=(x-\alpha)(x-\alpha-1)(x-\alpha-2)$ where α is an integer.

126 (continued). If A is positive and integral, then

$$A(x-a_1)^4(x-a_2)^4\ldots(x-a_n)^4+1$$

is reducible only if $A/4$ is the fourth power of an integer.

127. Let $P(x)$ be a polynomial with integral coefficients and assume that there exists an integer n such that the following three conditions are satisfied:

(1) The zeros of $P(x)$ lie in the half-plane $\Re x < n - \frac{1}{2}$.
(2) $P(n-1) \neq 0$.
(3) $P(n)$ is a prime number.

Then $P(x)$ is irreducible.

128. Let

$$p = a_0 a_1 \ldots a_m = a_0 10^m + a_1 10^{m-1} + \cdots + a_m$$

be a prime number expressed in the decimal system, $0 \leqq a_\nu \leqq 9$, $\nu = 0, 1, \ldots, m$; $a_0 \geqq 1$. Then the polynomial

$$a_0 x^m + a_1 x^{m-1} + \cdots + a_{m-1} x + a_m$$

is irreducible. [**127**, III **24**.]

129. Let r and s be odd prime numbers, such that

$$r \equiv 1 \pmod 8 \quad \text{and} \quad \left(\frac{r}{s}\right) = \left(\frac{s}{r}\right) = 1.$$

(For definition of the Legendre symbol $\left(\frac{n}{p}\right)$ see V **45**.) Such prime numbers are for example $r=41$ and $s=5$. The polynomial $P(x)=$

$$(x-\sqrt{r}-\sqrt{s})(x-\sqrt{r}+\sqrt{s})(x+\sqrt{r}-\sqrt{s})(x+\sqrt{r}+\sqrt{s})$$
$$=x^4-2(r+s)x^2+(r-s)^2$$
$$=(x^2+r-s)^2-4rx^2 \tag{1}$$
$$=(x^2-r+s)^2-4sx^2 \tag{2}$$
$$=(x^2-r-s)^2-4rs \tag{3}$$

is irreducible but it is "reducible" (it can be expressed as the product of two factors of the 2nd degree) *modulo m, where m is an arbitrary integer*, as can be shown in an elementary manner by formulae (1), (2) and (3).

Chapter 3. Arithmetical Aspects of Power Series

§ 1. Preparatory Problems on Binomial Coefficients

130. The product of any m consecutive integers is divisible by the product of the first m positive integers.

131. The product of m integers in arithmetic progression is divisible by $m!$ if the common difference of the progression is prime to $m!$.

132. The product of the differences of any m integers is divisible by the product of the differences of the first m positive integers. (The product of the differences of m variables $x_1, x_2, \ldots, x_m$, $\prod_{j<k}^{1,2,\ldots,m}(x_k-x_j)$, consists of $\frac{1}{2}m(m-1)$ factors.) [Solution V **96**.]

133. For what integers n is $(n-1)!$ *not* divisible by n? For what integers n is $(n-1)!$ *not* divisible by n^2?

134. What is the exponent of the highest power of a prime number p that is a divisor of $n!$?

135. In how many zeros does 1000! end when expressed in the decimal system?

136. Let a and b be positive integers. Then the integer $2a!\ 2b!$ is divisible by $a!\ (a+b)!\ b!$.

137. If h and n are positive integers, then $(hn)!/[(h!)^n n!]$ is an integer.

§ 2. On Eisenstein's Theorem

A power series in ascending powers of z

$$a_0+a_1z+a_2z^2+\cdots+a_nz^n+\cdots$$

is termed a power series *with rational coefficients*, if the coefficients $a_0, a_1, a_2, \ldots, a_n, \ldots$ are rational numbers, and it is termed a power series *with integral coefficients* if its coefficients $a_0, a_1, a_2, \ldots, a_n, \ldots$ are all rational integers. A power series with rational coefficients may be expressed in the form

$$\frac{s_0}{t_0}+\frac{s_1}{t_1}z+\frac{s_2}{t_2}z^2+\cdots+\frac{s_n}{t_n}z^n+\cdots$$

where s_n, t_n are rational integers, $(s_n, t_n)=1$, $t_n \geqq 1$; s_n is the numerator and t_n the denominator of the nth coefficient.

138. Let s be an integer, t a positive integer, $(s, t)=1$. Show that only prime factors of t divide the denominators of the coefficients of the power series with rational coefficients

$$(1+z)^{\frac{s}{t}} = \sum_{n=0}^{\infty} \frac{s}{t}\left(\frac{s}{t}-1\right)\cdots\left(\frac{s}{t}-n+1\right)\frac{z^n}{n!}.$$

139. Let p be a prime factor of t and let $p^{\alpha}, p^{\alpha_0}, p^{\alpha_1}, \ldots, p^{\alpha_n}, \ldots$ respectively be the highest powers of p that divide $t, t_0, t_1, \ldots, t_n, \ldots$, where t_n denotes the denominator of the nth coefficient in the power series expansion of $(1+z)^{\frac{s}{t}}$; $(s, t)=1$. Calculate $\lim_{n\to\infty}(\alpha_n/n)$.

We term the function $f(z)$ an *algebraic function*, if it satisfies an equation of the form

$$P_0(z)[f(z)]^l + P_1(z)[f(z)]^{l-1} + \cdots + P_{l-1}(z)f(z) + P_l(z) = 0$$

where $P_0(z), P_1(z), \ldots, P_l(z)$ are polynomials, $P_0(z) \not\equiv 0$.

The proper perspective for evaluating the particular result contained in **138**, **139** is given by the following general theorem due to *Eisenstein*:

"If the power series with rational coefficients

$$a_1z + a_2z^2 + a_3z^3 + \cdots + a_nz^n + \cdots$$

represents an algebraic function of z, then there exists an integer T, such that

$$a_1Tz + a_2T^2z^2 + a_3T^3z^3 + \cdots + a_nT^nz^n \cdots$$

has integral coefficients." [**153**.]

140. Find the smallest integer T with the property that all the coefficients of the power series expansion $(1+Tz)^{\frac{s}{t}}$ are integers.

141. Prove Eisenstein's theorem in the simplest special case, namely for *rational* functions.

We say that the power series

$$\frac{s_0}{t_0} + \frac{s_1}{t_1}z + \frac{s_2}{t_2}z^2 + \cdots + \frac{s_n}{t_n}z^n + \cdots$$

with rational coefficients (s_n, t_n integers, $t_n \geqq 1$, $(s_n, t_n)=1$) satisfies the *Eisenstein condition*, if there exists an integer T such that T^n is divisible by t_n for $n=1, 2, 3, \ldots$.

142. If two power series $f(z)$ and $g(z)$ with rational coefficients satisfy the Eisenstein condition, then the series $f(z)+g(z)$, $f(z)-g(z)$, $f(z)g(z)$, also satisfy the condition, and also $f(z)/g(z)$ (assuming $g(0) \gtrless 0$) and $f(g(z))$ (assuming $g(0)=0$).

143. If the two power series with rational coefficients

$$a_0 + a_1z + a_2z^2 + \cdots + a_nz^n + \cdots, \quad b_0 + b_1z + b_2z^2 + \cdots + b_nz^n + \cdots$$

satisfy the Eisenstein condition, then the series

$$a_0b_0 + a_1b_1z + a_2b_2z^2 + \cdots + a_nb_nz^n + \cdots$$

also satisfies the condition.

144. Show that the Eisenstein condition is equivalent to the following pair of simultaneous conditions:

(1) The denominators $t_1, t_2, t_3, \ldots, t_n, \ldots$ contain only finitely many different prime numbers.

(2) $(\log t_n)/n$ is bounded, $n=1, 2, 3, \ldots$.

145. Expand the two functions

$$\frac{1}{1-z}+\frac{1}{2-z}+\frac{1}{4-z}+\cdots+\frac{1}{2^n-z}+\cdots,$$

$$\frac{z}{2-z}+\frac{z^2}{2^4-z^2}+\frac{z^3}{2^9-z^3}+\cdots+\frac{z^n}{2^{n^2}-z^n}+\cdots.$$

in powers of z. Do the power series obtained satisfy the two conditions (1), (2) in **144**? Are the functions represented algebraic?

146. By a *hypergeometric* series we mean a series of the form

$$F(\alpha, \beta, \gamma; z)=\sum_{n=0}^{\infty} \frac{\alpha(\alpha+1)\ldots(\alpha+n-1)\cdot\beta(\beta+1)\ldots(\beta+n-1)}{1\cdot 2 \quad \ldots \quad n \quad \cdot\gamma(\gamma+1)\ldots(\gamma+n-1)} z^n.$$

If α and β are rational numbers but not both integers, and if γ is a positive integer, then it is possible to choose a positive integer T such that all the coefficients in the hypergeometric series $F(\alpha, \beta, \gamma; Tz)$ become integers. (The function

$$\frac{2}{\pi}\int_0^1 \frac{dx}{\sqrt{(1-x^2)(1-zx^2)}}=\sum_{n=0}^{\infty}\left(\frac{\frac{1}{2}(\frac{1}{2}+1)\ldots(\frac{1}{2}+n-1)}{1\cdot 2\ldots n}\right)^2 z^n$$

for example is not an algebraic function [I **90**]: the Eisenstein condition is necessary but *not sufficient* for a function to be algebraic.)

147. Assume that $\alpha\neq\gamma$, $\beta\neq\gamma$ and α, β, γ are not all rational, but all the coefficients in the hypergeometric series $F(\alpha, \beta, \gamma; z)$ are rational. Then α and β are the roots of an irreducible quadratic equation with rational coefficients.

148. In the case discussed in **147**, $F(\alpha, \beta, \gamma; z)$ does not represent an algebraic function of z. Prove this from Eisenstein's theorem!

§ 3. On the Proof of Eisenstein's Theorem

149. A *rational function* (quotient of two polynomials), whose expansion in ascending powers of z has only rational numbers as coefficients, is the quotient of two polynomials with rational coefficients.

150. Let the power series $f(z)$ with rational coefficients satisfy an equation of the form

$$P_0(z)[f(z)]^l+P_1(z)[f(z)]^{l-1}+\cdots+P_{l-1}(z)f(z)+P_l(z)=0;$$

where $P_0(z), P_1(z), \ldots, P_l(z)$ are polynomials, $P_0(z)\not\equiv 0$. Show that $f(z)$ also satisfies an equation of the same form, where the polynomials occurring in it have coefficients that are *rational integers*.

151. Let the power series with rational coefficients

$$y=a_0+a_1z+a_2z^2+\cdots+a_nz^n+\cdots$$

satisfy an algebraic differential equation, i.e. an equation of the form

$$R(z, y, y', y'', \ldots, y^{(r)})=0,$$

where R denotes a rational entire function (multinomial) of its $r+2$ arguments. Show that y must satisfy an equation of the form

$$R^*(z, y, y', y'', \ldots, y^{(r)})=0,$$

where R^* denotes a rational entire function with *integral* coefficients.

152. Let the functions $f(z)=c_0+c_1z+c_2z^2+\cdots$, $F_0(z), F_1(z), F_2(z), \ldots, F_l(z)$ $(F_0(z)\not\equiv 0)$ be regular in a neighbourhood of the origin and related by the equation

$$F_0(z)[f(z)]^l+F_1(z)[f(z)]^{l-1}+\cdots+F_{l-1}(z)f(z)+F_l(z)=0.$$

Then the power series

$$f^*(z)=c_m+c_{m+1}z+c_{m+2}z^2+\cdots,$$

if m is suitably chosen, satisfies an equation of the form

$$G_0(z)[f^*(z)]^k+G_1(z)[f^*(z)]^{k-1}+\cdots+G_{k-1}(z)f^*(z)+G_k(z)=0,$$

where $k\leqq l$, $G_0(0)=G_1(0)=\cdots=G_{k-2}(0)=0$, $G_{k-1}(0)\neq 0$, $G_0(z)\not\equiv 0$ and $G_\kappa(z)$, $\kappa=0, 1, 2, \ldots, k$, is a rational entire expression with integral coefficients in $F_0(z)$, $F_1(z), \ldots, F_l(z), z, c_0, c_1, \ldots, c_{m-1}$, possibly divided by a power of z. (Function-theoretic preparation for **153**, **154**.)

153. If the power series $P_0(z), P_1(z), \ldots, P_l(z)$ with rational coefficients satisfy the Eisenstein condition, and if the power series $f(z)$ with rational coefficients satisfies the equation

$$P_0(z)[f(z)]^l+P_1(z)[f(z)]^{l-1}+\cdots+P_{l-1}(z)f(z)+P_l(z)=0,$$

then $f(z)$ also satisfies the Eisenstein condition.

154. Let s_n and t_n be integers, $(s_n, t_n)=1$, $t_n\geqq 1$, and denote the greatest prime factor that divides t_n by P_n. We say that the power series

$$\frac{s_0}{t_0}+\frac{s_1}{t_1}z+\frac{s_2}{t_2}z^2+\cdots+\frac{s_n}{t_n}z^n+\cdots$$

satisfies the Tchebychev condition if P_n/n is bounded, and the Hurwitz condition if $\log P_n/\log n$ is bounded, $n=2, 3, 4, \ldots$. For the Tchebychev condition as well as for the Hurwitz condition analogous theorems to **142**, **143**, and **153** hold.

§ 4. Power Series with Integral Coefficients Associated with Rational Functions

155. We term the power series with integral coefficients

$$a_0+a_1z+a_2z^2+\cdots+a_nz^n+\cdots$$

primitive if the numbers $a_0, a_1, a_2, \ldots, a_n, \ldots$ have no common divisor apart from ± 1. Prove that the product of two primitive power series is a primitive power series.

156. If the power series with integral coefficients

$$a_0 + a_1 z + a_2 z^2 + \cdots + a_n z^n + \cdots$$

represents a rational function, then it can be expressed in the form $P(z)/Q(z)$, where $P(z)$ and $Q(z)$ are polynomials with integral coefficients and $Q(0) = 1$. [**155**.]

157. Let

$$\vartheta = 0.\, a_1 a_2 a_3, \ldots a_n \ldots = \frac{a_1}{10} + \frac{a_2}{10^2} + \frac{a_3}{10^3} + \cdots + \frac{a_n}{10^n} + \cdots$$

be a positive number $\leqq 1$ expressed in the decimal system, $0 \leqq a_\nu \leqq 9$, $\nu = 1, 2, 3, \ldots$. The power series with integral coefficients

$$a_1 z + a_2 z^2 + a_3 z^3 + \cdots + a_n z^n + \cdots$$

represents a rational function if ϑ is rational, and does not represent a rational function if ϑ is irrational.

158. Let there be only finitely many distinct numbers in the infinite sequence of numbers $a_0, a_1, a_2, \ldots, a_n, \ldots$. Then the power series

$$a_0 + a_1 z + a_2 z^2 + \cdots + a_n z^n + \cdots$$

represents a rational function if and only if the sequence of coefficients after a certain number of terms is periodic.

159. Let l be an integer, $l \geqq 0$, and $P(z)$ a polynomial with integral coefficients. In the power series expansion of $P(z)(1-z)^{-l-1}$ the coefficients are integers that form a *periodic* sequence modulo any prime number p; the length of the period is a power of p.

160. Let D be an integer and not divisible by the odd prime number p. The coefficients in the power series expansion of $[(D-1)z]/[(1-Dz)(1-z)]$ are periodic mod p, where the length of the period is a proper or an improper divisor of $p-1$.

161. Let the integers D and p be as in **160**. In the power series expansion of $(1-Dz^2)^{-1}$ the coefficients are periodic mod p. The length of the period is always a divisor of $2(p-1)$. It is or is not a divisor of $p-1$ according to $\left(\frac{D}{p}\right) = +1$ or -1.

162. The sequence of Fibonacci numbers (cf. VII **53**) is periodic mod m, where m is any integer. Calculate the length of the period for all prime numbers less than 30.

163. In a power series expansion with integral coefficients that represents a rational function the coefficients, after a certain number of terms, form a *periodic* sequence mod m (m arbitrary).

164. The algebraic function $1/\sqrt{1-4z}$ yields on expansion in powers of z a power series with integral coefficients. These coefficients are for no odd prime number p periodic mod p. (**163** cannot be extended to algebraic functions.)

§ 5. Function-Theoretic Aspects of Power Series with Integral Coefficients

165. The radius of convergence of a non-terminating power series with integral coefficients is $\leqq 1$.

166. A non-terminating power series with integral coefficients that is convergent in the interior of the unit circle cannot represent a bounded function there.

167. If a power series with integral coefficients represents an algebraic function that is not a rational function, then its radius of convergence is <1.

168. In **167** the upper bound 1 is the best possible. In other words: If $\varepsilon>0$, then there exists a power series with integral coefficients that represents an algebraic function that is not a rational function and whose radius of convergence is $>1-\varepsilon$.

169. Let the coefficients of the polynomial

$$P(x)=a_0x^r+a_1x^{r-1}+\cdots+a_{r-1}x+a_r, \qquad a_0\neq 0, \qquad r\geqq 1$$

be real. The analytic function defined by the power series

$$[P(0)]+[P(1)]z+[P(2)]z^2+\cdots+[P(n)]z^n+\cdots$$

is

(1) a rational function, if the r numbers $a_0, a_1, \ldots, a_{r-1}$ are all rational,

(2) not a rational function, if the r numbers $a_0, a_1, \ldots, a_{r-1}$, are not all rational. (Cf. II **168**.) [I **85**.]

170. If the sequence of non-negative integers

$$a_1, \; a_2, \; a_3, \; \ldots, \; a_n, \; \ldots$$

is bounded and contains infinitely many terms that are different from 0, then the series

$$a_1\frac{z}{1-z}+a_2\frac{z^2}{1-z^2}+a_3\frac{z^3}{1-z^3}+\cdots+a_n\frac{z^n}{1-z^n}+\cdots$$

does not represent a rational function.

171. If the sequence

$$a_1, \; a_2, \; a_3, \; \ldots, \; a_n, \; \ldots$$

satisfies the same assumptions as in **170**, then the series

$$a_1\frac{z}{1+z}+a_2\frac{z^2}{1+z^2}+a_3\frac{z^3}{1+z^3}+\cdots+a_n\frac{z^n}{1+z^n}+\cdots$$

does not represent a rational function.

172. Denote by Q_n the "cross sum" of the number n, i.e. the sum of its digits. E.g. we have $Q_{137}=1+3+7=11$. The power series

$$Q_1z+Q_2z^2+Q_3z^3+\cdots+Q_nz^n+\cdots$$

has the unit circle as its natural boundary.

173. In an infinite table of logarithms let the logarithms of all the positive integers 1, 2, 3, ... be written underneath each other in their natural order, in such a fashion that decimal places of the same order appear in the same vertical column. If we interpret the digits in any vertical column as the coefficients of a power series, then we obtain a power series that cannot be analytically continued.

In a different formulation we are required to prove the following theorem:

Let j be an integer and in the decimal representation of $\log 1, \log 2, \log 3, \ldots,$ $\log n, \ldots$ let the jth digit after the decimal point be denoted by $d_1, d_2, d_3, \ldots,$ $d_n, \ldots$ respectively. Then the analytic function defined by the power series

$$d_1 z + d_2 z^2 + d_3 z^3 + \cdots + d_n z^n + \cdots$$

has as its natural boundary the circle $|z| = 1$.

§ 6. Power Series with Integral Coefficients in the Sense of Hurwitz

The power series

$$a_0 + \frac{a_1}{1!} z + \frac{a_2}{2!} z^2 + \cdots + \frac{a_n}{n!} z^n + \cdots$$

is termed a power series with integral coefficients in the sense of *Hurwitz*, or with *H-integral coefficients* for short, if $a_0, a_1, a_2, \ldots, a_n, \ldots$ are rational integers. [A. Hurwitz, Math. Ann. Vol. 51, pp. 196–226 (1899).]

174. If $f(z)$ is a power series with H-integral coefficients, then

$$\frac{df(z)}{dz} \quad \text{and} \quad \int_0^z f(z)\, dz$$

are also power series with H-integral coefficients.

175. If $f(z)$ and $g(z)$ are power series with H-integral coefficients, then

$$f(z) + g(z), \quad f(z) - g(z), \quad f(z) g(z)$$

are also power-series with H-integral coefficients, and under the further assumption that $g(0) = \pm 1$ also

$$\frac{f(z)}{g(z)}.$$

176. If $f(z)$ is a power series with H-integral coefficients and $f(0) = 0$, then

$$\frac{[f(z)]^m}{m!}$$

is also a power series with H-integral coefficients, $n = 1, 2, 3, \ldots$ [**174**].

177. If $f(z)$ and $g(z)$ are power series with H-integral coefficients and $f(0) = 0$, then $g[f(z)]$ is also a power series with H-integral coefficients.

178. The function $\varphi(z)$, well defined by the differential equation

$$\left(\frac{d\varphi(z)}{dz}\right)^2 = 1 - [\varphi(z)]^4$$

and the initial conditions $\varphi(0) = 0$, $\varphi'(0) > 0$, may be expanded in a power series with H-integral coefficients.

By a *congruence* between two power series with H-integral coefficients

$$a_0 + \frac{a_1}{1!} z + \frac{a_2}{2!} z^2 + \cdots + \frac{a_n}{n!} z^n + \cdots \equiv b_0 + \frac{b_1}{1!} z + \frac{b_2}{2!} z^2 + \cdots + \frac{b_n}{n!} z^n + \cdots \pmod{m}$$

we mean the infinite set of congruences between the coefficients

$$a_0 \equiv b_0 \pmod m, \qquad a_1 \equiv b_1 \pmod m, \ldots, \qquad a_n \equiv b_n \pmod m, \ldots$$

179.

$$(e^z-1)^3 \equiv 2\left(\frac{z^3}{3!}+\frac{z^5}{5!}+\frac{z^7}{7!}+\cdots\right) \pmod 4.$$

180. For every prime number p, we have

$$(e^z-1)^{p-1} \equiv -\left(\frac{z^{p-1}}{(p-1)!}+\frac{z^{2(p-1)}}{(2p-2)!}+\frac{z^{3(p-1)}}{(3p-3)!}+\cdots\right) \pmod p.$$

181. For every composite number m greater than 4 we have

$$(e^z-1)^{m-1} \equiv 0 \pmod m.$$

182. The *Bernoulli* numbers B_n are defined as the coefficients in the power series expansion

$$\frac{z}{e^z-1}=1-\frac{z}{2}+\sum_{n=1}^{\infty}\frac{(-1)^{n-1}B_n}{(2n)!}z^{2n}$$

(I **154**). We have

$$B_1=\frac{1}{6}=1-\frac{1}{2}-\frac{1}{3}, \qquad B_2=\frac{1}{30}=-1+\frac{1}{2}+\frac{1}{3}+\frac{1}{5},$$

$$B_3=\frac{1}{42}=1-\frac{1}{2}-\frac{1}{3}-\frac{1}{7}, \qquad B_4=\frac{1}{30}=-1+\frac{1}{2}+\frac{1}{3}+\frac{1}{5},$$

$$B_5=\frac{5}{66}=1-\frac{1}{2}-\frac{1}{3}-\frac{1}{11}, \qquad B_6=\frac{691}{2730}=-1+\frac{1}{2}+\frac{1}{3}+\frac{1}{5}+\frac{1}{7}+\frac{1}{13},$$

$$B_7=\frac{7}{6}=2-\frac{1}{2}-\frac{1}{3}, \qquad B_8=\frac{3617}{510}=6+\frac{1}{2}+\frac{1}{3}+\frac{1}{5}+\frac{1}{17}.$$

In general we have

$$B_n=G_n+(-1)^n\left(\frac{1}{2}+\frac{1}{3}+\frac{1}{\alpha}+\frac{1}{\beta}+\frac{1}{\gamma}+\cdots\right),$$

where G_n is an integer and $2, 3, \alpha, \beta, \gamma, \ldots$ denote those prime numbers that *exceed by one the divisors of 2n.* [Expand first z and then $z/(e^z-1)$ in ascending powers of (e^z-1) and apply **179–181**.]

183. The coefficients C_n in the series

$$\frac{z}{\varphi(z)}=\sum_{n=0}^{\infty}\frac{C_n}{(4n)!}z^{4n},$$

where $\varphi(z)$ denotes the function defined in **178**, are rational numbers and moreover the denominator of C_n is not divisible by any square other than 1 and is divisible only by such prime numbers as are $\equiv 1 \pmod 4$. [Expand first z and then $z/\varphi(z)$ in ascending powers of $\varphi(z)$.]

184. The differential equation

$$\left(\frac{d\wp(z)}{dz}\right)^2 = 4[\wp(z)]^3 - 4\wp(z)$$

has a well-defined solution of the form

$$\wp(z) = \frac{1}{z^2} + \sum_{n=1}^{\infty} \frac{D_n}{(4n)!} z^{4n-2}.$$

($\wp(z)$ is a particular elliptic function, termed a "lemniscate function"; its period-parallelogram is a square.) The numbers D_n are rational, and moreover the denominator of D_n is not divisible by any square other than 1, and is divisible only by such prime numbers as are $\equiv 1 \pmod 4$. [Show that $\wp(z) = [\varphi(z)]^{-2}$, cf. **183**. Expand first z^2 and then $z^2[\varphi(z)]^{-2}$ in ascending powers of $\varphi(z)$; z^2 as a function of $\varphi(z)$ satisfies a linear differential equation of second order.]

185. If a power series with H-integral coefficients $\sum_{n=0}^{\infty} (a_n/n!)z^n$ satisfies a homogeneous linear differential equation with constant coefficients, then the sequence of coefficients $a_0, a_1, a_2, \ldots, a_n, \ldots$ is periodic mod m after a certain number of terms, where m is an arbitrary integer. [Examples of this are found already in **179, 180**.]

186. The power series

$$y = 1 + \frac{2x}{1!} + \frac{6x^2}{2!} + \frac{20x^3}{3!} + \cdots + \binom{2n}{n}\frac{x^n}{n!} + \cdots$$

shows that theorem **185** cannot without further assumptions be extended to homogeneous linear differential equations whose coefficients are polynomials.

187. If a transcendental entire function $g(z)$ has a power series expansion with H-integral coefficients, then we have for the maximum modulus (IV Chap. 1) of $g(z)$

$$\limsup_{r\to\infty} M(r)e^{-r}\sqrt{r} \geqq \frac{1}{\sqrt{2\pi}}.$$

§ 7. The Values at the Integers of Power Series that Converge about $z=\infty$

188. Let the power series in descending powers of z

$$b_m z^m + b_{m-1}z^{m-1} + \cdots + b_1 z + b_0 + \frac{b_{-1}}{z} + \frac{b_{-2}}{z^2} + \frac{b_{-3}}{z^3} + \cdots$$

converge outside some circle. Let the numbers $b_1, b_2, \ldots, b_m$ be rational. If the series assumes integral values for infinitely many integral values of z, then b_0 is rational and

$$b_{-1} = b_{-2} = b_{-3} = \cdots = 0.$$

(Generalization of **93**.) [Hurwitz-Courant, p. 30.]

189. The series

$$\sqrt{2z^2+1} = \sqrt{2}z + \frac{\sqrt{2}}{4z} - \frac{\sqrt{2}}{32z^3} + \frac{\sqrt{2}}{128z^5} - \frac{5\sqrt{2}}{2048z^7} + \cdots$$

represents an integer for infinitely many integral values of z.

190. Deduce **114** from **188**.

191. Let the power series

$$F(z)=b_m z^m + b_{m-1} z^{m-1} + \cdots + b_1 z + b_0 + \frac{b_{-1}}{z} + \frac{b_{-2}}{z^2} + \frac{b_{-3}}{z^3} + \cdots$$

that is not everywhere divergent represent integers for all sufficiently large integral values of z. Then $F(z)$ is an integral-valued polynomial.

192. If two polynomials are such that they assume integral values at the same points of the complex plane, then either their sum or their difference is a constant. (This constant is naturally an integer.)

193. If a polynomial $g(x)$ assumes real values for all values of x for which a second polynomial $f(x)$ assumes values belonging to a set of real numbers that is unbounded from above and from below, then $g(x)$ is in fact always real whenever $f(x)$ is real. (If $f(x)$ is of odd degree, then the assumption that the set of numbers is unbounded from above or from below is sufficient.) For we then have the identity

$$g=\varphi(f),$$

where $\varphi(y)$ is a polynomial in y with real coefficients. (From this result we easily obtain **192**.)

Chapter 4. Some Problems on Algebraic Integers

§ 1. Algebraic Integers. Fields

A real or complex number α is called an *algebraic number* if it is the zero of a polynomial with integral coefficients (introduction to chapter 2), i.e. if there exist rational integers $a_0, a_1, a_2, \ldots, a_n$, $a_0 \neq 0$, such that

$$\text{(E)} \qquad a_0\alpha^n + a_1\alpha^{n-1} + a_2\alpha^{n-2} + \cdots + a_{n-1}\alpha + a_n = 0.$$

If $a_0 = 1$, then α is called an *algebraic integer*, or an "integer" for short, as we shall use the term in **194–237**. There are thus irrational and also complex integers. The rational integers are the usual integers

$$\ldots, -3, -2, -1, 0, 1, 2, 3, \ldots.$$

If an algebraic integer happens to be a rational number it is necessarily a rational (ordinary) integer (Hardy-Wright, p. 178). If α and β are algebraic numbers, then

$$\alpha+\beta, \quad \alpha-\beta, \quad \alpha\beta, \quad \frac{\alpha}{\beta}$$

are also algebraic numbers, where in the last case we suppose $\beta \neq 0$. The algebraic numbers form a *field*. If α and β are algebraic integers, then

$$\alpha+\beta, \quad \alpha-\beta, \quad \alpha\beta$$

are also algebraic integers. The algebraic integers form an *integral domain*.

194. If α is an integer, then $\sqrt{\alpha}$ is also an integer.

195. If r and s are rational numbers and $r+s\sqrt{-1}$ is an integer, then r and s must be rational *integers*.

196. If r and s are rational numbers and $r+s\sqrt{-5}$ is an integer, then r and s must be rational integers.

197. If r and s are rational numbers and $r+s\sqrt{-3}$ is an integer, then $2r$ and $2s$ must be rational integers and $2r \equiv 2s \pmod 2$; r and s need not necessarily be rational integers.

197.1. If both ϑ/π and $\cos\vartheta$ are rational, then $\cos\vartheta$ must have one of the following five values:

$$1,\ \tfrac{1}{2},\ 0,\ -\tfrac{1}{2},\ -1.$$

197.2. Consider the interior angle included between two contiguous faces of a regular solid and call it

$$T,\quad H,\quad O,\quad D,\quad \text{or}\quad I$$

according as the solid is a

tetra-, hexa-, octa-, dodeca, or icosa-

hedron (a hexahedron is a cube). Show that the ratios

$$\frac{T}{H},\ \frac{O}{H},\ \frac{D}{H},\ \frac{I}{H}$$

are irrational.

[Show first, using appropriate spherical triangles, that $\cos T=\frac{1}{3}$, $\cos H=0$, $\cos O=-\frac{1}{3}$, $\cos 2D=-\frac{3}{5}$, $\cos 2I=\frac{1}{9}$.]

The numbers $\alpha_1, \alpha_2, \ldots, \alpha_m$ are called *linearly dependent* if there exist rational integers $a_1, a_2, \ldots, a_m$ *not all zero* such that

$$a_1\alpha_1+a_2\alpha_2+\cdots+a_m\alpha_m=0.$$

If no relation of this kind exists then $\alpha_1, \alpha_2, \ldots, \alpha_m$ are called *linearly independent.*

Thus, if $\beta\neq 0$, the numbers α and β are linearly dependent or independent according as α/β is rational or irrational. The number α is algebraic if there exists an n (a positive rational integer) such that $1, \alpha, \alpha^2, \ldots, \alpha^n$ are linearly dependent.

197.3. The numbers (angles) T, H and O are linearly dependent; in fact

$$T-2H+O=0$$

(see **197.2**; it can be also proved by elementary geometry). Yet T and O are linearly independent.

197.4 (continued). Show that T, H, D and I are linearly independent.

Also H, O, D and I are linearly independent, and any linear homogeneous relation with rational coefficients between T, H, O, D and I can differ only trivially (that is, only by a numerical factor) from the one given in **197.3**.

197.5. If both ϑ/π and $\tan^2\vartheta$ are rational then $\tan^2\vartheta$ must have one of the following five values (regarding $\infty=1/0$ as "rational".)

$$0,\ \tfrac{1}{3},\ 1,\ 3,\ \tfrac{1}{0}.$$

197.6. With the exception of the four cases $n=1, 2, 4$, and 8, $\tan 2\pi/n$ is irrational for all positive rational integers n.

Algebraic numbers that are zeros of the same *irreducible* polynomial [introduction to Chapter 2, § 3] are termed *conjugate* to one another [Hecke p. 65, Hardy-Wright pp. 230–1]. Algebraic numbers that are zeros of an irreducible polynomial of degree n are themselves said to be *of degree n*. The rational numbers are of degree 1 and conjugate only to themselves.

198. There are only finitely many integers of given degree n that together with their conjugates lie in a fixed circle $|z| < k$ of the complex plane.

199. The only integer that together with all its conjugates lies in the open unit circle is the number 0.

200. Let all the zeros of a polynomial with rational integral coefficients and leading coefficient 1 lie in the closed unit circle. Such a zero must lie either at the center or on the boundary of the unit circle and in the latter case it must be a root of unity. [I.e. it must satisfy an equation of the form $x^h = x^k$, where h, k are rational integers, $0 < h < k$; **198**.]

201. If the integer α together with all its conjugates is real and in absolute value < 2, then $\alpha = 2 \cos 2\pi p/q$, where p and q are rational integers.

The set of all numbers that can be expressed as rational functions of an algebraic number ϑ with rational coefficients is called an *algebraic number field*, or simply "a field", or more precisely the field *generated by* ϑ. The degree of the generating number ϑ is termed *the degree of the field*. Thus for example the field generated by a number of the second degree is called a quadratic field, or more precisely a real quadratic field or a complex quadratic field according as the generating number is real or complex.

202. The irrational integers

$$\sqrt{-1}, \quad \sqrt{-3}, \quad \sqrt{-5}$$

generate three distinct complex quadratic number fields, in which the integers have the forms

$$a + b\sqrt{-1}, \quad a + b\,\frac{1+\sqrt{-3}}{2}, \quad a + b\sqrt{-5}$$

respectively, where a and b are *rational* integers.

203. A set $\mathfrak{S}$ of complex (or real) numbers is called a "discontinuous integral domain" if it has the following two properties:

(1) If ζ' and ζ'' belong to the set $\mathfrak{S}$, then $\zeta' + \zeta''$, $\zeta' - \zeta''$, and $\zeta'\zeta''$ also belong to the set $\mathfrak{S}$.

(2) The point 0 is not a point of accumulation of the set of numbers $\mathfrak{S}$. (By (1) the number $0 = \zeta' - \zeta'$ belongs to the set $\mathfrak{S}$.)

A discontinuous integral domain consists either of all or of some of the integers of a complex quadratic number field. (E.g. only of the numbers 0, ± 6, ± 12, $\pm 18, \ldots$ or only of the single number 0.)

§ 2. Greatest Common Divisor

We say that the integer α is *divisible* by the integer ϑ if there exists an integer κ (the quotient) such that $\alpha=\kappa\vartheta$. We also say that "ϑ is a divisor of α" or "ϑ divides α", etc. Integers that are divisors of 1 are called *units*. ϑ is termed a *greatest common divisor* (abbreviated to g.c. divisor) of the m integers $\alpha_1, \alpha_2, \ldots, \alpha_m$ if there exist $2m$ integers $\kappa_1, \kappa_2, \ldots, \kappa_m, \lambda_1, \lambda_2, \ldots, \lambda_m$ such that

$$\alpha_1=\kappa_1\vartheta, \qquad \alpha_2=\kappa_2\vartheta, \ldots, \qquad \alpha_m=\kappa_m\vartheta, \qquad \alpha_1\lambda_1+\alpha_2\lambda_2+\cdots+\alpha_m\lambda_m=\vartheta.$$

Two integers α, β whose g.c divisor $=1$ are termed *relatively prime.*

204. If we were to designate as "integers" not all integers but only those that belong to the field generated by $\sqrt{-5}$, then the two numbers 3 and $1+2\sqrt{-5}$ would have *no* g.c. divisor in the sense of the above definition. I.e. it is impossible to find in the field generated by $\sqrt{-5}$, to which the numbers 3 and $1+2\sqrt{-5}$ belong, five integers α, β, γ, δ, ϑ such that the equations

$$3=\alpha\vartheta, \qquad 1+2\sqrt{-5}=\beta\vartheta, \qquad 3\gamma+(1+2\sqrt{-5})\delta=\vartheta$$

hold simultaneously. [Investigate the least values of $(a+b\sqrt{-5})(a-b\sqrt{-5})$, for a, b rational integers; **202.**]

205 (continued). The squares $3^2=9$ and $(1+2\sqrt{-5})^2=-19+4\sqrt{-5}$ have a g.c. divisor, even under the restricted definition of "integer" given in **204**.

206 (continued). Find the g.c. divisor of 3 and $1+2\sqrt{-5}$ (which is not contained in the field generated by $\sqrt{-5}$!). [**194.**]

The deep theorem that any two integers *have* a g.c. divisor [Hecke, p. 121] is *not* to be used in the problems which follow.

207. If ϑ is a g.c. divisor of $\alpha_1, \alpha_2, \ldots, \alpha_m$, then ϑ is divisible by every common divisor of $\alpha_1, \alpha_2, \ldots, \alpha_m$.

208. Let ϑ be a g.c. divisor of the numbers $\alpha_1, \alpha_2, \ldots, \alpha_m$, and ϑ' another g.c. divisor of the same numbers $\alpha_1, \alpha_2, \ldots, \alpha_m$. Then ϑ'/ϑ is a unit.

209. If ϑ is a g.c. divisor of $\alpha_1, \alpha_2, \ldots, \alpha_m$, then $\gamma\vartheta$ is a g.c. divisor of $\gamma\alpha_1, \gamma\alpha_2, \ldots, \gamma\alpha_m$.

210. If α is relatively prime to $\beta\gamma$, then α is relatively prime to β as well as to γ.

211. If α is relatively prime to β as well as to γ, then α is also relatively prime to $\beta\gamma$.

212. If α is relatively prime to β, then a g.c. divisor of α and γ is also a g.c. divisor of α and $\beta\gamma$.

213. If δ is a g.c. divisor of α and β, then δ^n is a g.c. divisor of α^n and β^n, $n=1, 2, 3, \ldots$.

214. If δ is a g.c. divisor of α and β, then $\sqrt[n]{\delta}$ is a g.c. divisor of $\sqrt[n]{\alpha}$ and $\sqrt[n]{\beta}$, $n=1, 2, 3, \ldots$.

215. If the m numbers $\alpha_1, \alpha_2, \ldots, \alpha_m$ are *coprime* (*pairwise relatively prime*), that is if the g.c. divisor of any two of them is 1, and if $\mu=\alpha_1\alpha_2\ldots\alpha_m$, then 1 is a g.c. divisor of the m numbers

$$\frac{\mu}{\alpha_1}, \quad \frac{\mu}{\alpha_2}, \quad \ldots, \quad \frac{\mu}{\alpha_m}.$$

216. All rational prime numbers, with the exception of finitely many, are relatively prime to a given integer α, $\alpha\neq 0$.

217. Let the integers $a_1, a_2, \ldots, a_n$ be rational and let $\alpha_1, \alpha_2, \ldots, \alpha_n$ be the n zeros of the polynomial $x^n+a_1x^{n-1}+a_2x^{n-2}+\cdots+a_{n-1}x+a_n$. Then $\alpha_1, \alpha_2, \ldots, \alpha_n$ *have* a g.c. divisor δ; in fact $\delta=\sqrt[N]{d}$, where $N=n!$ and d is the g.c. divisor of

$$a_1^{\frac{N}{1}}, \quad a_2^{\frac{N}{2}}, \quad a_3^{\frac{N}{3}}, \quad \ldots, \quad a_n^{\frac{N}{n}}.$$

By the *norm* of α, denoted by $N(\alpha)$, where α is a number belonging to a given field K of nth degree, we mean the product of its n *conjugates*, each distinct conjugate being represented q times if the degree of α is n/q [Hecke, p. 81].The term "conjugate" is used here in a different sense to that of the explanatory comments preceding **198** [Hecke, p. 65, 70].

218. If α, β belong to K and α is a divisor of β, then $N(\alpha)$ is a divisor of $N(\beta)$.

219. Let the field K be such that every pair of integers α and β belonging to K have a g.c. divisor belonging to K, in the sense that there exist five more integers α', β', γ, δ, ϑ in K such that

$$\alpha=\alpha'\vartheta, \qquad \beta=\beta'\vartheta, \qquad \alpha\gamma+\beta\delta=\vartheta.$$

(Not always the case! Cf. **204**.) For this to be the case we have the following necessary and sufficient condition: If α and β are integers in K and such that α is not a divisor of β nor β a divisor of α, then there exist in K two integers ξ, η such that we have simultaneously

$$0<|N(\alpha\xi+\beta\eta)|<|N(\alpha)|, \qquad 0<|N(\alpha\xi+\beta\eta)|<|N(\beta)|.$$

(If a and b are rational integers satisfying $|a|>|b|$, and b is not a divisor of a, then the remainder on dividing a by b, i.e. $r=a\cdot 1-b[a/b]$, has the property required here of $\alpha\xi+\beta\eta$.)

220. Any two numbers of the field generated by $\sqrt{-1}$ have a g.c. divisor belonging to the same field.

§ 3. Congruences

Let α, β and μ be integers. That $\alpha-\beta$ is divisible by μ is expressed as in the theory of rational numbers in terms of a *congruence* relation by writing

$$\alpha\equiv\beta \pmod{\mu}.$$

221. The congruence relation between two algebraic integers with respect to a given modulus is *symmetric* and *transitive*. That is we have the following:

From $\alpha\equiv\beta \pmod{\mu}$ it follows that $\beta\equiv\alpha \pmod{\mu}$.
From $\alpha\equiv\beta, \beta\equiv\gamma \pmod{\mu}$ it follows that $\alpha\equiv\gamma \pmod{\mu}$.

222. From

$$\alpha \equiv \beta, \qquad \gamma \equiv \delta \qquad (\text{mod } \mu)$$

it follows that

$$\alpha+\gamma \equiv \beta+\delta, \qquad \alpha-\gamma \equiv \beta-\delta, \qquad \alpha\gamma \equiv \beta\delta \qquad (\text{mod } \mu).$$

223. If 1 is the g.c. divisor of α and μ, and μ is not a unit, then

$$\alpha \not\equiv 0 \qquad (\text{mod } \mu).$$

224. Let $\alpha, \beta, \ldots$ be integers, $f(x, y, \ldots)$ a polynomial with rational integral coefficients and p a rational prime number. Then we have

$$(f(\alpha, \beta, \ldots))^p \equiv f(\alpha^p, \beta^p, \ldots) \qquad (\text{mod } p).$$

225. Let

$$\alpha_1, \alpha_2, \ldots, \alpha_m; \quad \omega_1, \omega_2, \ldots, \omega_m$$

($\alpha_k \neq 0$, $\omega_k \neq 0$, $\omega_k \neq \omega_l$ for $k \gtrless l$; $k, l = 1, 2, 3, \ldots, m$) be integers. Then not all the numbers

$$\frac{\alpha_1\omega_1^n + \alpha_2\omega_2^n + \cdots + \alpha_m\omega_m^n}{n}, \qquad n = 1, 2, 3 \ldots$$

can be integers. [Consider suitable prime numbers p and set $n = p, 2p, 3p, \ldots, rp$; **216**, **224**.]

226. The zeros of the cyclotomic polynomial $K_m(x)$ [**36**] are also zeros of $x^m - 1$, and thus integers. They are given by $\alpha^{r_1}, \alpha^{r_2}, \ldots, \alpha^{r_h}$ if we set $e^{\frac{2\pi i}{m}} = \alpha$, $\varphi(m) = h$ (cf. introduction to chapter 1, § 5), and $r_1, r_2, \ldots, r_h$ form a reduced remainder system mod m, i.e. if the rational integers $r_1, r_2, \ldots, r_h$ are all relatively prime to m and pair-wise not congruent (mod m).—Is $K_m(x)$ reducible or irreducible?

Solution: Let us consider that factor $f(x)$ among the irreducible factors of $K_m(x)$ with integral coefficients [**116**] that has α among its zeros. The zeros of $f(x)$, i.e. the integers conjugate to α, are all of the form α^n, where n is an integer, and their number is $\leqq h$ [in fact $< h$ if $K_m(x)$ is reducible]. In any case for an arbitrary rational integer n we have

$$(*) \qquad |f(\alpha^n)| \leqq 2^h,$$

since the absolute value of each factor of $f(\alpha^n)$, being a chord of the unit circle, is $\leqq 2$. Now let p be a prime number. Since $f(\alpha) = 0$, we obtain [**224**]

$$f(\alpha^p) \equiv (f(\alpha))^p \equiv 0 \qquad (\text{mod } p).$$

Thus $f(\alpha^p)/p$ is an integer. Its conjugates are also integers of the form $f(\alpha^{np})/p$. Hence if $p > 2^h$, all the conjugates of $f(\alpha^p)/p$ are by (*) in absolute value < 1, and hence [**199**] $f(\alpha^p) = 0$, and we have shown that α^p is also a zero of $f(x)$.

By a theorem of Dirichlet [footnote to **110**] all the $\varphi(m) = h$ arithmetical progressions

$$\begin{array}{l} r_1, \; r_1+m, \; r_1+2m, \; \ldots; \\ r_2, \; r_2+m, \; r_2+2m, \; \ldots; \\ \ldots\ldots\ldots\ldots\ldots\ldots\ldots \\ r_h, \; r_h+m, \; r_h+2m, \; \ldots \end{array}$$

contain infinitely many prime numbers. Hence in particular we can construct a reduced system of remainders mod m, consisting of prime numbers which are all $>2^h$. Then, by the preceding, the factor $f(x)$ of $K_m(x)$ must have all the zeros $\alpha^{r_1}, \alpha^{r_2}, \ldots, \alpha^{r_h}$ of $K_m(x)$ as zeros and the same degree as $K_m(x)$ and in fact must be identical with $K_m(x)$. Thus $K_m(x)$ is irreducible.

227. By modifying the preceding argument give a proof of the irreducibility of the cyclotomic polynomial $K_m(x)$ without using the Dirichlet theorem concerning arithmetical progressions. It can be avoided by a more laborious but also more elementary construction of the reduced system of remainders mod m.

227.1. If the rational integers m and n are relatively prime and $n \geqq 3$, then $2\cos(2\pi m/n)$ is an algebraic integer of degree $\varphi(n)/2$.

227.2 (continued). From this result derive **197.1** [**38.1**].

§ 4. Arithmetical Aspects of Power Series

228. If the expansion of a rational function both in ascending as well as in descending powers of z is a series with rational integral coefficients, then its poles different from 0 and ∞ lie at units.

229. Let a power series (in ascending powers) be convergent in the unit circle and represent a non-integral rational function. If its coefficients are rational integers, then its poles lie at roots of unity. [**156**, **200**.]

230. If the power series

$$\alpha_0 + \frac{\alpha_1}{z} + \frac{\alpha_2}{z^2} + \cdots + \frac{\alpha_n}{z^n} + \cdots$$

represents a rational function and its coefficients $\alpha_0, \alpha_1, \alpha_2, \ldots$ are algebraic integers, then the poles of the function are also algebraic integers.

231. If

$$\alpha_1, \alpha_2, \ldots, \alpha_m, \qquad \omega_1, \omega_2, \ldots, \omega_m$$
$$(\alpha_k \neq 0, \quad \omega_k \neq 0, \quad \omega_k \neq \omega_l \ \text{ for } \ k \gtrless l; \quad k, l = 1, 2, \ldots, m)$$

are algebraic integers, then the coefficients in the power series expansion of

$$\frac{\alpha_1}{1-\omega_1 z} + \frac{\alpha_2}{1-\omega_2 z} + \cdots + \frac{\alpha_m}{1-\omega_m z}$$

are also algebraic integers. However, the same is not true of the expansion that is obtained from it by term-by-term integration.

232. Let $f(z)$ be represented by a power series with rational coefficients and let its derivative $f'(z)$ be a rational function. $f(z)$ is rational if its power series satisfies the Eisenstein condition (cf. introduction to **142**) and transcendental if it does not.

233. Let the algebraic function $f(z)$ be defined by the equation

$$P_0(z)[f(z)]^l + P_1(z)[f(z)]^{l-1} + \cdots + P_{l-1}(z)f(z) + P_l(z) = 0,$$

where $P_0(z), P_1(z), \ldots, P_l(z)$ are polynomials with algebraic coefficients. If α is an algebraic number, then the coefficients in the expansion of $f(z)$ in powers of $(z-\alpha)$ are also algebraic numbers. [Special case: If $z=\alpha$ is a regular point of $f(z)$, then $f(\alpha)$ is an algebraic number.]

234. Let a rational entire function $F(z, y)$ of the two variables z, y with rational coefficients be termed *irreducible* if it cannot be factorized into the product of two entire functions with rational coefficients that are both of lower degree with respect to y.

If $F(z, y)$ is irreducible, then either no solution y of the equation

$$F(z, y)=0$$

is a rational function of z, or all solutions are rational functions of z.

[Moreover: either no solution is a rational *entire* function of z or all solutions are rational entire functions of z.—Note **233** and the chief characteristic of irreducible equations: the roots of an irreducible equation have the bad habit of always bringing the whole family on a visit. Cf. Hecke, p. 64, Theorem 49.]

235. If the coefficients in the power series expansion of an algebraic function are algebraic numbers, then these coefficients all belong to a *finite* field (i.e. to a field generated by a single algebraic number, cf. introduction to **202**). [**151**, **152**.]

236 (continued). If the power series in question is

$$\alpha_0+\alpha_1 z+\alpha_2 z^2+\cdots,$$

then there exists an integer τ such that all the numbers $\alpha_1\tau, \alpha_2\tau^2, \alpha_3\tau^3, \ldots$ are integers. (Generalization of Eisenstein's theorem.)

237. Let the series in descending powers of z

$$a_m z^m+a_{m-1}z^{m-1}+\cdots+a_1 z+a_0+\frac{a_{-1}}{z}+\frac{a_{-2}}{z^2}+\frac{a_{-3}}{z^3}+\cdots$$

be convergent outside some circle, and let its values for infinitely many rational integral values of z be rational integral. If the coefficients $a_m, a_{m-1}, \ldots, a_1, a_0, a_{-1}, a_{-2}, \ldots$ are all rational numbers, then the series represents a rational entire function (it terminates, **188**); if the coefficients all belong to a finite field, then the power series represents an algebraic function. (Example: **189**.)

Chapter 5. Miscellaneous Problems

§ 1. Lattice Points in Two and Three Dimensions

In problems **237.1–244** we shall consider a rectangular coordinate system in the plane, and we shall term a point of the plane a *lattice point* if both its coordinates are rational integers.

237.1. The three sides of a triangle are of lengths l, m, and n respectively. The numbers l, m and n are positive integers satisfying

$$l \leqq m \leqq n.$$

Find the number of different triangles of this kind with longest side of given length n. [Count lattice points.]

237.2. The number of lattice points on the periphery of a circle whose radius is $\sqrt{n}$ and whose center is at the origin depends on the nature of the prime factors of n. Verify the proposition of Gauss quoted in **58.7** for $n \leqq 10$ and for some larger integers.

238. All three vertices of an equilateral triangle cannot be lattice points.

238.1. A polygon is equiangular, has n sides, and all its vertices are lattice points. Then n is either 4 or 8.

238.2. A regular polygon has n sides and all its vertices are lattice points. Then $n=4$.

239. How thick must the trunks of the trees in a regularly spaced circular forest grow if they are to block completely the view from the center?

Let s be a given positive integer. Let every lattice point p, q that satisfies the inequality $1 \leqq p^2+q^2 \leqq s^2$ be the center of a circle of radius r. If r is sufficiently small, there exist rays running from the point 0, 0 to infinity without meeting the circles described above (the forest is transparent); such rays no longer exist if r is sufficiently large (for $r=\frac{1}{2}$ the circles touch). Let $r=\rho$ be the value of r which divides the two cases (limit of transparency). Then we have

$$\frac{1}{\sqrt{s^2+1}} \leqq \rho < \frac{1}{s}.$$

240. In the plane draw the *widest possible* infinite straight "path", contained between two parallels and avoiding all the lattice points, that makes an angle equal to arctan x with the y-axis. Denoting the width of this path by $\varphi(x)$, determine $f(x)=\varphi(x)\sqrt{1+x^2}$.

241. Two lattice points x, y and x', y' are termed congruent mod n, if

$$x \equiv x' \pmod{n}, \qquad y \equiv y' \pmod{n}.$$

Among any kn^2+1 distinct lattice points there are always $k+1$ that are congruent mod n.

242. In the plane consider a domain of area (Jordan content) F. It may contain no lattice points, but can always be moved parallel to itself so that $[F]+1$ lattice points lie in it.

243. The lattice points in the closed first quadrant are *enumerable*. I.e. there exist functions $f(x, y)$ defined for ordered pairs of non-negative integers x, y that combine the following two properties:

(1) The values assumed by $f(x, y)$ form the set of numbers $1, 2, 3, 4, 5, \ldots$.

(2) The function $f(x, y)$ takes different values at distinct points; i.e. if x, y, x', y' are integers, $x \geqq 0$, $y \geqq 0$, $x' \geqq 0$, $y' \geqq 0$, $(x-x')^2+(y-y')^2>0$, then $f(x, y) \neq f(x', y')$.

There even exist *rational entire* functions $f(x, y)$ that enumerate the lattice points, i.e. that have the properties (1) (2). An example is the function

$$\begin{aligned} f(x, y) &= \tfrac{1}{2}(x^2+2xy+y^2+3x+y+2) \\ &= \binom{x+y+1}{2}+x+1 \end{aligned}$$

and the function arising from it by interchanging x and y [the lattice points are counted successively on the line segments

$$x+y=0, \qquad x+y=1, \qquad x+y=2, \qquad \ldots, \qquad x\geqq 0, \qquad y\geqq 0$$

Let $f(x, y)$ be a rational entire function of degree m

$$f(x, y)=\varphi_0(x, y)+\varphi_1(x, y)+\cdots+\varphi_m(x, y),$$

where $\varphi_\mu(x, y)$ denotes a rational entire homogeneous function of degree μ. If $f(x, y)$ enumerates the lattice points, then clearly $\varphi_m(x, y)\geqq 0$ for $x\geqq 0$, $y\geqq 0$. If $\varphi_m(x, y)>0$ for $x\geqq 0$, $y\geqq 0$, $x+y>0$ and $f(x, y)$ enumerates the lattice points, then the degree of $f(x, y)$ is

$$m=2.$$

[The number of lattice points contained inside the contour line $f(x, y)=$const. is related to the enclosed area.]

244. Consider the square domain $\frac{1}{2}\leqq x\leqq n+\frac{1}{2}$, $\frac{1}{2}\leqq y\leqq n+\frac{1}{2}$ as a "chess board with n^2 squares", i.e. by means of straight lines parallel to the axis divide it into n^2 squares of unit area. The "problem of the n queens" requires us to choose from the n^2 lattice points that constitute the centers of the n^2 squares n lattice points $(x_1, y_1), (x_2, y_2), \ldots, (x_n, y_n)$ that satisfy the $2n(n-1)$ inequalities

$$x_\mu \neq x_\nu, \qquad y_\mu \neq y_\nu, \qquad x_\mu - x_\nu \neq y_\mu - y_\nu, \qquad x_\mu - x_\nu \neq -(y_\mu - y_\nu),$$

where $\mu\neq\nu$; $\mu, \nu=1, 2, \ldots, n$. Replace the inequalities by the more restrictive "incongruence relations"

$$x_\mu \not\equiv x_\nu, \qquad y_\mu \not\equiv y_\nu, \qquad x_\mu - x_\nu \not\equiv y_\mu - y_\nu, \qquad x_\mu - x_\nu \not\equiv -(y_\mu - y_\nu)$$

(mod n) and show that these have a solution if and only if n is relatively prime to 6.

In problems **244.1–244.4**, we shall consider a rectangular system of coordinates in three-dimensional space and we shall term a point a *lattice point* if all three of its coordinates are rational integers.

244.1. A regular polygon has n sides and all its vertices are lattice points. Then n is equal to 3, 4 or 6.

244.2 (continued). Give examples showing that the three cases $n=3$, 4 and 6 can actually occur.

244.3. All the vertices of a regular polyhedron are lattice points. Then the polyhedron is either a cube or a tetrahedron or an octahedron.

244.4 (continued). Give examples showing that the three cases mentioned above can actually occur.

§ 2. Miscellaneous Problems

245. Let q denote an odd prime number and let $r_1, r_2, \ldots, r_q$ and $s_1, s_2, \ldots, s_q$ be two complete systems of residues mod q. Then the q numbers $r_1s_1, r_2s_2, \ldots, r_qs_q$ do *not* form a complete system of residues mod q.

246. Let the highest power of the odd prime number p that divides the number n be p^α, $\alpha \geqq 1$. Then we have

$$1^\lambda + 2^\lambda + 3^\lambda + \cdots + n^\lambda \equiv -\frac{n}{p} \quad \text{or} \quad 0 \qquad (\operatorname{mod} p^\alpha),$$

according as λ is or is not divisible by $p-1$, $\lambda = 1, 2, 3, \ldots$.

247. Let p be the smallest prime number that divides n. Then there exist two complete systems of residues mod n

$$\begin{array}{llll} r_1, & r_2, & \ldots, & r_n, \\ s_1, & s_2, & \ldots, & s_n \end{array}$$

such that each of the $p-2$ lines

$$\begin{array}{llll} r_1+s_1, & r_2+s_2, & \ldots, & r_n+s_n, \\ r_1+2s_1, & r_2+2s_2, & \ldots, & r_n+2s_n, \\ \cdots & \cdots & \cdots & \cdots \\ r_1+(p-2)s_1, & r_2+(p-2)s_2, & \ldots, & r_n+(p-2)s_n \end{array}$$

represents a complete system of residues mod n. However then

$$r_1+(p-1)s_1, \quad r_2+(p-1)s_2, \quad \ldots, \quad r_n+(p-1)s_n$$

is not a complete system of residues mod n.

247.1. If p is a prime number, then

$$\binom{p}{1}, \quad \binom{p}{2}, \quad \binom{p}{3}, \quad \ldots, \quad \binom{p}{p-1},$$

$$\begin{array}{llll} s_2^p, & s_3^p, & \ldots, & s_{p-1}^p, \\ S_2^p, & S_3^p, & \ldots, & S_{p-1}^p \end{array}$$

are divisible by p. (For the definitions of s_k^n and S_k^n see the introductions to I **197** and I **186**, respectively.)

247.2. If p is an odd prime, the coefficients of the Tchebychev polynomial $T_p(x)$ (defined at the beginning of Part Six) are all divisible by p, with the exception of the leading coefficient which is equal to 2^{p-1}.

248. Every integral power of a number n can be represented as the sum of n consecutive odd integers.

249. A number in the sequence $2, 3, 4, \ldots, n$ $(n > 2)$ is relatively prime to all the other numbers of the sequence if and only if it is a prime number that exceeds $n/2$. (That such a prime number always exists for $n > 2$ was proved by Tchebychev. Cf. Oeuvres, Vol. 1, p. 63, St. Petersbourg 1899.)

250. The partial sums of the harmonic series

$$\frac{1}{1} + \frac{1}{2} + \frac{1}{3} + \cdots + \frac{1}{n}$$

for $n > 1$ are not equal to an integer. This follows directly from the Tchebychev theorem [**249**], but is to be proved *without* using that theorem.

251. The sum of two or more *consecutive* terms of the harmonic series, i.e. a sum of the form

$$\frac{1}{n}+\frac{1}{n+1}+\cdots+\frac{1}{m}, \qquad n=1, 2, 3, \ldots; \quad n<m,$$

cannot be an integer. If it is written in the form of a fraction in its lowest terms, then its denominator is even and its numerator is odd.

252. If the positive integer n is divisible by all numbers that are $\leqq \sqrt{n}$, then n is either 24 or a divisor of 24.

More generally show by an elementary argument the following: If $0<\alpha<1$, then there exist only a finite number of positive integers n such that n is divisible by $1, 2, 3, \ldots, [n^\alpha]$.

253. Let Q be a prime number and relatively prime to $10P$. The arithmetic mean of the digits in a period of the expression of P/Q as a decimal fraction is 4.5 if and only if the length of the period is an even number.

254. The number $n=2^h+1$, $h\geqq 2$ is a prime number if and only if

$$3^{\frac{n-1}{2}} \equiv -1 \qquad (\text{mod } n).$$

255. The following theorem given by *Euler* as a conjecture is true:

The Diophantine equation

$$4xyz-x-y-t^2=0$$

has no solution in *positive* integers x, y, z, t.

256. If the prime number $q \geqq 11$, then there always exist positive odd prime numbers p_1, p_2, p_3, p_4, not necessarily distinct, that are all smaller than q, and satisfy the equations

$$\left(\frac{p_1}{q}\right) = +1, \qquad \left(\frac{p_2}{q}\right) = -1,$$

$$\left(\frac{q}{p_3}\right) = +1, \qquad \left(\frac{q}{p_4}\right) = -1$$

respectively. $\left[\left(\frac{p_1}{q}\right)\right.$ etc., are *Legendre* symbols defined in V **45.**$\left.\right]$,

257. The decimal fraction

$$0.23571113171923\ldots$$

(formed by writing consecutively all the prime numbers) is irrational. [**249**, **110**.]

258. The number

$$e=1+\frac{1}{1!}+\frac{1}{2!}+\frac{1}{3!}+\frac{1}{4!}+\cdots$$

is irrational.

259. The number e is not only irrational, but it is also not an algebraic number of the second degree, i.e. it cannot satisfy an equation of the form

$$ae+be^{-1}+c=0,$$

where a, b, c are integers and are not all $=0$.

260. If the Euler (Mascheroni) constant $C=-\Gamma'(1)$ were a rational number, then $\Gamma'(n+1)$ would have to be an integer for all integers n from some point onward.

260.1. The number

$$\lambda=\frac{\log 3}{\log 2}$$

is irrational.

260.2. The numbers a and b are positive; a may be rational or irrational and the same holds for b, and so there are four possible cases. Show by means of examples that a^b can be rational or irrational in each of these four cases (eight examples are required).

261. The question whether the number π has the property that the arithmetic mean of its first n decimal places converges to 4.5 must be left unresolved. If however the number π has this property then the number $4-\pi$ has the same property.

262. Denote by q_n the denominator of the nth Bernoulli number B_n [**182**]. Then we have

$$\lim_{n\to\infty}\sqrt[n]{\frac{q_1q_2q_3\cdots q_n}{(2n)!}}=\frac{1}{2}.$$

262.1. If $\zeta(s)$ denotes the Riemann zeta-function defined in the introduction to **39**,

$$\lim_{n\to\infty}\left(\zeta(2)+\zeta(3)+\cdots+\zeta(n+1)-n\right)=1.$$

262.2.

$$\lim_{n\to\infty}\left(\frac{1}{2}\,\zeta(2)+\frac{1}{3}\,\zeta(3)+\cdots+\frac{1}{n}\,\zeta(n)-\log n\right)=0.$$

263. Let the arithmetic function $f(n)$ be multiplicative [introduction to Ch. 1, § 6] and converge to 0 as n tends to infinity *through the prime numbers and the powers of prime numbers.* Then $f(n)$ also satisfies the stronger requirement that

$$\lim_{n\to\infty} f(n)=0.$$

as n tends to infinity through *all* positive integers.

264. For every given positive number δ,

$$\lim_{n\to\infty}\frac{n^{1-\delta}}{\varphi(n)}=0,\qquad \lim_{n\to\infty}\frac{\tau(n)}{n^\delta}=0.$$

265. Let the quadratic polynomial ax^2+bx+c with integral coefficients be represented by the lattice point a, b, c of a three-dimensional cubic lattice. Let the number of those lattice points that lie in the cube

$$-n\leqq a\leqq n,\qquad -n\leqq b\leqq n,\qquad -n\leqq c\leqq n$$

and correspond to a *reducible* polynomial be r_n. Prove that

$$\lim_{n\to\infty}\frac{r_n}{(2n+1)^3}=0.$$

(It is in a certain sense the "normal case" for a quadratic polynomial to be irreducible.)

266. The probability that a polynomial with integral coefficients of given degree is reducible is equal to 0. More precisely: Let h be an integer, $h \geqq 2$, and let r_n denote the number of those lattice points of $(h+1)$-dimensional space that lie in the cube

$$-n \leqq a_0 \leqq n, \qquad -n \leqq a_1 \leqq n, \qquad \ldots, \qquad -n \leqq a_h \leqq n$$

and correspond to a *reducible* polynomial $a_0 x^h + a_1 x^{h-1} + \cdots + a_h$. Then we have

$$r_n = O(n^h \log^2 n).$$

(This result generalizes and improves **265**.) [II **46**.]

Part Nine. Geometric Problems

§ 1. Some Geometric Problems

1. If we throw a heavy convex polyhedron with arbitrary interior mass distribution onto a horizontal floor then it will come to rest in a stable position on one of its faces. That is there exists for an arbitrary point P lying in the interior of the convex polyhedron one face F (at least) with the following property: The perpendicular dropped from P onto the plane in which F lies has its foot in the *interior* of the face F. Give a purely geometrical proof free from mechanical considerations for the existence of the face F.

1.1. The intersection of a plane π with a circular cone is an ellipse. Let S denote the area of the part of the conical surface between the ellipse and the vertex of the cone, p and q the distances of the vertex from the two endpoints of the major axis of the ellipse, and finally α the angle included between the axis of the cone and any generator. Prove that

$$S=\pi\frac{p+q}{2}\sqrt{pq}\sin\alpha.$$

1.2. Two spheres, one of radius r, the other of radius R, are externally tangent to each other. We denote by B the smallest convex solid containing both spheres (a tight bag containing two balls). Let V stand for the volume of B, S for the area of its surface, and M for its Minkowski constant (surface integral of the mean curvature). Prove that

$$V=\frac{4\pi}{3}\frac{R^5-r^5}{R^2-r^2},$$

$$S=4\pi\frac{R^4-r^4}{R^2-r^2},$$

$$M=4\pi\frac{R^3-r^3}{R^2-r^2}.$$

1.3. A lampshade has the shape of a frustum of a right circular cone. Its perimeter is P at the bottom, p at the top, its slant height is s, and the inclination i of the slant height to the bottom plane is $\geqq 60°$; $P\geqq p$. Show that such a lampshade can be cut out in one piece from a rectangular sheet of paper of dimensions P and $s+[p(P-p)]/8s$. There will even be paper left over for a flap to glue the ends together, except in the extreme case where $P=p$, $i=\pi/2$, and no paper will be wasted.

1.4. The numbers a and b are positive and $f(x)$ is a non-linear function such that

$$f(0)=0, \quad f(a)=b \qquad \text{and} \qquad f(x)\geqq 0, \quad f''(x)\geqq 0$$

in the interval $[0, a]$. Prove that

$$2\pi\int_0^a f(x)[1+(f'(x))^2]^{1/2}\,dx < \pi b(a^2+b^2)^{1/2}.$$

(The inequality becomes intuitively obvious when both sides are interpreted as areas of surfaces of revolution.)

2. If on a circular field a given quantity of wheat is piled up, then the maximum slope of the resulting heap is smallest when it forms a right circular cone. Prove the analytic formulation of this fact, i.e. the following theorem:

Let the function $f(x, y)$ possess both the partial derivatives

$$\frac{\partial f}{\partial x}=f_x'(x, y), \qquad \frac{\partial f}{\partial y}=f_y'(x, y).$$

If $f(x, y)=0$ on the boundary of the unit circle $x^2+y^2=1$, then there exists a point ξ, η in the unit circle such that

$$\sqrt{[f_x'(\xi, \eta)]^2+[f_y'(\xi, \eta)]^2} > \frac{3}{\pi}\iint f(x, y)\,dx\,dy,$$

where the double integral is extended over the unit circle.

3. If a point mass traverses a unit length in unit time moving from rest to rest, then at some point between the two rest points it must experience an acceleration of magnitude greater than 4. Demonstrate this fact analytically.

4. If a curve viewed from the point O appears everywhere convex, then it subtends a visual angle $<180°$; if it appears everywhere concave and extends to infinity it subtends a visual angle $>180°$.

Let O be the origin of a system of polar coordinates r, φ and $r=1/f(\varphi)$ the equation of the curve, where $f(x)$ is twice differentiable. In calculating the radius of curvature we are led to the following two theorems:

(1) If the function $f(x)>0$ in the interval $a\leqq x\leqq a+\pi$, then there exists a point ξ, $a<\xi<a+\pi$, such that

$$f(\xi)+f''(\xi)>0.$$

(2) If $f(x)>0$, $f(x)+f''(x)>0$ for $a<x<b$ and $f(a)=f(b)=0$, then $b-a>\pi$.

Prove these theorems by purely analytical methods.

Let $\mathfrak{C}$ be a closed convex curve with continuous curvature in the plane. Let the *support function* of $\mathfrak{C}$ (i.e. the support function of the convex domain $\mathfrak{D}$ enclosed by $\mathfrak{C}$, cf. III preceding **112**) be denoted by $h(\varphi)$ and let the radius of curvature of $\mathfrak{C}$ at the point where the direction of the tangent is $\varphi+\pi/2$ be denoted by $r(\varphi)$. The function $h(\varphi)$ is twice continuously differentiable and $r(\varphi)$ is continuous. We have

$$r(\varphi)=h(\varphi)+h''(\varphi).$$

For the area a of $\mathfrak{C}$ (i.e. the area of $\mathfrak{D}$) and for the length l of $\mathfrak{C}$ we have the formulae:

$$a=\frac{1}{2}\int_0^{2\pi} h(\varphi)r(\varphi)\,d\varphi, \qquad l=\int_0^{2\pi} h(\varphi)\,d\varphi.$$

5. On a closed convex curve with continuous curvature there are always at least three different pairs of points with the following property: The tangents at the two points of a pair are parallel and the curvature of the curve at the two points is the same.

6. If a convex curve of circumference l and of area a can roll unimpeded in the interior of another convex curve of circumference L and area A, then we have

$$lL\leqq 2\pi(a+A).$$

In the x, y, z space let $\mathfrak{S}$ be a closed, convex surface with continuous curvature. The expression $x\cos\alpha+y\cos\beta+z\cos\gamma$ where $\cos^2\alpha+\cos^2\beta+\cos^2\gamma=1$, has in the convex domain $\mathfrak{D}$ enclosed by $\mathfrak{S}$ a precisely determined maximum $h(\alpha,\beta,\gamma)$. The quantity $h(\alpha,\beta,\gamma)$ is a function of the direction α, β, γ and is called the *support function* of $\mathfrak{D}$. The plane

$$x\cos\alpha+y\cos\beta+z\cos\gamma-h(\alpha,\beta,\gamma)=0$$

is a tangent plane (*plane of support*) of the surface $\mathfrak{S}$, and in fact that tangent plane whose normal outward from $\mathfrak{S}$ has direction cosines $\cos\alpha$, $\cos\beta$, $\cos\gamma$. Let r and r' be the two principal radii of curvature at the point of tangency. We may regard the quantities h, r, r' as functions of a variable point ω on the unit sphere. They are continuous functions of ω. If V, S and M denote the volume, the surface area and the integral of the mean curvature of $\mathfrak{S}$, respectively, then we have the formulae

$$V=\frac{1}{3}\iint hrr'\,d\omega,$$
$$S=\iint rr'\,d\omega=\frac{1}{2}\iint h(r+r')\,d\omega,$$
$$M=\frac{1}{2}\iint (r^{-1}+r'^{-1})rr'\,d\omega=\iint h\,d\omega,$$

where the integration is extended over the unit sphere whose element of area is denoted by $d\omega$. M is customarily termed the *Minkowski constant* of $\mathfrak{S}$, or of the solid enclosed by $\mathfrak{S}$.

7. Denote the volume, the surface area and the integral of the mean curvature of two convex surfaces by v, s, m and V, S, M respectively. If the first body can roll unimpeded in the interior of the second, then we have the inequalities

$$mS+sM\leqq 12\pi(V+v),$$
$$mS-sM\leqq 4\pi(V-v).$$

8. Every closed surface is in equilibrium under the action of uniform pressure applied from all sides. In order to show this, prove that the following formulae hold for every closed surface:

$$\iint\cos\alpha\,dS=0, \qquad \iint\cos\beta\,dS=0, \qquad \iint\cos\gamma\,dS=0,$$
$$\iint(y\cos\gamma-z\cos\beta)\,dS=0, \qquad \iint(z\cos\alpha-x\cos\gamma)\,dS=0,$$
$$\iint(x\cos\beta-y\cos\alpha)\,dS=0.$$

Here $\cos\alpha$, $\cos\beta$, $\cos\gamma$ are the direction cosines of the outward normal and dS is the element of surface area.

9. If on every element dS of a closed, rigid surface with continuous curvature there acts a force of magnitude $\frac{1}{2}(1/R_1+1/R_2)dS$ in the direction of the inward normal, where R_1 and R_2 are the two principal radii of curvature, then the surface is in equilibrium.

10 (continued)**.** If the magnitude of the force acting on dS in the direction of the inward normal is $(1/R_1R_2)\,dS$ (i.e. proportional to the Gaussian curvature rather than to the mean curvature as in the preceding problem), then the surface is also in equilibrium.

11. At every edge e of a rigid polyhedron there acts a force F. The point of application of F is the mid-point of e. F is perpendicular to e and lies in the plane that bisects the interior angle included by the two faces of the polyhedron meeting in e. The magnitude of F is $l|\cos(\alpha/2)|$, where l denotes the length of e and α the interior angle at the edge e considered above. Finally F is directed inwards or outwards according as the edge e is convex or re-entrant, i.e. according to the sign of $\cos(\alpha/2)$. Under the system of all the forces F the polyhedron is in equilibrium.

12. If at every element ds of a closed, rigid space curve with continuous curvature there acts a force in the direction of the positive principal normal of magnitude ds/r, where r denotes the radius of curvature, then the curve is in equilibrium.

13. Imagine a unit vector drawn from the origin of a coordinate system in the direction of the tangent at each point of a continuously differentiable space curve. The end-points of these unit vectors generate the *spherical image* of the space curve. If the space curve is closed, then its spherical image will be intersected by every great circle of the unit sphere.

14. Let the function $f(x)$ be defined in the interval $a\leqq x\leqq b$ and satisfy the following conditions:

(1) $f(x)$ is positive in the interior and $=0$ at the end-points of the interval a, b;

(2) $f(x)$ is twice continuously differentiable for $a<x<b$;

(3) $f'(x)$ converges to $+\infty$ as $x\to a$, and to $-\infty$ as $x\to b$.

Under these conditions the function

$$\frac{f''(x)}{f(x)(1+[f'(x)]^2)^2}=F(x)$$

cannot be monotonic in the whole interval $a<x<b$, except in the case that

$$f(x)=\sqrt{(x-a)(b-x)},\qquad F(x)=-\left(\frac{2}{b-a}\right)^2.$$

15. On a surface of revolution with continuous curvature there are always two different parallel circles with the same Gaussian curvature.

16. Let a, b, c be the sides, and α, β, γ the angles of a triangle in radian measure, a lying opposite to α, b to β and c to γ. Then we have

$$\frac{\pi}{3}\leqq\frac{a\alpha+b\beta+c\gamma}{a+b+c}\leqq\frac{\pi}{2}.$$

Equality holds in the first inequality if and only if the triangle is equilateral, and in the second inequality if and only if the triangle degenerates into a straight line counted twice.

17. Let $k_1, k_2, \ldots, k_6$ be the lengths of the six edges of a tetrahedron and $\alpha_1, \alpha_2, \ldots, \alpha_6$ the radian measures of the angles formed by the faces of the tetrahedron that meet at the respective edges. Then we have

$$\frac{\pi}{3} < \frac{k_1\alpha_1 + k_2\alpha_2 + \cdots + k_6\alpha_6}{k_1 + k_2 + \cdots + k_6} < \frac{\pi}{2}.$$

The given bounds are the best possible.

17.1. Consider a triangle. Let a, b and c denote its sides and A its area. Prove that

$$A \leqq \frac{\sqrt{3}}{4}(abc)^{2/3}$$

with equality only for equilateral triangles. [The shape of a triangle is specified by two parameters.]

17.2 (continued). If $p > 0$

$$A \leqq \frac{\sqrt{3}}{4}\left(\frac{a^p + b^p + c^p}{3}\right)^{\frac{2}{p}},$$

with equality only for equilateral triangles.

17.3 (continued). Let r and R denote the radii of the inscribed and the circumscribed circle of a triangle, respectively. Prove that

$$2r \leqq R,$$

with equality only for equilateral triangles.

17.4. Formulate a proposition that has an analogous relation to **17.1** as **17** has to **16**.

17.5 (continued). Try to find simple particular cases for which the proposition formulated above can be conveniently tested.

18. Let $P_0P_1P_2$ be a triangle right-angled at P_2, P_2P_3 the perpendicular from P_2 to P_0P_1, similarly $P_3P_4 \perp P_1P_2$, $P_4P_5 \perp P_2P_3$, etc. Find the point $\lim_{n\to\infty} P_n$.

19. Let K and K' be two circles in space, each given as the intersection of a sphere and a plane

$$(K) \qquad x^2 + y^2 + z^2 + ax + by + cz + d = 0, \qquad Ax + By + Cz + D = 0,$$
$$(K') \qquad x^2 + y^2 + z^2 + a'x + b'y + c'z + d' = 0, \qquad A'x + B'y + C'z + D' = 0$$

Find a rational entire function of the 16 real coefficients $a, b, \ldots, d, A, \ldots, D, a', \ldots, d', A', \ldots, D'$, that is < 0 if and only if the two circles are *interlocked* i.e. their relative position is such that they have no common point, and, if conceived of as being made of wire, say, they cannot be taken apart without tearing.

20. Let X_1, X_2, X_3 be the projective coordinates of a variable point in the plane and let X_0 be defined by $X_0 + X_1 + X_2 + X_3 = 0$. In the triply infinite family of conics

$$\lambda_0 X_0^2 + \lambda_1 X_1^2 + \lambda_2 X_2^2 + \lambda_3 X_3^2 = 0$$

there are doubly infinitely many that degenerate to pairs of straight lines; and moreover every point of the plane is the singular point (intersection) of such a pair of straight lines. Find the integral curves of these pairs of straight lines. (At every point P of the required curve the tangent coincides with one straight line of the pair of straight lines whose point of intersection is P.)

21. Let the point P describe a plane curve. Let ρ be the radius of curvature at P. Let the segment of the normal at P lying between the two rectangular coordinate axes be of length ν. Determine those curves for which the ratio of ρ and ν is fixed:

$$\rho = 2n\nu.$$

The constant n may be supposed positive, since n is transformed into $-n$ if x and y are interchanged.—For integral n there is a particular solution in the form of a *rational* curve of order $2n$ that has a point of intersection of multiplicity $2n$ with the line at infinity. [Introduce as a parameter the angle τ between the x-axis and the tangent at P.]

22. The family of surfaces of second order

$$F(x_1, x_2, x_3, t) \equiv \frac{x_1^2}{a_1 - t} + \frac{x_2^2}{a_2 - t} + \frac{x_3^2}{a_3 - t} - t = 0 \qquad (0 < a_1 < a_2 < a_3)$$

with parameter t has an enveloping surface H of order 10 and class 4. Determine the lines of curvature of H. [Find the differential equation for one radius of curvature of H in terms of the parameter t.]

23 (continued). Deduce a parametric representation of H with the lines of curvature as parameter lines. The latter are algebraic curves of 12th order.

24 (continued). The central surface and the parallel surfaces of H may be regarded as envelopes of a suitably chosen family of surfaces of second order.

25. Let $F(x, y)$ be a continuous function, periodic with period 1 in the two variables x, y; i.e.

$$F(x+1, y) = F(x, y+1) = F(x, y).$$

Furthermore let $F(x, y)$ have the property (e.g. let it satisfy a *Lipschitz* condition) that through every point of the plane there passes one and only one infinitely continuable integral curve of the differential equation

$$\frac{dy}{dx} = F(x, y).$$

Then there exists a number ω associated with the function $F(x, y)$ such that for every integral curve $y = f(x)$ the difference $f(x) - \omega x$ is bounded for all values of x.

Appendix

§ 1. Additional Problems to Part One

A I **9.1.** In how many different ways can you pay the sum of n cents with l different kinds of coins of value $a_1, a_2, \ldots, a_l$ cents respectively, using just k coins?

[We are required to find $A_{n,k}$, the number of those solutions of the Diophantine equation

$$a_1x_1+a_2x_2+\cdots+a_lx_l=n$$

in non-negative integers $x_1, x_2, \ldots, x_l$, for which

$$x_1+x_2+\cdots+x_l=k.$$

The function of the variables z and q

$$\frac{1}{(1-q^{a_1}z)(1-q^{a_2}z)\ldots(1-q^{a_l}z)},$$

which for $z=1$, $q=\zeta$ solves the change problem as presented in I **9**, may be useful.]

A I **9.2.** Generalizing A I **9.1**, find $A_{n_1n_2\ldots n_k}$, the number of solutions in non-negative integers $x_1, x_2, \ldots, x_l$ of the following system of k Diophantine equations:

$$\begin{aligned} a_{11}x_1+a_{12}x_2+\cdots+a_{1l}x_l&=n_1,\\ a_{21}x_1+a_{22}x_2+\cdots+a_{2l}x_l&=n_2,\\ \cdots\cdots\cdots\cdots\cdots&\cdots\\ a_{k1}x_1+a_{k2}x_2+\cdots+a_{kl}x_l&=n_k; \end{aligned}$$

$a_{\kappa\lambda}$ and n_κ ($\kappa=1, 2, \ldots, k$; $\lambda=1, 2, \ldots, l$) are given non-negative integers.

A I **20.1.** Define $p(n)$ (the number of *partitions* of n) as the number of different decompositions of the positive integer n into a sum of positive integers. E.g.

$$\begin{aligned} 5=4+1=3+2 \quad &=2+2+1 \quad =1+1+1+1+1\\ =3+1+1&=2+1+1+1 \end{aligned}$$

and so $p(5)=7$. It is convenient to also define $p(0)=1$.

Compute $p(n)$ for $0\leqq n\leqq 10$.

A I **20.2.** Prove

$$\sum_{n=0}^{\infty} p(n)x^n=\frac{1}{(1-x)(1-x^2)(1-x^3)\ldots}.$$

A I **54.1.**

$$\left\{\prod_{n=1}^{\infty}(1-q^n)\right\}^3=\sum_{n=0}^{\infty}(-1)^n(2n+1)q^{n(n+1)/2}.$$

A I **60.12.** There is another approach to problem I **60.10.**

As in I **60.11**, cut the area under the zig-zag path into r vertical strips of width 1 of which

$$x_0,\ x_1,\ x_2,\ \ldots,\ x_s$$

are of height

$$0,\ 1,\ 2,\ \ldots,\ s,$$

respectively; $s=n-r$. Then the area is

$$\begin{aligned} 0x_0+1x_1+2x_2+\cdots+sx_s&=\alpha,\\ x_0+\ x_1+\ x_2+\cdots+\ x_s&=r. \end{aligned}$$

Find $c_{r+s,r,\alpha}$, the number of non-negative integral solutions of this system of two Diophantine equations. [A I **9.1**, cf. I **51**.]

A I **191.1.** If n is an integer, $n>1$, then

$$0!S_1^n-1!S_2^n+2!S_3^n-\cdots+(-1)^{n-1}(n-1)!S_n^n=0.$$

A I **191.2.** If n is a positive integer, then

$$1!S_1^n-2!S_2^n+3!S_3^n-\cdots+(-1)^{n-1}n!S_n^n=(-1)^{n-1}.$$

A I **203.1.** There are $4n$ players who wish to play bridge at n tables. Each player must have another player as partner and each pair of partners must have another pair as opponents. Show that the choice of partners and opponents can be made in exactly

$$\frac{(4n)!}{n!8^n}$$

different ways (3, 315, 155925, ... ways when $n=1, 2, 3, \ldots$; always an odd number).

Solutions

Part Four. Functions of One Complex Variable Special Part

1. [For **1–76** cf. A. Wiman: Acta Math. **37**, pp. 305–326 (1914). There is additional material in G. Valiron: General Theory of Integral Functions. Toulouse: É. Privat 1923 and in R. P. Boas: Entire Functions. New York: Academic Press 1954. Cf. also references given for I **110**.] We are concerned with the maximum of the sequence

$$1,\quad \frac{r}{1},\quad \frac{r}{1}\frac{r}{2},\quad \frac{r}{1}\frac{r}{2}\frac{r}{3},\quad \dots,\quad \frac{r}{1}\frac{r}{2}\cdots\frac{r}{n},\quad \dots$$

The factors $r/1, r/2, \dots, r/n, \dots$ are decreasing from left to right. If

$$\frac{r}{n} \geqq 1 > \frac{r}{n+1},\quad \text{i.e.}\quad n \leqq r < n+1,$$

then the nth term $r^n/n!$ is the maximum term. Hence

$$\mu(r) = \frac{r^{[r]}}{[r]!},\qquad \nu(r) = [r].$$

2. Since all the coefficients are positive, $M(r) = e^r$; $N(r) = 0$.

3. $\mu(r) = r^n/(2n+1)!$ for $2n(2n+1) \leqq r \leqq (2n+2)(2n+3)$,

$$\nu(r) = \left[\frac{\sqrt{1+4r}-1}{4}\right] \sim \frac{\sqrt{r}}{2}.$$

4. Since the coefficients have alternating signs, the maximum modulus is attained for negative z.

$$M(r) = \frac{e^{\sqrt{r}} - e^{-\sqrt{r}}}{2\sqrt{r}},\qquad N(r) = \left[\frac{\sqrt{r}}{\pi}\right] \sim \frac{\sqrt{r}}{\pi}.$$

5. $\nu(r) = 0$.

6. $N(r) = 0$.

7. $|a_n| r^n$ is the maximum term when r exceeds each of the numbers $|a_0/a_n|^{\frac{1}{n}}$, $|a_1/a_n|^{\frac{1}{n-1}}, \dots, |a_{n-1}/a_n|$.

8. By the fundamental theorem of algebra and since the absolute value of the polynomial $\sim |a_n||z|^n$ as $|z| \to \infty$.

9. From I **119**, I **120** it follows that the second limit $= \infty$. From $\mu(r) \geqq |a_n| r^n$ it follows that if $a_n \neq 0$

$$\liminf_{r\to\infty} \frac{\log \mu(r)}{\log r} \geqq n, \quad \text{i.e.} \quad \lim_{r\to\infty} \frac{\log \mu(r)}{\log r} = \infty.$$

10. The first part of the assertion is proved in a similar way as the first part of **9**. For the second part cf. for example **2**.

11. $\mu_k(r) = \mu_1(r^k)$, $\nu_k(r) = k\nu_1(r^k)$. The exponent nk in the term $a_n z^{nk}$ is of course taken as the value of $\nu_k(r)$ when that term has the maximum absolute value.

12. $M_k(r) = M_1(r^k)$, $N_k(r) = kN_1(r^k)$.

13. (1) For $(2n-1)2n \leqq r^2 < (2n+1)(2n+2)$ we have

$$\nu(r) = 2n \sim r, \qquad \mu(r) = \frac{r^{2n}}{(2n)!},$$

$$\frac{\nu(r)}{\log \mu(r)} = -\frac{2n}{r} \frac{1}{\frac{1}{r}\left(\log \frac{1}{r} + \log \frac{2}{r} + \cdots + \log \frac{2n}{r}\right)} \sim -\frac{1}{\int_0^1 \log x \, dx} = 1.$$

(2) For $\frac{(2n-1)2n}{4} \leqq r^2 < \frac{(2n+1)(2n+2)}{4}$ we have

$$\nu(r) = 2n \sim 2r, \qquad \mu(r) = \frac{1}{2} \frac{(2r)^{2n}}{(2n)!},$$

$$\frac{\nu(r)}{\log \mu(r)} = -\frac{2n}{2r} \frac{1}{\frac{1}{2r}\left(\log \frac{1}{2r} + \log \frac{2}{2r} + \cdots + \log \frac{2n}{2r}\right) + \frac{\log 2}{2r}} \sim -\frac{1}{\int_0^1 \log x \, dx} = 1.$$

One would like to have a theorem which contains this example and, on the analogy of **14**, reads somewhat as follows: "If $f(z)$ is an entire function and if (in contrast to **11**) $\mu_k(r)$ and $\nu_k(r)$ denote the maximal term and the central index of the expansion of $(f(z))^k$ in ascending powers of z, $k = 1, 2, 3, \ldots$, respectively, then the limit

$$\lim_{r\to\infty} \frac{\nu_k(r)}{\log \mu_k(r)}$$

is independent of k." This theorem is in general not true (the limit need not even exist) but it does hold under special assumptions, e.g. under the assumption of **67** [**68**]. Cf. also **59**, **60**.

14. $M_k(r) = (M_1(r))^k$, $N_k(r) = kN_1(r)$.

15. [**19**.]

16. $N(r)$ is a counting function, cf. II Ch. 4, § 1.

17. Let $\nu(r_1) = l \geqq 1$. Then

$$\mu(r_2) \geqq |a_l| r_2^l \geqq |a_l| r_1^{l-1} r_2 = \mu(r_1) \frac{r_2}{r_1}.$$

18. [Solution III **280**.]

19. Since a maximum term exists, *there exist* points which lie in every half-plane $\eta \geqq n\xi + \log |a_n|$, $n = 0, 1, 2, \ldots$. Their aggregate $\mathfrak{G}$ extends to infinity and, being the intersection of infinitely many convex regions (half-planes), it is itself convex. The boundary of $\mathfrak{G}$ from below forms a polygon whose equation is $\eta = \log \mu(e^\xi)$. The derivative *from the right*

$$\frac{d \log \mu(e^\xi)}{d\xi}$$

exists also at the corners and always $=\nu(e^\xi)$. This derivative is piece-wise constant and, because of the convexity, it is non-decreasing. From this **15** follows.

20. [III **304**.] The exceptional case arises when there is just one term different from 0 in the power series [III **305**].

21. Let $0<r<r'$. The two pairs of points

$$(\log \alpha r, \log \mu(\alpha r)), \qquad (\log r', \log \mu(r'))$$

and

$$(\log r, \log \mu(r)), \qquad (\log \alpha r', \log \mu(\alpha r'))$$

both lie on the boundary of $\mathfrak{G}$ [solution **19**] and the second pair, in fact, lies between the first pair. The centers of the connecting lines (secants) have the same abscissa

$$\frac{\log \alpha r + \log r'}{2} = \frac{\log r + \log \alpha r'}{2};$$

for the ordinates we have because of the convexity [**19**]

$$\frac{\log \mu(\alpha r) + \log \mu(r')}{2} \geqq \frac{\log \mu(r) + \log \mu(\alpha r')}{2}.$$

22. From **20**, as **21** from **19**. Note the exception.

23. For polynomials the assertion is obvious [solution **7**]. For power series with an infinite radius of convergence $\nu(r)$ must tend to infinity, since if $m<n$ the term $|a_m|r^m$ is exceeded by $|a_n|r^n$ for $r>(|a_m|/|a_n|)^{\frac{1}{n-m}}$. If $\mu(\alpha r)=|a_m|(\alpha r)^m$ then $\mu(r)\geqq |a_m|r^m$ and hence $\mu(\alpha r)/\mu(r)\leqq \alpha^m$; $m\to\infty$ as r increases.

24. The limit in question exists and is <1 for every entire function which is not constant [**22**]. If it is positive and equal to α^k, where $k(k>0)$ is chosen suitably, then

$$\frac{M(\alpha^{-n+1})}{M(\alpha^{-n})} \geqq \alpha^k, \quad \text{hence} \quad M(\alpha^{-n}) \leqq \alpha^{-nk} M(1), \qquad n=1, 2, 3, \ldots,$$

Hence it follows that the entire function is a polynomial of degree $\leqq k$ [**8**, **10**]. For polynomials the assertion is obvious.

25. Among the numbers $\rho_1, \rho_2, \ldots$ there are $\nu(0)$ which vanish, and hence none if $\nu(0)=0$.—If ξ_0 is the abscissa of a corner of the polygon forming the boundary of $\mathfrak{G}$ [solution **19**] at which the two sides with direction coefficients m, n $(m<n)$ meet, then $\rho_{m+1}=\rho_{m+2}=\cdots=\rho_{n-1}=\rho_n=e^{\xi_0}$. Between the corner in question and the adjoining corner to the right we have $|a_n|r^n$ as the greatest term.

26. Among the numbers $r_1, r_2, \ldots$ there are $N(0)$ which vanish, and hence none if $N(0)=0$.—Cf. **16**.

27.

$$\rho_n = n; \qquad \rho_n = 2n(2n+1); \qquad \rho_{2n-1} = \rho_{2n} = \sqrt{(2n-1)2n}.$$

28.

$$w_1 = -\frac{\pi i}{2}, \qquad w_2 = \frac{3\pi i}{2}, \qquad w_3 = -\frac{5\pi i}{2}, \ldots, \qquad r_n = (2n-1)\frac{\pi}{2};$$

$$w_n = r_n = n^2\pi^2; \qquad r_{2n-1} = r_{2n} = (2n-1)\frac{\pi}{2}.$$

29. $\rho_n \leqq R$, $n = 1, 2, 3, \ldots$. In the interval $[0, r]$, $r < R$, we have a finite number, namely $\nu(r)$, members of the sequence $\rho_1, \rho_2, \rho_3, \ldots$, which can hence have a point of accumulation only at the right-hand endpoint of the interval $[0, R]$.

30. The zeros cannot have a point of accumulation in the interior of the circle of convergence.

31. The proof is by mathematical induction on the intervals into which $\rho_1, \rho_2, \ldots$ divide the real axis. For $0 \leqq r < \rho_1$ we have $\mu(r) = |a_0|$. Let $0 \leqq m < n$, and let $|a_n| r^n$ succeed $|a_m| r^m$ as the maximum term. Assume that

$$\mu(r) = \frac{|a_0| r^m}{\rho_1 \rho_2 \cdots \rho_m} = |a_m| r^m \quad \text{for} \quad \rho_m \leqq r < \rho_{m+1},$$

and hence that $|a_m| = |a_0|/(\rho_1 \rho_2 \ldots \rho_m)$. $|a_n| r^n$ becomes the maximum term at the point $r = \rho_{m+1} = \rho_{m+2} = \cdots = \rho_{n-1} = \rho_n$ [**25**]. Hence

$$\mu(\rho_{m+1}) = |a_m| \rho_{m+1}^m = |a_n| \rho_n^n = |a_n| \rho_{m+1}^m \rho_{m+1} \rho_{m+2} \cdots \rho_n,$$

$$|a_n| = \frac{|a_m|}{\rho_{m+1} \rho_{m+2} \cdots \rho_n} = \frac{|a_0|}{\rho_1 \rho_2 \cdots \rho_m \rho_{m+1} \cdots \rho_n},$$

and $\mu(r) = |a_n| r^n$ for $\rho_n \leqq r < \rho_{n+1}$, q.e.d.

32. *Jensen's* inequality [III **120**]. Equality is not attained for any entire function differing from a constant.

33. By careful integration from

$$r \frac{d \log \mu(r)}{dr} = \nu(r),$$

which was proved in solution **19**.

34. Equivalent to **32**, since by II **147**

$$\int_0^r \frac{N(t)}{t} \, dt = N(r) \log r - \int_0^r \log r \, dN(r).$$

35. Denote the limits on the left and on the right by α and β, respectively. Consider first α finite. Let $\varepsilon > 0$; for r sufficiently large we have $\nu(r) < r^{\alpha+\varepsilon}$ and $|a_{\nu(r)}| < 1$ and hence

$$\mu(r) < r^{\nu(r)} < r^{r^{\alpha+\varepsilon}}, \qquad \frac{\log \log \mu(r)}{\log r} < \alpha + \varepsilon + \frac{\log \log r}{\log r}.$$

From this it follows that $\beta \leqq \alpha$. On the other hand [**33**]

$$\int_r^{2r} \frac{\nu(t)}{t} \, dt = \log \mu(2r) - \log \mu(r).$$

For r sufficiently large $\mu(r)>1$, hence $\nu(r)\log 2<\log\mu(2r)$, i.e. $\alpha\leqq\beta$. If α is infinite the first half of the proof is obvious.

36. Cf. the second half of the proof of **35**. [**34**.]

37. Assume $a_0=1$. We find [**33**, II **147**]

$$-r^{-k}\nu(r)+\sum_{0<\rho_\nu\leqq r}\rho_\nu^{-k}=k\int_0^r t^{-k-1}\nu(t)\,dt=k\int_0^r t^{-k}\,d\log\mu(t)$$
$$=kr^{-k}\log\mu(r)+k^2\int_0^r t^{-k-1}\log\mu(t)\,dt.$$

If $\int_1^\infty t^{-k-1}\log\mu(t)\,dt$ is convergent, then we have $\lim_{r\to\infty}r^{-k}\log\mu(r)=0$ [II **113**] and hence $\sum_{\nu=1}^n(\rho_\nu^{-k}-\rho_n^{-k})$ is bounded [I **78**]. If $\sum_{n=1}^\infty\rho_n^{-k}$ is convergent the conclusion follows more readily.

38. [G. Valiron: Darboux Bull. Series 2, **45**, pp. 258–270 (1921).] Assuming $f(0)=1$ and with the notation of III **121** we obtain, as in **37**,

$$-r^{-k}N(r)+\sum_{0<r_\nu\leqq r}r_\nu^k=kr^{-k}\log\mathfrak{G}(r)+k^2\int_0^r t^{-k-1}\mathfrak{G}(t)\,dt.$$

From $\mathfrak{G}(r)\leqq M(r)$ one argues as in **37**.

39.

$$-(1-r)^{k+1}\nu(r)+\sum_{\rho_\nu\leqq r}(1-\rho_\nu)^{k+1}=(k+1)\int_0^r(1-t)^k\nu(t)\,dt\qquad\text{[II 147]}.$$

$$(1-r)^k\log\mu(r)+k\int_0^r(1-t)^{k-1}\log\mu(t)\,dt=\int_0^r(1-t)^kt^{-1}\nu(t)\,dt\qquad\text{[33]}.$$

Let $a_0=1$. The right-hand sides behave in the same manner for $r=1$. If the integral on the left in the second line exists then we have $\lim_{r\to1-0}(1-r)^k\log\mu(r)=0$ [II **112**] and $\sum_{\nu=1}^n[(1-\rho_\nu)^{k+1}-(1-\rho_n)^{k+1}]$ is bounded [I **78**].

40. [F. and R. Nevanlinna: Acta Soc. Sc. Fennicae, **50**, No. 5 (1922).] Related to **39** as **38** is to **37**.

41. [J. Hadamard: J. Math. Pures Appl., Series 4, **9**, p. 174 (1893).] The straight line through $(n,-\log|a_n|)$ with slope $\log r$ has the equation

$$\eta=-\log|a_n|+\log r(\xi-n)$$

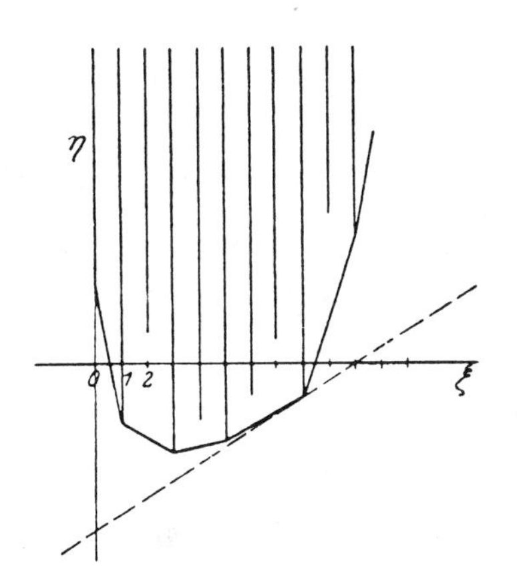

and cuts off a segment of length $-\log|a_n|r^n$ from the η-axis. Of all these straight lines the line of support has the lowest point of intersection. Hence the segment cut from the vertical axis by the supporting line $=-\log\mu(r)$, the abscissa of the corner in which the supporting line cuts $\mathfrak{K}$ is $\nu(r)$ (taking the corner furthest to the right if there are several), and the slope of the portion of the boundary of $\mathfrak{K}$ between the abscissae $n-1$ and n equals $\log\rho_n$. From the figure it can be seen that $\log\rho_{n+1}\geqq\log\rho_n$ and that, if $a_0\neq 0$,

$$-\log|a_n|=-\log|a_0|+\log\rho_1+\log\rho_2+\cdots+\log\rho_n,$$

if $(n, -\log|a_n|)$ lies on the boundary of $\mathfrak{K}$ [**31**]. If the supporting line rotates from slope $\log\rho_n$ to slope $\log\rho_{n+1}$ at the corner with abscissa n, then $\log\mu(\rho_{n+1})-\log\mu(\rho_n)=n(\log\rho_{n+1}-\log\rho_n)$, etc.

42. Taking for r the smallest number for which the inequalities

$$|a_0|\leqq|a_m|r^m,\qquad |a_1|r\leqq|a_m|r^m,\qquad \ldots,\qquad |a_{m-1}|r^{m-1}\leqq|a_m|r^m$$

are all simultaneously satisfied, we obtain $|a_m|r^m$ as the maximum term. This value of r is therefore ρ_m.—In figure **41** the slopes

$$\frac{-\log|a_n|+\log|a_0|}{n-0},\quad \frac{-\log|a_n|+\log|a_1|}{n-1},\quad \ldots,\quad \frac{-\log|a_n|+\log|a_{n-1}|}{n-(n-1)}$$

are all $\leqq$ the slope $\log\rho_n$.

43. If a_0 is to be the maximum term for a value of $r>0$ then $a_0\neq 0$. Then by the proof of **31** we have

$$|a_n|=\frac{|a_0|}{\rho_1\rho_2\cdots\rho_n},\qquad \left|\frac{a_n}{a_{n+1}}\right|=\rho_{n+1}\geqq\rho_n=\left|\frac{a_{n-1}}{a_n}\right|,\qquad n=1,2,3,\ldots.$$

In figure **41** the points $(n, -\log|a_n|)$ all lie on the boundary in the present case and $\dfrac{-\log|a_{n-1}|-\log|a_{n+1}|}{2}\geqq-\log|a_n|$; this expresses the convexity.

44. The curve $y=\alpha^{-1}x^\alpha$, $x>0$, is concave since $y''=(\alpha-1)x^{\alpha-2}<0$. From this the assertion concerning $\nu(r)$ follows [**43**]. The maximum of the expression $\alpha^{-1}x^\alpha+x\log r$, where we consider x as variable, $x>0$, and r as constant, is attained at

$$(*)\qquad x=(-\log r)^{-\frac{1}{1-\alpha}}$$

and equals $(\alpha^{-1}-1)x^\alpha=(\alpha^{-1}-1)(-\log r)^{-\frac{\alpha}{1-\alpha}}$. If the number (*) is an integer $=n$, then the nth term $=e^{(\alpha^{-1}-1)n^\alpha}$ and is the maximum term [a second proof of the assertion concerning $\nu(r)$] and the value given for $\mu(r)$ is attained exactly (it is otherwise somewhat too large). That it is asymptotically correct follows from

$$\lim_{n\to\infty}\frac{e^{(\alpha^{-1}-1)(n+1)^\alpha}}{e^{(\alpha^{-1}-1)n^\alpha}}=1.$$

45. Set $-\log r=\tau$ and compare the series in question with the integral

$$\int_0^\infty e^{\alpha-1}x^{\alpha-\tau x}\,dx \sim \sqrt{\frac{2\pi}{1-\alpha}}\,\tau^{-\frac{\alpha}{2(1-\alpha)}-1}\exp\left(\frac{1-\alpha}{\alpha}\tau^{-\frac{\alpha}{1-\alpha}}\right) \qquad \text{[II **208**]}.$$

Passing from the series to the integral introduces an error of the order of magnitude of the maximal term of the series [cf. II **8**]. The ratio of this to the magnitude of the integral is of order $\tau^{\frac{\alpha}{2(1-\alpha)}+1}$ [**44**], and this justifies the transition.

46. The curve $y=\alpha x\log x$ is convex since $y''=\alpha/x>0$. The maximum of $-\alpha x\log x+x\log r$ for variable x, $x>0$, and fixed r is attained for $\alpha(1+\log x)=\log r$. If the corresponding value of x is an integer n then the nth term is the maximum term. If the value of x in question is given by $x=n+t$, $0\leqq t\leqq 1$, then

$$r=(e(n+t))^\alpha, \qquad n^{-n\alpha}r^n=e^{n\alpha}(1+tn^{-1})^{n\alpha}\sim e^{\alpha(n+t)}=\exp\left(\alpha e^{-1}r^{\frac{1}{\alpha}}\right),$$

$$\rho_n=\frac{n^{n\alpha}}{(n-1)^{(n-1)\alpha}}\sim(ne)^\alpha \qquad \text{[**43**]}.$$

47. [**46**, **45**, II **209**.]

48. It is sufficient to investigate the series in question at a single point only, at $z=0$ say [III **251**]. Set $g(z)=\sum_{n=0}^\infty (a_n/n!)z^n$. Then we are concerned with the convergence of the series $\sum_{n=0}^\infty a_n$. We have

$$|a_n|<n!\,\frac{M(r)}{r^n}.$$

(a) Assume $l<1$, and hence $M(r)<Ae^{(l+\varepsilon)r}$, $A>0$, $\varepsilon>0$, $l+\varepsilon<1$. For $r=n/(l+\varepsilon)$ we obtain $|a_n|<An!(e/n)^n(l+\varepsilon)^n$, and hence [I **69**] that $\sum_{n=0}^\infty a_n$ is convergent.—The optimum value of r used above is obtained by differentiating $An!\,r^{-n}\,e^{(l+\varepsilon)r}$ with respect to r and *minimizing* it.

(b) Assume $l>1$. Then $\sum_{n=0}^\infty a_n$ cannot be convergent, for otherwise we should have $|a_n|<K$, and hence $|f(z)|<Ke^{|z|}$ in contradiction to $l>1$.

(c) Assume $l=1$. Then $\sum_{n=0}^\infty a_n$ may be either convergent or divergent. Choose for example a_n such that $a_n>0$, $\lim_{n\to\infty}\sqrt[n]{a_n}=1$ and $\sum_{n=0}^\infty a_n$ converges. Then $l\leqq 1[(b)]$, and also $M(r)>a_nr^n/n!$ for arbitrary n. For $n=[r]$ we obtain that $l\geqq 1$, and hence $l=1$. On the other hand choose a_n such that $\lim_{n\to\infty}a_n=0$, $\sum_{n=0}^\infty a_n$ diverges. Then $l\leqq 1[(b)]$, and also $l\geqq 1[(a)]$, hence $l=1$.

49. $N(r)$ is the number of lattice points in the circle $|z|\leqq r$; the primitive periods are $\omega_1=1$, $\omega_2=i$. If w is any period so is iw, hence

$$(*) \qquad \sigma(iu)=i\sigma(u).$$

Also

$$\sigma(u+1)=-e^{\eta_1(u+\frac{1}{2})}\sigma(u), \qquad \sigma(u+i)=-e^{\eta_2(u+\frac{1}{2}i)}\sigma(u).$$

In the first equation replace u by $iu-1$ and multiply the resulting equation by the second equation. Then

$$\sigma(iu)\sigma(u+i)=e^{(\eta_1 i+\eta_2)u+\frac{1}{2}(-\eta_1+i\eta_2)}\sigma[i(u+i)]\sigma(u).$$

Hence from (*) and then $\eta_1\omega_2-\eta_2\omega_1=2\pi i$ we obtain successively

$$\eta_1=i\eta_2=\pi,$$

and for arbitrary integers m_1, m_2

$$\sigma(u+m_1+im_2)=\pm\sigma(u)e^{\pi(m_1-im_2)u+\frac{\pi}{2}(m_1{}^2+m_2{}^2)}.$$

If we restrict $u(u\neq 0)$ to the parallelogram of the periods surrounding the point $u=0$, set $z=u+m_1+im_2$ and let $m_1^2+m_2^2$ tend to infinity we obtain

$$|z|^2\sim m_1^2+m_2^2, \qquad \log|\sigma(z)|\sim\frac{\pi}{2}|z|^2.$$

49.1. By substituting $-z$ for z we obtain from I **50**

$$(1+qz)(1+q^2z)(1+q^3z)\cdots=1+\sum_{n=1}^{\infty}\frac{q}{1-q}\frac{q^2}{1-q^2}\cdots\frac{q^n}{1-q^n}z^n.$$

Let q here be positive, $0<q<1$. Then

$$r_n=-w_n=\frac{1}{q^n}, \qquad \rho_n=\frac{1-q^n}{q^n}=r_n-1.$$

Now, by substituting $-q$ for q, we obtain

$$(1-qz)(1+q^2z)(1-q^3z)\cdots=1+\sum_{n=1}^{\infty}(-1)^{\left[\frac{n+1}{2}\right]}\frac{q}{1+q}\frac{q^2}{1-q^2}\cdots\frac{q^n}{1+(-1)^{n-1}q^n}z^n.$$

Let q here be positive, $0<q\leqq\frac{1}{2}$. Then

$$r_n=\frac{1}{q^n}=(-1)^{n-1}w_n, \qquad \rho_n=\frac{1+(-1)^{n-1}q^n}{q^n}=r_n+(-1)^{n-1}.$$

Observe that, for $n\geqq 2$,

$$\rho_n-\rho_{n-1}=\frac{1}{q^n}+(-1)^{n-1}-\left(\frac{1}{q^{n-1}}+(-1)^n\right)=\frac{1}{q^n}(1-q)+2(-1)^{n-1}$$
$$\geqq 4.2^{-1}-2=0.$$

50. In order to compute $\mu(r)$ for $\sin\sqrt{z}/\sqrt{z}$, note that if we associate the integer n with r as $r\to\infty$ by the double inequality

$$(*)\qquad 2n\sqrt{1+\frac{1}{2n}}\leqq\sqrt{r}<2n\sqrt{\left(1+\frac{1}{n}\right)\left(1+\frac{3}{2n}\right)}$$

and set $\sqrt{r}=2n+t$, then t remains bounded. In the interval (*) we have by **3**

$$\mu(r)=\frac{\sqrt{r}^{\,2n}}{(2n+1)\cdot(2n)!}\sim\frac{\sqrt{r}^{\,2n}}{2n\cdot\sqrt{2\pi}(2n)^{2n+\frac{1}{2}}e^{-2n}}$$
$$=\frac{1}{\sqrt{2\pi}}(2n)^{-\frac{3}{2}}e^{\sqrt{r}}\left(1+\frac{t}{2n}\right)^{2n}e^{-t};$$

$$\mu(r)\sim\frac{1}{\sqrt{2\pi}}r^{-\frac{3}{4}}e^{\sqrt{r}}, \qquad \log\mu(r)\sim\sqrt{r}.$$

	$\frac{\sin\sqrt{z}}{\sqrt{z}}$	$\cos z$	$(\cos z)^2$	e^z+i	e^z	$\sigma(z)$
$\frac{r_n}{\rho_n}$	$\frac{\pi^2}{4}$	$\frac{\pi}{2}$	$\frac{\pi}{2}$	π	—	—
$\frac{N(r)}{\nu(r)}$	$\frac{2}{\pi}$	$\frac{2}{\pi}$	$\frac{2}{\pi}$	$\frac{1}{\pi}$	0	—
$\frac{N(r)}{\log M(r)}$	$\frac{1}{\pi}$	$\frac{2}{\pi}$	$\frac{2}{\pi}$	$\frac{1}{\pi}$	0	2
$\frac{\nu(r)}{\log \mu(r)}$	$\frac{1}{2}$	1	1	1	1	—
$\frac{M(r)}{\sqrt{2\pi \log \mu(r)}\mu(r)}$	$\frac{1}{2}$	$\frac{1}{2}$	$\frac{1}{2}$	1	1	—
$\frac{\log N(r)}{\log r}$	$\frac{1}{2}$	1	1	1	—	2

The relations which can be observed in this table will be explained in part by the subsequent theorems.

51. $\mu(r) \leqq M(r)$ [definition]. If

$$\limsup_{r\to\infty} \frac{\log\log \mu(r)}{\log r} = \lambda, \qquad \lambda \text{ finite}, \quad \lambda < \beta,$$

then for ρ sufficiently large

$$|a_n|\rho^n \leqq \mu(\rho) < e^{\rho^\beta},$$

and hence [solution **48**, (a)] choosing $\rho = (n/\beta)^{\frac{1}{\beta}}$, we have for n sufficiently large

$$|a_n| < \left(\frac{e\beta}{n}\right)^{\frac{n}{\beta}}, \qquad |a_n| r^n < \left(\frac{e\beta r^\beta}{n}\right)^{\frac{n}{\beta}}.$$

Setting $2e\beta = k$, we have[1]

$$M(r) \leqq \sum_{n=0}^{kr^\beta} |a_n| r^n + \sum_{n=kr^\beta+1}^{\infty} |a_n| r^n < (kr^\beta + 1)\mu(r) + \sum_{n=kr^\beta}^{\infty} 2^{-\frac{n}{\beta}}.$$

[See R. P. Boas, Entire Functions, New York: Academic Press 1954, for more on the concept of the order of an entire function, as well as for other topics, newer

---

[1] For integral as well as for non-integral a and b, $a \leqslant b$, $\sum_{n=a}^{b} c_n$ is to be interpreted in this chapter as

$$\sum_{n=[a]}^{[b]} c_n = c_{[a]} + c_{[a]+1} + \cdots + c_{[b]-1} + c_{[b]}.$$

developments and bibliography.]

52. [**51**, **35**, II **149**, I **113**.]

53. If the order is λ, $\varepsilon>0$, then in accordance with solution **51** we have for r sufficiently large

$$rf'(r)=\left(\sum_{n=1}^{kr^{\lambda+\varepsilon}}+\sum_{n=kr^{\lambda+\varepsilon}+1}^{\infty}\right)na_nr^n<kr^{\lambda+\varepsilon}\sum_{n=0}^{kr^{\lambda+\varepsilon}}a_nr^n$$

$$+\sum_{n=kr^{\lambda+\varepsilon}+1}^{\infty}n\cdot 2^{-\frac{n}{\beta}}<kr^{\lambda+\varepsilon}f(r)+k',$$

where k, k' are constants, and hence

$$\limsup_{r\to\infty}\frac{\log(rf'(r)f(r)^{-1})}{\log r}\leqq\lambda+\varepsilon,$$

for ε arbitrarily close to 0. On the other hand, if for $r>r_0$ we have $rf'(r)f(r)^{-1}<r^\beta$, then by integration we obtain

$$\log f(r)<\frac{r^\beta}{\beta}+C$$

for $r>r_0$, where C is a constant.

54. If the order is λ and $\varepsilon>0$ then we have for r sufficiently large [solution **51**]

$$\mu(r)<M(r)<kr^{\lambda+\varepsilon}\mu(r).$$

55. Let σ be chosen so that $\sigma(l-1)>1$. For t sufficiently large $M(t)>t^\sigma$ [**10**], and hence

$$\mu(t)M(t)^{-l}\leqq M(t)^{1-l}<t^{-\sigma(l-1)}.$$

For the case $l=1$ consider for example the special function $f(z)=e^z$. The integral in III **156** is divergent in this case.

56. [A. Wiman, loc. cit. **1**.] For every $\varepsilon>0$ there exist power series $1+b_1r+b_2r^2+\cdots+b_nr^n+\cdots$ with finite radius of convergence and positive coefficients b_n such that, denoting the central index by n, we have

$$\sum_{\nu=0}^{\infty}\frac{b_\nu y^\nu}{b_n y^n}<(\log b_n y^n)^{\frac{1}{2}+\varepsilon},$$

for all positive variable y sufficiently close to the circle of convergence [**45**]. For an entire function $\sum_{n=0}^{\infty}a_nz^n$ (assuming $a_0=1$) we have at a point r, with which has been associated a positive number y in the sense of I **122**,

$$\frac{|a_\nu|r^\nu}{|a_n|r^n}\leqq\frac{b_\nu y^\nu}{b_n y^n}\quad\text{for}\quad \nu=0,1,2,\ldots.$$

In particular for $\nu=0$ we have $b_ny^n\leqq|a_n|r^n$. At such a point r we have the required inequality. We can assume $a_0=1$ without loss of generality.

57. [A. Wiman, loc. cit. **1**.] Let $\lambda<\beta<\lambda+\varepsilon=1/\alpha$. Then [**52**] for n sufficiently large we have $\log n<\beta\log\rho_n$ and hence

$$\rho_n\cdot n^{-n\alpha}(n-1)^{(n-1)\alpha}\sim\rho_ne^{-\alpha}n^{-\alpha}>n^{\frac{1}{\beta}-\alpha}e^{-\alpha}\to\infty.$$

The assertion now follows from I **118** and **47** in a similar fashion as **56** from I **122** and **45**.

58. Let

$$\limsup_{r\to\infty}\frac{\log N(r)}{\log r}=\limsup_{n\to\infty}\frac{\log\ n}{\log r_n}=\lambda \qquad \text{[II 149]}.$$

Then $\lambda\leqq 1$. By **51**, **36** it is clear that the order $\geqq\lambda$. That the order $\leqq\lambda$ is shown as follows: We have

$$M(r)\leqq|c|r^q\prod_{n=q+1}^{\infty}\left(1+\frac{r}{r_n}\right).$$

If $\lambda=1$ choose, for any given positive number ε, an integer N such that $N>q$, $\sum_{n=N+1}^{\infty}(1/r_n)<\varepsilon$. Then

$$M(r)<|c|r^q\prod_{n=q+1}^{N}\left(1+\frac{r}{r_n}\right)\cdot e^{\varepsilon r},\qquad \limsup_{r\to\infty}\frac{\log\log M(r)}{\log r}\leqq 1.$$

If $\lambda<1$ choose the positive number η so small that $\lambda+\eta<1$. For sufficiently large n we have [I **113**]: $1/r_n<1/n^{1/(\lambda+\eta)}$. Apply II **37** to the logarithm of the function $\prod_{n=1}^{\infty}\{1+[r/n^{1/(\lambda+\eta)}]\}$.

59. [For **59–63** cf. also G. Pólya: Math. Ann. **88**, pp. 177–183 (1923).] Assume without loss of generality that $f(0)=1$, also in **60** to **65**. We have [**31**, II **159**]

$$\frac{\log\mu(r)}{\nu(r)}=\frac{1}{\nu(r)}\sum_{\rho_n\leqq r}\log\frac{r}{\rho_n}\sim\int_0^1\log x^{-\frac{1}{\lambda}}\,dx.$$

Examples in **13**.

60. Similarly to **59**, making use of II **160** instead of II **159**.

61. $$\frac{1}{N(r)}\sum_{n=1}^{\infty}\log\left(1+\frac{re^{i\vartheta}}{r_n}\right)\sim\int_0^{\infty}\log\left(1+x^{-\frac{1}{\lambda}}e^{i\vartheta}\right)dx.$$

62. Similarly to **61**, making use of II **161** instead of II **159**:

$$\frac{\log M(r)}{N(r)}\leqq\frac{1}{N(r)}\sum_{n=1}^{\infty}\log\left(1+\frac{r}{r_n}\right).$$

63. According to **32** we have

$$\limsup_{r\to\infty}\frac{\log M(r)}{N(r)}\geqq\limsup_{r\to\infty}\frac{1}{N(r)}\sum_{r_n\leqq r}\log\frac{r}{r_n}\geqq\int_0^1\log x^{-\frac{1}{\lambda}}\,dx \qquad \text{[II 160]}.$$

64. According to III **121** we have

$$\left(\frac{\mathfrak{g}(r)}{\mathfrak{G}(r)}\right)^{\frac{1}{N(r)}}=e^{-\frac{1}{2}+\frac{1}{2N(r)}\sum\limits_{r_n\leqq r}\left(\frac{r_n}{r}\right)^2}.$$

By II **159** we have

$$\lim_{r\to\infty}\frac{1}{N(r)}\sum_{r_n\leqq r}\left(\frac{r_n}{r}\right)^2=\int_0^1 x^{\frac{2}{\lambda}}\,dx=\frac{\lambda}{\lambda+2}.$$

For polynomials $\sum_{r_n \leqq r} r_n^2$ remains bounded.

65. Similarly to **64**, making use of II **160** instead of II **159**.

66. Setting $f(z)=\sum_{n=0}^{\infty} a_n z^n$ we have [III **122**, III **130**]

$$\mathfrak{M}(r)=\sum_{n=0}^{\infty} |a_n|^2 r^{2n}, \qquad \mathfrak{m}(r)=\sum_{n=0}^{\infty} \frac{|a_n|^2 r^{2n}}{n+1}.$$

Set $g(z)=\sum_{n=0}^{\infty} [|a_n|^2/(n+1)]z^{2n+2}$; $g(z)$ has the same order λ as $f(z)$ since

$$\frac{r^2[\mu(r)]^2}{\nu(r)+1}<g(r)<r^2[M(r)]^2$$

[**35**, **51**]. Apply **53** to

$$\frac{\mathfrak{M}(r)}{\mathfrak{m}(r)}=\frac{rg'(r)}{2g(r)}.$$

67. (1) $$|a_n|<r^{-n}M(r)<r^{-n}e^{br^\alpha},$$

when $r>r_0$, where r_0 is independent of n. The right-hand side attains its minimum

$$\left(\frac{\alpha be}{n}\right)^{\frac{n}{\alpha}} \qquad \text{for} \qquad r=\left(\frac{n}{\alpha b}\right)^{\frac{1}{\alpha}}.$$

(2) We prove the assertion for arbitrary positive k and for ε sufficiently small. (This is sufficient!) By assumption we have

(*) $$e^{a(1-\varepsilon^3)r^\alpha}<M(r)<e^{a(1+\varepsilon^4)r^\alpha}$$

for r sufficiently large. From this and from [III **122**]

$$|a_0|^2+|a_1|^2r^2+|a_2|^2r^4+\cdots \leqq [M(r)]^2$$

we deduce [II **80**]

$$\left(\sum_{n=1}^{m} |a_n|r^n\right)^2 \leqq \sum_{n=1}^{m} 1 \sum_{n=1}^{m} |a_n|^2 r^{2n}<me^{2a(1+\varepsilon^4)r^\alpha},$$

(**) $$\sum_{n=1}^{m} |a_n|n^k r^n<m^{k+\frac{1}{2}}e^{a(1+\varepsilon^4)r^\alpha}.$$

We introduce the abbreviations

$$\sum_{1}^{\alpha ar^\alpha(1-\varepsilon)}=\sum\nolimits^{I}, \qquad \sum_{\alpha ar^\alpha(1+\varepsilon)}^{3\alpha ar^\alpha}=\sum\nolimits^{II}, \qquad \sum_{3\alpha ar^\alpha}^{\infty}=\sum\nolimits^{III}.$$

From this definition and from (**) for r sufficiently large, if $(6\alpha ar^\alpha)^{k+1/2}<\exp[a(\varepsilon^3-\varepsilon^4)r^\alpha]$ we have

$$\sum\nolimits^{I} |a_n|n^k r^n<e^{a(1+\varepsilon^3)r^\alpha}, \qquad \sum\nolimits^{II} |a_n|n^k r^n<e^{a(1+\varepsilon^3)r^\alpha}.$$

We could take the sum on the left from any lower limit $\geqq 0$ to any upper limit $\leqq 6\alpha ar^\alpha$. Therefore we can take the essential step of replacing r by $r(1-\lambda)$ and by $r(1+\lambda)$ (λ fixed, $0<\lambda<1$) on the right-hand side and under the summation sign on the left-hand side of the first and second inequalities, respectively, *without altering the limits of summation.* We deduce that

$$(1-\lambda)^{\alpha a r^\alpha(1-\varepsilon)} \sum^I |a_n| n^k r^n < e^{a(1+\varepsilon^3)(1-\lambda)^\alpha r^\alpha},$$
$$(1+\lambda)^{\alpha a r^\alpha(1+\varepsilon)} \sum^{II} |a_n| n^k r^n < e^{a(1+\varepsilon^3)(1+\lambda)^\alpha r^\alpha}.$$

Hence, with the aid of the first half of (*) (not used till now),

$$[M(r)]^{-1} \sum^I |a_n| n^k r^n < \exp\left[-a(1-\varepsilon^3)r^\alpha + a(1+\varepsilon^3)(1-\lambda)^\alpha r^\alpha - \alpha a r^\alpha(1-\varepsilon)\log(1-\lambda)\right],$$
$$[M(r)]^{-1} \sum^{II} |a_n| n^k r^n < \exp\left[-a(1-\varepsilon^3)r^\alpha + a(1+\varepsilon^3)(1+\lambda)^\alpha r^\alpha - \alpha a r^\alpha(1+\varepsilon)\log(1+\lambda)\right].$$

Expanding in ascending powers of ε and λ we obtain

$$-(1-\varepsilon^3)+(1+\varepsilon^3)(1\mp\lambda)^\alpha - \alpha(1\mp\varepsilon)\log(1\mp\lambda) = \frac{\alpha^2}{2}\lambda^2 - \alpha\lambda\varepsilon + \cdots$$
$$= -\frac{\varepsilon^2}{2} + \cdots < 0,$$

setting $\alpha\lambda=\varepsilon$ and choosing ε sufficiently small to make the expansion valid and also to ensure that its sign is determined by the term of lowest order.

This constitutes the proof as far as $\sum^I$ and $\sum^{II}$ are concerned. $\sum^{III}$ is of no account; for if $e<3h<3$, $3ah=be$ then, by (1),

$$\sum^{III} n^k |a_n| r^n < \sum^{III} n^k \left(\frac{\alpha b e r^\alpha}{n}\right)^{\frac{n}{\alpha}}$$
$$= \sum^{III} n^k \left(\frac{3\alpha a r^\alpha}{n} h\right)^n < \sum^{III} n^k h^n,$$

which is the remainder of a convergent series and hence tends to 0 as $r\to\infty$.

68. Assume (1). Then the function is of finite order [**51**] and we deduce (2) [**54**]. Also (3) must hold. For, if $\nu(r)$ for arbitrarily large r were outside the bounds $\alpha a r^\alpha(1-\varepsilon)$ and $\alpha a r^\alpha(1+\varepsilon)$ then for these values of r we would also have $\log\mu(r) < \log M(r) - \delta r^\alpha$ [**67**], in contradiction to (2).

Assume (2). Then the function is of finite order [**51**], and hence we deduce (1) [**54**].

Assume (3). Take $a_0 \neq 0$ (without loss of generality) and deduce (2) with the help of **33**.

69. (1) a_n is decreasing. If $0<\varepsilon<1$ and

$$m-1 \leqq \alpha a r^\alpha(1-\varepsilon) < m < n < \alpha a r^\alpha(1+\varepsilon) \leqq n+1,$$

then by **67** (2) we have for r sufficiently large

$$0 < f(r) - \sum_{\nu=m}^{n} a_\nu r^\nu < \varepsilon f(r),$$

$$a_m r^n \cdot 2\varepsilon\alpha a r^\alpha > f(r)(1-\varepsilon), \qquad a_n r^m \cdot 2\varepsilon\alpha a r^\alpha < f(r)(1+\varepsilon).$$

Hence, noting that $\log r \sim (1/\alpha) \log m \sim (1/\alpha) \log n$ and $m/(1-\varepsilon) \sim n/(1+\varepsilon)$, we deduce

$$\liminf_{m\to\infty} \frac{\log a_m}{m \log m} \geqq -\frac{1}{\alpha}\frac{1+\varepsilon}{1-\varepsilon}, \qquad \limsup_{n\to\infty} \frac{\log a_n}{n \log n} \leqq -\frac{1}{\alpha}\frac{1-\varepsilon}{1+\varepsilon}.$$

(2) If $a_1/a_0 \geqq a_2/a_1 \geqq a_3/a_2 \geqq \cdots$, then $a_n r^n$ becomes the *maximum term* for a suitable value of r [**43**]. It follows that we may replace the number n by $\nu(r)$ in the expression under consideration, i.e. in $\log n^{\frac{1}{\alpha}} a_n^{\frac{1}{n}} = (1/\alpha) \log n + (1/n) \log a_n$. Since $a_{\nu(r)} r^{\nu(r)} = \mu(r)$ we have by **68**

$$\frac{1}{\alpha} \log \nu(r) + \frac{1}{\nu(r)} \log a_{\nu(r)} = \frac{1}{\alpha} \log \frac{\nu(r)}{r^\alpha} + \frac{\log \mu(r)}{\nu(r)} \to \frac{1}{\alpha} \log \alpha a + \frac{1}{\alpha}.$$

70. In $\sum_{n=0}^{\infty} a_n r^n$ as well as in $\sum_{n=1}^{\infty} n^k a_n r^n$ the center dominates, i.e. the sum of the terms whose subscript lies between $\alpha a r^\alpha(1-\varepsilon)$ and $\alpha a r^\alpha(1+\varepsilon)$ outweighs the other parts of the series [**67**].

71. Special case of **70**: $k=1$. With no regularity assumptions we have **53**.

72. From I **94** and **70**:

$$\frac{\sum_{n=0}^{\infty} a_n r^n}{(\beta b r^\beta)^k \sum_{n=0}^{\infty} b_n r^n} = \frac{\sum_{n=1}^{\infty} a_n r^n}{\sum_{n=1}^{\infty} n^k b_n r^n} \cdot \frac{\sum_{n=1}^{\infty} n^k b_n r^n}{(\beta b r^\beta)^k \sum_{n=0}^{\infty} b_n r^n}.$$

73. Replacing r by $\sqrt{r}$ we obtain the situation discussed in **72**:

$$a_n = \frac{1}{n!n!}\frac{1}{2^{2n}}, \qquad b_n = \frac{1}{(2n)!}, \qquad f(r) = \cos i\sqrt{r} \sim \frac{e^{\sqrt{r}}}{2},$$

$$\beta = \frac{1}{2}, \qquad b=1, \qquad k = -\frac{1}{2}, \qquad s = \frac{1}{\sqrt{\pi}}.$$

For by Stirling's formula [II **205**, also II **202**]

$$\frac{a_n}{b_n} = \frac{(2n)!}{n!^2 2^{2n}} \sim \frac{1}{\sqrt{\pi n}}.$$

74. Letting $\omega = e^{\frac{2\pi i}{l}}$ we have

$$1 + \frac{x^l}{l!} + \frac{x^{2l}}{(2l)!} + \frac{x^{3l}}{(3l)!} + \cdots = \frac{e^x + e^{\omega x} + e^{\omega^2 x} + \cdots + e^{\omega^{l-1} x}}{l} \sim \frac{1}{l} e^x$$

as $x \to +\infty$. Let $x = l r^{\frac{1}{l}}$.—We have [II **31**]

$$a(a+1)(a+2)\ldots(a+n-1) \sim \frac{1}{\Gamma(a)} n^{a-1} n!,$$

and hence

$$\frac{P(1)P(2)\ldots P(n)(nl)!}{Q(1)Q(2)\ldots Q(n) l^{nl}} \sim \frac{\Gamma(b_1)\Gamma(b_2)\ldots\Gamma(b_q)}{\Gamma(a_1)\Gamma(a_2)\ldots\Gamma(a_p)} (2\pi)^{\frac{1-l}{2}} l^{\frac{1}{2}} n^{\Delta + \frac{l+1}{2}},$$

using Stirling's formula. Now apply **72**.

75. Using the calculations in solution **74**, Stirling's formula and I **94**, we obtain

$$1+\sum_{n=1}^{\infty}\frac{P(1)P(2)\ldots P(n)}{Q(1)Q(2)\ldots Q(n)}r^n \sim \frac{\Gamma(b_1)\Gamma(b_2)\ldots\Gamma(b_q)}{\Gamma(a_1)\Gamma(a_2)\ldots\Gamma(a_p)}(2\pi)^{\frac{l}{2}}\sum_{n=1}^{\infty}n^{\Delta+\frac{l}{2}}\left(\frac{er^{\frac{1}{l}}}{n}\right)^{ln}.$$

Replacing the last series by the integral

$$\int_1^{\infty}x^{\Delta+\frac{l}{2}}\left(\frac{er^{\frac{1}{l}}}{x}\right)^{lx}dx=l^{-\frac{l}{2}-\Delta-1}\int_l^{\infty}x^{\frac{l}{2}+\Delta+1}\left(\frac{elr^{\frac{1}{l}}}{x}\right)^{x}\frac{dx}{x}$$

and this integral on the basis of II **207** by

$$l^{-\frac{l}{2}-\Delta-1}\sqrt{2\pi}\left(lr^{\frac{1}{l}}\right)^{\frac{l+1}{2}+\Delta}e^{lr^{\frac{1}{l}}},$$

we obtain the desired formula. Passing from the series to the integral introduces an error of the order of magnitude of the maximum term of the series [cf. II **8**]. The maximum of $(er^{\frac{1}{l}}x^{-1})^{lx}$ for fixed r is $\exp(lr^{\frac{1}{l}})$. The maximum term is of order $r^{\frac{2\Delta+l}{2l}}\exp(lr^{\frac{1}{l}})$ and hence its ratio to the integral is of order $r^{-\frac{1}{2l}}$. Thus this step is justified. [**45**, **47**.]

76. Changing finitely many coefficients if necessary we can assume $a_0=1$, $a_n^2>a_{n-1}a_{n+1}$ for $n=1, 2, 3, \ldots$. Then [**43**, **31**] we have $\rho_n=a_{n-1}/a_n$ and

$$\lim_{n\to\infty}n\log\frac{\rho_{n+1}}{\rho_n}=\lim_{n\to\infty}n\log\left[1+\left(\frac{a_n^2}{a_{n-1}a_{n+1}}-1\right)\right]=\frac{1}{\lambda}.$$

(1)
$$\frac{\log\rho_n}{\log n}\sim\frac{\log\rho_1+\log\frac{\rho_2}{\rho_1}+\log\frac{\rho_3}{\rho_2}+\cdots+\log\frac{\rho_n}{\rho_{n-1}}}{1+\frac{1}{2}+\frac{1}{3}+\cdots+\frac{1}{n}}\to\frac{1}{\lambda}$$
[I **70**; **52**].

(2) If $\rho_n\leqq r<\rho_{n+1}$ then $\nu(r)=n$, $\mu(r)=r^n/(\rho_1\rho_2\ldots\rho_n)$ [**25**, **31**] and

$$\frac{\log\mu(r)}{\nu(r)}=\log\frac{r}{\rho_n}+\frac{(n-1)\log\frac{\rho_n}{\rho_{n-1}}+(n-2)\log\frac{\rho_{n-1}}{\rho_{n-2}}+\cdots+1\log\frac{\rho_2}{\rho_1}}{n}$$
$$\sim 0+\frac{1}{\lambda}$$
[I **67**].

(3) Let $\rho_n\leqq r<\rho_{n+1}$, $l>0$ ($l<0$ similarly). Then

$$\log\frac{a_{n+l}r^{n+l}}{a_nr^n}=\log\frac{r^l}{\rho_{n+1}\rho_{n+2}\cdots\rho_{n+l}}$$
$$=l\log\frac{r}{\rho_{n+1}}-(l-1)\log\frac{\rho_{n+2}}{\rho_{n+1}}-\cdots-\log\frac{\rho_{n+l}}{\rho_{n+l-1}}$$
$$\sim 0-\frac{(l-1)+(l-2)+\cdots+1}{n}\cdot\frac{1}{\lambda}\sim-\frac{x^2}{2\lambda}.$$

The sum of the rectangles $=\dfrac{M(r)}{\mu(r)\sqrt{\nu(r)}}\sim\dfrac{M(r)}{\mu(r)\sqrt{\log\mu(r)}}$, the area $=\sqrt{2\pi}\lambda$; cf. **57**.

77. Since the derivative of a function which is schlicht in a region is everywhere different from zero in this region, all the zeros of the polynomial $g(z)=1+2a_2z+3a_3z^2+\cdots+na_nz^{n-1}$ lie outside the unit circle. Hence the absolute value of their product is $\geqq 1$.

[For newer developments and bibliography on the topic of this chapter see G. M. Goluzin: Geometric Theory of Functions of a Complex Variable. Amer. Math. Soc. 1969, especially the Supplement, and L. V. Ahlfors: Conformal Invariants. McGraw-Hill: New York 1973.]

78. Let z_1 and z_2 be arbitrary points inside the unit circle and set $w_1=f(z_1)$, $w_2=f(z_2)$. If $\varphi[f(z_1)]=\varphi[f(z_2)]$, i.e. $\varphi(w_1)=\varphi(w_2)$, then we must have $w_1=w_2$, and hence $z_1=z_2$.

79. $\varphi(z)$ is evidently regular inside the unit circle $|z|<1$ since in this region $f(z^2)/z^2$ is regular and different from zero. Besides $\varphi(z)$ is an odd function, $\varphi(-z)=-\varphi(z)$. Now let z_1 and z_2 be arbitrary points inside the unit circle $|z|<1$, such that $\varphi(z_1)=\varphi(z_2)$. Then $f(z_1^2)=f(z_2^2)$, hence $z_1^2=z_2^2$. The conclusion that $z_1=-z_2\neq 0$ is not possible since it would imply $\varphi(z_1)=-\varphi(z_2)\neq 0$.

80. $[\varphi(z)]^2$ is an even function, i.e. a function of $z^2=\zeta$; set $[\varphi(z)]^2=f(\zeta)$. If ζ_1, ζ_2 are two complex numbers, $|\zeta_1|<1$, $|\zeta_2|<1$, for which $f(\zeta_1)=f(\zeta_2)$, choose z_1, z_2 such that $z_1^2=\zeta_1$, $z_2^2=\zeta_2$. Then $[\varphi(z_1)]^2=[\varphi(z_2)]^2$, $\varphi(z_1)=\pm\varphi(z_2)=\phi(\pm z_2)$, and hence $z_1=\pm z_2$, $\zeta_1=\zeta_2$.

81. Assume that the common center of $\mathfrak{C}$ and $\mathfrak{c}$ is the point $z=0$, the radius of $\mathfrak{C}$ is R and of $\mathfrak{c}$ is r, and finally that the mapping is given by the function $w=f(z)=a_0+a_1z+a_2z^2+\cdots+a_nz^n+\cdots$. Then we have [III **124**]

$$|\mathfrak{R}|=\pi\sum_{n=1}^{\infty} n|a_n|^2R^{2n},\qquad |\mathfrak{r}|=\pi\sum_{n=1}^{\infty} n|a_n|^2r^{2n},\qquad |\mathfrak{C}|=\pi R^2,\qquad |\mathfrak{c}|=\pi r^2,$$

$$\frac{|\mathfrak{R}|}{|\mathfrak{r}|}-\frac{|\mathfrak{C}|}{|\mathfrak{c}|}=\frac{r^2R^2\sum_{n=2}^{\infty} n|a_n|^2(R^{2n-2}-r^{2n-2})}{\sum_{n=1}^{\infty} n|a_n|^2r^{2n+2}}\geqq 0.$$

The equality sign holds only if $a_2=a_3=a_4=\cdots=0$. The expression used for $\mathfrak{R}$ is not contained in III **124** since there it was assumed that the mapping function is regular on the boundary of $\mathfrak{C}$. However by taking limits one can show that this formula represents in general the so-called *inner measure* of $\mathfrak{R}$. Using any other concept of measure the result is true *a fortiori*.

82. $|\mathfrak{R}|=\pi(|a_1|^2R^2+2|a_2|^2R^4+3|a_3|^2R^6+\cdots+n|a_n|^2R^{2n}+\cdots)\geqq\pi|a_1|^2R^2=\pi a^2R^2$. Notation and further considerations as in **81**.

83. We consider the annulus $0<r<|z|<R$, and let the mapping be given by the function $w=\sum_{n=-\infty}^{\infty} a_nz^n$. The images of *all* the circles $|z|=\text{const.}$ are described in the positive sense at the same time as the circles themselves [III **190**] and contain $\mathfrak{r}$ in their interior [solution III **188**]. From III **127** we can deduce (the areas are positive!)

$$|\mathfrak{R}|=\pi\sum_{n=-\infty}^{\infty} n|a_n|^2R^{2n},\qquad |\mathfrak{r}|=\pi\sum_{n=-\infty}^{\infty} n|a_n|^2r^{2n},\qquad |\mathfrak{C}|=\pi R^2,\qquad |\mathfrak{c}|=\pi r^2,$$

where we interpret $|\mathfrak{R}|$ as the inner measure of $\mathfrak{R}$ and $|\mathfrak{r}|$ as the outer measure of $\mathfrak{r}$

[solution **81**; for other definitions of the measures the required inequality holds *a fortiori*]. Excluding the case $|\mathfrak{r}|=0$, we have

$$\frac{|\mathfrak{R}|}{|\mathfrak{r}|}-\frac{|\mathfrak{C}|}{|\mathfrak{c}|}=\frac{r^2R^2\sum_{n=-\infty}^{\infty} n|a_n|^2(R^{2n-2}-r^{2n-2})}{\sum_{n=-\infty}^{\infty} n|a_n|^2 r^{2n+2}}\geqq 0,$$

since $n(R^{2n-2}-r^{2n-2})>0$ for integral n, with the exception of 0 and 1.

84. Let the center of the circle in question be $z=0$ and the mapping be given by the function $w=f(z)=a_1z+a_2z^2+\cdots+a_nz^n+\cdots$.

(1) Assume that $f(z)$ is continuous on the boundary of the circle. Since the function $f(z)/z$ is regular in the interior of the circle and (because the mapping is one–one) is different from 0 and of constant absolute value on the boundary, it follows that it is constant [III **142**, III **274**]. (2) Without the assumption of continuity on the boundary we argue as follows: By **82** we have at the center

$$\left|\left(\frac{dw}{dz}\right)_0\right|^2=|f'(0)|^2\leqq 1,$$

and since the relation between z and w is reciprocal also

$$\left|\left(\frac{dz}{dw}\right)_0\right|^2=\frac{1}{|f'(0)|^2}\leqq 1,$$

i.e. $|f'(0)|^2=1$, hence [**82**] $f(z)=f'(0)z$.

85. Since the relation between the two annuli is reciprocal we have the equality sign when applying the inequality of **83**. The mapping is thus a rotation and dilatation which because of the coincidence of the two outer boundaries must be a pure rotation. The assumption concerning the sense of description of the curves is essential: The mapping of the annulus $\frac{1}{2}<|z|<2$ onto the annulus $\frac{1}{2}<|w|<2$ by means of $zw=1$ is not a mere rotation.

86. Assume that there are two mappings. First by means of one mapping go from the unit circle onto $\mathfrak{R}$ and then by means of the inverse of the second mapping from $\mathfrak{R}$ back onto the unit circle. This results in a mapping of the unit circle onto itself with the center mapped onto itself, and this must be a rotation [**84**] and in fact the rotation through an angle 0 since $f'(0)>0$.—The result proved above also holds if one prescribes, instead of the condition $f'(0)>0$, any fixed argument for $f'(0)$.

87. Let the function $w=f(z)$ map the unit circle $|z|<1$ schlicht onto itself, $f(0)=w_0$, $\arg f'(0)=\alpha$. The linear function

$$\frac{w_0+e^{i\alpha}z}{1+\bar{w}_0e^{i\alpha}z}=w_0+(1-|w_0|^2)e^{i\alpha}z+\cdots$$

yields a mapping with the same properties and hence [**86**] coincides with $f(z)$. Set

$$-w_0e^{-i\alpha}=z_0.$$

88. [For **88–96** cf. P. Koebe: J. reine angew. Math., **145**, pp. 177–225 (1915); cf. also E. Lindelöf: Quatrième congrès des math. scandinaves à Stockholm, 1916. Upsala: Almqvist and Wicksells 1920, pp. 59–75.] The points $z=a$

and $z=1/\bar{a}$ are the exceptional points (branch points) to each of which only one point corresponds, namely $w=\sqrt{a}\eta^{-1}$ and $w=1/(\overline{\sqrt{a}\eta})$ respectively. Solving for z we obtain

$$z=w\eta^2\frac{2\eta^{-1}\sqrt{a}-(1+|a|)w}{1+|a|-2\eta\overline{\sqrt{a}}w}.$$

Thus η is to be chosen such that $\eta\sqrt{a}>0$. Substituting $-\sqrt{a}$ for $\sqrt{a}$ and simultaneously $-\eta$ for η leaves the relation unaltered.

89. It is assumed that $|a|<1$ i.e. that $\mathfrak{R}$ does not coincide with $|z|<1$. The two images of z determined by **88** lie inside the unit circle if z lies inside the unit circle [III **5**] and are distinct if $z\neq a$. However the image of $z=a$, i.e. $w=\sqrt{a}\eta^{-1}$ [**88**] is not a point of $\mathfrak{R}^*$ or of $\mathfrak{R}^{**}$ but a common boundary point of both.

90. Now a has been chosen positive, $a>0$, and hence $\sqrt{a}>0$, $\eta=1$. The two images w^* and w^{**} of z are roots of the equation

$$(1+a)w^2-2\sqrt{a}(1+z)w+(1+a)z=0.$$

If z lies on the cut, i.e. if z is real, $a<z<1$, then $(1+a)^2z-(1+z)^2a>0$, hence w^*, w^{**} are conjugate complex numbers, and in view of $w^*w^{**}=z$ we have $|w^*|=\sqrt{z}$. Thus the point in the image of the cut closest to the origin is $\sqrt{a}$. This is the required nearest boundary point; for the images of boundary points lying on $|z|=1$ lie on $|w|=1$. The Koebe image region of the slit region is "crescent shaped"; to the angle of 360° around the end-point of the cut in the slit region there corresponds an angle of 180° in the Koebe image region.

91. If the images w_1 and w_2 of two points z_1 and z_2 lying on the boundary of the circle $|z|=a$ are equidistant from the origin, i.e. if they lie on the same circle $|w|=\text{const.}$, then from

$$\frac{w}{z}=\frac{1+a-2\sqrt{a}w}{2\sqrt{a}-(1+a)w},$$

we deduce that w_1 and w_2 lie on the same circle

$$\left|\frac{1+a-2\sqrt{a}w}{2\sqrt{a}-(1+a)w}\right|=\text{const.},$$

i.e. that they are intersection points of these two circles, and hence that w_2 and w_1 and also z_2 and z_1 are situated symmetrically with respect to the real axis. Consequently in the image of the semicircle $|z|=a$, $\mathfrak{J}z\geqq 0$ the distance from the origin varies monotonically as z runs from $+a$ to $-a$, and in fact monotonically decreasing, as we see from the two extreme situations: $w=\sqrt{a}$, the image of $z=a$, is the furthest and $w=\{\sqrt{a}/(1+a)\}[1-a-\sqrt{2(1+a^2)}]$, the image of $z=-a$, the nearest boundary point of the Koebe image region.

92. According to the formula in solution **88**, z/w is a function of w which maps the circle $|w|<1$ onto itself [III **5**]. That is we have for $|w|<1$: $|z/w|<1$, $|z|<|w|$.

93. The circular disc $|z|<|a|$, which is completely contained in the region $\mathfrak{R}$, is mapped into a region which contains a circular disc of radius $>|a|$ **[92]**; hence $|a_1|>|a|$. One can also show directly **[91]** that

$$\frac{|\sqrt{a}|}{1+|a|}\left(-1+|a|+\sqrt{2(1+|a|^2)}\right)>|a|.$$

94. If $z, z_1, z_2, \ldots, z_n$ are variable points in $\mathfrak{R}, \mathfrak{R}_1, \mathfrak{R}_2, \ldots, \mathfrak{R}_n$ respectively, then, according to solution **88**, we have the first term of the power series expansion

$$z_1=\frac{1+|a|}{2|\sqrt{a}|}z+\cdots$$

and similarly

$$z_2=\frac{1+|a_1|}{2|\sqrt{a_1}|}z_1+\cdots,\quad z_3=\frac{1+|a_2|}{2|\sqrt{a_2}|}z_2+\cdots,\quad \ldots,\quad z_n=\frac{1+|a_{n-1}|}{2|\sqrt{a_{n-1}}|}z_{n-1}+\cdots$$

and hence

$$f_n'(0)=\frac{1+|a|}{2|\sqrt{a}|}\cdot\frac{1+|a_1|}{2|\sqrt{a_1}|}\cdots\frac{1+|a_{n-1}|}{2|\sqrt{a_{n-1}}|}.$$

Since $f_n(z)$ is regular inside the circle $|z|<|a|$ and satisfies $|f_n(z)|<1$ there, we have

$$|f_n'(0)||a|\leqq 1.$$

95. In accordance with **94** the infinite product

$$\left(1+\frac{(1-|\sqrt{a}|)^2}{2|\sqrt{a}|}\right)\left(1+\frac{(1-|\sqrt{a_1}|)^2}{2|\sqrt{a_1}|}\right)\cdots\left(1+\frac{(1-|\sqrt{a_n}|)^2}{2|\sqrt{a_n}|}\right)\cdots\leqq\frac{1}{|a|},$$

converges, and hence

$$\lim_{n\to\infty}(1-|\sqrt{a_n}|)=0.$$

96. Since the mapping is one-to-one and the point $z=0$ is mapped onto itself, the functions

$$\frac{f_1(z)}{z},\quad \frac{f_2(z)}{z},\quad \ldots,\quad \frac{f_n(z)}{z},\quad \ldots$$

are regular and different from 0 in $\mathfrak{R}$. Their absolute values increase with increasing index at every point z **[92]**. Hence the functions

$$\psi_1(z)=\log\frac{f_1(z)}{z},\qquad \psi_n(z)=\log\frac{f_n(z)}{f_{n-1}(z)},\qquad n=2,3,4,\ldots$$

are regular in $\mathfrak{R}$ and have *positive* real part. The determination of logarithm is to be made such that $\psi_n(0)$ becomes real. We have further that in the whole region $\mathfrak{R}$

$$\left|\frac{f_n(z)}{z}\right|\leqq\frac{1}{|a|}$$

[III **278**], and hence

$$\Re\psi_1(z)+\Re\psi_2(z)+\cdots+\Re\psi_n(z)\leqq -\log|a|.$$

From this one concludes immediately that the real part of the series $\sum_{n=1}^{\infty}\psi_n(z)$ converges. Application of III **257**, III **258** (attention to the point $z=0$) shows that the same holds for the imaginary part.—If $|w|<1$, the value w is assumed by $f_n(z)$ as soon as $|a_n|>|w|$. Because of **95** it follows that $f(z)=\lim_{n\to\infty} f_n(z)$ also assumes the value w [III **201**]. That $f(z)$ is schlicht follows from the fact that $f_n(z)$ is schlicht [III **202**].

97. The normed mapping function is [III **76**]:

$$f(a;z)=(\rho^2-|a|^2)\frac{z-a}{\rho^2-\bar{a}z}.$$

We have

$$r_a=\frac{\rho^2-|a|^2}{\rho},\qquad \bar{r}=\rho.$$

98.

$$r_a=\frac{|a|^2-\rho^2}{\rho};\qquad \lim_{a\to\infty} r_a=\infty.$$

99. The auxiliary mapping $\zeta=z^{\frac{\pi}{\vartheta_0}}$ takes the given sector into the upper half-plane $\Im\zeta>0$, and the point a into $\zeta_0=a^{\frac{\pi}{\vartheta_0}}$. The further mapping

$$w=r_a\frac{\zeta-\zeta_0}{\zeta-\bar{\zeta}_0}$$

maps the upper half-plane onto the interior of the circle $|w|<r_a$. Here r_a is determined by the equation

$$\left|\frac{dw}{dz}\right|_{z=a}=\left|\frac{dw}{d\zeta}\right|_{\zeta=\zeta_0}\left|\frac{d\zeta}{dz}\right|_{z=a}=1.$$

We have, setting $a=|a|e^{i\alpha}$, $0<\alpha<\vartheta_0$,

$$r_a=\frac{2\vartheta_0}{\pi}|a|\sin\frac{\alpha\pi}{\vartheta_0}.$$

100. Let $f(z)$ be the normed mapping function of $\mathfrak{R}$ associated with the central point a and $\varphi(z')$ the normed mapping function of $\mathfrak{R}'$ associated with a'. Then $h^{-1}\varphi(hz+k)$ has all the characteristic properties of $f(z)$, and hence we have $h^{-1}\varphi(hz+k)=f(z)$.—Similarly for the outer radius.

101. We may consider the real line segment $-2\leqq z\leqq 2$ [**100**]. The normed mapping function associated with the point at infinity as central point is then [III **79**]

$$w=f(z)=\frac{z+\sqrt{z^2-4}}{2}=z-\frac{1}{z}-\frac{1}{z^3}-\frac{2}{z^5}-\cdots.$$

For $z=2\cos\vartheta$ we have $w=e^{i\vartheta}$, $0\leqq\vartheta\leqq 2\pi$, i.e. $|w|=1$, $\bar{r}=1$.

102. In the case of the mapping used in **101**, the ellipse with foci -2, 2 for which the sum of the lengths of the axes is $4R=l$ goes over into the circle $|w|=R=l/4$ [III **80**].

103. $r_a=2d$. [III **6** or a special case of **99**.]

104. [**100**.]

105. The mapping $\zeta=z+(1/z)$ takes the exterior of the curve in question into the slit region whose boundary is the real line segment $-(a_2+a_2^{-1})\leqq z\leqq a_1+a_1^{-1}$. Now the outer radius is left invariant [**104**]; i.e. [**101**]

$$\bar{r}=\frac{a_1+a_1^{-1}+a_2+a_2^{-1}}{4}=\frac{(a_1+a_2)(1+a_1a_2)}{4a_1a_2}.$$

106. $r_a=(2D/\pi)\sin(\pi d/D)$. In the proof it may be assumed that we have the strip

$$|\mathfrak{J}z|<\frac{\pi}{2}, \qquad D=\pi,$$

and that $a=i(\pi/2-d)$. The mapping $z'=e^z$ takes this strip onto the half-plane $\mathfrak{R}z'>0$ and a into $i\,e^{-ia}$[**103, 104**].

107. Definition.

108. In view of **107**, the transformation $\zeta=1/z$ reduces the problem to that of **105**: Setting $a_1=b_1^{-1}$, $a_2=b_2^{-1}$, we obtain

$$r_0=\frac{4b_1^{-1}b_2^{-1}}{(b_1^{-1}+b_2^{-1})(1+b_1^{-1}b_2^{-1})}=\frac{4b_1b_2}{(b_1+b_2)(1+b_1b_2)}.$$

109. In the mapping $\zeta=(z-z_1)/(z-z_2)$ the exterior of $\mathfrak{C}$ goes over into the sector

$$\vartheta_2-2\pi<\arg\zeta<\vartheta_1,$$

and the point $z=\infty$ into $\zeta=1$. For the inner radius r_1 of this sector with respect to $\zeta=1$ we have [**100, 107**]

$$r_1=\frac{|z_2-z_1|}{\bar{r}}.$$

By **99** we have also

$$r_1=\frac{2(\vartheta_1+2\pi-\vartheta_2)}{\pi}\sin\frac{\vartheta_1\pi}{\vartheta_1+2\pi-\vartheta_2}.$$

Hence we have

$$\bar{r}=\frac{|z_2-z_1|}{2(2-\delta)\sin\dfrac{\vartheta_1}{2-\delta}}, \qquad \text{where} \quad \pi\delta=\vartheta_2-\vartheta_1.$$

If $\vartheta_1+\vartheta_2=2\pi$, then $\mathfrak{C}$ is "symmetric",

$$\bar{r}=|z_2-z_1|\,\frac{\pi}{4\vartheta_1};$$

cf. the special cases $\vartheta_1=\pi$ [**101**], $\vartheta_1=\pi/2$ [**97**].

For $\vartheta_2=\vartheta_1$, $\mathfrak{C}$ is a circular arc,

$$\bar{r}=\frac{|z_2-z_1|}{4\sin\dfrac{\vartheta_1}{2}};$$

this result takes a different form if we introduce the radius of curvature ρ and the angle φ subtended by $\mathfrak{C}$ at the center, namely $\bar{r}=\rho\sin(\varphi/4)$.—For $\vartheta_1=\pi/2$, $\vartheta_2=\pi$ we obtain the outer radius of a semicircular disc of diameter d, $\bar{r}=2d/(3\sqrt{3})$.

110. The normed mapping function associated with the central point a is [III **76**]

$$f(a;z)=c\,\frac{\dfrac{f(z)}{r_b}-\dfrac{f(a)}{r_b}}{1-\dfrac{\overline{f(a)}}{r_b}\dfrac{f(z)}{r_b}}=cr_b\,\frac{f(z)-f(a)}{r_b^2-\overline{f(a)}f(z)},$$

where the constant c is to be determined from

$$\left(\frac{df(a;z)}{dz}\right)_{z=a}=\frac{cr_bf'(a)}{r_b^2-|f(a)|^2}=1.\qquad \text{Thus}\quad r_a=|c|.$$

111. By the mapping $\zeta=(1/z)-2$ we obtain the ζ-plane slit along the interval -2, 2. Therefore, the required mapping function is

$$w=f(z)=\frac{\zeta-\sqrt{\zeta^2-4}}{2}=\frac{1-2z-\sqrt{1-4z}}{2z},\qquad \frac{w}{(1+w)^2}=z.$$

By **110** we have for real a

$$r_a=\left(\frac{1-f(a)}{1+f(a)}\right)^2=1-4a.$$

In particular $r_0=1$. As a runs through the interval $(-\infty,\frac{1}{4})$, r_a decreases monotonically from ∞ to 0.

112. By **110**; $|f'(a)|$ remains greater than a fixed positive number as a approaches a point of L from the interior. At the same time $|f(a)|$ converges to r_b.

113. Replace in **110** b by a and a by $a+\varepsilon$, where ε is an arbitrary vector whose magnitude tends to 0. If $((\varepsilon^2))$ denotes a quantity that goes to 0 at least as rapidly as ε^2, then

$$|f'(a+\varepsilon)|=|1+2c_2\varepsilon+((\varepsilon^2))|=1+2\Re c_2\varepsilon+((\varepsilon^2)),$$

and hence

$$r_{a+\varepsilon}=r_a(1-2\Re c_2\varepsilon)+((\varepsilon^2)),$$

that is, $\Re c_2\varepsilon\geqq 0$ for $|\varepsilon|$ sufficiently small, independently of $\arg\varepsilon$. From this we conclude that $c_2=0$.

114. The points in question lie on the straight line $\Im a=\text{const.}$ [**103**].

115. From the expression (*) for the normed mapping function given at the beginning of § 4 we obtain by inversion

$$z-a=\varphi(w)=w+c_2'w^2+c_3'w^3+\cdots,$$

where $\varphi(w)$ is schlicht in the interior of the circle $|w|<r_a$ and maps it onto $\Re$. The function $\sqrt[n]{\varphi(w^n)}=w+c_2''w^{1+n}+c_3''w^{2+n}+\cdots$ is then schlicht in the circle $|w|<r_a^{\frac{1}{n}}$ [**79**] and takes this circle precisely into $\Re'$.

116. We have [solution **115**]

$$w=f(z)=\left(\frac{1-2z^n-\sqrt{1-4z^n}}{2z^n}\right)^{\frac{1}{n}}, \qquad \frac{w}{(1+w^n)^{\frac{2}{n}}}=z,$$

from which we easily obtain [**110**]

$$r_a=(1-|f(a)|^2)\left|\frac{f(a)}{a}\right|^{n-1}\frac{|1-[f(a)]^n|}{|1+[f(a)]^n|^3}.$$

In particular we have $r_0=1$. As a runs along the ray $\arg z=(2\pi\nu)/n$, $0\leqq|a|<\sqrt[n]{\tfrac{1}{4}}$, $f(a)$ describes the radius of the unit circle that makes the angle $(2\pi\nu)/n$ with the positive real axis. The required limit is equal to 0 for all values of ν.

117. Apply **115** in a suitable manner to the slit region whose boundary is the real line segment $-\beta^2$, α^2. This region has the outer radius $(\alpha^2+\beta^2)/4$ [**101**], and hence

$$\bar{r}=\frac{\sqrt{\alpha^2+\beta^2}}{2}.$$

118. The function $\varphi(z)=F(z)/f(z)$ is regular in $\mathfrak{R}$ including the point $z=a$. We have $\varphi(a)=1$. Let $0<\varepsilon<r_a$. About any boundary point of $\mathfrak{R}$ we can determine a neighbourhood so small that for all points of $\mathfrak{R}$ lying within it we have

$$|f(z)|>r_a-\varepsilon,$$

that is

$$|\varphi(z)|<\frac{M}{r_a-\varepsilon}.$$

Hence in $\mathfrak{R}$ [III **278**] we have $|\varphi(z)|\leqq M/r_a$, in particular $1=|\varphi(a)|\leqq M/r_a$. The equality sign holds if and only if $\varphi(z)\equiv 1$, i.e. $F(z)\equiv f(z)$.

119. If

$$f(z)=z-a+c_2(z-a)^2+\cdots+c_n(z-a)^n+\cdots$$

is the normed mapping function, then the function formed with the conjugate complex coefficients

$$\bar{f}(z)=z-a+\bar{c}_2(z-a)^2+\cdots+\bar{c}_n(z-a)^n+\cdots$$

is also regular in $\mathfrak{R}$. For, let z_0 be an arbitrary point of $\mathfrak{R}$ and let $F_0(z-a)=f(z)$, $F_1(z-a_1),\ldots,F_{l-1}(z-a_{l-1})$, $F_l(z-a_l)$, where $a_1,a_2,\ldots,a_{l-1}$ are suitable points of $\mathfrak{R}$, $a_l=z_0$, be the successive expansions which effect the continuation of $f(z)$ to the point z_0. Then from $\bar{f}(z)$ we reach the point $\bar{a}_l=\bar{z}_0$ by means of the successive expansions $\bar{F}_0(z-a)=\bar{f}(z)$, $\bar{F}_1(z-\bar{a}_1),\ldots,\bar{F}_{l-1}(z-\bar{a}_{l-1})$, $\bar{F}_l(z-\bar{a}_l)$. $\bar{f}(z)$ is hence regular in the mirror image of $\mathfrak{R}$ with respect to the real axis, i.e. in $\mathfrak{R}$ itself. Moreover $\bar{f}(\bar{z})=\overline{f(z)}$ and hence $|\bar{f}(z)|<r_a$ in $\mathfrak{R}$. By **118** we have hence $\bar{f}(z)=f(z)$.

120. $-f(-z)$ is regular in $\mathfrak{R}$, and $|-f(-z)|<r_a$. By **118** we have hence $-f(-z)=f(z)$.

121. If $f(z)$ and $f^*(z)$ are the normed mapping functions belonging to the central point a of $\mathfrak{R}$ and $\mathfrak{R}^*$ respectively, then we have [**118**]

$$r_a=\max_{(\mathfrak{R})}|f(z)|\geqq\max_{(\mathfrak{R}^*)}|f(z)|\geqq\max_{(\mathfrak{R}^*)}|f^*(z)|=r_a^*.$$

Both equality signs can be attained, however not simultaneously.

122. From **121** with the aid of **97**.

123. [**118**, **121**, **122**.]

***124.** [III **309**.]

124.1. (a) By **102** and **124**

$$2\pi \frac{2a+2b}{4} < l.$$

(b) By calculus and II **80**

$$l = \int_0^{2\pi} \sqrt{a^2 \sin^2 \varphi + b^2 \cos^2 \varphi} \sqrt{\sin^2 \varphi + \cos^2 \varphi}\, d\varphi$$

$$> \int_0^{2\pi} \sqrt{(a \sin^2 \varphi + b \cos^2 \varphi)^2}\, d\varphi = \pi(a+b).$$

125. [**82**.]

126. III **126**, cf. also solution **83**.

127. [G. Pólya: Jber. deutsch. Math. Verein. **31** (1922) footnote, p. 111.] Denoting the outer measure of the interior of L by $|L|_e$ and the inner measure of the same region by $|L|_i$, we have [**125**, **126**]

$$\pi r_a^2 \leqq |L|_i \leqq |L|_e \leqq \pi \bar{r}^2.$$

128. If $f(z)$ is the normed mapping function belonging to the origin as central point then $P(z)/[f(z)]^k$ is regular in the interior of L and is equal to a_k for $z=0$. Similarly for the second inequality.

129. Let $z=\varphi(\zeta)$ be a function which maps the interior of the circle $|\zeta|<1$ into the interior of L. The mapping is one–one and continuous for $|\zeta| \leqq 1$ [last remark preceding **97**]. Apply III **233** to $f[\varphi(\zeta)]$.

130.

$$w = \frac{e^{\frac{\pi z}{D}} - e^{\frac{\pi i}{D}}}{e^{\frac{\pi z}{D}} - e^{-\frac{\pi i}{D}}}.$$

The length of the arc in question is equal to $2\pi(1-D^{-1})$; it steadily increases as D increases.

131. [K. Löwner.] We may assume that O as well as the center of the image circle are at the origin. Let the radius of the image circle be r. Denote by $f_1(z)$ and $f_2(z)$ the mapping functions of L_1 and L_2 respectively. Then

$$F(z) = \frac{f_1(z)}{f_2(z)}$$

is regular and different from 0 in the interior of L_1. Furthermore, on the common arcs of L_1 and L_2 we have

$$|f_1(z)| = |f_2(z)| = r, \qquad |F(z)| = 1,$$

while on the complementary part of L_1

$$|f_1(z)| = r, \qquad |f_2(z)| \leqq r, \qquad |F(z)| \geqq 1.$$

We deduce that $|F(z)| > 1$ in the interior of L_1. The regular branch of $\log F(z)$

which is real for $z=0$ satisfies the assumptions of **129**. Describing positively a common arc of L_1 and L_2 involves a negative change in $\mathfrak{J} \log F(z) = \mathfrak{J} \log f_1(z) - \mathfrak{J} \log f_2(z)$.

132. Let the interior conductor of an insulated cylindrical condenser be a straight line (wire) conductor. If the exterior conductor is deformed, without jeopardizing the insulation, in such a manner that the cross-sectional surface contains the old one as a part, extending beyond it at some places and not at others, then at those places where the condenser wall has remained undisturbed the density of electrostatic charge will increase—as can also be made plausible by means of physical considerations.

133. $\mathfrak{T}^*$ is simply connected (simple closed curves in the interior of $\mathfrak{T}^*$ belong both to the interior of L_1 and to the interior of L_2). Map also $\mathfrak{T}^*$ onto the unit circle, and again let the center of the circle be the image of O. The following table refers to the total length of those arcs which lie on the boundary of the unit circle in the mappings I, II, III (interior of L_1, L_2, $\mathfrak{T}^*$ respectively) as images of

	I	II	III
visible arcs of L_1	σ_1	—	σ_1^*
hidden arcs of L_1	τ_1	—	—
visible arcs of L_2	—	σ_2	σ_2^*
hidden arcs of L_2	—	τ_2	— .

Obviously

$$\sigma_1 + \tau_1 = 2\pi, \qquad \sigma_2 + \tau_2 = 2\pi, \qquad \sigma_1^* + \sigma_2^* = 2\pi.$$

Using **131** we have

$$\sigma_1^* \leqq \sigma_1, \qquad \sigma_2^* \leqq \sigma_2.$$

Hence we have

$$\tau_2 = 2\pi - \sigma_2 \leqq 2\pi - \sigma_2^* = \sigma_1^* \leqq \sigma_1,$$
$$\tau_1 = 2\pi - \sigma_1 \leqq 2\pi - \sigma_1^* = \sigma_2^* \leqq \sigma_2.$$

If we take L_2 to be a circle with center O, then we can express the result imprecisely but graphically as follows: In mapping a region onto a circle the parts of the boundary curve lying close to the point which is mapped onto the center of the circle are expanded more than the parts of the boundary lying further away.

134. The proof will be given under the assumption that the boundary of $\mathfrak{D}$ is a simple closed curve which cuts the circle $|z-\zeta| = \rho$ at only finitely many points. By deleting the boundary curve, we obtain from $\mathfrak{D}$ a simply connected region whose intersection with the circular disc $|z-\zeta| < \rho$ has a simply connected part $\mathfrak{T}$ which contains the point O in its interior [solution **133**]. Let $\rho\gamma$ be the total length of the *common* boundary of $\mathfrak{T}$ and of the circle $|z-\zeta| < \rho$. Then we have

$$\gamma + \Omega \leqq 2\pi.$$

In the mapping of $\mathfrak{T}$ onto the unit circle $|w| < 1$ let O go over into the center of the circle and the set of arcs counted in the calculation of $\rho\gamma$ into a set of circular arcs of total length δ. By **131** we then have

$$\delta \leqq \gamma.$$

Define a function $\varphi(w)$ regular in the circle $|w|<1$, which at interior points of the arcs counted in δ satisfies the equation

$$\Re\varphi(w)=\log A,$$

and at interior points of the remaining arcs of the circle $|w|=1$ satisfies the equation

$$\Re\varphi(w)=\log a$$

[III **231**]. We have

$$2\pi\Re\varphi(0)=\delta\log A+(2\pi-\delta)\log a$$

[III **118**]. Set

$$e^{\varphi(w)}=\Phi(w).$$

Then

$$|\Phi(0)|^{2\pi}=a^{2\pi}\left(\frac{A}{a}\right)^{\delta}\leqq a^{2\pi}\left(\frac{A}{a}\right)^{\gamma}\leqq a^{2\pi}\left(\frac{A}{a}\right)^{2\pi-\Omega}.$$

If the mapping in question of $\mathfrak{T}$ onto $|w|<1$ is effected by the function

$$\psi(z)=w,\qquad \psi(\zeta)=0$$

then $\Phi[\psi(z)]$ is different from 0 in the interior of $\mathfrak{T}$ and at all points of the boundary of $\mathfrak{T}$, with the exception of finitely many points, its modulus is not less than that of $f(z)$. We deduce [III **335**]

$$|f(\zeta)|\leqq|\Phi[\psi(\zeta)]|=|\Phi(0)|.$$

By reversing the argument we obtain a part of **131**, **133** from III **276**. Cf. also solution III **177**.

135. $f_n(z)$ vanishes only for $z=0$ (a one-to-one mapping). Applying III **278** to the two functions $z^{-1}f_n(z)$ and $z(f_n(z))^{-1}$ which are regular in $\Re_n$ we obtain in $\Re_n$

$$\frac{1}{1}<\left|\frac{z}{f_n(z)}\right|<\frac{a}{1}.$$

Let ε_n be a positive number satisfying

$$\left|\frac{e^{i\alpha_n}-1}{2}\right|^{\varepsilon_n}=\frac{1}{a}.$$

Then $\lim_{n\to\infty}\varepsilon_n=0$. Set

$$\frac{z}{f_n(z)}\left(\frac{z-1}{2}\right)^{\varepsilon_n}=\varphi_n(z).$$

We have $|\varphi_n(z)|<1$ for $|z|<1$ [III **278**]. From

$$|z||\tfrac{1}{2}(z-1)|^{\varepsilon_n}<|f_n(z)|<|z|$$

we obtain $|f_n(z)|\to|z|$ as $n\to\infty$ in the interior of the circle $|z|<1$. Note that $f_n'(0)>0$ and apply III **258** to $\log(f_n(z)z^{-1})$.

136. [L. Bieberbach: Berlin. Ber. pp. 940–955 (1916); G. Faber: München. Ber. pp. 39–42 (1916).] From III **126** we have for every $R>1$

$$\frac{|b_1|^2}{R^2}+\frac{2|b_2|^2}{R^4}+\frac{3|b_3|^2}{R^6}+\cdots<1.$$

Hence the sum of the first n terms <1. For $R\to 1$ we deduce

$$|b_1|^2+2|b_2|^2+\cdots+n|b_n|^2\leqq 1.$$

This is true for every value of n, so that $\sum_{n=1}^{\infty} n|b_n|^2\leqq 1$. If $|b_1|=1$, $b_1=e^{i\beta}$, then $b_2=b_3=\cdots=0$, i.e. $g(z)=z+b_0+(e^{i\beta}/z)$ [III **79**].

137. [K. Löwner: Math. Z. 3, pp. 69–72 (1919).] By applying Cauchy's inequality [II **80**] we obtain [**136**]

$$|g'(z)|=\left|1-\sum_{n=1}^{\infty}\frac{nb_n}{z^{n+1}}\right|\leqq 1+\sum_{n=1}^{\infty}\frac{\sqrt{n}}{|z|^{n+1}}\sqrt{n}|b_n|\leqq 1+\sqrt{\sum_{n=1}^{\infty}\frac{n}{|z|^{2n+2}}}$$

$$=\frac{1}{1-\dfrac{1}{|z|^2}}.$$

If $g'(\rho\bar{\varepsilon})=1/(1-\rho^{-2})$, $\rho>1$, $|\varepsilon|=1$, then $-b_n\varepsilon^{n+1}$ must be real and non-negative, and $\sqrt{n}|b_n|=\lambda(\sqrt{n}/\rho^{n+1})$, where the factor λ which is independent of n can be determined from the equation $\sum_{n=1}^{\infty} n|b_n|^2=1$. We have

$$\lambda=\rho^2-1,\qquad b_n=-\frac{\rho^2-1}{(\rho\varepsilon)^{n+1}},\qquad g(z)=z+b_0-\frac{1}{\varepsilon}\left(\rho-\frac{1}{\rho}\right)\frac{1}{\rho\varepsilon z-1}.$$

This function is schlicht for $|z|>1$, since

$$\left|\frac{1-\rho^2}{\rho\varepsilon z-1}+1\right|\leqq\rho$$

for $|z|>1$, and hence

$$\frac{1-\rho^2}{\rho\varepsilon z_1-1}+1\neq\rho\varepsilon z_2,\quad g(z_2)-g(z_1)\neq 0;\qquad |z_1|>1,\quad |z_2|>1,\quad z_1\neq z_2.$$

The identity

$$\varepsilon[g(z)-b_0]-\left(\rho-\frac{1}{\rho}\right)=\frac{\rho\varepsilon z(\varepsilon z-\rho)}{\rho\varepsilon z-1}.$$

now shows that the image of the unit circle is a circular arc with center $b_0+\bar{\varepsilon}(\rho-(1/\rho))$ and of radius ρ.

138. See **120**.

139. [G. Faber, *ibid.* **136.**] By **79**

$$\sqrt{g(z^2)}=z+\frac{\frac{1}{2}b_0}{z}+\frac{\frac{1}{2}b_1-\frac{1}{8}b_0^2}{z^3}+\cdots$$

is regular and schlicht for $|z|>1$, and hence [**136**] $|\frac{1}{2}b_0|\leqq 1$. We have $|\frac{1}{2}b_0|=1$ only if $\sqrt{g(z^2)}=z+(e^{i\beta}/z)$, β real, i.e. $g(z)=z+2e^{i\beta}+(e^{2i\beta}/z)$.

140. [L. Bieberbach, *ibid.* **136.**] Denoting by h an arbitrary boundary point of $\mathfrak{R}$ apply **139** to $g(z)-h$. The origin is then a boundary point of the image region. The conformal center of gravity is b_0-h, hence $|b_0-h|\leqq 2$.

141. [Cf. L. Bieberbach: Math. Ann. 77, pp. 153–172 (1916).] The lower estimate for D follows by a similar argument as in III **239**, the upper estimate from **140** as follows: If h_1 and h_2 are two boundary points of $\mathfrak{R}$, then

$$|b_0-h_1|\leqq 2, \qquad |b_0-h_2|\leqq 2, \qquad \text{and} \qquad |h_1-h_2|\leqq 4.$$

142. Denoting by d the distance between the two points, and by D the diameter of the arc joining them, we have [**141**]

$$d\leqq D\leqq 4\bar{r} \qquad [\mathbf{101}].$$

143. [K. Löwner, *ibid.* 137, pp. 74–75.] It is sufficient to prove that $|g(\rho)|\leqq \rho+(1/\rho)$, $\rho>1$. If we remove from the region $\mathfrak{R}$ the two arcs in $\mathfrak{R}$ which correspond to the two segments $1<z\leqq\rho$ and $-\rho\leqq z<-1$ according to the mapping $g(z)=w$, then we obtain a region $\mathfrak{R}^*$. Cut the exterior of the circle $|z|>1$ from 1 to ρ and from -1 to $-\rho$. Then the resulting region can be mapped onto the exterior of the circle $|\zeta|>\frac{1}{2}[\rho+(1/\rho)]$ [**105**] and has [**138**] the conformal center of gravity 0. Now apply **140** to the resulting mapping of the exterior of the circle $|\zeta|>\frac{1}{2}[\rho+(1/\rho)]$ onto $\mathfrak{R}^*$.

144. [G. Faber: München. Ber. pp. 49–64 (1920).] Since $g(z)-z$ is regular for $|z|>1$, and since the boundary values of $g(z)-z$ are all $\leqq 3$ in modulus [**140**], we have $|g(z)-z|\leqq 3$, $|z|>1$. Hence from III **280** we obtain $|g(z)-z|\leqq 3/|z|$. The equality sign can hold, if at all, only for $g(z)-z=e^{i\alpha}/z$, α real, but then $|g(z)-z|=1/|z|$. Hence it can never hold.

145. [K. Löwner.] The outer radius $\bar{r}$ of the curve described in the problem is greater than that of a straight line segment of length $2a$ [**121**] and smaller than that of an ellipse which contains it and has the sum of the lengths of its axes converging to 4 as $a\to 2$, $\delta\to 0$. From this we deduce [**101**, **102**] that $\bar{r}\to 1$ as $a\to 2$, $\delta\to 0$. In addition from **119** we obtain that in the case of the mapping in question the two boundary points lying on opposite sides of the slit at $z=-a$ are mapped into the points $z=+1$ and $z=-1$, so that one of the boundary points undergoes a displacement of $1+a$. As $a\to 2$, $1+a$ approaches arbitrarily closely to 3.

146. [L. Bieberbach, *ibid.*, **136**.] Setting $\zeta=1/z$, we have

$$g(\zeta)=[f(\zeta^{-1})]^{-1}=\zeta-a_2+\frac{a_2^2-a_3}{\zeta}+\cdots\neq 0 \quad \text{for} \quad |\zeta|>1; \qquad \mathbf{139}.$$

We have $|a_2|=2$ if and only if

$$g(\zeta)=\zeta+2e^{i\alpha}+\frac{e^{2i\alpha}}{\zeta}, \qquad f(z)=\frac{z}{(1+e^{i\alpha}z)^2}, \qquad \alpha \text{ real.}$$

147. [G. Faber, *ibid.* **136**, **144**; L. Bieberbach, *ibid.* **136**. The theorem without the precise constants is found already in P. Koebe: Gött. Nachr. pp. 197–210 (1907).] Let $R=1$. The function

$$\frac{f(z)}{1-h^{-1}f(z)}=z+\left(a_2+\frac{1}{h}\right)z^2+\cdots$$

is schlicht for $|z|<1$, hence [**146**] $|a_2+(1/h)|\leqq 2$, $|1/h|\leqq|a_2|+2\leqq 4$. [This variant of the proof stems from F. Hausdorff.] The equality sign applies only for

$$f(z)=\frac{z}{(1+e^{i\alpha}z)^2}, \qquad \alpha \text{ real.}$$

148. Let $\zeta=1/z$, $g(\zeta)=[f(\zeta^{-1})]^{-1}$ [solution **146**]. Applying **140** to $g(\zeta)$ yields for an arbitrary boundary point h

$$\left|a_2+\frac{1}{h}\right|\leqq 2, \qquad \left|\frac{1}{h}\right|\leqq |a_2|+2.$$

[Cf. also solution **147**.]

149. [Cf. G. Szegö: Jber. deutsch. Math. Verein. **31**, p. *42* (1922); **32**, p. *45* (1923).] Let h_1 and h_2 be two boundary points of $\mathfrak{R}$ with $\arg h_1-\arg h_2=\pi$. Applying **147** to $f(z)/[1-h_1^{-1}f(z)]$ yields $|1/(h_2^{-1}-h_1^{-1})|\geqq R/4$, i.e.

$$\frac{2}{\dfrac{1}{|h_1|}+\dfrac{1}{|h_2|}}\geqq\frac{R}{2}.$$

By the theorem about the arithmetic and the harmonic mean (II, Vol. I, p. 63) we then have in particular $|h_1|+|h_2|\geqq R$. If we have the equality sign, then we must have firstly $|h_1|=|h_2|=R/2$, $h_1=-h_2=(R/2)e^{i\gamma}$, and furthermore

$$\frac{f(z)}{1-h_1^{-1}f(z)}=\frac{R^2z}{(R+e^{i\alpha}z)^2}, \qquad \gamma,\ \alpha \text{ real.}$$

This function has the absolute value $R/4$ in the circle $|z|\leqq R$ only for $z=R\,e^{-i\alpha}$. We deduce that $e^{-i\alpha}=-e^{-i\gamma}$ and $f(z)=R^2z/(R^2+e^{2i\alpha}z^2)$. [In the reference above we find only $\max(|h|_1, |h_2|)\geqq R$. The observation that in fact $|h_1|+|h_2|\geqq R$ is due to T. Radó.]

150. [Cf. G. Pick, *ibid.* **151**; also L. Bieberbach: Math. Z. **4**, pp. 295–305 (1919); R. Nevanlinna: Översikt av Finska Vetenskaps-Soc. Förh. **62**(*A*), No. 7 (1919–1920).] Let $|z_0|<1$ and apply **146** to the function

$$\varphi(z)=A+Bf\left(\frac{z+z_0}{1+\bar{z}_0z}\right)$$

which is also schlicht inside the unit circle $|z|<1$. The constants A and B are to be determined by the conditions $\varphi(0)=0$, $\varphi'(0)=1$.

151. [With respect to **151**, **152**, **156**, **157** cf. P. Koebe, *ibid.* **147**; G. Plemelj: Verhandl. d. deutschen Naturforsch. **85**, III, p. 163 (1913); G. Pick: Leipz. Ber. **68**, pp. 58–64 (1916); G. Faber, L. Bieberbach, *ibid.* **136**; also T. H. Gronwall: C. R. Acad. Sci. (Paris), **162**, pp. 249–252 (1916).] We have

$$\log f'(z)+\log(1-|z|^2)=\int_0^z\left(\frac{f''(z)}{f'(z)}-\frac{2\bar{z}}{1-|z|^2}\right)dz,$$

where the integration is carried out along the straight line segment from 0 to z, $|z|=r<1$. We deduce [**150**]

$$|\log f'(z)+\log(1-|z|^2)|\leqq\int_0^r\frac{4}{1-r^2}\,dr=2\log\frac{1+r}{1-r}.$$

Considering the real part we obtain

$$-2\log\frac{1+r}{1-r}\leqq\log|f'(z)|+\log(1-r^2)\leqq 2\log\frac{1+r}{1-r}, \qquad \text{q.e.d.}$$

Considering the imaginary part we obtain in addition

$$|\mathfrak{I} \log f'(z)| \leqq 2 \log \frac{1+r}{1-r}.$$

(The rotation theorem.) The equality sign can never occur. [L. Bieberbach, *ibid.* **150**.]

152. From $f(z)=\int_0^z f'(z)\,dz$ we obtain [**151**]

$$|f(z)| \leqq \int_0^r \frac{1+r}{(1-r)^3}\,dr = \frac{r}{(1-r)^2}.$$

We obtain the lower estimate for $|f(z)|$ in two ways as follows:

(1) If w is the point of the image of the circle $|z|=r$ which is closest to the origin, and if L is the curve of the z-plane that corresponds to the line segment joining 0 and w, then from $|w|=\int_L |f'(z)|\,|dz|$ we obtain, since $|dz| \geqq dr$, that

$$|w| \geqq \int_0^r |f'(z)|\,dr \geqq \int_0^r \frac{1-r}{(1+r)^3}\,dr = \frac{r}{(1+r)^2}.$$

(2) Map the interior of the circle $|z|<1$ slit along the real line segment from r to 1 onto the interior of the circle $|\zeta|<4r/(1+r)^2$ by means of the function $z=g(\zeta)=\zeta+b_2\zeta^2+b_3\zeta^3+\cdots$ [**108**] and then apply **147** to the boundary point $f(r)$ of the mapping $w=f[g(\zeta)]$.

153. [T. H. Gronwall: Nat. Acad. Proc. **6**, pp. 300–302 (1920); R. Nevanlinna, *ibid.* **161**, p. 17, footnote.] Setting $\zeta=1/z$, apply **143** to $[f(\zeta^{-1})]^{-1}+a_2$.

154. [T. Radó, problem: Jber. deutsch. Math. Verein. **32**, p. *15* (1923).] By **80**, $f(z)=\sqrt{g(z^2)}$ where $g(z)$ is regular and schlicht inside the unit circle $|z|<1$. Hence for $|z|=r$, we have [**152**]

$$\frac{r^2}{(1+r^2)^2} \leqq |g(z^2)| \leqq \frac{r^2}{(1-r^2)^2}.$$

155. With the notation of III **128** we have

$$J(\rho) \leqq \pi\left(\frac{\rho}{1-\rho^2}\right)^2 \qquad [\mathbf{154}],$$

and hence, since $f(0)=0$,

$$\int_0^{2\pi} |f(r\,e^{i\vartheta})|^2\,d\vartheta \leqq 4\pi \int_0^r \frac{\rho}{(1-\rho^2)^2}\,d\rho.$$

156. By Cauchy's theorem we have

$$f^{(n)}(z)=\frac{n!}{2\pi i}\oint \frac{f(\zeta)}{(\zeta-z)^{n+1}}\,d\zeta,$$

where the integration is carried out along a circle $|\zeta-z|=\rho$, $0<\rho<1-|z|=1-r$. If we take $\omega_n(r)$ to be the *smallest* function of the kind considered here, we have

$$\omega_n(r) < \frac{n!}{\rho^n}\frac{r+\rho}{(1-r-\rho)^2}.$$

157. [L. Bieberbach: Math. Z. **2**, pp. 161–162, footnote 5 (1918).] We have

$$\omega_n = \frac{\omega_n(0)}{n!} \qquad \textbf{[156]},$$

and hence

$$\omega_n < \frac{1}{\rho^{n-1}(1-\rho)^2}, \qquad 0<\rho<1.$$

Now set $\rho=(n-1)/(n+1)$.

158. In **155** replace $f(z)$ by $\sqrt{f(z^2)}$ [**80**] and r by $\sqrt{r}$. We obtain

$$\frac{1}{2\pi}\int_0^{2\pi} |f(r\,e^{2i\vartheta})|\,d\vartheta = \frac{1}{2\pi}\int_0^{2\pi} |f(r\,e^{i\varphi})|\,d\varphi \leqq \frac{r}{1-r}.$$

159. [J. E. Littlewood. Cf. Proc. Lond. Math. Soc. Series 2, **23**, pp. 481–519 (1925).] By considering the special function $f(z)=z/(1-z)^2=\sum_{n=1}^{\infty} nz^n$, we obtain that $\omega_n \geqq n$. The upper estimate of ω_n follows from

$$|a_n| \leqq \frac{1}{2\pi r^n}\int_0^{2\pi} |f(r\,e^{i\vartheta})|\,d\vartheta < \frac{1}{r^{n-1}(1-r)} \qquad \textbf{[158]}$$

if we choose the optimum value of r, $r=(n-1)/n$. Then we have

$$|a_n| < \frac{n^n}{(n-1)^{n-1}} = \left(1+\frac{1}{n-1}\right)^{n-1} n < en \qquad [\text{I } \textbf{170}].$$

Since the date of publication of the original German edition, much research has been devoted to the determination of ω_n. See *loc. cit.* solution **77**.

160. [G. Pick: Wien. Ber. **126**, pp. 247–263 (1917).] The assertion concerning M follows from **122** or from III **280**. From

$$\frac{f(z)}{[1+e^{i\alpha}M^{-1}f(z)]^2} = z + (a_2 - 2e^{i\alpha}M^{-1})z^2 + \cdots$$

we now deduce $|a_2 - 2e^{i\alpha}M^{-1}| \leqq 2$ [**146**]. With the optimum choice of α, $\alpha = \pi + \arg a_2$, we obtain the required inequality. The equality sign holds if and only if

$$\frac{f(z)}{[1+e^{i\alpha}M^{-1}f(z)]^2} = \frac{z}{(1+e^{i\beta}z)^2}, \qquad \beta \text{ real}.$$

Then the values of $w=e^{i\alpha}M^{-1}f(z)$ fall in that part of the unit circle to which in the mapping $w/(1+w)^2=W$ [**111**] the slit region with the bounding ray $\arg W = \alpha-\beta$, $1/(4M) \leqq |W| < +\infty$ corresponds. Therefore we must have $\alpha=\beta$. To the slit segment $1/(4M) \leqq W \leqq \frac{1}{4}$ there corresponds the line segment $2M-1-2\sqrt{M(M-1)} \leqq w \leqq 1$. The extreme case thus occurs if $f(z)$ effects a schlicht mapping of the unit circle on that slit region which is formed from the circle of radius M by cutting from an arbitrary point of the circumference along a straight line of length $2M\sqrt{M-1}\,(\sqrt{M}-\sqrt{M-1})$ at right angles into the interior of the circle.

161. [R. Nevanlinna: Översikt av Finska Vetenskaps-Soc. Förh. **63(A)**, no. 6 (1920–1921).] By III **109** we have that

$$z\frac{f'(z)}{f(z)} = 1 + \gamma_1 z + \gamma_2 z^2 + \cdots$$

is regular for $|z|<1$ and has positive real part there. Hence we have [III **235**]

$$z\frac{f'(z)}{f(z)} \ll \frac{1+z}{1-z}.$$

We deduce [I **63**]

$$|a_n| \leqq n, \qquad n=2, 3, 4, \ldots$$

We can have $|a_n|=n$ only if

$$z\frac{f'(z)}{f(z)}=\frac{1+e^{i\alpha}z}{1-e^{i\alpha}z}, \qquad \alpha \text{ real},$$

(in which case the equality holds for all n.)

162. [T. H. Gronwall, *ibid.* **151**; K. Löwner: Leipz. Ber. 69, pp. 89–106 (1917).] If $f(z)$ is a convex mapping then the image in the mapping $w=zf'(z)$ is star-shaped [III **110**]. Hence we have [**161**]

$$zf'(z) \ll \frac{z}{(1-z)^2}, \qquad \text{i.e.} \quad |a_n| \leqq 1, \quad n=2, 3, 4, \ldots.$$

We have $|a_n|=1$ if and only if $zf'(z)=z/(1-e^{i\alpha}z)^2$, $f(z)=z/(1-e^{i\alpha}z)$, α real (in which case the equality holds for all n).

163. [Cf. E. Study: Vorlesungen über ausgewählte Gegenstände der Geometrie, Part 2, Leipzig: B. G. Teubner 1913. R. Nevanlinna, *ibid.* **161**.] From **150** we deduce for $|z|<r$

$$\Re z\frac{f''(z)}{f'(z)} > \frac{-4r+2r^2}{1-r^2}.$$

The right side is $\geqq -1$, if r is not greater than the smaller root of the equation $r^2-4r+1=0$. [III **108**.]

164. By means of the mapping $(e^{iz}-i)/(e^{iz}+i)=\zeta$ the question may be reduced to that of III **297.** If $f(z)$ is not identically zero, then the series

$$\sum_{n=1}^{\infty}\left(1-\left|\frac{e^{i(x_n+iy_n)}-i}{e^{i(x_n+iy_n)}+i}\right|\right)$$

must be convergent. Furthermore, $y_n \to \infty$. But for positive y tending to infinity we have

$$1-\left|\frac{e^{i(x+iy)}-i}{e^{i(x+iy)}+i}\right|=1-\left(\frac{1+e^{-2y}-2e^{-y}\sin x}{1+e^{-2y}+2e^{-y}\sin x}\right)^{\frac{1}{2}} \sim 2e^{-y}\sin x.$$

In a similar manner we can replace the general term of the above series for negative y by $2e^{y}\sin x$.

165. That the condition is sufficient follows from the general existence theorem for linear differential equations. On the other hand it is also necessary. We let $u_1, u_2, \ldots, u_n$ be n linearly independent entire functions that satisfy the given equation. From the equations

$$-u_j^{(n)}=f_1(z)u_j^{(n-1)}+f_2(z)u_j^{(n-2)}+\cdots+f_n(z)u_j, \qquad j=1, 2, \ldots, n,$$

the coefficients $f_\nu(z)$ may be calculated as fractions whose numerators are entire functions. The common denominator of these functions is the Wronksian determinant of $u_1, u_2, \ldots, u_n$, that is $e^{-\int f_1(z)\,dz}$ [VII § 5], and thus an entire function which has no zeros.

166. For sufficiently small values of z, different from 0, we have

$$\varphi(z)=\gamma_{-k}z^{-\frac{k}{q}}+\gamma_{-k+1}z^{-\frac{k-1}{q}}+\cdots+\gamma_0+\gamma_1 z^{\frac{1}{q}}+\gamma_2 z^{\frac{2}{q}}+\cdots,$$

where q is a positive integer and $z^{\frac{1}{q}}$ denotes a fixed branch. From the assumptions of the problem it follows firstly that $\varphi(z)$ remains bounded as $z\to 0$, and hence $\gamma_{-k}=\gamma_{-k+1}=\cdots=\gamma_{-1}=0$. Furthermore $\gamma_0=c_0$. If it were possible that $q>1$ and not all the coefficients $\gamma_1,\gamma_2,\ldots,\gamma_{q-1}$ were equal to 0, say $\gamma_1=\gamma_2=\cdots=\gamma_{s-1}=0$, $\gamma_s\neq 0$, $1\leqq s\leqq q-1$, then the function

$$z^{-\frac{s}{q}}(\varphi(z)-c_0)$$

would be greater in magnitude than a fixed positive number as $z\to 0$. From this we could deduce that

$$\lim_{z\to 0} z^{-1}(\varphi(z)-c_0)=\infty:$$

Contradiction. Hence $\gamma_1=\gamma_2=\cdots=\gamma_{q-1}=0$. Further $\gamma_q=c_1$, and then again $\gamma_{q+1}=\gamma_{q+2}\cdots=\gamma_{2q-1}=0$ and so on.

167. If the function $f(z)$ is regular and single-valued in the half-open annular region $R\leqq|z|<\infty$, then its zeros in the region can have a point of accumulation only at infinity. We may therefore construct an entire function $g(z)$ that has in this region the same zeros as $f(z)$. Then the function $\log[f(z)/g(z)]$ is regular but not necessarily single-valued for $R\leqq|z|<\infty$. Its real part is single-valued and its imaginary part increases by $2\pi im$ in traversing the circumference of the circle $|z|=r$, $r>R$ in the positive sense, where m is an integer *independent of* r [III **190**, $a=0$]. Hence we have that

$$\log\frac{f(z)}{g(z)}-m\log z$$

is single-valued and regular for $R\leqq|z|<\infty$ and may there be represented by a Laurent series:

$$\log\frac{f(z)}{g(z)}-m\log z=\sum_{n=-\infty}^{\infty} c_n z^n=\gamma(z)+\sum_{n=-\infty}^{-1} c_n z^n=\gamma(z)+\psi(z),$$

where $\gamma(z)$ is an entire function and the series $\psi(z)$ is convergent for $|z|\geqq R$. We may set $z^m g(z)\,e^{\gamma(z)}=z^{-p}G(z)$.—If $f(z)$ is merely assumed to be regular in the open annular region $R<|z|<\infty$, then its zeros may also have points of accumulation on the circumference of the circle $|z|=R$.

168. Set $e^{-iz}=w$, and hence $e^y=|w|$. Because of the periodicity of $f(z)$, the function $f(i\log w)=F(w)$ is a single-valued function of w that has only finitely many zeros and poles in the annular region $0<|w|<\infty$. We may therefore determine the polynomials $P(w)$, $Q(w)$ and a power-series $\sum_{n=-\infty}^{\infty} c_n w^n=\psi(w)$ convergent for $0<|w|<\infty$ [solution **167**] so that in this region we have

$$f(w)=\frac{P(w)}{Q(w)}\,e^{\psi(w)}.$$

Denote by $A(r)$ the maximum of $\Re\psi(w)$ on the circumference of the circle $|w|=r$ and by w_0 the point at which $A(r)$ is attained. We have

$$\log M(y) - \log |P(w_0)| + \log |Q(w_0)| \geqq A(r).$$

From the assumption it follows furthermore that there exists a number θ, $0<\theta<1$, such that

$$\log M(y) < e^{\theta|y|} = \begin{cases} |w|^{\theta} \\ |w|^{-\theta}, \end{cases}$$

according as $|w| \geqq 1$ or $|w| \leqq 1$. From this we deduce that $A(r)r^{-\theta}$ is bounded for $r>1$ and $A(r)r^{\theta}$ is bounded for $r<1$. From solution III **237** it follows that $\psi(w)=$ const.

169. [H. A. Schwarz, oral communication.] Let

$$f(z) = a_0 + a_1 z + a_2 z^2 + \cdots + a_n z^n + \cdots$$

be regular and not constant in a circle $|z|<\rho$ about the origin. Then in this circle

$$F(z) = \overline{a_0} + \overline{a_1} z + \overline{a_2} z^2 + \cdots + \overline{a_n} z^n + \cdots$$

is also regular and not constant. Furthermore for real values of x and y satisfying $x^2+y^2<\rho^2$ we have

$$\varphi(x, y) = f(x+iy)F(x-iy) = \sum_{k=0}^{\infty} \sum_{l=0}^{\infty} a_k \bar{a}_l (x+iy)^k (x-iy)^l.$$

This series is also convergent for *complex* x and y of sufficiently small absolute value. As it represents an algebraic function of x and y, we have identically in x and y

$$\sum_{\nu=0}^{n} \Phi_\nu(x, y)[\varphi(x, y)]^\nu = \sum_{\nu=0}^{n} \Phi_\nu(x, y)[f(x+iy)F(x-iy)]^\nu = 0,$$

where $\Phi_\nu(x, y)$, are polynomials in x and y, $\Phi_n(x, y) \not\equiv 0$. We may assume that $\Phi_n(0, 0) \neq 0$, for otherwise we could have considered $f(z-a)$ with suitable a instead of $f(z)$ from the beginning.

Let a variable ζ be restricted to the unit circle and let a constant t be chosen as follows:

(1) $t \neq 0$;

(2) the power series for the function $\varphi(x, y)$ considered above is convergent for $|x| \leqq t$, $|y| \leqq t$;

(3) $F(t) \neq 0$;

(4) $\Phi_n\left(\frac{(\zeta+1)t}{2}, \frac{(\zeta-1)t}{2i}\right)$ does not vanish identically in ζ. This condition can always be satisfied, since the constant term of this polynomial in ζ and t coincides with that of $\Phi_n(x, y)$, and thus is different from 0.

We now have identically for $|\zeta|<1$

$$\sum_{\nu=0}^{n} \Phi_\nu\left(\frac{(\zeta+1)t}{2}, \frac{(\zeta-1)t}{2i}\right)[F(t)]^\nu [f(\zeta t)]^\nu = 0.$$

170. [E. Landau.] Assume that the circle in the problem is the unit circle. It contains the real interval $-1<w<1$. Consider the set of those real points z at which the value of the function $w=f(z)$ is real and of absolute value <1. These points z have at least one finite limit point z_0. For otherwise their number would be countable and hence $f(z)$ could not assume all values in the interval $-1<w<1$ whose cardinality is that of the continuum. $f(z_0)$ is real, and so is

$$f'(z_0)=\lim \frac{f(z)-f(z_0)}{z-z_0},$$

where we are free to choose for z real values for which $f(z)$ is also real. Similarly we can demonstrate that all the coefficients in the expansion of $f(z)$ in powers of $z-z_0$ are real, and by continuation of the power series it follows that $f(z)$ is real for all real z: Contradiction.

171. Such functions do not exist. For assume that the relation in question is satisfied. If $f(z)$ is identically 0, then obviously also $f_1(z), f_2(z), \ldots, f_n(z)$ are identically 0. If $f(z)$ is not identically 0, we may assume that $f(z)$ does not vanish anywhere in $\mathfrak{D}$ (if necessary we may replace $\mathfrak{D}$ by a sub-domain). Hence $f_\nu(z)f(z)^{-1}$ is regular in $\mathfrak{D}$ and the function

$$\left|\frac{f_1(z)}{f(z)}\right|+\left|\frac{f_2(z)}{f(z)}\right|+\cdots+\left|\frac{f_n(z)}{f(z)}\right|=1$$

also obtains its maximum in the interior of $\mathfrak{D}$. Hence [III **300**]

$$f_\nu(z)f(z)^{-1}=\text{const.}, \qquad \nu=1, 2, \ldots, n.$$

172. Such functions do not exist. For if we confine ourselves to a sub-domain of $\mathfrak{D}$ where $g(z)\neq 0$ and $\sqrt{g(z)}=\varphi(z)$ is regular, then we have $|g(z)|=|\varphi(z)|^2$. By assumption $|\varphi(z)|^2$ is a regular harmonic function, i.e. [III **87**]

$$\left(\frac{\partial^2}{\partial x^2}+\frac{\partial^2}{\partial y^2}\right)|\varphi(x+iy)|^2=0.$$

By III **58** we hence deduce that $\varphi(z)=\text{const.}$, $g(z)=\text{const.}$, $f(z)=\text{const.}$

172.1. [Cf. G. Pólya, problem: Jber. deutsch. Math. Ver., Section 2, **35**, p. 48 (1926). Solution by H. Kneser, op. cit. **35**, p. 119 (1926), and by N. Tchebotarev, op. cit. **36**, pp. 40–41 (1927).]

(1) If $z_1\neq z_2$ and $g(z_1)=g(z_2)=0$, the convexity implies that $g(z)=0$ along the line segment that joins z_1 and z_2, and so $g(z)=0$ identically.

(2) If $g(z)$ has a unique zero z_0, let δ be the minimum of $|g(z)|$ on the circle $|z-z_0|=1$. The convexity implies that $|g(z)|>\delta$ for $|z-z_0|>1$ and so $g(z)$ must be a polynomial having a unique, possibly multiple, zero at the point z_0.

(3) If $g(z)$ never vanishes, $\log g(z)$ is an entire function. The points where $|g(z)|=e^\alpha$ form the boundary line L of a convex domain where $|g(z)|\leqq e^\alpha$. By III **191**, $\mathfrak{J}\log g(z)$ varies monotonically when z moves along L. Hence $\log g(z)$ can take no value more than once and so it must be a linear function of z.

173. Let $h(z)$ be an entire function. $f(z)=e^{h(z)}$ and $g(z)=e^{-h(z)}$ satisfy the conditions with $a=1$, $b=-1$, $c=0$ for

$$\frac{e^h-1}{e^{-h}-1}=-e^h, \qquad \frac{e^h+1}{e^{-h}+1}=e^h, \qquad \frac{e^h-0}{e^{-h}-0}=e^{2h}$$

are entire functions with no zeros.—Note that there do not exist two distinct entire functions of finite order that have the same $a-$, $b-$, $c-$ and $d-$ points (a, b, c, d all distinct). [G. Pólya: Nyt Tidsskr. for Math. (B) **32**, p. 21 (1921).]

174. There exists such a function $G(z)$. An even more stringent condition can be satisfied: Not only $G^{(n)}(n)$, but also the lower order derivatives $G(n)$, $G'(n)$, $G''(n), \ldots, G^{(n-1)}(n)$ can be arbitrarily prescribed, for $n=1, 2, 3, \ldots$. In fact it is more appropriate to discuss the solution to a more general interpolation problem for entire functions; see **174.1**.

174.1. There exists an entire function $h(z)$ that has zeros of order m_n at the points z_n, $n=1, 2, \ldots$, and no other zeros (see Hurwitz-Courant, pp. 120–124, or Ahlfors, p. 157). It has an expansion of the form

$$h(z)=(z-z_n)^{m_n}(h_{n,0}+h_{n,1}(z-z_n)+h_{n,2}(z-z_n)^2+\cdots),$$

where $h_{n,0}\neq 0$ for $n=1, 2, 3, \ldots$. According to the condition that we are required to satisfy

$$\frac{f(z)}{h(z)}=\frac{1}{(z-z_n)^{m_n}}\frac{w_{n,0}+w_{n,1}(z-z_n)+\cdots+w_{n,m_n-1}(z-z_n)^{m_n-1}+\cdots}{h_{n,0}+h_{n,1}(z-z_n)+\cdots+h_{n,m_n-1}(z-z_n)^{m_n-1}+\cdots}$$

$$=P_n\left(\frac{1}{z-z_n}\right)+f_n(z),$$

where $f_n(z)$ is an analytic function, regular at the point z_n, and $P_n(t)$ is a polynomial without constant term, of degree not exceeding m_n, that is completely determined by the numbers $w_{n,k}$ and $h_{n,k}$, where $0\leqq k<m_n$. There exists a meromorphic function $m(z)$ that has no poles except at the points $z_1, z_2, z_3, \ldots$ and whose singular part at the point z_n is precisely $P_n[1/(z-z_n)]$ (see Hurwitz-Courant, pp. 108–110, or Ahlfors, pp. 149–150). The required entire function is

$$f(z)=m(z)h(z).$$

174.2. Let $G(z)$ have the zeros z_n of order m_n, $n=1, 2, 3, \ldots$, and no other zeros. Therefore $F(z_n)\neq 0$. The proposition requires that $1-F(z)f(z)$ should have a zero of order not less than m_n at the point z_n, that is

$$(1-F(z)f(z))^{(k)}=0$$

for $z=z_n$ and $0\leqq k<m_n$. These m_n equations determine uniquely

$$f(z_n)=1/F(z_n),\ f'(z_n),\ \ldots,\ f^{(m_n-1)}(z_n)$$

for $n=1, 2, 3, \ldots$. But there exists an entire function $f(z)$ that interpolates the values so determined [**174.1**]. Now

$$\frac{1-F(z)f(z)}{G(z)}=g(z)$$

is an entire function.

175. The first inequality is obvious. Further from

$$|a_n|\leqq\frac{M(r+\delta)}{(r+\delta)^n},\qquad n=0, 1, 2, \ldots$$

it follows that

$$\mathfrak{M}(r) \leqq M(r+\delta) \sum_{n=0}^{\infty} \left(\frac{r}{r+\delta}\right)^n = \frac{r+\delta}{\delta} M(r+\delta).$$

176. [E. Landau.] We have [**175**]

$$[M(r)]^n \leqq \mathfrak{M}_n(r) \leqq \frac{r+\delta}{\delta} [M(r+\delta)]^n,$$

and hence

$$M(r) \leqq [\mathfrak{M}_n(r)]^{\frac{1}{n}} \leqq \left(\frac{r+\delta}{\delta}\right)^{\frac{1}{n}} M(r+\delta).$$

Let n tend to ∞; $\delta > 0$ is arbitrarily small, $M(r)$ is a continuous function of r as can easily be seen from the definition.

177. [E. Landau.] $(1/n) \log \mathfrak{M}_n(r)$ is a convex function of $\log r$ [II **123**]. The limit of convex curves is convex [**176**].

178. [I. Schur.] The function

$$\varepsilon^{-1} M f[\varepsilon M^{-1} f(z)] = z + A_2 z^2 + A_3 z^3 + \cdots + A_n z^n + \cdots$$

is schlicht and its absolute value $< M^2$ for $|z| < 1$. The coefficient A_n is a polynomial of degree $(n-1)$ in ε, and the constant term of this polynomial $= a_n$ and its leading coefficient $= M^{-n+1} a_n$. If we substitute successively in place of ε the $(n-1)$th roots of unity, we obtain by addition and by applying the inequality $|A_n| \leqq \omega_n[M^2]$,

$$|a_n + M^{-n+1} a_n| \leqq \omega_n[M^2], \qquad \text{i.e.} \quad \omega_n[M] \leqq \frac{\omega_n[M^2]}{1 + M^{-n+1}}.$$

Here $\omega_n[M]$ denotes the upper bound of $|a_n|$ under the assumptions of **160**. Applying this result repeatedly, we obtain, since $\omega_n[M] \leqq \omega_n$,

$$\omega_n[M] \leqq \omega_n \prod_{\nu=0}^{\infty} \frac{1}{1 + M^{(-n+1)2^\nu}} = \omega_n(1 - M^{-n+1})$$

[solution I **14**].

179. Let m be an arbitrary positive integer. Then $\varphi(m\vartheta)$ is a trigonometric polynomial of mnth order that satisfies the assumptions. Thus we have for all values of ϑ

$$|m\varphi'(m\vartheta)| \leqq mn + K, \qquad \text{i.e.} \quad |\varphi'(\vartheta)| \leqq n + \frac{K}{m}.$$

Let m tend to ∞.

180. [H. Poincaré: Amer. J. Math. **14**, p. 214 (1892), cf. H. Bohr: Nyt Tidsskr. for Math. (B) **27**, pp. 73–78 (1916).] Determine positive integers $\lambda_1, \lambda_2, \ldots, \lambda_n, \ldots$ satisfying

$$\lambda_1 = 1, \qquad \lambda_n > \lambda_{n-1}, \qquad \left(\frac{n}{n-1}\right)^{\lambda_n} > \varphi(n+1), \qquad n = 2, 3, 4, \ldots$$

The function

$$g(z) = \varphi(2) + 1 + \sum_{n=2}^{\infty} \left(\frac{z}{n-1}\right)^{\lambda_n}$$

is an entire function. For $n \leqq x \leqq n+1$, $n \geqq 2$, we have

$$g(x) \geqq g(n) > \left(\frac{n}{n-1}\right)^{\lambda_n} > \varphi(n+1) \geqq \varphi(x).$$

181. Example: $\sin \sqrt{z}/\sqrt{z} = 1 - z/3! + z^2/5! - \cdots$; rational entire functions (that are not constant) tend to ∞ in all directions.

182. [H. von Koch: Ark. för Mat., Astron. och Fys. **1**, pp. 627–641 (1903).] Example: $e^z + z$. In the proof we must distinguish between the two cases: $-\pi/2 < \arg z < \pi/2$, $\pi/2 \leqq \arg z \leqq 3\pi/2$. Uniform convergence is not possible. [Hurwitz-Courant, p. 96.]

183. [Cf. J. Malmquist: Acta Math. **29**, pp. 203–215 (1905).] [III **158**, III **160**.]—An entire function of finite order cannot behave in this manner [III **330**].

184. [G. Mittag-Leffler: Proceedings of the 3rd International Congress of Mathematicians in Heidelberg 1904. Leipzig: B. G. Teubner 1905, pp. 258–264. Atti del IV. congr. internaz. dei mat. Roma 1908, **1**. Rome: Tip. della Acc. dei Lincei 1909, pp. 67–85.]

$$e^{-E(z)} - e^{-E(2z)} \quad \text{or} \quad E(z)e^{-E(z)},$$

where $E(z)$ denotes the function considered in **183**.

185. The function $E(z)$ [III **158**] has either no negative zeros or else only finitely many [III **160**]. In the first case set

$$F(z) = E(z)$$

and in the second case

$$F(z) = \frac{E(z)}{(z+\alpha_1)(z+\alpha_2)\ldots(z+\alpha_l)},$$

where $-\alpha_1, -\alpha_2, \ldots, -\alpha_l$ denote all the negative zeros of $E(z)$. In any case $F(z)$ is real and of constant sign for real negative z. Hence we have

$$i \int_0^\infty F(-t^2)\, dt = C \neq 0.$$

The existence of the integral follows from III **160**. The odd entire function

$$g(z) = C^{-1} \int_0^z F(z^2)\, dz$$

furnishes an example of a function with the required properties, as can be shown by displacing the path of integration [III **160**].

186. [G. Pólya: Interméd. des math. Series 2, **1**, pp. 81–82 (1922).] Assume $0 < \alpha < 2\pi$ and let $g(z)$ be the odd entire function constructed in **185**. The function

$$\frac{1 + g(\sqrt{z})g(\sqrt{e^{i\alpha}z})}{2}$$

is an entire function and, if the sign of the two square roots is appropriately chosen, tends to 1 in the sector $0 < \arg z < 2\pi - \alpha$ and to 0 in the sector $2\pi - \alpha < \arg z < 2\pi$. By means of a suitable linear combination of such functions one obtains an example of a function with the required properties.

187. No. For the set of all possible divisions has a higher cardinality ($2^{\mathfrak{c}}$) than that of the continuum ($\mathfrak{c}$), while the set of all entire functions has cardinality $\mathfrak{c}^{\aleph_0}=\mathfrak{c}$, i.e. that of the continuum. [What limits along the rays emanating from the origin can be prescribed for an entire function? See A. Roth: Commentarii Math. Helvet. **11**, pp. 77–125 (1938), especially theorem IV(b).]

188. If there existed a continuous curve L going to infinity, along which e^z approached a limit different from 0 and ∞, then for every ε, $\varepsilon>0$, there would exist a point z_0 on L, such that $|\arg e^z-\arg e^{z_0}|<\varepsilon$ for all points z past the point z_0 on the curve L, where the determination of the argument is made appropriately.

If $\varepsilon<\pi$, then, since the argument depends continuously on z, it follows that $|\Im z-\Im z_0|<\varepsilon$, and hence L, from the point z_0 on, would have to lie in a strip of width 2ε parallel to the real axis. But in such a strip z could approach ∞ only in one of two ways, either such that e^z tends to ∞ or such that e^z tends to 0.

189. The function $\int_0^z e^{-\frac{x^2}{2}}\,dx=\sum_{n=0}^{\infty}\frac{(-1)^n}{n!}\,\frac{z^{2n+1}}{2^n(2n+1)}$ tends to ∞, as z tends to infinity along the positive or negative imaginary axis. Further it converges uniformly to $\sqrt{\pi/2}$ and $-\sqrt{\pi/2}$ as z tends to infinity in the sectors $-\pi/4\leqq\arg z\leqq\pi/4$ and $3\pi/4\leqq\arg z\leqq 5\pi/4$, respectively. For we have for $z=re^{i\vartheta}$, $r>0$, $-\pi/4\leqq\vartheta\leqq\pi/4$,

$$\int_0^{re^{i\vartheta}} e^{-\frac{x^2}{2}}\,dx=\int_0^r e^{-\frac{x^2}{2}}\,dx+\int_r^{re^{i\vartheta}} e^{-\frac{x^2}{2}}\,dx,$$

and the absolute value of the second term is less than $r\int_0^{\pi/4} e^{-\frac{r^2}{2}\cos 2\vartheta}\,d\vartheta$ [solution III **151**]. If there were a continuous curve going to infinity along which the function in question converged to a finite limit different from $\pm\sqrt{\pi/2}$, then this curve at a sufficient distance from the origin would have no point in the above sectors and also no point on the imaginary axis. We may thus assume for example that it lies entirely in the sector $\pi/4<\arg z<\pi/2$. However in that case the function would be bounded in the region between the curve and the ray $\arg z=\pi/4$ [use a generalization of III **330** which bears the same relation to it as III **324** bears to III **322**], and hence [III **340**] it would also have to converge to $\sqrt{\pi/2}$ along the above curve. [Cf. A. Hurwitz: C.R. Acad. Sci. (Paris) **143**, p. 879 (1906) and **144**, p. 65 (1907).]

190. The given function converges to ∞ along the $2n$ rays $\arg z=(2k-1)[\pi/(2n)]$, $k=1,2,\ldots,2n$, and to finite limits along the bisectors of the angles of the $2n$ sectors determined by these rays. More precisely, it converges along the ray $\arg z=k(\pi/n)$ to the limit

$$e^{ik\frac{\pi}{n}}\int_0^{\infty}\frac{\sin(x^n)}{x^n}\,dx=e^{ik\frac{\pi}{n}}\frac{1}{n-1}\,\Gamma\left(\frac{1}{n}\right)\sin\left(\frac{n-1}{n}\,\frac{\pi}{2}\right),\qquad k=1,2,\ldots,2n.$$

[III **152**]. The proof then proceeds similarly as in solution **189**.

191. [F. Iversen: Öfversikt av Finska Vetenskaps-Soc. Förh. **58**(A), No. 3 (1915–1916).] Set $\delta_m=2(1-\varepsilon_m)$. Then $\delta_m=0$ or 4 according as $\varepsilon_m=+1$ or $\varepsilon_m=-1$ respectively. If z tends to infinity along the ray

$$\arg z=2\pi\sum_{m=0}^{\infty}\frac{\delta_m}{8^{m+1}}$$

then we have for $m=0, 1, 2, \ldots$

$$\arg z^{8^m}=2\pi\left(\frac{\delta_m}{8}+\frac{\delta_{m+1}}{8^2}+\cdots\right).$$

Since δ_m can only assume the values 0 and 4, it follows that z^{8^m} lies either in the sector $(0, \pi/4)$ or in the sector $(\pi, 5\pi/4)$ according as $\varepsilon_m=+1$ or $\varepsilon_m=-1$, respectively. Hence [**189**] we have, setting $\gamma(z)=\sqrt{2/\pi}\int_0^z e^{-\frac{x^2}{2}}\,dx$,

$$\lim \gamma(z^{8^m})=\varepsilon_m.$$

Moreover [solution **189**] there exists a number G such that for all z on the above ray and for all values of m we have

$$|\gamma(z^{8^m})|<G$$

Now take $\varepsilon>0$ and m so large that we have $a_{m+1}+a_{m+2}+\cdots<\varepsilon$. As z tends to infinity in the special manner described above, the function

$$a_0\gamma(z)+a_1\gamma(z^8)+\cdots+a_m\gamma(z^{8^m})$$

converges to the value $\varepsilon_0 a_0+\varepsilon_1 a_1+\cdots+\varepsilon_m a_m$. The remainder is in absolute value $<G\varepsilon$ for all z.

192. Let the entire function $g(z)$ in question be of order λ, where λ is finite, $\lambda>0$, i.e. $|g(z)|<A\, e^{B|z|^{\lambda+\varepsilon}}$, where ε, A, B, are constants, $\varepsilon>0$, $A>0$, $B>0$. Furthermore let $\lim g(z)=a$ as z tends to infinity along a ray emanating from the origin, and let $\lim g(z)=b\neq a$ as z tends to infinity along another ray emanating from the origin. Let the angle between the two rays be γ. It follows from III **330** $[\gamma=\beta-\alpha]$ that, if $(\lambda+\varepsilon)\gamma<\pi$, $g(z)$ remains bounded in the sector between the two rays. However, by III **340** we would then have $a=b$: Contradiction. Thus we have $(\lambda+\varepsilon)\gamma\geqq\pi$ for arbitrary positive ε, i.e. $\lambda\gamma\geqq\pi$, $\lambda\geqq\pi/\gamma$. By assumption there exists at least one number γ satisfying $\gamma\leqq 2\pi/n$, i.e. $\lambda\geqq n/2$. [For further results see T. Carleman: Ark. för Mat. Astron. och Fys. **15**, No. 10 (1920) and L. Ahlfors: Ann. Soc. Sci. Fenn., Nova Series A, **1**, No. 9 (1930) (proof of a conjecture by A. Denjoy).]

193. [Cf. F. Iversen: Thèse, Helsingfors 1914.] Let $g(z)$ be the given entire function and let $z^{-1}[g(z)-g(0)]=h(z)$ be not constant. Denote by $\mathfrak{R}$ a connected region at whose finite boundary points $|h(z)|=1$ and in whose interior $|h(z)|>1$. The point $z=\infty$ belongs to the boundary of $\mathfrak{R}$ [III **338**]. Along a line that terminates at $z=\infty$ and lies in $\mathfrak{R}$ or runs along the boundary of $\mathfrak{R}$, we have

$$|g(z)|\geqq|zh(z)|-|g(0)|\geqq|z|-|g(0)|.$$

194. Let m be the number of zeros of $g(z)-a$. Determine the polynomial $P(z)$ of degree m such that $[P(z)]/[g(z)-a]$ is regular in the whole z-plane, and then the polynomial $Q(z)$ of degree at most m such that also

$$\frac{1}{z^{m+1}}\left(\frac{P(z)}{g(z)-a}-Q(z)\right)$$

is regular in the whole z-plane. There hence exists [**193**] a continuous curve going to infinity such that at points sufficiently far along it we have

$$\left|\frac{P(z)}{g(z)-a}-Q(z)\right|>|z|^{m+1}.$$

We have there $|Q(z)|<A|z|^m$, $|P(z)|<B|z|^m$, where A and B are positive constants, and hence

$$\left|\frac{P(z)}{g(z)-a}\right|>|z|^{m+1}-A|z|^m, \qquad \left|\frac{1}{g(z)-a}\right|>\frac{|z|-A}{B}.$$

195. [T. Carleman.] Assume $f(z)\leqq M$ in the annular region $0<|z|<1$. If $|z_0|=\frac{1}{2}$, then by Cauchy's theorem we obtain

$$\left|\frac{f^{(n)}(z_0)}{n!}\right|\leqq\frac{M}{(\frac{1}{2})^n}.$$

Since $f^{(n)}(z)$ was assumed to be *bounded*, we have [III **337**] $|f^{(n)}(z)|\leqq n!2^nM$ in the annular region $0<|z|\leqq\frac{1}{2}$, and in particular for $z=\frac{1}{4}$. Hence we have that

$$f(z)=f(\tfrac{1}{4})+\frac{f'(\frac{1}{4})}{1!}(z-\tfrac{1}{4})+\cdots+\frac{f^{(n)}(\frac{1}{4})}{n!}(z-\tfrac{1}{4})^n+\cdots$$

is convergent for $|z-\frac{1}{4}|<\frac{1}{2}$, and hence in a circular disc covering the point $z=0$.

196. Let $w_\nu\neq 0$, $\nu=1,2,3,\ldots$ and

$$g(z)=\prod_{\nu=1}^{\infty}\left(1-\frac{z}{w_\nu}\right), \qquad g^*(z)=\prod_{\nu=1}^{\infty}\left(1+\frac{z}{|w_\nu|}\right), \qquad M^*(r)=g^*(r),$$
$$m^*(r)=|g^*(-r)|.$$

Then we clearly have for $|z|=r$

$$m^*(r)\leqq|g(z)|\leqq M^*(r).$$

Let us take the most unfavourable case and then we may assume without loss of generality that all the zeros of $g(z)$ are real and negative, i.e.

$$g(z)=\prod_{\nu=1}^{\infty}\left(1+\frac{z}{|w_\nu|}\right).$$

The assertion then may be stated as follows: For all sufficiently large r: $g(r)<e^{\varepsilon r}$ and for arbitrarily large r: $|g(-r)|>e^{-\varepsilon r}$ where ε is a given positive number. The first inequality is obtained as in the case $\lambda=1$ of solution **58**. To prove the second inequality, apply theorem III **332** to $g(z)g(-z)$ considered as a function of z^2. We have for arbitrarily large r

$$|g(r)g(-r)|>1, \qquad |g(-r)|>\frac{1}{g(r)}>e^{-\varepsilon r}.$$

197. [A. Wiman: Ark. för Mat., Astron. och. Fys. **2**, No. 14 (1905); cf. E. Lindelöf: Palermo Rend. **25**, p. 228 (1908).] We assume that the zeros of the entire function $g(z)$ considered, which according to Hadamard must be of genus zero [cf. remarks preceding **58**], are real and negative [solution **196**]. We consider the function $g(z)\,e^{-z^{\lambda-\varepsilon}}$ in the sector $-\pi<\vartheta<+\pi$ in which it is analytic.

This function is continuous for $-\pi \leqq \vartheta \leqq +\pi$ and assumes arbitrarily large values on the positive real axis [assumption concerning $M(r)$]. If it were bounded on the real negative axis, for example if it were $\leqq 1$, then it would be bounded everywhere [III **332**]. This is not the case as we have just noted, and hence we must have for certain arbitrarily large values of r

$$m(r)\, e^{-r^{\lambda-\varepsilon} \cos \pi(\lambda-\varepsilon)} = |g(-r)\, e^{-(-r)^{\lambda-\varepsilon}}| > 1,$$

q.e.d. [A more precise and more comprehensive result (encompassing the case $\lambda < 1$), stated by J. E. Littlewood and first proved by A. Wiman and G. Valiron, was also proved by G. Pólya: J. Lond. Math. Soc. **1**, pp. 78–86 (1926).]

198. [Ch. H. Müntz, cf. Math. Abhd. dedicated to H. A. Schwarz. Berlin: J. Springer 1914, pp. 303–312. T. Carleman: Ark. för Mat., Astron. och Fys. **17**, No. 9, p. 15 (1923).] The integral

$$\int_0^1 t^z h(t)\, dt = f(z)$$

is convergent, if z lies in the half-plane $\Re z > -1$, and $f(z)$ is analytic there. For $\Re z \geqq 0$ the integral is proper and $|f(z)|$ is bounded, $|f(z)| \leqq \int_0^1 |h(t)\, dt$. If there are infinitely many λ_ν which are $\leqq 1$, then they must have a point of accumulation on the line segment $0 \leqq z \leqq 1$, and $f(z)$ is identically equal to 0. If all the λ_ν with the exception of a finite number are $\geqq 1$, then from the divergence of $\sum_{n=1}^{\infty} \lambda_n^{-1}$ there follows again that $f(z)$ is identically equal to 0 [III **298**]. At any rate we have that $f(n) = \int_0^1 t^n h(t)\, dt = 0$, $n = 0, 1, 2, \ldots$ [II **139**].

199. [T. Carleman.] The function $\int_0^1 g(zt)h(t)\, dt$ is entire and of order $\leqq \lambda$, and hence by the assumptions the function

$$\gamma(z) = \frac{\int_0^1 g(zt)h(t)\, dt}{g(z)}$$

is also an entire function and of order $\leqq \lambda$. Let $M(r)$ be the maximum of $|g(z)|$ on the circumference of the circle $|z| = r$. The minimum of $|\gamma(z)|$ on the circumference of the circle $|z| = r$ is

$$\leqq \frac{M(\alpha r)}{M(r)} \int_0^\alpha |h(t)|\, dt + \int_\alpha^1 |h(t)|\, dt,$$

where $0 < \alpha < 1$, and α is arbitrarily close to 1, and hence it converges to 0 as $r \to \infty$ [**24**], although it should be unbounded if the order of $\gamma(z)$ were > 0 [cf. **197**]. Hence $\gamma(z)$ is a constant and moreover equal to 0. Hence all the coefficients of $\int_0^1 g(zt)h(t)\, dt$ are equal to 0. Hence by assumption $\int_0^1 t^n h(t)\, dt = 0$, $n = 0, 1, 2, \ldots$ [II **139**].

200. [T. Carleman.] Let $g(z) = \prod_{\nu=1}^{\infty} [1 - (z/w_\nu)]$, $w_\nu \neq 0$ (it is sufficient to investigate this case). Then we have $|g(iy)|^2 = \prod_{\nu=1}^{\infty} [1 + (y^2/w_\nu^2)]$. From this formula it follows that $\lim_{y \to \infty} |g(i\alpha y)|/|g(iy)| = 0$, where α is a fixed number, $0 < \alpha < 1$ [**24**]. Hence the function

$$\gamma(z) = \frac{\int_0^1 g(zt)h(t)\, dt}{g(z)}$$

tends to 0 along the positive and along the negative imaginary axis. Moreover it is entire by assumption. On all sufficiently large circles $|z|=r$, we have $|g(z)|<e^{\varepsilon r}$ and on certain arbitrarily large circles $|g(z)|>e^{-\varepsilon r}$ [**196**]. Moreover $\left|g\left(r\,e^{\frac{i\pi}{4}}\right)\right|> e^{-\varepsilon r}$, $\left|g\left(r\,e^{\frac{3i\pi}{4}}\right)\right|>e^{-\varepsilon r}$ for all sufficiently large r. For we have

$$|g(z)g(-z)|^2=\prod_{\nu=1}^{\infty}\left|1-\frac{z^2}{w_\nu^2}\right|^2=\prod_{\nu=1}^{\infty}\left(1-\frac{r^2\cos 2\vartheta}{w_\nu^2}+\frac{r^4}{w_\nu^4}\right)\geqq 1,$$

for $\vartheta=\pi/4$, $3\pi/4$.

Apply III **325**, slightly modified, to the function $\gamma(z)$ in the two half-planes $\Re z\geqq 0$ and $\Re z\leqq 0$. Assumption (1) must be extended in the sense of III **323**. The ray $\vartheta=0$ is to be replaced by $\vartheta=\pi/4$ and $\vartheta=3\pi/4$ respectively in the two cases. [Concluding remark in solution III **325**.] We deduce that $|\gamma(z)|\leqq$const. in the whole plane, and hence $\gamma(z)\equiv$const., $\gamma(z)\equiv 0$. Hence we have $\int_0^1 t^n h(t)\,dt=0$, if $g(z)=\sum_{n=0}^{\infty} a_n z^n$ and $a_n\gtreqless 0$. However since the roots are real, at least one of a_n, a_{n+1} is different from 0 [V **166**]. Finally apply **198**.

201. [S. Bernstein: C.R. Acad. Sci. (Paris) **176**, pp. 1603–1605 (1923).] From $|a_n|\leqq K\rho^n$ it follows that

$$|F(z)|\leqq\sum_{n=0}^{\infty}\frac{K\rho^n|z|^n}{n!}=K\,e^{\rho|z|}.$$

The function $F(z)\,e^{i\rho z}$ satisfies conditions analogous to those given in III **322** in the two sectors $0\leqq\vartheta\leqq\pi/2$, $\pi/2\leqq\vartheta\leqq\pi$, for we have (for x, y real, $y>0$)

$$|F(x)\,e^{i\rho x}|\leqq M,\qquad |F(iy)\,e^{-\rho y}|\leqq K.$$

If we now consider $F(z)\,e^{-i\rho z}$ in the lower half-plane, we find in view of III **322** that in the whole plane

$$|F(x+iy)|\leqq L\,e^{\rho|y|},$$

where $L=\max(M, K)$. It follows [III **165**] that, for real x,

$$(*)\qquad \left|\frac{F'(x)}{\sin\rho x}\right|=\left|\frac{\rho F(x)\cos\rho x}{\sin^2\rho x}-\rho\sum_{n=-\infty}^{\infty}\frac{(-1)^n F\left(\frac{n\pi}{\rho}\right)}{(\rho x-n\pi)^2}\right|$$

$$\leqq\frac{\rho|F(x)\cos\rho x|}{\sin^2\rho x}+\rho M\sum_{n=-\infty}^{\infty}\frac{1}{(\rho x-n\pi)^2}$$

$$=\frac{\rho|F(x)\cos\rho x|+\rho M}{\sin^2\rho x}.$$

From this we firstly deduce that $|F'[\pi/(2\rho)]|\leqq\rho M$. Now we may apply the same proof to the function $F(z+x_0-[\pi/(2\rho)])$, where x_0 is a real constant, since

$$\left|F\left(z+x_0-\frac{\pi}{2\rho}\right)\right|\leqq K\,e^{\rho\left|x_0-\frac{\pi}{2\rho}\right|}e^{\rho|z|}.$$

If in (*) we have the equality sign, then we obtain $(-1)^n F(n\pi)=A$ (independent of n) and

$$\frac{d}{dz}\left(\frac{F(z)}{\sin\rho z}\right)=-\frac{\rho A}{\sin^2\rho z}.$$

202. $F'(z)$ is a function of the same kind as $F(z)$. By repeated application of **201** it follows that for real values of x

$$|F'(x)| \leqq M\rho, \qquad |F''(x)| \leqq M\rho^2, \ldots, \qquad |F^{(n)}(x)| \leqq M\rho^n, \ldots,$$

and hence

$$|F(x+iy)| \leqq \sum_{n=0}^{\infty} \frac{|F^{(n)}(x)|}{n!} |y|^n \leqq M\, e^{\rho|y|}.$$

203. The conditions of III **166** are satisfied [solution **201**]. We have for real z, $\rho|z| < \pi/2$,

$$\frac{|G(z)|}{2\rho|z| \cos \rho z} \leqq \sum_{n=0}^{\infty} \frac{M}{((n+\frac{1}{2})\pi)^2 - \rho^2 z^2} = M \frac{\sin \rho z}{2\rho z \cos \rho z} \leqq M \frac{\rho|z|}{2\rho|z| \cos \rho z}.$$

(For $z=0$ we set $G(z)/z = G'(0)$, $(\sin \rho z)/\rho z = 1$).) If in this inequality we have equality, then it follows that

$$(-1)^n G\left(\frac{(n+\frac{1}{2})\pi}{\rho}\right) = cM, \qquad |c| = 1, \quad n = 0, 1, 2, \ldots, \qquad z = 0.$$

For $\rho|z| \geqq \pi/2$ the above inequality (even with $<$) is trivial.

204. Applying **203** to

$$G(z) = \frac{F(z_0+z) - F(z_0-z)}{2},$$

where z_0 is real, gives **201**. (Cf. VI **82**.)

205. Let $\lambda < \mu < 2\lambda$. The function $f(z)e^{iz^\mu}$ is bounded on the two rays $\arg z = 0$ and $\arg z = \pi/(2\mu)$, and hence [III **330**] also in the sector of aperture $\pi/(2\mu) < \pi/(2\lambda) \leqq \pi$ lying between them. We deduce that the function $f(z)(\sin z^\mu)^{-1}$ is bounded on the two rays $\arg z = -\pi/(2\mu)$ and $\arg z = \pi/(2\mu)$, and also on those arcs of the circles $|z|^\mu = (n+\frac{1}{2})\pi$, $n = 1, 2, 3, \ldots$, that lie in the sector $-\pi/(2\mu) \leqq \arg z \leqq \pi/(2\mu)$. Hence we have

$$\frac{1}{2\pi i} \oint \frac{f(\zeta)}{\sin \zeta^\mu} \frac{d\zeta}{(\zeta - z)^2} = \frac{d}{dz}\left(\frac{f(z)}{\sin z^\mu}\right) + \sum_{n=1}^{\infty} \frac{(-1)^n f\left((n\pi)^{\frac{1}{\mu}}\right)}{\mu (n\pi)^{1-\frac{1}{\mu}} \left(z - (n\pi)^{\frac{1}{\mu}}\right)^2},$$

where the integral is taken in the positive sense along the boundary of the infinite sector

$$-\frac{\pi}{2\mu} < \arg z < \frac{\pi}{2\mu}, \qquad |z| > \rho \quad \text{and} \quad 0 < \rho < \pi^{\frac{1}{\mu}}.$$

Since the series $\sum_{n=1}^{\infty} n^{-1-\frac{1}{\mu}}$ is convergent and $f(x)$ is bounded for $x > 0$, we deduce in view of I **182** that $f'(x) = O(x^{\mu-1})$, as x tends to infinity through the intervals $((n+\frac{1}{4})\pi)^{\frac{1}{\mu}} \leqq x \leqq ((n+1-\frac{1}{4})\pi)^{\frac{1}{\mu}}$, $n = 1, 2, 3, \ldots$. With the aid of an analogous formula in which $\sin z^\mu$ is replaced by $\cos z^\mu$, we can show that $f'(x) = O(x^{\mu-1})$ also in the remaining intervals.

***206.** [G. Pólya, problem: Jber. deutsch. Math. Verein. Section 2, **35**, p. 48 (1926). Solution by L. Tchakalov, op. cit. **37**, pp. 30–33 (1928).]

***207.** [R. Nevalinna and G. Pólya, problem: Jber. deutsch. Math. Verein. Section 2, **40**, p. 80 (1931). Solution by H. Schmidt, op. cit. **43**, pp. 6–7 (1933).]

***208.** [G. Pólya, problem: Jber. deutsch. Math. Verein., Section 2, **34**, pp. 97–98 (1925). Solution by G. Szegö, op. cit. **35**, pp. 86–89 (1926).]

***209.** [G. Pólya, problem: Jber. deutsch. Math. Verein., Section 2, **40**, pp. 80–81 (1931). Solutions op. cit. **43** (1933) by L. Tchakalov, pp. 10–11 and G. Szegö, pp. 11–13; comments by G. Pólya, pp. 67–69. See also R. E. A. C. Paley and Norbert Wiener: Fourier transforms in the complex domain, Amer. Math. Soc. Colloquium Publ. **19**, New York 1934, pp. 81–83.]

***210.** [G. Pólya, problem: Jber. deutsch. Math. Verein., Section 2, **40**, p. 80 (1931). Solution by G. Szegö, op. cit. **43**, pp. 7–9 (1933).] See *Picard's theorem*, for example in Ahlfors pp. 297–298, Hille Vol. II pp. 219–221 or Titchmarsh pp. 277–284. Cf. **194**.

***211.** [G. Pólya, problem: Jber. deutsch. Math. Verein., Section 2, **40**, p. 81 (1931). Solutions op. cit. **43** by N. Obreschkov, pp. 13–15, G. Szegö, pp. 15–16, and N. Tchebotarev, pp. 16–17.]

***212.** [G. Pólya, problem: Jber. deutsch. Math. Verein., Section 2, **40**, p. 81 (1931). Solution by O. Szász, op. cit. **43**, pp. 20–23 (1933). For a stronger result, see Paley and Wiener, op. cit. **209**, pp. 83–85. See also Norman Levinson: Gap and Density Theorems. Amer. Math. Soc. Colloquium Publications **26**, New York 1940, pp. 3–5.]

Part Five. The Location of Zeros

1. [For this whole chapter cf. E. Laguerre: Oeuvres, **1**. Paris: Gauthier-Villars 1898.] Obvious.

2. Consider the deletion of a term $a_\mu \neq 0$ and take for example the case that

$$a_{\mu-k} \neq 0, \qquad a_{\mu-k+1} = \cdots = a_{\mu-1} = a_{\mu+1} = \cdots = a_{\mu+l-1} = 0, \qquad a_{\mu+l} \neq 0.$$

The number of changes of sign in the sequence $a_0, a_1, \ldots, a_{\mu-1}, a_{\mu+1}, \ldots$ is less by two than the number of changes of sign in the sequence $a_0, a_1, \ldots, a_{\mu-1}, a_\mu, a_{\mu+1}, \ldots$ if

$$\operatorname{sgn} a_{\mu-k} = -\operatorname{sgn} a_\mu = \operatorname{sgn} a_{\mu+l},$$

while the two numbers are equal for all other combinations of signs. In case a_μ is the first or the last non-vanishing term, with its deletion one or no change of sign of the sequence is lost.

3. The two sequences $a_0, a_1, \ldots, a_\mu, a_{\mu+1}, \ldots$ and $a_0, a_1, \ldots, a_\mu, b, a_{\mu+1}, \ldots$ have the same number of changes of sign in the following three cases: (1) $b=0$, (2) $\operatorname{sgn} b = \operatorname{sgn} a_\mu$, (3) $\operatorname{sgn} b = \operatorname{sgn} a_{\mu+1}$: Obvious!

4. $a_0, a_0+a_1, a_1, a_1+a_2, a_2, a_2+a_3, \ldots, a_\mu, a_\mu+a_{\mu+1}, a_{\mu+1}, \ldots$ has the same number of changes of sign as $a_0, a_1, a_2, \ldots, a_\mu, \ldots$ since, if we set $a_\mu + a_{\mu+1} = b$, one of the three cases mentioned in solution **3** must occur. Make use of **2**.

5. [A. Hurwitz.] The number of changes of sign does not increase as we consider consecutively the sequences

$$\begin{array}{llllll}
a_0, & a_1, & a_2, & a_3, & \ldots, & a_l, \quad a_{l+1}, \quad \ldots \\
a_0, & a_0+a_1, & a_1+a_2, & a_2+a_3, & \ldots, & a_{l-1}+a_l, \quad a_l+a_{l+1}, \quad \ldots \\
a_0, & a_0+a_1, & a_0+2a_1+a_2, & a_1+2a_2+a_3, & \ldots, & a_{l-1}+2a_l+a_{l+1}, \quad \ldots \\
a_0, & a_0+a_1, & a_0+2a_1+a_2, & a_0+3a_1+3a_2+a_3, \ldots & &
\end{array}$$

from the top down. [Solution **4**.] But the nth sequence constructed in this manner agrees in its first n terms with the sequence to be discussed. Hence *these* terms cannot have more than C changes of sign. Since n is arbitrary we obtain the required result.

6. Obvious, even in the case of multiple zeros.

7. Obvious.

8. Since $f(x)$ is analytic, the interval (a, b) contains only finitely many zeros. Passing through a zero the sign of $f(x)$ does or does not reverse, according as the zero in question is of odd or even multiplicity, respectively.

9. Obvious. Cf. **8**.

10. Take $\varepsilon > 0$, ε sufficiently small.

$$f(a+\varepsilon) = f(a+\varepsilon) - f(a) = \varepsilon f'(a+\varepsilon_1), \qquad 0 < \varepsilon_1 < \varepsilon,$$
$$-f(b-\varepsilon) = f(b) - f(b-\varepsilon) = \varepsilon f'(b-\varepsilon_2), \qquad 0 < \varepsilon_2 < \varepsilon.$$

From $\operatorname{sgn} f(a+\varepsilon)=\operatorname{sgn} f(b-\varepsilon)\neq 0$ it follows that $\operatorname{sgn} f'(a+\varepsilon_1)=-\operatorname{sgn} f'(b-\varepsilon_2)\neq 0$. Apply **8** to $f'(x)$ in the interval $a+\varepsilon_1, b-\varepsilon_2$.—The formulation of Rolle's theorem we have given offers a more precise statement about a narrower class of functions than the usual formulation in differential calculus.

11. By the assumptions we have

$$\operatorname{sgn} a_{j+1}=\operatorname{sgn}(a_{j+1}-a_j), \qquad \operatorname{sgn} a_{k+1}=\operatorname{sgn}(a_{k+1}-a_k).$$

From $\operatorname{sgn} a_{j+1}=-\operatorname{sgn} a_{k+1}\neq 0$ we obtain $\operatorname{sgn}(a_{j+1}-a_j)=-\operatorname{sgn}(a_{k+1}-a_k)\neq 0$. Apply **9** to the sequence of differences.

12. (α) If $f(x)$ has an N-fold zero at the point $x=x_1$, $N>0$, then $f'(x)$ has an $(N-1)$-fold zero there. This applies to the case in which the interval reduces to a point.—(β) If $a<b$, let the points in which $f(x)$ vanishes be $x_1, x_2, \ldots, x_l$, $a\leqq x_1< x_2<\cdots<x_l\leqq b$. We divide the closed interval $x_1\leqq x\leqq x_l$ into l parts: The point x_1 and the half-open intervals $x_1<x\leqq x_2$, $x_2<x\leqq x_3, \ldots, x_{l-1}<x\leqq x_l$. Passing from $f(x)$ to $f'(x)$ at the point x_1 *one* zero is lost [case (α)]; in the half-open interval $x_1<x\leqq x_2$ none is lost [**10** and case (α)]; similarly for the other half-open intervals.

13. We denote the indices of change in question by $\nu_1+1, \nu_2+1, \ldots, \nu_C+1$, $0\leqq \nu_1<\nu_2<\cdots<\nu_C\leqq n-1$. Each of the $C-1$ subsequences

$$\begin{array}{llll} a_{\nu_1+1}-a_{\nu_1}, & a_{\nu_1+2}-a_{\nu_1+1}, & \ldots, & a_{\nu_2+1}-a_{\nu_2}, \\ a_{\nu_2+1}-a_{\nu_2}, & a_{\nu_2+2}-a_{\nu_2+1}, & \ldots, & a_{\nu_3+1}-a_{\nu_3}, \\ \cdots & \cdots & \cdots & \cdots \\ a_{\nu_{C-1}+1}-a_{\nu_{C-1}}, & a_{\nu_{C-1}+2}-a_{\nu_{C-1}+1}, & \ldots, & a_{\nu_C+1}-a_{\nu_C} \end{array}$$

contains at least one change of sign [**11**].

14. Let $\operatorname{sgn} f(a)=\operatorname{sgn} f'(a)\neq 0$. Let the points in which $f(x)$ vanishes be $x_1, x_2, \ldots, x_l$, $a<x_1<x_2<\cdots<x_l<b$. We divide the interval $a<x\leqq x_l$ into the subintervals

$$a<x\leqq x_1, \qquad x_1<x\leqq x_2, \qquad \ldots, \qquad x_{l-1}<x\leqq x_l.$$

Take $\varepsilon>0$, ε sufficiently small. From

$$-f(x_1-\varepsilon)=f(x_1)-f(x_1-\varepsilon)=\varepsilon f'(x_1-\eta), \qquad 0<\eta<\varepsilon$$

it follows by the assumptions that

$$\operatorname{sgn} f'(a)=\operatorname{sgn} f(a)=\operatorname{sgn} f(x_1-\varepsilon)=-\operatorname{sgn} f'(x_1-\eta)\neq 0,$$

and hence that $f'(x)$ has at least one zero between a and $x_1-\eta$ [**8**]. With regard to the point x_1 and the other sub-intervals cf. solution **12**. Similarly we prove: If $\operatorname{sgn} f(b)= -\operatorname{sgn} f'(b)\neq 0$, then there is an additional zero of $f'(x)$ in the interval $x_l<x<b$.

15. We shall keep the notation of **13**. Apart from the $C-1$ subsequences considered there, the following two subsequences

$$\begin{array}{lllll} a_0, & a_1-a_0, & \ldots, & a_{\nu_1+1}-a_{\nu_1}, & \\ a_{\nu_C+1}-a_{\nu_C}, & a_{\nu_C+2}-a_{\nu_C+1}, & \ldots, & a_n-a_{n-1}, & -a_n \end{array}$$

each have a change of sign. For if a_α is the first non-vanishing term of the first sequence, then we have $0\leqq\alpha\leqq\nu_1$ and

$$\operatorname{sgn}(a_\alpha-a_{\alpha-1})=\operatorname{sgn} a_\alpha=-\operatorname{sgn} a_{\nu_1+1}=-\operatorname{sgn}(a_{\nu_1+1}-a_{\nu_1});$$

now apply **9**. Similarly in the case of the second subsequence above.

16. In the case of infinitely many zeros the assertion is obvious [**10**].—Let x_l be the last zero of $f(x)$. In the interval $a \leqq x \leqq x_l$, $f'(x)$ has at most one less zero than $f(x)$ [**12**]. Since $\lim_{x\to\infty} f(x) = \int_{x_l}^{\infty} f'(t)\,dt = 0$, $f'(x)$ cannot be of constant sign in the interval $x_l < x < \infty$.

17. In the case of infinitely many indices of change the assertion is obvious [**11**]. Let the sequence $a_0, a_1, a_2, \ldots$ have C changes of sign and let $\nu_C + 1$ be the last index of change. The sequence

$$a_0,\ a_1 - a_0,\ \cdots,\ a_{\nu_C+1} - a_{\nu_C}$$

has at least C changes of sign [solution **13, 15**]. Moreover

$$\operatorname{sgn} a_{\nu_C+1} = \operatorname{sgn}(a_{\nu_C+1} - a_{\nu_C}) \neq 0.$$

Since the infinite series

$$a_{\nu_C+1} + (a_{\nu_C+2} - a_{\nu_C+1}) + (a_{\nu_C+3} - a_{\nu_C+2}) + \cdots = 0$$

its terms different from zero cannot all have the same sign. (It is understood that there is, as in **15**, one trivial exceptional case.)

18. Apply **12** or **16** to the function $e^{\alpha x} f(x)$ [**6**]; its derivative is $e^{\alpha x}[\alpha f(x) + f'(x)]$.

19. Consider the sequences

$$\begin{array}{llll} a_0, & a_1, & a_2, & \ldots,\ a_n, \ldots \\ a_0, & a_1\alpha, & a_2\alpha^2, & \ldots,\ a_n\alpha^n, \ldots \\ a_0, & a_1\alpha - a_0, & a_2\alpha^2 - a_1\alpha, & \ldots,\ a_n\alpha^n - a_{n-1}\alpha^{n-1}, \ldots \\ a_0\alpha, & a_1\alpha - a_0, & a_2\alpha - a_1, & \ldots,\ a_n\alpha - a_{n-1}, \ldots . \end{array}$$

Cf. **7**, then solution **15** and **17** and then again **7**.

20. Set $\int_0^x f(t)\,dt = F(x)$; then $F(x)$ and $F'(x) = f(x)$ have the same sign in an appropriate neighbourhood of $x = 0$. [**14**.]

21. Set $a_0 + a_1 + \cdots + a_n = A_n$, $n = 0, 1, 2, \ldots$ and apply the first half of **19** with $\alpha = 1$ to the sequence $A_0, A_1, A_2, \ldots, A_n, \ldots$.

22. Obvious.

23. The continuous curve $y = \int_0^x f(t)\,dt$ consists of $R + 1$ monotonic sections. The first section begins at the point $x = 0$, $y = 0$ and does not cause any changes of sign. Each of the remaining S sections can cross at most once from one side of the x-axis to the other. (A more precise formulation can easily be given.) If $f(x)$ is analytic cf. also **20, 24**.

24. Let the interval be $a < x < b$. If we assume that $f(x)$ is analytic then there exists an ε, $\varepsilon > 0$, such that $f(x) \neq 0$ for $a < x < a + \varepsilon$ and $b - \varepsilon < x < b$ [**8, 22**].

25. Cf. **10, 24**.

26. $f(x) = P(x)/Q(x)$, where $Q(x) = (x - a_1)(x - a_2)\ldots(x - a_n)$ and $P(x)$ is a real polynomial of degree $\leqq n - 1$.

(1) If $\varepsilon > 0$, and ε sufficiently small, then $\operatorname{sgn} f(a_1 + \varepsilon) = +1$, $\operatorname{sgn} f(a_2 - \varepsilon) = -1$. Hence $f(x)$ and thus $P(x)$ has a zero in the interval $a_1 < x < a_2$ [**8**]. Proceeding similarly we can demonstrate that there are $n - 2$ zeros of $P(x)$ (for $a_1 < x < a_2$, $a_2 < x < a_3, \ldots, a_{n-2} < x < a_{n-1}$). The single possible remaining zero cannot be non-real, for the non-real zeros of real polynomials occur in complex conjugate *pairs.*

(2) Show, as in (1), that $f(x)$ has a zero in each of the $n-3$ intervals (a_1, a_2), $(a_2, a_3), \ldots, (a_{k-2}, a_{k-1}), (a_{k+1}, a_{k+2}), \ldots, (a_{n-1}, a_n)$. If ε is sufficiently small and ω sufficiently large, $\varepsilon>0$, $\omega>0$, then

$$\operatorname{sgn} f(-\omega) = -\operatorname{sgn} f(a_1-\varepsilon) = \operatorname{sgn} f(a_n+\varepsilon) = -\operatorname{sgn} f(\omega) = +1.$$

Hence there exists a zero in each of the intervals $(-\infty, a_1)$ and $(a_n, +\infty)$ [**8**].

27.

$$f(0)>0, \quad f\left(\frac{\pi}{n}\right)<0, \quad f\left(\frac{2\pi}{n}\right)>0, \quad \ldots, \quad f\left(\frac{2n\pi}{n}\right)>0.$$

Hence [**8**] $f(x)$ has $2n$ real zeros in the strip $0<\Re x<2\pi$. $f(x)$ has no other zeros there [VI **14**].

28. Cf. **7**.

29. Assume $a_0=a_1=\cdots=a_{\alpha-1}=0$, $a_\alpha \neq 0$. Denote by $a_k x^k$ and $a_{\mathfrak{l}} x^{\mathfrak{l}}$ two terms such that

$$a_k\neq 0, \qquad a_{k+1}=a_{k+2}=\cdots=a_{\mathfrak{l}-1}=0, \qquad a_{\mathfrak{l}}\neq 0,$$
$$P(x)=a_\alpha x^\alpha+\cdots+a_k x^k+a_{\mathfrak{l}}\alpha^{\mathfrak{l}}+\cdots.$$

α must be regarded as a special value of k. If $\mathfrak{l}-k$ is odd, then the pair of terms $a_k x^k + a_{\mathfrak{l}} x^{\mathfrak{l}}$ contributes a unit either to C^+ or to C^-. If $\mathfrak{l}-k$ is even, then either both C^+ and C^- gain a unit from $a_k x^k + a_{\mathfrak{l}} x^{\mathfrak{l}}$ or neither gains a unit. From this it follows, assuming $a_n \neq 0$, that

$$(n-\alpha)-(C^+ + C^-) = \textstyle\sum^{I} (\mathfrak{l}-k-1) + \sum^{II} [\mathfrak{l}-k-(1-\operatorname{sgn} a_k a_{\mathfrak{l}})].$$

$\sum^{I}$ extends over odd $\mathfrak{l}-k$, $\sum^{II}$ over even $\mathfrak{l}-k$, for we have $n-\alpha=\sum^{I}(\mathfrak{l}-k)+\sum^{II}(\mathfrak{l}-k)$. Both in $\sum^{I}$ and in $\sum^{II}$ every term $\geqq 0$.

30. Replace x by αx [**28**] and apply **15**.

31. [G. Pólya: Arch. Math. Phys. Ser. 3, **23**, p. 22 (1914).] From **19** there follows even more: for the second part instead of the convergence of $\sum_{n=0}^{\infty} a_n \alpha^n$ we need only that $\lim_{n\to\infty} a_n\alpha^n=0$.

32. For $\alpha=1$ the result follows from **4**. Cf. **28**.

33. Equivalent to **21**. Also follows from **31**.

34. [Laguerre, op. cit. **1**, p. 22. The proof has gaps.] For $\alpha=1$ the result follows from **5** [**28**].

35. If of the numbers $a_0, a_1, a_2, \ldots, a_n$ only one, a_α say, is different from zero, then we are considering the product

$$\frac{a_\alpha}{p_1 p_2 \cdots p_\alpha}\left(1+\frac{x}{p_1}+\frac{x^2}{p_1^2}+\cdots\right)\left(1+\frac{x}{p_2}+\frac{x^2}{p_2^2}+\cdots\right)\cdots\left(1+\frac{x}{p_\alpha}+\frac{x^2}{p_\alpha^2}+\cdots\right),$$

which, when multiplied out, clearly has no changes of sign. Assume that the statement has been proved for all sequences of numbers that contain one fewer non-vanishing term than the given sequence $a_0, a_1, a_2, \ldots, a_n$. Let a_α be the first non-vanishing term, $0 \leqq \alpha < n$. Set

$$0+\frac{a_{\alpha+1}x}{p_{\alpha+1}-x}+\frac{a_{\alpha+2}x^2}{(p_{\alpha+1}-x)(p_{\alpha+2}-x)}+\cdots+\frac{a_n x^{n-\alpha}}{(p_{\alpha+1}-x)\cdots(p_n-x)}$$
$$=0+B_1x+B_2x^2+\cdots.$$

Denote the number of changes of sign in the sequence

$$\begin{array}{ll} a_0,\ a_1,\ \ldots,\ a_\alpha,\ a_{\alpha+1},\ \ldots,\ a_n & \text{by } \{a\}, \\ 0,\ a_{\alpha+1},\ a_{\alpha+2},\ \ldots,\ a_n & \text{by } \{b\}, \\ A_0,\ A_1,\ A_2,\ A_3,\ \ldots & \text{by } \{A\}, \\ 0,\ B_1,\ B_2,\ B_3,\ \ldots & \text{by } \{B\}. \end{array}$$

By the assumption of the mathematical induction we have

$$\{B\} \leqq \{b\}.$$

Let a_β be the first non-vanishing term of the sequence $0, a_{\alpha+1}, a_{\alpha+2}, \ldots, a_n$. Then we have

$$\{a\} = \{b\} + \frac{1 - \operatorname{sgn} a_\alpha a_\beta}{2}.$$

The first non-vanishing term in $0, B_1\ B_2, \ldots$ has the same sign as a_β. Hence the number of changes of sign in the sequence

$$a_\alpha,\ B_1,\ B_2,\ B_3,\ \ldots \text{ is equal to } \{B\} + \frac{1 - \operatorname{sgn} a_\alpha a_\beta}{2}.$$

Now we have

$$(A_0 + A_1 x + A_2 x^2 + \cdots)(p_1 - x)(p_2 - x)\ldots(p_\alpha - x)$$
$$= x^\alpha (a_\alpha + B_1 x + B_2 x^2 + \cdots),$$

and hence [**31**]

$$\{A\} \leqq \{B\} + \frac{1 - \operatorname{sgn} a_\alpha a_\beta}{2}.$$

From the three relations established above it follows that $\{A\} \leqq \{a\}$, q.e.d.

36. Let $\alpha_1, \alpha_2, \ldots, \alpha_Z$ be the positive zeros of $P(x) = a_0 + a_1 x + \cdots + a_n x^n$. Then we have $P(x) = Q(x)(\alpha_1 - x)(\alpha_2 - x)\ldots(\alpha_Z - x)$, where $Q(x)$ is a polynomial of degree $(n - Z)$ with real coefficients. The number of changes of sign of $Q(x)$ is $\geqq 0$, that of $Q(x)(\alpha_1 - x)$ is $\geqq 1$ [**30**], that of $Q(x)(\alpha_1 - x)(\alpha_2 - x)$ is $\geqq 2, \ldots$, that of $Q(x)(\alpha_1 - x)(\alpha_2 - x)\ldots(\alpha_Z - x)$ is $\geqq Z$, q.e.d.

37. Let a_α be the first and a_ω the last non-vanishing coefficient of $P(x)$, $\alpha \leqq \omega$ (only $\alpha < \omega$ is interesting), $0 < \xi < \alpha_1 \leqq \alpha_2 \leqq \cdots \leqq \alpha_Z < X < \infty$. If ξ is sufficiently close to 0 and X to ∞ then it can be deduced that

$$\operatorname{sgn} P(\xi) = \operatorname{sgn} a_\alpha, \qquad \operatorname{sgn} P(X) = \operatorname{sgn} a_\omega \qquad [\mathbf{8, 9}].$$

With the aid of **36** we obtain the following: If $C = 1$, then also $Z = 1$. This may also easily be seen directly [III **16**].

38. [Laguerre, op. cit. **1**, p. 5; cf. op. cit. **31**, p. 24.] If C is finite (this case only is of interest here), then all the non-vanishing coefficients from some a_ω on have the same sign. Hence the ωth derivative of the power series does not vanish for $0 < x < \rho$. Thus the power series has only finitely many positive zeros [**12**]; denote them by $\alpha_1, \alpha_2, \ldots, \alpha_Z$. The power series

$$\frac{a_0 + a_1 x + a_2 x^2 + \cdots}{(\alpha_1 - x)(\alpha_2 - x)\ldots(\alpha_Z - x)} = b_0 + b_1 x + b_2 x^2 + \cdots$$

is regular in the circle of convergence of $a_0+a_1x+a_2x^2+\cdots$, and hence convergent there [Hurwitz-Courant, p. 49, p. 266]. The number of changes of sign of $b_0+b_1x+b_2x^2+\cdots$ is $\geqq 0$. Now the theorem follows in a similar fashion from the second case of **31** as **36** from **30**.

39. The radius of convergence is $=1$. The power series takes the value 2 for $x=0$ and the value $2-(1-\frac{1}{2})-(\frac{1}{2}-\frac{1}{3})-\cdots=1$ for $x=1$. The number of zeros in the interval $0<x<1$ is thus even [**8**], but on the other hand it is $\leqq 1$ [**38**], and hence $=0$. We can also see directly that for complex z, $|z|<1$, we have

$$\left|2-\frac{z}{1\cdot 2}-\frac{z^2}{2\cdot 3}-\cdots\right|\geqq 2-\frac{|z|}{1\cdot 2}-\frac{|z|^2}{2\cdot 3}-\cdots>2-\frac{1}{1\cdot 2}-\frac{1}{2\cdot 3}-\cdots=1.$$

40. Let a_α be the first non-vanishing coefficient and ω the last index of change. We have

$$\lim_{x\to 0} x^{-\alpha}(a_0+a_1x+a_2x^2+\cdots)=a_\alpha,$$

$$\lim_{x\to\rho-0} (a_0+a_1x+a_2x^2+\cdots)=+\infty\cdot\operatorname{sgn} a_\omega \qquad [\mathbf{8,\ 9}].$$

41. [C. Runge, cf. op. cit. **31**, p. 25.] Replacing x by $-x$, the case of the zeros $<\xi_\alpha$ reduces to the case of the zeros $>\xi_\omega$ and this case in turn, by replacing x by $x+\text{const.}$, reduces to the special case in which $\xi_\alpha>0$. In this special case set

$$\frac{a_0+a_1(x-\xi_1)+a_2(x-\xi_1)(x-\xi_2)+\cdots+a_n(x-\xi_1)(x-\xi_2)\ldots(x-\xi_n)}{(x-\xi_1)(x-\xi_2)\ldots(x-\xi_n)}$$

$$=a_n+\frac{a_{n-1}}{\xi_n}\frac{1}{x}\frac{1}{\frac{1}{\xi_n}-\frac{1}{x}}+\cdots+\frac{a_0}{\xi_1\,\xi_2\ldots\xi_n}\frac{1}{x^n}\frac{1}{\left(\frac{1}{\xi_1}-\frac{1}{x}\right)\left(\frac{1}{\xi_2}-\frac{1}{x}\right)\cdots\left(\frac{1}{\xi_n}-\frac{1}{x}\right)}$$

$$=A_0+\frac{A_1}{x}+\frac{A_2}{x^2}+\frac{A_3}{x^3}+\cdots.$$

The power series converges for $x>\xi_\omega$ and has all the required zeros whose total number is $\leqq$ the number of changes of sign of the sequence $A_0, A_1, A_2, \ldots$ [**38**], which itself is $\leqq$ the number of changes of sign of the sequence $a_0, a_1, a_2, \ldots, a_{n-1}, a_n$ [**35, 6, 7**], q.e.d. That the difference between the two numbers cannot be odd may be proved as in **37**.

42. Notation as in **38, 40**. For the power series

$$f_n(x)=1+\frac{x}{1!}+\frac{x^2}{2!}+\cdots+\frac{x^n}{n!}-\lambda\, e^x$$

$$=(1-\lambda)+\frac{(1-\lambda)}{1!}x+\cdots+\frac{(1-\lambda)}{n!}x^n-\frac{\lambda}{(n+1)!}x^{n+1}-\cdots$$

we have $C=1$ and hence $Z\leqq 1$ [**38**]. Moreover $\rho=\infty$, hence $1-Z$ is even [**40**], and thus $1-Z=0$. Let x_n be the unique positive zero of $f_n(x)$. As x passes through the value x_n, $\operatorname{sgn} f_n(x)$ changes from $+1$ to -1 [solution **40**]. Hence if $f_n(a)>0$, we have that also $x_n - a > 0$. Now for fixed a we have $\lim_{n\to\infty} f_n(a) = (1-\lambda)\, e^a$, whence it follows ($a$ arbitrary!) that $\lim_{n\to\infty} x_n = \infty$. Finally we have

$$f_n(x_{n-1}) = f_{n-1}(x_{n-1}) + \frac{x_{n-1}^n}{n!} = \frac{x_{n-1}^n}{n!} > 0, \qquad \text{i.e.} \quad x_n > x_{n-1}.$$

43. $x^{-5}(e^x - 1)$ has a minimum and no maximum since

$$\frac{d}{dx}\,[x^{-5}\,(e^x - 1)] = -x^{-6}(5\,e^x - 5 - x\,e^x) = -x^{-6} \sum_{n=1}^{\infty} \frac{(5-n)x^n}{n!}$$

and this series has only one index of change, namely $n = 6$ [**38**, **40**]. (*Planck's* law of radiation.)

43.1. [G. Pólya and I. J. Schoenberg: Pacific J. Math. **8**, p. 322 (1958)]. Let $x(1-x)^{-1} = t$. Then

$$\frac{K_n(x)}{(1-x)^n} = \sum_{\nu=0}^{n} f\left(\frac{\nu}{n}\right)\binom{n}{\nu} t^\nu.$$

The number of changes of sign in the sequence of coefficients of this polynomial in t is obviously $\leqq R(f)$. Apply **36**.

44. The zeros in question are also zeros of the series

$$(1-x)^{-1}(a_0 + a_1 x + a_2 x^2 \ldots) = a_0 + (a_0 + a_1)x + (a_0 + a_1 + a_2)x^2 + \cdots$$

[**38**].

44.1. [J. Steinig. Cf. **83.1**.] Since

$$a_0 + a_1 + a_2 + \cdots + a_{n-1} + a_n = 0,$$
$$P(x)(1-x)^{-1} = a_0 + (a_0 + a_1)x + \cdots + (a_0 + a_1 + \cdots + a_{n-1})x^{n-1},$$

and this polynomial has at most C positive zeros [**36**]. That the case of equality can be attained in the inequality just proved is shown by the example $(1-x)^n$ where $C = n-1$.

45. [Cf. M. Fekete: Palermo Rend. **34**, p. 89 (1912).] $x = 1$ is a zero of the polynomial and multiplication by $(1-x)^{-2}$ yields a power series with no negative coefficients.

46. [G. Pólya: Jber. deutsch. Math. Verein. **28**, p. 37 (1919).] Setting

$$f(x) = \sum_{\nu=1}^{163} \left(\frac{\nu}{163}\right) x^\nu$$

we find that

$$(1-x)^{-2} f(x)$$
$$= x + x^2 - x^5 - x^6 - \cdots - x^{65} + 7x^{66} + 14x^{67} + \cdots + 163x^{161} + 163x^{162}(1-x)^{-1}$$

has two indices of change. The number of zeros in the interval $0 < x < 1$ is thus $= 0$ or $= 2$ [**38**, **40**]. It is actually $= 2$, since for $0 < x < 1$ we have

$$x^{-1} f(x) = \sum_{\nu=1}^{10} \left(\frac{\nu}{163}\right) x^{\nu-1} - x^{10} - x^{11} - x^{12} + x^{13} + x^{14} + x^{15} + \cdots$$
$$< \sum_{\nu=1}^{10} \left(\frac{\nu}{163}\right) x^{\nu-1} + x^{16}(1-x)^{-1},$$

and the right-hand side is $= -0.00995\ldots$ if $x = 0.7$.

47. Let the number of negative zeros be Z^-, the number of positive zeros Z^+ and thus $n = Z^- + \alpha + Z^+$ with the notation of solution **29**. We have [solution **29**]

$$n - (C^- + \alpha + C^+) = (Z^- - C^-) + (Z^+ - C^+) \geqq 0$$

and [**36**]

$$Z^- - C^- \leqq 0, \qquad Z^+ - C^+ \leqq 0,$$

and hence

$$Z^- - C^- = 0, \qquad Z^+ - C^+ = 0, \qquad \text{q.e.d.}$$

48. If the determinant were to vanish, then one could find a solution, not identically zero, of the corresponding homogeneous system, i.e. n real numbers $c_1, c_2, \ldots, c_n$ satisfying $c_1^2 + c_2^2 + \cdots + c_n^2 > 0$, such that the polynomial

$$c_1 x^{\nu_1} + c_2 x^{\nu_2} + \cdots + c_n x^{\nu_n}$$

vanishes for $x = \alpha_1, \alpha_2, \ldots, \alpha_n$: Contradiction to **36**! For the sequence of coefficients contains not more than n terms different from 0, and therefore cannot have n changes of sign and thus the polynomial certainly cannot have n positive zeros $\alpha_1, \alpha_2, \ldots, \alpha_n$. The determinant is thus $\neq 0$. It is > 0 if $n = 1$. Assume that it is > 0 if it has $n-1$ rows and regard α_n as variable. As α_n increases from α_{n-1} to $+\infty$, the determinant $\neq 0$ and thus has constant sign. As $\alpha_n \to +\infty$ this appears as the sign of a minor of order $n-1$ of the same structure and hence is $= +1$.

49. By **36** the number of real zeros, with the notation of solution **29**, is $\leqq C^- + \alpha + C^+$, and hence the number of non-real zeros is $\geqq n - (C^- + \alpha + C^+) = \sum^{I} (\mathfrak{l} - k - 1) + \sum^{II} [\mathfrak{l} - k - (1 - \operatorname{sgn} a_k a_{\mathfrak{l}})]$. If a_k is $\neq 0$, $a_{k+1} = a_{k+2} = \cdots = a_{k+2m} = 0$, then the corresponding k and $\mathfrak{l}$ are such that $\mathfrak{l} - k$ is either odd, $\geqq 2m+1$, or even, $\geqq 2m+2$.

50. [Laguerre, op. cit. **1**, p. 111.]

$$P(x)(1 + b_1 x + b_2 x^2 + \cdots + b_{2m} x^{2m}) = 1 - b_{2m+1} x^{2m+1} + \cdots.$$

The expression on the right is not a constant since $P(x)$ is not constant, and hence it is a polynomial that has at least $2m$ non-real zeros [**49**]. These must by assumption all be zeros of the factor $1 + b_1 x + b_2 x^2 + \cdots + b_{2m} x^{2m}$.

51. [J. Grommer: J. reine angew. Math. **144**, pp. 130–131 (1914).] If $P(x) = (1 - x_1 x)(1 - x_2 x) \ldots (1 - x_n x)$, then we have [**50**] $b_{2m} = S(x_1, x_2, \ldots, x_n)$. If we had $b_{2m} \leqq 0$, then $1 + b_1 x + b_2 x^2 + \cdots + b_{2m} x^{2m}$ would not have $2m$ non-real zeros.

52. *First proof.* Obvious from the representation in terms of the zeros of $P(x)$ [**51**].

Second proof. For simplicity assume

$$P(x) = (x - a_1)(x - a_2) \ldots (x - a_n) \qquad \text{where} \quad 0 < a_1 < a_2 < \cdots < a_n.$$

Then we have [VI, § 9]

$$1 = \sum_{\nu=1}^{n} \frac{1}{P'(a_\nu)} \frac{P(x)}{x - a_\nu}, \quad \text{hence} \quad B_k = \sum_{\nu=1}^{n} \frac{a_\nu^k}{P'(a_\nu)}.$$

Consider the polynomial of degree p, $p \geqq n$,

$$x^p - \sum_{\nu=1}^{n} \frac{a_\nu^p}{P'(a_\nu)} \frac{P(x)}{x - a_\nu} = x^p - B_p x^{n-1} + \cdots = L_p(x).$$

$L_p(x)$ vanishes for $x=a_1, a_2, \ldots, a_n$, and thus **[36]** has at least n changes of sign. The number of its non-vanishing coefficients is however $\leqq n+1$, and hence we have $B_p>0$.

53. If the degree, the number of real zeros and the number of non-real zeros of the polynomial are $=n$, r, and $n-r$ respectively, then for the derivative they are $=n-1$, $\geqq r-1$ [**12**], and $\leqq n-1-(r-1)$ respectively.

54. Let the polynomial be of degree n. Apart from the $(n-1)$ zeros demonstrated in solution **12**, the derivative has no other zeros.

55. [**53, 54.**]

56. If we set $(d^\nu/dx^\nu)(1+x^2)^{-1/2}=Q_\nu(x)(1+x^2)^{-\nu-(1/2)}$, then $Q_0(x)=1$, $Q_{\nu+1}(x)=-(2\nu+1)xQ_\nu(x)+(1+x^2)Q_\nu'(x)$. Hence $Q_\nu(x)$ is a polynomial of degree ν. Assume that $Q_\nu(x)$ has ν real distinct zeros. Then $Q_{\nu+1}(x)$ has a zero between every two zeros of $Q_\nu(x)$ [**10**], a zero between $-\infty$ and the smallest zero of $Q_\nu(x)$ and another zero between the largest zero of $Q_\nu(x)$ and $+\infty$ [**16**], and apart from these $(\nu-1)+2$ zeros no other zeros.

57. Proof as in **56**. Direct verification is also possible since the n roots of the equation

$$\frac{d^{n-1}}{dx^{n-1}}\frac{x}{1+x^2}=\frac{(-1)^{n-1}(n-1)!}{2}\left(\frac{1}{(x+i)^n}+\frac{1}{(x-i)^n}\right)=0$$

have the following values:

$$\cot\frac{\pi}{2n},\quad \cot\frac{3\pi}{2n},\quad \ldots,\quad \cot\frac{(2n-1)\pi}{2n}.$$

58. In the case of $H_n(x)$ the method of proof in **56** may be followed unchanged; in the other two cases it can be followed with slight changes [**16**].

The functions	$(1-x^2)^n$;	$e^{-x}x^n$;	e^{-x^2},
vanish for $x=$	$-1, +1$;	$0, +\infty$;	$-\infty, +\infty$,
with order	n, n;	n, ∞;	∞, ∞.

59. The degree of $Q_n(x)$ is $=n(q-1)$ [recursion formula, analogously to **56**]. Let the number of positive zeros be $=p_n$, and the number of zeros at the origin $=v_n$. From

$$(1+x^q)^n\frac{d^{n-1}}{dx^{n-1}}(x^{q-1}-x^{2q-1}+x^{3q-1}-\cdots)=Q_n(x)$$

it follows that

$$v_1=q-1,\qquad v_2=q-2,\qquad v_3=q-3,\quad\ldots,\quad v_q=0,\qquad v_{q+1}=q-1,\quad\ldots,$$

etc. with period q. Since moreover, setting $\omega=e^{2\pi i/q}$, we have $Q_n(\omega x)=\omega^{v_n}Q_n(x)$, we see that to every zero lying in the interior of the positive real axis there corresponds a similarly situated zero on each of the other $q-1$ rays described. If we assume that the statement to be proved is true for a certain integer n (mathematical induction), we then obtain

$$v_n+qp_n=n(q-1).$$

Since the $(n-1)$th derivative of $x^{q-1}(1+x^q)^{-1}$ vanishes for $x=+\infty$, we have [**16**]

$$p_{n+1} \geqq p_n, \quad \text{if} \quad v_n = 0; \qquad p_{n+1} \geqq p_n + 1, \quad \text{if} \quad v_n > 0.$$

In all cases we have

$$v_{n+1} + qp_{n+1} \leqq n(q-1) + q - 1 = v_n + qp_n + q - 1,$$

$$p_{n+1} \leqq p_n + \frac{v_n - v_{n+1} + q - 1}{q} = \begin{cases} p_n & \text{if} \quad v_n = 0, \\ p_n + 1, & \text{if} \quad v_n > 0. \end{cases}$$

Hence all $\leqq$ signs should be replaced by $=$,

$$v_{n+1} + qp_{n+1} = (n+1)(q-1).$$

With regard to the simplicity of the zeros cf. **55**, **56**. The proof for $R_n(x)$ is similar. The q half-rays in question are the bisectors of the segments joining neighbouring poles of the function $1/(1+z^q)$, distinguished from the other half-rays through the origin by the property that along them the function e^{-z^q} decreases most rapidly. (Cf. G. Pólya: Math. Z. **12**, p. 38 (1922).)

60. Passing from

$$a_0 + \binom{n}{1} a_1 x + \binom{n}{2} a_2 x^2 + \cdots + a_n x^n$$

to

$$a_0 x^n + \binom{n}{1} a_1 x^{n-1} + \binom{n}{2} a_2 x^{n-2} + \cdots + a_n$$

the number of non-real zeros does not alter, while passing to

$$n \left[a_1 + \binom{n-1}{1} a_2 x + \binom{n-1}{2} a_3 x^2 + \cdots + a_n x^{n-1} \right]$$

the number of non-real zeros does not increase [**53**].

61. Sharpening the result of **60** [cf. **54**] we find that the polynomials

$$x^2 + 2m_1 x + m_2^2, \quad m_1 x^2 + 2m_2^2 x + m_3^3, \quad \ldots, \quad m_{n-2}^{n-2} x^2 + 2m_{n-1}^{n-1} x + m_n^n$$

have real *simple* zeros. Therefore we have

$$m_1^2 > m_2^2, \qquad m_2^4 > m_1 m_3^3, \quad \ldots, \quad m_{n-1}^{2n-2} > m_{n-2}^{n-2} m_n^n,$$

from which we obtain successively

$$m_1 > m_2, \qquad m_2^4 > m_2 m_3^3, \quad \ldots, \quad m_{n-1}^{2n-2} > m_{n-1}^{n-2} m_n^n.$$

62. We are concerned with the zeros of $e^{-\alpha x} (d/dx) [e^{\alpha x} P(x)]$ [**6**, **16**].

62.1. For the first polynomial it is obvious, see I **199**. Denote the second polynomial by $P_n(x)$. Then, by I **187**,

$$P_{n+1}(x) = x(P_n(x) + P_n'(x)).$$

Use mathematical induction based on a somewhat sharper version of **62** (simple zeros!).

63. Without loss of generality we may assume $a_n \neq 0$. Setting

$$a_0 + a_1 x + a_2 x^2 + \cdots + a_n x^n = a_n (x + \alpha_1)(x + \alpha_2) \ldots (x + \alpha_n),$$

we are concerned with the zeros of

$$a_n\, e^{-\alpha_n x} \frac{d}{dx}\, e^{(\alpha_n-\alpha_{n-1})x} \frac{d}{dx} \cdots \frac{d}{dx}\, e^{(\alpha_2-\alpha_1)x} \frac{d}{dx}\, e^{\alpha_1 x} P(x).$$

Iterated application of **62**.

64. Limiting case of **63**. Observe III **201** and set $n=2m$,

$$a_0+a_1x+a_2x^2+\cdots+a_{2m}x^{2m}=\left(1-\frac{x^2}{m}\right)^m \to e^{-x^2} \qquad \text{as} \quad m\to\infty.$$

65. Also $a_0x^n+a_1x^{n-1}+\cdots+a_n$ has only real zeros and hence also [**63**]

$$\frac{1}{n!}\left(a_nx^n+a_{n-1}\frac{dx^n}{dx}+a_{n-2}\frac{d^2x^n}{dx^2}+\cdots\right)$$

$$=\frac{a_n}{n!}\,x^n+\frac{a_{n-1}}{(n-1)!}\,x^{n-1}+\frac{a_{n-2}}{(n-2)!}\,x^{n-2}+\cdots.$$

65.1. Use the symbol $D=d/dx$. We require to consider

$$(b_0+b_1D+b_2D^2+\cdots+b_kD^k+\cdots)P(x)=Q(x).$$

Use **63**:

$$\begin{aligned}&(a_0+a_1D+\cdots+a_nD^n)Q(x)\\&=(a_0+a_1D+\cdots+a_nD^n)(b_0+b_1D+b_2D^2+\cdots)P(x)\\&=P(x).\end{aligned}$$

66.

$$\alpha P(x)+xP'(x)=x^{-\alpha+1}\frac{d}{dx}\,[x^\alpha P(x)]=-(-x)^{1-\alpha}\frac{d}{dx}\,[(-x)^\alpha P(x)]$$

has neither fewer vanishing nor fewer positive nor fewer negative zeros than $P(x)$. For $\lim_{x\to\infty} x^\alpha P(x)=0$, if $\alpha<-n$ [**16**] and $\lim_{x\to 0} x^\alpha P(x)=0$, if $\alpha>0$.

67. [Laguerre, op. cit. **1**, p. 200.]

$$a_0(0+\alpha)+a_1(1+\alpha)x+a_2(2+\alpha)x^2+\cdots+a_n(n+\alpha)x^n=0$$

has not more non-real roots than

$$a_0+a_1x+a_2x^2+\cdots+a_nx^n=0 \qquad [\mathbf{66}].$$

This is the case $Q(x)=x+\alpha$; iterate.

68. Limiting case of **67**. Choose m sufficiently large and set

$$Q(x)=\left(1+\frac{x^2\log q}{m}\right)^m \to q^{x^2}.$$

69. [E. Laguerre, problem; solution by G. Pólya: Interméd. des math. **20**, p. 127 (1913).] Let α be real. The polynomial

$$Q(x)=\left(1+\frac{\alpha\sqrt{x}}{m}\right)^m+\left(1-\frac{\alpha\sqrt{x}}{m}\right)^m$$

has only real negative zeros, namely [**57**]

$$x=-\left(\frac{m}{\alpha}\tan\frac{\pi}{2m}\right)^2,\quad -\left(\frac{m}{\alpha}\tan\frac{3\pi}{2m}\right)^2,\quad \ldots,\quad -\left(\frac{m}{\alpha}\tan\frac{(2m-1)\pi}{2m}\right)^2.$$

Setting $q=e^{\alpha}$, we are concerned with the limiting case of the equation

$$a_0Q(0)+a_1Q(1)x+a_2Q(2)x^2+\cdots+a_nQ(n)x^n=0$$

as $m\to\infty$ [**67**, III **201**].

70. If $\Phi(x)=f(x)-a-bx$, $\Phi(x_1)=\Phi(x_2)=\Phi(x_3)=0$, $x_1<x_2<x_3$, then $\Phi'(x)$ has a zero in the interior of each of the intervals $x_1\leqq x\leqq x_2$ and $x_2\leqq x\leqq x_3$ [**10**] and hence $\Phi''(x)=f''(x)$ has a reversal of sign between x_1 and x_3 [**25**].

71. By assumption

$$\Phi(x)=f(x)-a_0-a_1x-\cdots-a_{n-1}x^{n-1},$$

where $a_0, a_1, \ldots, a_{n-1}$ are constants, has $n+1$ zeros in the interval $a\leqq x\leqq b$. By application of **10** n times, the existence of an interior point ξ at which $\Phi^{(n)}(\xi)=f^{(n)}(\xi)=0$ may be deduced.

72. The difference has n coincident zeros at the point $x=0$. If it had in addition an $(n+1)$th zero in the interval $-1<x<\infty$, then $(1+x)^{\alpha-n}$ would have to vanish there [**71**], which is not the case.

73.

$$e^x-1-\frac{x}{1!}-\frac{x^2}{2!}-\cdots-\frac{x^{n-1}}{(n-1)!}$$

has n coincident zeros at the point $x=0$, while e^x has no zeros. The assertion follows from **71** similarly as in **72.** It is, in fact, essentially equivalent to the fact that the function e^x is enveloped by its Maclaurin series for $x<0$ [I **141**].

74. [J. J. Sylvester: Mathematical Papers, **2**. Cambridge University Press 1908, p. 516.] It is sufficient to show that

$$1+\frac{x}{1!}+\frac{x^2}{2!}+\cdots+\frac{x^n}{n!}=f_n(x)$$

does not have two *consecutive negative* zeros. If a and b were such zeros, then we would have

$$f_n(a)=f_n'(a)+\frac{a^n}{n!}=0, \qquad f_n(b)=f_n'(b)+\frac{b^n}{n!}=0, \qquad \operatorname{sgn} f_n'(a)=\operatorname{sgn} f_n'(b)\neq 0;$$

$f_n'(x)$ would have by **8** an even number and by **10** an odd number of zeros in the interval $a<x<b$!

75. Obvious for $\mathfrak{l}=1$. Assume the statement to be true for the case of $\mathfrak{l}-1$ exponential functions. If $g(x)$ has Z and

$$g^*(x)=\frac{d^{m_\mathfrak{l}}}{dx^{m_\mathfrak{l}}}\,[e^{-a_\mathfrak{l}x}g(x)]=\frac{d^{m_\mathfrak{l}}}{dx^{m_\mathfrak{l}}}\,[e^{-a_\mathfrak{l}x}g(x)-P_\mathfrak{l}(x)]$$

Z^* real zeros, then we have

$$Z^*\geqq Z-m_\mathfrak{l} \qquad [\mathbf{12}].$$

On the other hand in $g^*(x)$ there occur only $\mathfrak{l}-1$ exponential functions [with the exponents $(a_1-a_\mathfrak{l})x, \ldots, (a_{\mathfrak{l}-1}-a_\mathfrak{l})x$; concerning the degrees of the polynomials that arise see **62**]. By assumption we hence have

$$m_1+m_2+\cdots+m_{\mathfrak{l}-1}-1\geqq Z^*.$$

76. [Cf. G. Pólya, problem: Arch. Math. Phys. Ser. 3, **28**, p. 173 (1920).] If the determinant were to vanish, there would exist a solution not identically zero of the corresponding homogeneous system, i.e. n real numbers $c_1, c_2, \ldots, c_n$ such that the entire function

$$c_1 e^{\beta_1 x} + c_2 e^{\beta_2 x} + \cdots + c_n e^{\beta_n x}$$

does not vanish identically but vanishes for $x = \alpha_1, \alpha_2, \ldots, \alpha_n$. Contradiction, since a function constructed in this manner has at most $n-1$ real zeros [**75**]. For the determination of the sign of the determinant cf. **48**.

77. [Laguerre, op. cit. **1**, p. 3.]

77.1. The given points are zeros since

$$F(m) = \Delta^n 0^m$$

for $m = 1, 2, 3, \ldots, n-1$ (notation used in III **220**). There are not more than $n-1$ real zeros [**77**, even **75** is sufficient].

78.[1] The "*Dirichlet* series" $\sum_{n=1}^{\infty} a_n e^{-\lambda_n s}$ may be differentiated term-by-term in the interior of its region of convergence. The method of proof of **77** can be followed without change.

79. [J. J. Sylvester, op. cit. **74**, p. 360, p. 401; cf. op. cit. **31**, p. 30.]

(1) If $a_1, a_2, \ldots, a_n$ have the same sign and m is even, then $C = 0$ and clearly also $Z = 0$.

(2) If $a_1, a_2, \ldots, a_n$ have the same sign and m is odd, then $C = 1$ and, since

$$P'(x) = m[a_1(x-\lambda_1)^{m-1} + a_2(x-\lambda_2)^{m-1} + \cdots + a_n(x-\lambda_n)^{m-1}]$$

never vanishes, cf. case (1), $Z \leqq 1$ [**10**].

(3) Let $a_\alpha a_{\alpha+1} < 0$, $1 \leqq \alpha < n$, and assume that the statement is true for $C-1$ changes of sign. Let $P(x) = b_0 x^m + b_1 x^{m-1} + \cdots + b_m$. Choose λ such that

$$\lambda_\alpha < \lambda < \lambda_{\alpha+1}, \qquad P(\lambda) \neq 0$$

and, if $b_0 \neq 0$, $b_1 + \lambda m b_0 \neq 0$. Set

$$P(x)(x-\lambda)^{-m} = F(x),$$

$$(x-\lambda)^{m+1} F'(x) = a_1^*(x-\lambda_1)^{m-1} + a_2^*(x-\lambda_2)^{m-1} + \cdots + a_n^*(x-\lambda_n)^{m-1} = P^*(x),$$

where $a_\nu^* = m(\lambda_\nu - \lambda) a_\nu$, $\nu = 1, 2, \ldots, n$. The number of changes of sign of the sequence $a_1^*, a_2^*, \ldots, a_n^*, (-1)^{m-1} a_1^*$ is

$$C^* = C - 1.$$

Denoting by Z^* the number of real zeros of $P^*(x)$, by assumption we have

$$C^* \geqq Z^*.$$

(3a) $$\lim_{x \to \pm\infty} \frac{F(x)}{x^2 F'(x)} = -\frac{b_0}{b_1 + \lambda m b_0} \neq 0, \qquad \text{if} \quad b_0 \neq 0.$$

Either sgn $F(x) = \operatorname{sgn} F'(x)$ in a neighbourhood of $x = -\infty$
or sgn $F(x) = -\operatorname{sgn} F'(x)$ in a neighbourhood of $x = +\infty$.

---

[1] In solutions **78**, **80–84** the necessary convergence considerations are not given, but their result is used. Cf. e.g. E. Landau: Münch. Ber. **36**, p. 151 (1906).

Applying **14** to $-\infty$, λ (appropriately!) and simultaneously **12** to λ, $+\infty$, or if necessary conversely, we find that in any case

$$Z^* \geqq Z-1.$$

$$\text{(3b)} \qquad \lim_{x\to\infty} \frac{F(x)}{xF'(x)} = -\frac{1}{s}, \qquad \text{if} \quad b_0=b_1=\cdots=b_{s-1}=0, \quad b_s\neq 0.$$

Now we have sgn $F(x)=$ sgn $F'(x)$ as $x\to-\infty$, sgn $F(x)=-$sgn $F'(x)$ as $x\to+\infty$, and hence by applying **14** to $-\infty$, λ and λ, $+\infty$ it follows that in fact $Z^*\geqq Z$.

80. [Laguerre, op. cit. **1**, p. 29; cf. G. Pólya: C.R. Acad. Sci. (Paris), **156**, p. 996 (1913).] If $R=0$, then clearly $Z=0$. If $R>0$, let λ_0 be the common end-point of two neighbouring intervals of opposite constant sign of the function $\varphi(\lambda)$ [definition preceding **22**]. The number of reversals of sign of

$$\varphi^*(\lambda)=(\lambda_0-\lambda)\varphi(\lambda)$$

is then $R^*=R-1$. The number of zeros in question of the function

$$F^*(x)=e^{-\lambda_0 x}\frac{d}{dx}[e^{\lambda_0 x}F(x)]=\int_0^\infty \varphi^*(\lambda)\, e^{-\lambda x}\, d\lambda$$

is $Z^*\geqq Z-1$ [**12**]. Mathematical induction as in **77, 79**.

81. [Cf. L. Fejér: C.R. Acad. Sci. (Paris), **158**, pp. 1328–1331 (1914).] The number of real zeros Z of the function

$$F(x)=\int_0^\infty f(t)t^x\, dt=\int_{-\infty}^\infty e^{-\lambda}f(e^{-\lambda})\, e^{-\lambda x}\, d\lambda$$

is not less than the number of changes of sign of the sequence $a_0, a_1, a_2, \ldots$, as can be seen by a complete survey of the various cases [**8, 9**]. We have $F(n)=a_n$ in case (1), $F(n)=0$ in cases (2), (3). The choice of the limits of integration ($0,\infty$ or $-\infty,\infty$) does not affect the proof of **80**.

82. Transformation by integration by parts yields

$$\int_0^\infty \varphi(\lambda)\, e^{-\lambda x}\, d\lambda = x\int_0^\infty \Phi(\lambda)\, e^{-\lambda x}\, d\lambda,$$

if the integral on the left is convergent and $x>0$ [**80**].

83. Transformation by partial summation yields

$$\sum_{n=1}^\infty a_n\, e^{-\lambda_n x}=\sum_{n=1}^\infty (a_1+a_2+\cdots+a_n)\,(e^{-\lambda_n x}-e^{-\lambda_{n+1}x}),$$

if the series on the left is convergent and $x>0$. Set

$$\varphi(\lambda)=a_1+a_2+\cdots+a_n, \qquad \text{if} \quad \lambda_n\leqq x<\lambda_{n+1}, \quad n=1,2,3,\ldots.$$

Then the series under consideration becomes

$$x\sum_{n=1}^\infty (a_1+a_2+\cdots+a_n)\int_{\lambda_n}^{\lambda_{n+1}} e^{-\lambda x}\, d\lambda = x\int_{\lambda_1}^\infty \varphi(\lambda)\, e^{-\lambda x}\, d\lambda.$$

The number of reversals of sign of the piecewise constant function $\varphi(\lambda)$ equals the number of changes of sign of the sequence $a_1, a_1+a_2, a_1+a_2+a_3, \ldots$ [**80**].

83.1. [J. Steinig: Rendiconti di Mat. Ser. 6, **4**, p. 633 (1971).] Keep the notation of **83** but add that $a_k=0$ for $k>n$. By our assumption

$$x^{-1} D(x)=\int_{\lambda_1}^{\lambda_n} \varphi(\lambda)\, e^{-\lambda x}\, d\lambda,$$

and this integral exists for all values of x [**80**].

84.

$$\sum_{n=1}^{\infty} \frac{n!a_n}{x(x+1)\ldots(x+n)}=\sum_{n=0}^{\infty} a_n \frac{\Gamma(x)\Gamma(n+1)}{\Gamma(x+n+1)}=\sum_{n=0}^{\infty} a_n \int_0^1 t^{x-1}(1-t)^n\, dt$$

$$=\int_0^1 t^{x-1}f(1-t)\, dt=\int_0^{\infty} e^{-\lambda x}f(1-e^{-\lambda})\, d\lambda. \qquad \text{[80, 24, 44.]}$$

85. [Laguerre, op. cit. **1**, p. 28.] If we set

$$f(x)=a_n\left(\frac{1}{\alpha_n}\right)^{-x}+a_{n-1}\left(\frac{1}{\alpha_{n-1}}\right)^{-x}+\cdots+a_1\left(\frac{1}{\alpha_1}\right)^{-x},$$

there are at most $n-1$ integers k such that $f(k)=0$ [**75**, also **48**].

Hence it is impossible that the series

$$\Phi(x)=\sum_{\nu=1}^{n} a_\nu F(\alpha_\nu x)=\sum_{k=0}^{\infty} p_k(a_1\alpha_1^k+a_2\alpha_2^k+\cdots+a_n\alpha_n^k)x^k=\sum_{k=0}^{\infty} p_k f(k)x^k$$

should vanish identically. The following four numbers, the positive zeros of $\Phi(x)$, the changes of sign of $p_0 f(0), p_1 f(1), p_2 f(2), \ldots$, the positive zeros of $f(x)$ and the changes of sign of $a_n, a_n+a_{n-1}, a_n+a_{n-1}+a_{n-2}, \ldots$ are ordered monotonically non-decreasing. Apply successively **38**, **8**, **83**.

86. A function of the form

$$c_1F(x\beta_1)+c_2F(x\beta_2)+\cdots+c_nF(x\beta_n)$$

cannot have more than $n-1$ positive zeros [**85**], and hence can certainly not vanish for $x=\alpha_1, \alpha_2, \ldots, \alpha_n$. The proof then proceeds as in **48**, **76**. The determinant moreover is >0; for in the case $\alpha_\nu=\beta_\nu$ its sign can be determined with the aid of the theory of quadratic forms from the fact that

$$\sum_{\lambda=1}^{n}\sum_{\mu=1}^{n} F(\alpha_\lambda\alpha_\mu)x_\lambda x_\mu=\sum_{k=0}^{\infty} p_k(\alpha_1^k x_1+\alpha_2^k x_2+\cdots+\alpha_n^k x_n)^2 \geqq 0.$$

87. We assume that the Descartes rule of signs holds.

(1) If

$$1 \leqq \nu_1<\nu_2<\nu_3<\cdots<\nu_l \leqq n,$$

then $W[h_{\nu_1}(x), \ldots, h_{\nu_l}(x)] \neq 0$ for $a<x<b$. For if this Wronskian determinant were to vanish for $x=x_0$, then there would exist constants $c_1, c_2, \ldots, c_l$, not all vanishing, satisfying the l simultaneous equations

$$\begin{aligned}
&c_1h_{\nu_1}(x_0) \;+c_2h_{\nu_2}(x_0) \;+\cdots+c_lh_{\nu_l}(x_0) \;=0,\\
&c_1h'_{\nu_1}(x_0) \;+c_2h'_{\nu_2}(x_0) \;+\cdots+c_lh'_{\nu_l}(x_0) \;=0,\\
&\cdots\cdots\cdots\cdots\cdots\cdots\cdots\cdots\cdots\cdots\\
&c_1h_{\nu_1}^{(l-1)}(x_0)+c_2h_{\nu_2}^{(l-1)}(x_0)+\cdots+c_lh_{\nu_l}^{(l-1)}(x_0)=0,
\end{aligned}$$

i.e. the function $c_1 h_{\nu_1}(x)+c_2 h_{\nu_2}(x)+\cdots+c_l h_{\nu_l}(x)$ would have l zeros (coinciding with x_0) in the interval $a<x<b$, while the sequence of coefficients $c_1, c_2, \ldots, c_l$ can have at most $l-1$ changes of sign: Contradiction.

(2) Let us prove for example that all the $(n-1)$-rowed Wronksian determinants have the same sign. Let $a<x_0<b$. Determine $a_1, a_2, \ldots, a_n$ from the system

$$\begin{aligned}
&a_1 h_1(x_0) \quad +a_2 h_2(x_0) \quad +\cdots+a_n h_n(x_0) \quad =0,\\
&a_1 h_1'(x_0) \quad +a_2 h_2'(x_0) \quad +\cdots+a_n h_n'(x_0) \quad =0,\\
&\cdots\cdots\cdots\cdots\cdots\cdots\cdots\cdots\cdots\cdots\cdots\\
&a_1 h_1^{(n-2)}(x_0)+a_2 h_2^{(n-2)}(x_0)+\cdots+a_n h_n^{(n-2)}(x_0)=0,\\
&a_1 h_1^{(n-1)}(x_0)+a_2 h_2^{(n-1)}(x_0)+\cdots+a_n h_n^{(n-1)}(x_0)=1,
\end{aligned}$$

whose determinant does not vanish (cf. (1)). The function

$$a_1 h_1(x)+a_2 h_2(x)+\cdots+a_n h_n(x)$$

has $n-1$ zeros in the interior of a, b (they are all $=x_0$). Hence the sequence of coefficients $a_1, a_2, \ldots, a_n$ must have exactly $n-1$ changes of sign, that is, setting $(-1)^{n-1} W[h_1(x_0), h_2(x_0), \ldots, h_n(x_0)]=W$, the numbers

$$a_1 W=W(h_2, h_3, \ldots, h_n), \qquad -a_2 W=W(h_1, h_3, \ldots, h_n),$$
$$a_3 W=W(h_1, h_2, h_4, \ldots, h_n), \ldots$$

(all taken at the point $x=x_0$) must all be of the same sign, q.e.d.

88. Applied to the special cases with number of rows $l=1, 2$, the criterion **87** says that on the one hand $h_1(x), h_2(x), \ldots, h_n(x)$, on the other hand

$$W(h_1, h_2)=h_1^2 \frac{d}{dx}\frac{h_2}{h_1}, \qquad W(h_2, h_3)=h_2^2 \frac{d}{dx}\frac{h_3}{h_2}, \ldots$$

must be of the same sign. May also be shown directly.

89. Let $l \leqq n-1$, $1 \leqq \nu_1<\cdots<\nu_j<\alpha<\nu_{j+1}<\cdots<\nu_l \leqq n$. We have [VII **57**]

$$\begin{aligned}
&W(h_{\nu_1}, h_{\nu_2}, \ldots, h_{\nu_j}, h_\alpha, h_{\nu_{j+1}}, \ldots, h_{\nu_l})=\\
&=h_\alpha^{l+1} W\left[\frac{h_{\nu_1}}{h_\alpha}, \frac{h_{\nu_2}}{h_\alpha}, \ldots, \frac{h_{\nu_j}}{h_\alpha}, 1, \frac{h_{\nu_{j+1}}}{h_\alpha}, \ldots, \frac{h_{\nu_l}}{h_\alpha}\right]\\
&=h_\alpha^{l+1} W\left[-\left(\frac{h_{\nu_1}}{h_\alpha}\right)', \ldots, -\left(\frac{h_{\nu_j}}{h_\alpha}\right)', \left(\frac{h_{\nu_{j+1}}}{h_\alpha}\right)', \ldots, \left(\frac{h_{\nu_l}}{h_\alpha}\right)'\right].
\end{aligned}$$

90. Let us assume that the Descartes rule of signs is valid for a system of any $n-1$ functions that satisfy the determinantal conditions. Let the function

$$F(x)=a_1 h_1(x)+a_2 h_2(x)+\cdots+a_n h_n(x)$$

have Z zeros in the interior of a, b and let the sequence of coefficients $a_1, a_2, \ldots, a_n$ have C changes of sign. The case $C=0$ is clear. Let now $\alpha+1$ be an index of change. With the notation of **89** we obtain

$$\begin{aligned}
\frac{d}{dx}\frac{F(x)}{h_\alpha(x)}&=a_1 \frac{d}{dx}\frac{h_1}{h_\alpha}+\cdots+a_{\alpha-1}\frac{d}{dx}\frac{h_{\alpha-1}}{h_\alpha}+a_{\alpha+1}\frac{d}{dx}\frac{h_{\alpha+1}}{h_\alpha}+\cdots+a_n \frac{d}{dx}\frac{h_n}{h_\alpha}\\
&=-a_1 H_1(x)-\cdots-a_{\alpha-1} H_{\alpha-1}(x)+a_{\alpha+1} H_\alpha(x)+\cdots+a_n H_{n-1}(x)\\
&=F^*(x).
\end{aligned}$$

Let Z^* be the number of zeros of $F^*(x)$ in the interval $a<x<b$ and C^* the number of changes of sign of the sequence

$$-a_1, \quad -a_2, \quad \ldots, \quad -a_{\alpha-1}, \quad a_{\alpha+1}, \quad \ldots, \quad a_n,$$

or, equivalently [**3**], the number of changes of sign of the sequence

$$-a_1, \quad -a_2, \quad \ldots, \quad -a_{\alpha-1}, \quad -a_\alpha, \quad a_{\alpha+1}, \quad \ldots, \quad a_n.$$

We have [**12**]

$$Z^* \geqq Z-1, \qquad C^* = C-1.$$

The functions $H_1(x), H_2(x), \ldots, H_{n-1}(x)$ satisfy the determinantal conditions **87** [**89**]. From the assumption of the mathematical induction proof we then have

$$Z^* \leqq C^*.$$

91.

$$W(e^{\lambda_1 x}, e^{\lambda_2 x}, \ldots, e^{\lambda_n x}) = e^{(\lambda_1+\lambda_2+\cdots+\lambda_n)x} \prod_{j<k}^{1,2,\ldots,n} (\lambda_k-\lambda_j) > 0.$$

92. With the notation of solution **63** we have

$$a_n\, e^{-\alpha_n x} \frac{d}{dx} e^{(\alpha_n-\alpha_{n-1})x} \frac{d}{dx} \cdots \frac{d}{dx} e^{(\alpha_2-\alpha_1)x} \frac{d}{dx} e^{\alpha_1 x} f(x)$$

$$= a_0 f(x) + a_1 f'(x) + \cdots + a_n f^{(n)}(x).$$

Apply **12** n times and **6** $n+1$ times.

93. [G. Pólya: Trans. Amer. Math. Soc. **24**, pp. 312–324 (1922); cf. H. Poincaré: Interméd. des math. **1**, pp. 141–144 (1894).] Using the formula VII **62**, the proof proceeds as in **92**.

94. We have thus assumed that

$$h^{(n)}(x) + \varphi_1(x) h^{(n-1)}(x) + \varphi_2(x) h^{(n-2)}(x) + \cdots + \varphi_n(x) h(x) = 0$$

identically, and that $f(x)-h(x)$ vanishes $n+1$ times in the interval. Apply **93** to $f(x)-h(x)$.

94.1. [See G. Pólya: C.R. Acad. Sci. (Paris), **199**, pp. 655–657 (1934); also for suggestions regarding some other differential operators.] If $f(x, y)$ vanishes identically, the assertion is obvious; therefore we exclude this trivial case. Hence, $f(x, y)$ attains both its minimum and its maximum in $\mathfrak{R}$ in the interior of $\mathfrak{R}$, at two different points, at one of which the inequalities

$$f(x, y) \leqq 0, \qquad \frac{\partial^2 f}{\partial x^2} + \frac{\partial^2 f}{\partial y^2} \geqq 0,$$

and at the other the opposite inequalities hold. Join the two points by a continuous path.

95. [Cf. H. A. Schwarz: Gesammelte Mathematische Abhandlungen, **2**, p. 296. Berlin: J. Springer, 1890. T. J. Stieltjes: Oeuvres, **2**, p. 110. Groningen: P. Noordhoff, 1918.] Assume $|\varphi_\lambda(x_\mu)| \neq 0$. Denoting the value of the quotient $|f_\lambda(x_\mu)| : |\varphi_\lambda(x_\mu)|$ by Q, the function of x

$$|f_k(x_1) f_k(x_2) \ldots f_k(x_{n-1}) f_k(x)| - Q |\varphi_k(x_1) \varphi_k(x_2) \ldots \varphi_k(x_{n-1}) \varphi_k(x)|$$

vanishes for $x=x_{n-1}$ and for $x=x_n$. By Rolle's theorem we obtain

$$|f_k(x_1)f_k(x_2)\dots f_k(x_{n-1})f_k'(\eta_n)|-Q|\varphi_k(x_1)\varphi_k(x_2)\dots\varphi_k(x_{n-1})\varphi_k'(\eta_n)|=0,$$

where $x_{n-1}<\eta_n<x_n$. Now replace x_{n-1} by x. The resulting function of x vanishes for $x=x_{n-2}$ and $x=x_{n-1}$ and so on. Then replace η_n by x and so on. In this manner after $(n-1)+(n-2)+\cdots+2+1$ applications of Rolle's theorem we arrive at the linear equation for Q which was to be demonstrated.

96. Special case of **95**:

$$\varphi_1(x)=1,\qquad \varphi_2(x)=x,\qquad \varphi_3(x)=x^2,\quad \dots,\quad \varphi_n(x)=x^{n-1}.$$

By adding rows we obtain

$$\frac{1}{1^{n-1}2^{n-2}3^{n-3}\dots(n-1)^1}\begin{vmatrix} 1 & 1 & 1 & \dots & 1\\ x_1 & x_2 & x_3 & \dots & x_n\\ \dots & \dots & \dots & \dots & \dots\\ x_1^{n-1} & x_2^{n-1} & x_3^{n-1} & \dots & x_n^{n-1}\end{vmatrix}$$

$$=\begin{vmatrix} 1 & 1 & \dots & 1\\ \binom{x_1}{1} & \binom{x_2}{1} & \dots & \binom{x_n}{1}\\ \dots & \dots & \dots & \dots\\ \binom{x_1}{n-1} & \binom{x_2}{n-1} & \dots & \binom{x_n}{n-1}\end{vmatrix}=\frac{\prod\limits_{j<k}^{1,2,\dots,n}(x_k-x_j)}{\prod\limits_{j<k}^{1,2,\dots,n}(k-j)}.$$

97. Special case of **96**:

$$f_1(x)=1,\qquad f_2(x)=x,\qquad f_3(x)=x^2,\quad \dots,\quad f_{n-1}(x)=x^{n-2}.$$

98. Special case of **97**:

$$x_1=x,\qquad x_2=x+h,\qquad x_3=x+2h,\quad \dots,\quad x_n=x+(n-1)h.$$

99. [Op. cit. **93**.] For $n=2$, **93** (**) yields only one condition: $h(x)>0$. By the mean value theorem we have

$$\frac{1}{h(x_1)h(x_2)}\begin{vmatrix} h(x_1) & h(x_2)\\ f(x_1) & f(x_2)\end{vmatrix}=\frac{f(x_2)}{h(x_2)}-\frac{f(x_1)}{h(x_1)}=\frac{x_2-x_1}{[h(\xi)]^2}\begin{vmatrix} h(\xi) & h'(\xi)\\ f(\xi) & f'(\xi)\end{vmatrix}.$$

Set

$$\frac{d}{dx}\frac{h_2(x)}{h_1(x)}=H_1(x),\quad \dots,\quad \frac{d}{dx}\frac{h_{n-1}(x)}{h_1(x)}=H_{n-2}(x),\qquad \frac{d}{dx}\frac{f(x)}{h_1(x)}=F(x).$$

We obtain by applying Rolle's theorem $n-1$ times [solution **95**]

$$\frac{1}{h_1(x_1)h_1(x_2)\dots h_1(x_n)}\begin{vmatrix} h_1(x_1) & h_1(x_2) & \dots & h_1(x_n)\\ h_2(x_1) & h_2(x_2) & \dots & h_2(x_n)\\ \dots & \dots & \dots & \dots\\ h_{n-1}(x_1) & h_{n-1}(x_2) & \dots & h_{n-1}(x_n)\\ f(x_1) & f(x_2) & \dots & f(x_n)\end{vmatrix}=$$

$$=(x_n-x_{n-1})(x_{n-1}-x_{n-2})\dots(x_2-x_1)\begin{vmatrix} 1 & 0 & \dots & 0\\ h_2(x_1)/h_1(x_1) & H_1(\xi_1) & \dots & H_1(\xi_{n-1})\\ \dots & \dots & \dots & \dots\\ h_{n-1}(x_1)/h_1(x_1) & H_{n-2}(\xi_1)\dots & & H_{n-2}(\xi_{n-1})\\ f(x_1)/h_1(x_1) & F(\xi_1) & \dots & F(\xi_{n-1})\end{vmatrix},$$

where $x_1 < \xi_1 < x_2 < \xi_2 < \cdots < x_{n-1} < \xi_{n-1} < x_n$. Transform the Wronskian determinant appearing on the right side of the equation to be proved by taking out the factor $[h_1(\xi)]^n$ in accordance with VII **58**. The conditions **93** (**) are satisfied by $H_1(x), H_2(x), \ldots, H_{n-2}(x)$. [**89**.] If we assume that the statement is proved for the $n-1$ functions $H_1(x), H_2(x), \ldots, H_{n-2}(x), F(x)$, then from the transformation we have carried out it follows for the n functions $h_1(x), h_2(x), \ldots, h_{n-1}(x), f(x)$.

100. Special case of **99**:

$$h_\nu(x) = e^{\beta_\nu x}, \qquad \nu = 1, 2, \ldots, n-1, \qquad f(x) = e^{\beta_n x}, \qquad x_\mu = \alpha_\mu, \qquad \mu = 1, 2, \ldots, n.$$

The Wronskian determinants in question are >0 [**91**].

101. [For the whole of this chapter cf. E. Laguerre, op. cit. **1**. See also M. Marden: The Geometry of the Zeros. Mathematical Surveys III, New York: Amer. Math. Soc. 1949.] The mappings considered of the Z-plane onto the Z'-plane have the form $Z' = aZ + b$. From

$$\zeta = m_1 z_1 + m_2 z_2 + \cdots + m_n z_n, \qquad m_1 + m_2 + \cdots + m_n = 1,$$
$$z_1' = a z_1 + b, \qquad z_2' = a z_2 + b, \ldots, \qquad z_n' = a z_n + b,$$
$$\zeta' = m_1 z_1' + m_2 z_2' + \cdots + m_n z_n'$$

it follows that

$$\zeta' = a\zeta + b.$$

102. First let $z, z_1, z_2, \ldots, z_n$ be finite. The transformations in question have the form

$$Z' = \frac{a}{Z - z} + b, \qquad a \neq 0.$$

From

$$z_1' = \frac{a}{z_1 - z} + b, \qquad z_2' = \frac{a}{z_2 - z} + b, \qquad \ldots, \qquad z_n' = \frac{a}{z_n - z} + b,$$
$$\zeta' = m_1 z_1' + m_2 z_2' + \cdots + m_n z_n', \qquad m_1 + m_2 + \cdots + m_n = 1,$$
$$\zeta' = \frac{a}{\zeta - z} + b$$

it follows that

$$(*) \qquad \frac{1}{\zeta - z} = \frac{m_1}{z_1 - z} + \frac{m_2}{z_2 - z} + \cdots + \frac{m_n}{z_n - z},$$

independently of a and b.—If $z_\nu = \infty$ for a certain ν, and thus z finite, then the νth term on the right-hand side of (*) is to be omitted. If $z = \infty$, and thus $z_1, z_2, \ldots, z_n$ finite, then the transformation is the one considered in **101** and ζ coincides with the ordinary center of gravity [**101**].

103. The facts are well known if the point of reference is $z = \infty$ [III **32**]. By a linear mapping we now obtain: the circle through z_i, z_k and z is the boundary between two circular domains ($z_i \neq z_k$); from the finitely many such circular domains delete the interior of all those that contain none of the points $z_1, z_2, \ldots, z_n$ in their interior. The closed part of the plane which has not been thus deleted is $\mathfrak{C}_z$. The construction must be modified if $z_1, z_2, \ldots, z_n$ and z all lie on a circle: in this case $\mathfrak{C}_z$ is that arc of this circle that contains $z_1, z_2, \ldots, z_n$ and does not contain z.

104. Cf. **103** or directly by means of a mapping from the case $z=\infty$.

105. [**103**.]

106. Generalization of the case $z=\infty$ by means of a linear mapping.

107. If z and ζ_z both lay outside C, then a circle passing through z and ζ_z but not through C would not separate $z_1, z_1, \ldots, z_n$ in contradiction to **106.** If z lies outside and ζ_z on the boundary of C, apply **106** to the circle passing through z and ζ_z and touching C from the outside.

108. By **102**, since now all the masses $=1/n$, the center of gravity ζ_z is given by

$$\frac{1}{\zeta_z - z} = \frac{1}{n}\left(\frac{1}{z_1 - z} + \frac{1}{z_2 - z} + \cdots + \frac{1}{z_n - z}\right)$$

$$= \frac{n_1}{n}\frac{1}{w_1 - z} + \frac{n_2}{n}\frac{1}{w_2 - z} + \cdots + \frac{n_k}{n}\frac{1}{w_k - z}$$

where $n_1, n_2, \ldots, n_k$ denote the number of those of the $z_1, z_2, \ldots, z_n$ that coincide with $w_1, w_2, \ldots, w_k$, respectively. By the rational masses $n_1/n, n_2/n, \ldots, n_k/n$ any k masses whose sum is 1 may be arbitrarily closely approximated. This is for finite $w_1, w_2, \ldots, w_k, z$. For $w_\nu=\infty$ or $z=\infty$ cf. solution **102**.

109. To the point z_1 considered, even if z_1 lies at infinity, cf. the formula in solution **102**.

110. By **102**, since now all the masses $=1/n$,

$$\zeta_z = z - \frac{n}{\dfrac{1}{z-z_1} + \dfrac{1}{z-z_2} + \cdots + \dfrac{1}{z-z_n}}$$

$$= \frac{z_1 + z_2 + \cdots + z_n}{n} + \frac{\sum_{\mu=1}^{n}\sum_{\nu=1}^{n}(z_\mu - z_\nu)^2}{2n^2}\frac{1}{z} + \cdots.$$

111. By solution **110** we have

$$\zeta = z - \frac{nf(z)}{f'(z)} = -\frac{a_0 + \binom{n-1}{1}a_1 z + \binom{n-1}{2}a_2 z^2 + \cdots + a_{n-1}z^{n-1}}{a_1 + \binom{n-1}{1}a_2 z + \binom{n-1}{2}a_3 z^2 + \cdots + a_n z^{n-1}},$$

if z is finite, $\zeta = -a_{n-1}/a_n$, if $z=\infty$; $f(z)$ is expressed by the formula in the remarks preceding **111**. For $f(z)=z^n$ we have $\zeta=0$, as is also clear from other considerations [**107**].—Laguerre terms ζ the "point dérivé du point z" [op. cit. **1**, p. 56].

112. [Laguerre, op. cit. **1**, p. 61.] If all the zeros of $f(z)$ are real, then they lie in the closed lower (upper) half-plane, in whose interior also ζ must lie, if the imaginary part of z is positive (negative), except in the case in which all the zeros of $f(z)$ coincide [**107**]. If $f(z)$ has a complex zero z_1, then z and ζ simultaneously tend to z_1 [**109**]; hence sufficiently close to z_1 their imaginary parts have the same sign.

113. By **111** we have, assuming $a_n \neq 0$, that $\zeta = \infty$ if and only if z is a zero of the derivative $f'(z)$ that does not coincide with a zero of $f(z)$. Every straight line through such a zero of $f'(z)$ separates the zeros of $f(z)$ [**106**], and every circular disc that covers all the zeros of $f(z)$, also covers those of $f'(z)$ [**107**]. From **106** as well as from **107** Gauss's theorem [III **31**] again follows, in particular from **107** by the following consideration: the largest common part (or "intersection") of all circular discs (circular domains not containing ∞) that cover the points $z_1, z_2, \ldots, z_n$, is the smallest convex polygon that contains $z_1, z_2, \ldots, z_n$.

114. If x is a zero of $e^{-\frac{z}{c}}\{-(f(z)/c)+f'(z)\}$ and $f(x) \neq 0$, and thus also $f'(x) \neq 0$, then the center of gravity of $f(z)$ with respect to x is [**111**]

$$= x - \frac{nf(x)}{f'(x)} = x - nc = \zeta.$$

If x were to lie neither in C nor in $C+nc$, then both x and ζ would lie outside C, which is however not possible by **107**.

115. [J. v. Sz. Nagy; cf. L. Fejér: Jber. deutsch. Math. Verein. **26**, p. 119 (1917).] Let z_1 be finite; setting $f(z)=(z-z_1)g(z)$ we have $(x-z_1)g'(x)+g(x)=0$. The center of gravity ζ of $g(z)$ with respect to x is [**111**]

$$\zeta = x - \frac{(n-1)g(x)}{g'(x)} = x-(n-1)(z_1-x). \qquad [\mathbf{106}]$$

For $z_1 = \infty$ cf. III **31**.

116. [Laguerre, op. cit. **1**, p. 56, 133.] Let z be one of the zeros of $\alpha_1 z f'(z) - \alpha_2 f(z)$ and $f(z) \neq 0$, hence z finite, $f'(z) \neq 0$. The center of gravity of $f(z)$ with respect to z is

$$\zeta = z - n\frac{\alpha_1}{\alpha_2} z = \left(1 - n\frac{\alpha_1}{\alpha_2}\right) z.$$

Thus if we had $|z| < \min(1, |1-n(\alpha_1/\alpha_2)|^{-1})$, then we would also have $|\zeta| < 1$, in contradiction to **107**.

117. [**116**.]

118. [Laguerre, op. cit. **1**, p. 161.] z_1 must be a simple zero so that the center of gravity considered makes sense. Setting $f(z)=(z-z_1)g(z)$, we have $f'(z_1)= g(z_1)$, $f''(z_1)=2g'(z_1)$ and the center of gravity [**111**] is

$$= z_1 - \frac{(n-1)g(z_1)}{g'(z_1)}.$$

For $z_1 = \infty$ the remaining zeros must be finite, $a_n = 0$, $a_{n-1} \neq 0$, The center of gravity is then [**111**]

$$= -\frac{1}{2}\frac{a_{n-2}}{a_{n-1}}.$$

119. [Laguerre, op. cit. **1**, p. 142.] With respect to the zero $z_1 = \alpha + i\beta$ with the largest imaginary part, the center of gravity of the remaining $n-1$ zeros must have imaginary part $< \beta$, if $\beta > 0$ [**107**]. In fact for $\beta > 0$ it follows from the differential equation that, if $f(z_1)=0$ and thus $f'(z_1) \neq 0$, $f''(z_1) \neq 0$ [**118**], we have

$$\Im\left(z_1 - \frac{2(n-1)f'(z_1)}{f''(z_1)}\right) = \Im\left(z_1 - \frac{2(n-1)}{z_1}\right) = \beta + \frac{2(n-1)\beta}{\alpha^2+\beta^2} > \beta.$$

120. For $f(z_1)=0$, $z_1=\alpha+i\beta$, $\beta>0$, it would follow that

$$\mathfrak{I}\left(z_1-\frac{2(n-1)f'(z_1)}{f''(z_1)}\right)=\mathfrak{I}\left(z_1+\frac{(n-1)(z_1^2-1)}{z_1}\right)$$
$$=\beta+(n-1)\left(\beta+\frac{\beta}{\alpha^2+\beta^2}\right)>\beta.$$

Cf. **119**.

121. False. Let $a\neq b$ and C_1, C_2 circles with centers a and b, respectively, whose radii are so small that they do not contain the point $(a+b)/2$. Now consider $f(z)=(z-a)(z-b)$.

122.

$$\frac{n_1z_2+n_2z_1}{n_1+n_2}-\frac{n_1z_2^{(0)}+n_2z_1^{(0)}}{n_1+n_2}=\frac{n_1(z_2-z_2^{(0)})+n_2(z_1-z_1^{(0)})}{n_1+n_2}.$$

We have $(z^{(0)}-z_1^{(0)}):(z_2^{(0)}-z^{(0)})=(r-r_1):(r_2-r)=n_1:n_2$.

123. We may assume that C_1 and C_2 are the half-planes $\mathfrak{R}z\geqq c_1$, $\mathfrak{R}z\geqq c_2$. Then we have

$$\mathfrak{R}\,\frac{n_1z_2+n_2z_1}{n_1+n_2}\geqq\frac{n_1c_2+n_2c_1}{n_1+n_2}.$$

Furthermore, choosing z_1 and z_2 suitably, the mean value $(n_1z_2+n_2z_1)/(n_1+n_2)$ can be made equal to an arbitrary number c with $\mathfrak{R}c\geqq(n_1c_2+n_2c_1)/(n_1+n_2)$.

124. [J. L. Walsh: Trans. Amer. Math. Soc. **22**, p. 115 (1921).] Let z be a zero of $f_1'(z)f_2(z)+f_1(z)f_2'(z)$ and let z lie outside C_1 and C_2. Thus we have $f_1(z)\neq 0$, $f_1'(z)\neq 0$, $f_2(z)\neq 0$, $f_2'(z)\neq 0$, z finite. Denoting the center of gravity of $f_1(z)$ with respect to z by ζ_1 and the center of gravity of $f_2(z)$ with respect to z by ζ_2, we have

$$\zeta_1=z-\frac{n_1f_1(z)}{f_1'(z)},\qquad \zeta_2=z-\frac{n_2f_2(z)}{f_2'(z)}.$$

ζ_1 lies in C_1, ζ_2 lies in C_2 [**107**]. From this it follows by multiplication by n_2 and n_1, respectively, and addition that

$$z=\frac{n_1\zeta_2+n_2\zeta_1}{n_1+n_2}.$$

125. [J. L. Walsh, op. cit. **124**.] Analogously to **124** set: $f(z)=f_1(z)/f_2(z)$. If z is a zero of $f'(z)$ lying outside C_1 and C_2, and ζ_1 and ζ_2 the centers of gravity with respect to z of $f_1(z)$ and $f_2(z)$, respectively, then we have [**124**]

$$z=\frac{n_1\zeta_2-n_2\zeta_1}{n_1-n_2},\qquad \text{if}\quad n_1\gtrless n_2.$$

If $n_1=n_2$, then it follows that $\zeta_1=\zeta_2$, which is impossible for circular domains that have no common points.

126. [R. Jentzsch: Arch. Math. Phys. Ser. 3, **25**, p. 196 (1917); cf. M. Fekete: Jber. deutsch. Math. Verein. **31**, pp. *42–48* (1922).] Let a and b be two numbers for which all the zeros of $f(z)-a$ and $f(z)-b$ lie in $\mathfrak{O}_1$. We must show that the zeros of $f(z)-c$ also lie in $\mathfrak{O}_1$, if c lies on the straight line joining a and b. [Definition of convexity!] The zeros of $F(z)=[f(z)-a]^m[f(z)-b]^n$, for positive integers m, n, certainly lie in $\mathfrak{O}_1$. We have

$$F'(z)=(m+n)[f(z)-a]^{m-1}[f(z)-b]^{n-1}f'(z)\left(f(z)-\frac{na+mb}{m+n}\right),$$

so that all the zeros of $f(z)-(na+mb)/(m+n)$ also lie in $\mathfrak{O}_1$ [III **31**]. But the numbers $(na+mb)/(m+n)$ lie everywhere dense on the line joining a and b, as m and n run through all the positive integers.

127. Apply **124** to $F(z)=[f(z)-a]^{n_1}[f(z)-b]^{n_2}$ [solution **126**].

128.

$$a_0+a_1\zeta,\quad a_1+a_2\zeta,\quad a_2+a_3\zeta,\quad \ldots,\quad a_{n-1}+a_n\zeta,$$

or

$$a_1,\ a_2,\ a_3,\ \ldots,\ a_n,$$

according as ζ is finite or infinite, respectively. I.e. we have, if ζ is finite,

$$(\zeta-z)f'(z)+nf(z)=n\sum_{\nu=0}^{n-1}\binom{n-1}{\nu}(a_\nu+a_{\nu+1}\zeta)z^\nu.$$

129. Definition.

130. Well known for $\zeta=\infty$. For $\zeta\neq\infty$

$$\begin{aligned}g(z)[(\zeta-z)h'(z)+lh(z)]+h(z)[(\zeta-z)g'(z)+kg(z)]\\ =(\zeta-z)[g(z)h'(z)+g'(z)h(z)]+(k+l)g(z)h(z).\end{aligned}$$

131. If ζ_1, ζ_2 are both $=\infty$, the result is well known. If both are finite, we have that

$$\begin{aligned}(\zeta_1-z)[(\zeta_2-z)f'(z)+nf(z)]'+(n-1)[(\zeta_2-z)f'(z)+nf(z)]\\ =(\zeta_1-z)(\zeta_2-z)f''(z)+(n-1)(\zeta_1+\zeta_2-2z)f'(z)+n(n-1)f(z)\end{aligned}$$

is symmetric in ζ_1, ζ_2. If $\zeta_1=\zeta$ is finite, $\zeta_2=\infty$,

$$[(\zeta-z)f'(z)+nf(z)]'=(\zeta-z)f''(z)+(n-1)f'(z).$$

May also be demonstrated by means of the formal calculation in **137.**

132. For $\zeta=\infty$, it follows from $f'(z)\equiv 0$ that $f(z)$ is a constant, i.e. that all the n zeros are $=\infty$. If ζ is finite, then the solution of the homogeneous linear differential equation of first order

$$(\zeta-z)f'(z)+nf(z)=0,$$

is $f(z)=c(z-\zeta)^n$, where c is a constant.

133. Let z' be a point of the derived system.

(1) z' finite, $f(z')\neq 0$, $z'\neq\zeta$, hence also $f'(z')\neq 0$; $\zeta=z'-[nf(z')]/[f'(z')]$ [**111**].

(2) z' finite, $f(z')=0$, $f(z)=(z-z')^k\varphi(z)$, $\varphi(z')\neq 0$, $z'\neq\zeta$; $(z-z')^{-k+1}A_\zeta f(z)$ for $z=z'$ reduces to $k(\zeta-z')\varphi(z')\neq 0$.

(3) $z'=\zeta$; $f(z)=(z-\zeta)^k\varphi(z)$, $\varphi(\zeta)\neq 0$, $k<n$; $(z-\zeta)^{-k}A_\zeta f(z)$ for $z=\zeta$ reduces to $(n-k)\varphi(\zeta)\neq 0$.

(4) If $a_n\neq 0$ and $a_{n-1}+a_n\zeta=0$ [**128**], then $\zeta=-a_{n-1}/a_n$ [**111**].

134. For $\zeta=\infty$, cf. III **31**. For finite ζ consider the cases listed in **133**. Either ζ is the center of gravity of $f(z)$ with respect to z' as in (1) and (4) and then **106**

applies, or z' is itself a zero of $f(z)$ and then z' lies in the circular domain, namely on its boundary.

135. Follows from **107** by means of the same considerations as were used to deduce **134** from **106**.

136. From **135**. The circular-arc polygon in question is conceived of as the largest common part of circular domains that contain $z_1, z_2, \ldots, z_n$ but do not contain ζ [solution **113**].

137. If $\zeta=\infty$, then from $f^{(n-1)}(z)=0$ we obtain the ordinary center of gravity of $f(z)$, or ∞ or nothing in particular according as the precise degree of $f(z)$ is equal to n, $n-1$ or is $\leqq n-2$.—Let ζ be finite. We shall say that the polynomial of degree m, $\sum_{\nu=0}^{m}\binom{m}{\nu}b_\nu z^\nu$, and the power series $\sum_{\nu=-\infty}^{\infty} c_\nu u^\nu$ are associated with each other, in symbols

$$b_0+\binom{m}{1}b_1z+\binom{m}{2}b_2z^2+\cdots+b_mz^m\wedge\cdots+c_{-1}u^{-1}+c_0+c_1u+\cdots$$

if $c_0=b_0,\ c_1=b_1,\ldots,\ c_m=b_m,\ c_{-1},\ c_{-2},\ldots,\ c_{m+1},\ c_{m+2},\ldots$ *arbitrary*. We then have [**128**]

$$f(z)=a_0+\binom{n}{1}a_1z+\binom{n}{2}a_2z^2+\cdots+a_nz^n$$

$$\wedge a_0+a_1u+a_2u^2+\cdots+a_nu^n,$$

$$f'(z)=n\left[a_1+\binom{n-1}{1}a_2z+\binom{n-1}{2}a_3z^2+\cdots+a_nz^{n-1}\right]$$

$$\wedge(a_0+a_1u+a_2u^2+\cdots+a_nu^n)\frac{n}{u},$$

$$A_\zeta f(z)=n\left[(a_0+a_1\zeta)+\binom{n-1}{1}(a_1+a_2\zeta)z+\cdots+(a_{n-1}+a_n\zeta)z^{n-1}\right]$$

$$\wedge(a_0+a_1u+\cdots+a_nu^n)n\left(1+\frac{\zeta}{u}\right),$$

$$A_\zeta^{n-1}f(z)\wedge(a_0+a_1u+\cdots+a_nu^n)n(n-1)\ldots2\left(1+\frac{\zeta}{u}\right)^{n-1},$$

$$A_\zeta^{n-1}f(z)=n!\left(a_0+\binom{n-1}{1}a_1\zeta+\binom{n-1}{2}a_2\zeta^2+\cdots+a_{n-1}\zeta^{n-1}\right.$$

$$\left.+\left[a_1+\binom{n-1}{1}a_2\zeta+\binom{n-1}{2}a_3\zeta^2+\cdots+a_n\zeta^{n-1}\right]z\right).$$

We deduce that the zero in question is the center of gravity $f(z)$ with respect to ζ [**111**].

138. By solution **137** we have

$$A_{\zeta_1}A_{\zeta_2}\ldots A_{\zeta_n}f(z)\wedge n!(a_0+a_1u+\cdots+a_nu^n)\left(1+\frac{\zeta_1}{u}\right)\left(1+\frac{\zeta_2}{u}\right)\cdots\left(1+\frac{\zeta_n}{u}\right),$$

if $\zeta_1, \zeta_2, \ldots, \zeta_n$ are finite, otherwise

$$A_{\zeta_1}A_{\zeta_2}\ldots A_{\zeta_n}f(z)\wedge n!(a_0+a_1u+\cdots+a_nu^n)\left(1+\frac{\zeta_1}{u}\right)\left(1+\frac{\zeta_2}{u}\right)\cdots\left(1+\frac{\zeta_{n-k}}{u}\right)\frac{1}{u^k}.$$

139. Let $\zeta_1, \zeta_2, \ldots, \zeta_n$ all be finite, i.e. $b_n \neq 0$. We then have [**138**] that

$$\Sigma_\nu = (-1)^\nu \binom{n}{n-\nu}\frac{b_{n-\nu}}{b_n}, \qquad \nu = 0, 1, 2, \ldots, n$$

and hence

$$A(\zeta_1, \zeta_2, \ldots, \zeta_n)f(z)$$
$$=\frac{1}{b_n}\left[a_0b_n - \binom{n}{1}a_1b_{n-1} + \binom{n}{2}a_2b_{n-2} - \cdots + (-1)^{n-1}\binom{n}{n-1}a_{n-1}b_1 + (-1)^n a_nb_0\right].$$

If the last k of $\zeta_1, \zeta_2, \ldots, \zeta_n$ and only these are infinite, then we have [**138**]

$$\Sigma_0 = \Sigma_1 = \cdots = \Sigma_{k-1} = 0,$$

$$\Sigma_\nu = (-1)^{\nu-k}\frac{\binom{n}{\nu}b_{n-\nu}}{\binom{n}{k}b_{n-k}}, \qquad \nu = k, k-1, \ldots, n, \quad b_{n-k} \neq 0,$$

and hence

$$A(\zeta_1, \zeta_2, \ldots, \zeta_n)f(z)$$
$$=\frac{(-1)^k}{\binom{n}{k}b_{n-k}}\left[(-1)^k\binom{n}{k}a_kb_{n-k} + (-1)^{k+1}\binom{n}{k+1}a_{k+1}b_{n-k-1} + \cdots\right.$$
$$\left. + (-1)^{n-1}\binom{n}{n-1}a_{n-1}b_1 + (-1)^n a_nb_0\right]$$

[for $k=n$ this is to be read as $(-1)^n a_n$]. In every case we hence have

$$A(\zeta_1, \zeta_2, \ldots, \zeta_n)f(z) = \lambda_f\left[a_0b_n - \binom{n}{1}a_1b_{n-1} + \binom{n}{2}a_2b_{n-2} - \cdots\right.$$
$$\left. + (-1)^{n-1}\binom{n}{n-1}a_{n-1}b_1 + (-1)^n a_nb_0\right],$$

where $\lambda_f^{-1} = (-1)^k\binom{n}{k}b_{n-k}$, if b_{n-k} denotes the highest coefficient of $g(z)$ that is different from zero. We have

$$\lambda_f^{-1}A(\zeta_1, \zeta_2, \ldots, \zeta_n)f(z) = \lambda_g^{-1}A(z_1, z_2, \ldots, z_n)g(z).$$

140. The apolarity of z_1 and ζ_1 means: $z_1 = \zeta_1$. The apolarity of z_1, z_2 and ζ_1, ζ_2 means (apart from easily discussed exceptional cases)

$$(z_1-\zeta_1)(z_2-\zeta_2) + (z_1-\zeta_2)(z_2-\zeta_1) = 0, \qquad \frac{(z_1-\zeta_1)(z_2-\zeta_2)}{(z_1-\zeta_2)(z_2-\zeta_1)} = -1,$$

i.e. the two pairs of points are situated harmonically on a circle.

141. By **139** we must have $\Sigma_0 - \Sigma_n = 0$, i.e. $\zeta_1, \zeta_2, \ldots, \zeta_n$ are all finite and

$$\zeta_1\zeta_2\ldots\zeta_n = 1.$$

142.

$$\zeta = z_1, z_2, \ldots, z_n.$$

143.

$$A(\zeta_1, \zeta_2, \ldots, \zeta_n)f(z) = \Sigma_0 - \frac{\Sigma_1}{n} + c\Sigma_n = 1 - \frac{\zeta_1 + \zeta_2 + \cdots + \zeta_n}{n} + c\zeta_1\zeta_2\ldots\zeta_n.$$

144. Apply **135** repeatedly.

145. [J. H. Grace: Proc. Camb. Phil. Soc. **11**, pp. 352–357 (1900–1902); cf. G. Szegö: Math. Z. **13**, p. 31 (1922); J. Egerváry: Acta Univ. Hung. Francisco-Josephinae, **1**, pp. 39–45; Math. és phys. lapok, **29**, pp. 21–43 (1922).] Let $z_1, z_2, \ldots, z_n$ lie inside and $\zeta_1, \zeta_2, \ldots, \zeta_k$ outside the circular domain C and let $A_{\zeta_1} A_{\zeta_2} \ldots A_{\zeta_k} A_{\zeta_{k+1}} f(z)$ but not $A_{\zeta_1} A_{\zeta_2} \ldots A_{\zeta_k} f(z)$, $k \leqq n-1$, vanish identically. By **144** the zeros of $A_{\zeta_1} A_{\zeta_2} \ldots A_{\zeta_k} f(z)$ also lie inside C and by **132** they must all coincide with the point ζ_{k+1}. Thus ζ_{k+1} lies inside C.

146. [**145**.] For if two convex polygons have no common point, then there are straight lines that separate the two polygons. E.g. the perpendicular bisector of the shortest distance is such a line.

147. [E. Landau: Ann. de l'Éc. Norm. **24**, p. 180 (1907).] [**148**.]

148. [A. Hurwitz: cf. E. Landau, op. cit. **147**.] By **143** the polynomial in question is apolar to the polynomial with the zeros

$$\zeta_\nu = 1 - e^{\frac{2\pi i\nu}{n}}, \qquad \nu = 1, 2, \ldots, n,$$

that lie in the circular domain $|z-1| \leqq 1$ [**145**].

149. [L. Fejér: C.R. Acad. Sci. (Paris) **145**, p. 459 (1907); Math. Ann. **65**, pp. 413–423 (1908); Jber. deutsch. Math. Verein. **26**, pp. 114–128 (1917). Cf. also S. Sarantopoulos: C.R. Acad. Sci. (Paris) **174**, p. 592 (1922); P. Montel: C.R. Acad. Sci. (Paris) **174**, p. 851, p. 1220 (1922); Ann de l'Éc. Norm. Series 3, **40**, pp. 1–34 (1923).] For $k=2$ use **135** with $\zeta=0$. We have

$$A_0(1 - z + c_2 z^{\nu_2}) = \nu_2 - (\nu_2 - 1)z,$$

the derived system consists of $\nu_2/(\nu_2-1)$ and of the point at infinity taken with multiplicity $\nu_2 - 2$. Thus the exterior of the circle $|z| > \nu_2/(\nu_2-1)$ cannot contain all the zeros of $1 - z + c_2 z^{\nu_2}$.

For $k > 2$ repeated application of **135**, i.e. mathematical induction, is required. Denoting the polynomial in question by $f(z)$ we have

$$A_0 f(z) = \nu_k - (\nu_k - 1)z + c_2(\nu_k - \nu_2)z^{\nu_2} + \cdots + c_{k-1}(\nu_k - \nu_{k-1})z^{\nu_{k-1}}.$$

At least one point of the derived system lies in the circle

$$|z| \leqq \left(\frac{\nu_2}{\nu_2 - 1}\frac{\nu_3}{\nu_3 - 1}\cdots\frac{\nu_{k-1}}{\nu_{k-1}-1}\right)\frac{\nu_k}{\nu_k - 1}.$$

To see this replace z by $[\nu_k/(\nu_k-1)]u$ and in the equation for u make use of the assumption of the mathematical induction proof.—We have

$$\left[\left(1-\frac{1}{\nu_2}\right)\left(1-\frac{1}{\nu_3}\right)\cdots\left(1-\frac{1}{\nu_k}\right)\right]^{-1}\leqq\left[\left(1-\frac{1}{2}\right)\left(1-\frac{1}{3}\right)\cdots\left(1-\frac{1}{k}\right)\right]^{-1}=k.$$

Note that the polynomial of $k+1$ terms

$$\left(1-\frac{z}{k}\right)^k=1-z+\cdots$$

has the single zero k.

150. [J. H. Grace, op. cit. **145**, p. 356; cf. also P. J. Heawood: Quart. J. Math. **38**, pp. 84–107 (1907).] Denoting by

$$f(z)=a_0+\binom{n-1}{1}a_1z+\binom{n-1}{2}a_2z^2+\cdots+a_{n-1}z^{n-1}$$

the derivative of the polynomial in question, we have

$$\int_a^b f(z)\,dz=a_0b_{n-1}-\binom{n-1}{1}a_1b_{n-2}+\binom{n-1}{2}a_2b_{n-3}+\cdots$$
$$+(-1)^{n-1}a_{n-1}b_0=0,$$

where $b_0, b_1, \ldots, b_{n-1}$ remain fixed for all polynomials $f(z)$, i.e. for variable $a_0, a_1, \ldots, a_{n-1}$. The two polynomials of degree $n-1$, $f(z)$ and

$$g(z)=b_0+\binom{n-1}{1}b_1z+\binom{n-1}{2}b_2z^2+\cdots+b_{n-1}z^{n-1}$$

are thus apolar. We obtain the explicit expression for $g(x)$ by choosing in particular $a_\nu=(-1)^\nu x^{n-1-\nu}$, i.e. $f(z)=(x-z)^{n-1}$. From

$$g(x)=\int_a^b (x-z)^{n-1}\,dz=\frac{(x-a)^n-(x-b)^n}{n}$$

the zeros of $g(x)$ are obtained as

$$\zeta_\nu=\frac{a+b}{2}+i\frac{a-b}{2}\cot\frac{\nu\pi}{n}, \qquad \nu=1, 2, \ldots, n-1.$$

Now apply **145**.

151. [Cf. G. Szegö, op. cit. **145**, p. 35.] If $\gamma=0$, then either among the zeros of $f(z)$ or among those of $g(z)$ there must occur $z=0$. The assertion is then evident. (For $\beta_1=\beta_2=\cdots=\beta_n=\infty$, set $k=0$; $0\cdot\infty$ is indeterminate.) We have a similar deduction for $\gamma=\infty$. Assume now that $\gamma\neq 0, \infty$. The two polynomials $f(z)$ and $z^ng(-\gamma z^{-1})$ are then apolar. Of the zeros of the latter, namely $-\gamma\beta_\nu^{-1}$, $\nu=1, 2, \ldots, n$ (setting $-\gamma\cdot 0^{-1}=\infty$, $-\gamma\cdot\infty^{-1}=0$) by **145** at least one lies in C, i.e. $-\gamma\beta_\nu^{-1}=k$.

152. [**151**.]

153. [I. Schur, cf. G. Szegö, op. cit. **145**, p. 37.] Consider $\mathfrak{C}$ as the intersection of closed circular discs C that contain the origin and all the zeros of $f(z)$. By **151** every zero γ of the polynomial obtained by composition has the form ϑk, where $0\leqq\vartheta\leqq 1$ and k lies in C. But then γ lies in all the C and hence also in $\mathfrak{C}$.

154. [For the statement of the problem cf. Laguerre, op. cit. **1**, pp. 199–200 (1898). G. Pólya: Interméd. des Math. **20**, pp. 145–146 (1913). For the heuristic background cf. MPR **2**, p. 47, ex. 15.—G. Szegö, op. cit. **145**, p. 38, and J. Egerváry, op. cit. **145**.] Replace $g(z)$ by $g(bz)$ or $g(-bz)$, according as the zeros of $g(z)$ lie in the intervals $-b, 0$ or $0, b$, respectively, and then apply **153**.

155. [E. Malo: J. math. spéc., Series 4, **4**, p. 7 (1895).] By VI **85** we have that

$$Q_n(z)=1+\binom{n}{1}^2 z+\binom{n}{2}^2 z^2+\cdots+\binom{n}{n-1}^2 z^{n-1}+z^n=(1-z)^n P_n\left(\frac{1+z}{1-z}\right),$$

where $P_n(x)$ denotes the nth Legendre polynomial. Hence $Q_n(z)$ has only negative zeros [VI **97**]. First compose the first polynomial with $Q_n(cz)$ and the resulting polynomial

$$a_0+\binom{n}{1}a_1 z+\binom{n}{2}a_2 z^2+\cdots+a_n z^n$$

with the second polynomial, in which if necessary z is replaced by dz. The constants c, d are chosen in such a manner that all the zeros of the respective polynomials lie in the interval $-1, 0$. Here in both cases apply **153**, but choosing for $\mathfrak{C}$ the upper and lower closed half-plane respectively. Then it follows that the zeros of the polynomial in question all lie in the upper and at the same time in the lower half-plane.—Also follows from **154** by two applications of **65**.

156. [I. Schur: J. reine angew. Math. **144**, pp. 75–88 (1914).] Similarly as in **155**, noting that the polynomials [VI **99**]

$$1+\binom{n}{1}\frac{z}{1!}+\binom{n}{2}\frac{z^2}{2!}+\cdots+\binom{n}{n-1}\frac{z^{n-1}}{(n-1)!}+\frac{z^n}{n!}$$

have only real, negative zeros.

157. Since the zeros of $[1+(iz/n)]^n$ lie in the half-plane $\mathfrak{I}z>0$, those of

$$1-\binom{n}{2}\frac{z^2}{n^2}+\binom{n}{4}\frac{z^4}{n^4}-\cdots\to\cos z$$

are real [III **25**]. Observe III **203**.

158. $d^\nu/dz^\nu[1-(z^2/n)]^n$ has only real zeros [**55**] and, for fixed ν, as $n\to\infty$ it tends in every finite domain to $(d^\nu e^{-z^2})/(dz^\nu)$ [Hurwitz-Courant, p. 63]. From this the assertion follows [III **203**].

159. (a) [A. Hurwitz: Math. Ges. Hamburg, Festschrift, p. 25 (1890).]

$$(-1)^n\left(1+\frac{z^2}{4n^2}\right)^n P_n\left(\frac{z^2-4n^2}{z^2+4n^2}\right)=1-\binom{n}{1}^2\left(\frac{z}{2n}\right)^2+\binom{n}{2}^2\left(\frac{z}{2n}\right)^4-\binom{n}{3}^2\left(\frac{z}{2n}\right)^6+\cdots;$$

$P_n(z)$ has only real zeros, lying in the interval $-1, +1$ [VI **97**].

(b)

$$L_n\left(\frac{z^2}{4n}\right)=1-\frac{z^2}{2\cdot 2}+\frac{z^4}{2\cdot 4\cdot 2\cdot 4}\left(1-\frac{1}{n}\right)-\frac{z^6}{2\cdot 4\cdot 6\cdot 2\cdot 4\cdot 6}\left(1-\frac{1}{n}\right)\left(1-\frac{2}{n}\right)+\cdots;$$

$L_n(z)$ has only real positive zeros [VI **99**, solution (i)].

(c) From

$$\int_{-\pi}^{\pi}\frac{d\vartheta}{\sin\vartheta-iz}=\frac{2\pi i}{\sqrt{1+z^2}} \qquad \text{[III } \mathbf{148}\text{]}$$

by differentiating n times and changing the variable we derive that

$$\frac{1}{2\pi}\int_{-\pi}^{\pi}\left(1+\frac{iz\sin\vartheta}{n}\right)^{-n-1}d\vartheta=\frac{1}{n!}\left(-\frac{z}{n}\right)^{n}Q_n\left(\frac{n}{z}\right)\left(1+\frac{z^2}{n^2}\right)^{-n-\frac{1}{2}}$$

$$\to\frac{1}{2\pi}\int_{-\pi}^{\pi}e^{-iz\sin\vartheta}\,d\vartheta$$

as $n\to\infty$ where $Q_n(z)$ is the polynomial considered in solution **56** that has only real zeros.

(d) From III **205**, setting $f(t)=(1-t^2)^{-1/2}$.—(a), (b), (c) depend directly, (d) indirectly, on III **203**.

160. [G. Pólya: Tôhoku Math. J. **19**, p. 241 (1921).] Setting

$$\frac{d^{n-1}}{dz^{n-1}}\left(\frac{z^{q-1}}{z^q-1}\right)=\frac{\bar{Q}_n(z)}{(z^q-1)^n},\qquad \frac{d^n e^{z^q}}{dz^n}=\bar{R}_n(z)\,e^{z^q},$$

the zeros of $\bar{Q}_n(z)$, $\bar{R}_n(z)$ lie on q rays emanating from the point $z=0$ that divide the plane into q equal sectors, one of which is bisected by the positive real axis [**59**]. Now the zeros of $F(z^q)$ lie on the same rays, for we have

$$\bar{R}_{qn}(z)\,e^{z^q}=\sum_{k=0}^{\infty}\frac{(qn+qk)!}{(n+k)!}\frac{z^{qk}}{(qk)!},\qquad \frac{n!}{(qn)!}\bar{R}_{qn}\left(zq^{-1}n^{-\frac{q-1}{q}}\right)e^{z^q q^{-q}n^{-q+1}}$$

$$=\sum_{k=0}^{\infty}\frac{qn+1}{q(n+1)}\frac{qn+2}{q(n+2)}\cdots\frac{qn+k}{q(n+k)}\left(1+\frac{k+1}{qn}\right)\left(1+\frac{k+2}{qn}\right)\cdots\left(1+\frac{qk}{qn}\right)\frac{z^{qk}}{(qk)!}$$

$$\to\sum_{k=0}^{\infty}\frac{z^{qk}}{(qk)!}=F(z^q)$$

as $n\to\infty$ [III **203**]. The second proof operates in an analogous manner on $\bar{Q}_n(z)$.

161. [G. Pólya and I. Schur: J. reine angew. Math. **144**, pp. 89–113 (1914).] Set

$$\left(1-\frac{\alpha z}{k}\right)^k\left(1-\frac{z}{\alpha_1}\right)\left(1-\frac{z}{\alpha_2}\right)\cdots\left(1-\frac{z}{\alpha_k}\right)=P_k(z).$$

We have, uniformly in every finite domain, $\lim_{k\to\infty}P_k(z)=g(z)$.

162. [J. L. W. V. Jensen: Acta Math. **36**, p. 181 (1912).] If we write the polynomial considered in solution **161** in the form

$$P_k(z)=a_{0k}+\frac{a_{1k}}{1!}z+\frac{a_{2k}}{2!}z^2+\cdots+\frac{a_{nk}}{n!}z^n+\cdots,$$

we have [Hurwitz-Courant, pp. 61–64]

$$\lim_{k\to\infty}a_{nk}=a_n,$$

and the polynomial

$$a_{0k}z^n+\frac{a_{1k}}{1!}\frac{dz^n}{dz}+\frac{a_{2k}}{2!}\frac{d^2z^n}{dz^2}+\cdots=a_{0k}z^n+\binom{n}{1}a_{1k}z^{n-1}+\binom{n}{2}a_{2k}z^{n-2}+\cdots+a_{nk}$$

has only real zeros [**63**]. By taking the limit $k\to\infty$ [III **203**] and replacing z

by $1/z$, it follows that the zeros of the polynomials in question are *real*. That they are also positive follows from the observation that the coefficients in the power series expansion

$$g(-z)=1-\frac{a_1}{1!}z+\frac{a_2}{2!}z^2-\cdots$$

are all positive, and hence

$$a_1<0, \qquad a_2>0, \qquad a_3<0, \quad \ldots \qquad \textbf{[47]}.$$

163. It follows from **161**, **55** (slightly modified) and III **201**, that no derivative of $g(x)$ vanishes in the interval $-\infty<x<\alpha_1$. From this we deduce, as in **72**, that the difference

$$g(x)-a_0-\frac{a_1}{1!}x-\cdots-\frac{a_{n-1}}{(n-1)!}x^{n-1}$$

except for $x=0$, vanishes nowhere in the interval $-\infty<x\leqq\alpha_1$. Its sign in the interval $0<x<\alpha_1$ thus agrees with that of a_n. [I **144**.]

164.

$$e^{-z^2}=\frac{2}{\sqrt{\pi}}\int_0^\infty e^{-t^2}\cos 2zt\,dt$$

can be regarded as the limit of polynomials with only real zeros [**158**]. From this it follows [**63**] that, if $p>0$,

$$e^{-z^2}-p\frac{d^2e^{-z^2}}{dz^2}=\frac{2}{\sqrt{\pi}}\int_0^\infty e^{-t^2}(1+4pt^2)\cos 2zt\,dt$$

has only real zeros. By iterating this procedure we arrive at polynomials and by a limiting process [**161**] to $g(z)$ under the integral sign. [For a generalization see G. Pólya: J. reine angew. Math. **158**, pp. 6–18 (1927).]

165. [Op. cit. **161**.] Determine the integer m_k such that, setting $\beta+\beta_1^{-1}+\beta_2^{-1}+\cdots+\beta_k^{-1}=B_k$, we have in the circle $|z|\leqq k$

$$\left|\left(1+\frac{B_kz}{m_k}\right)^{m_k}-e^{B_kz}\right|<\frac{1}{k}\exp\left(-k\left(\frac{1}{|\beta_1|}+\frac{1}{|\beta_2|}+\cdots+\frac{1}{|\beta_k|}\right)\right).$$

Then we have, uniformly in every finite domain,

$$\lim_{k\to 0}\left(1-\frac{\alpha z^2}{k}\right)^k\left(1+\frac{B_kz}{m_k}\right)^{m_k}\left(1-\frac{z}{\beta_1}\right)\cdots\left(1-\frac{z}{\beta_k}\right)=G(z).$$

166. Show, as in **162**, that the polynomial

$$1+\binom{n}{1}b_1z+\binom{n}{2}b_2z^2+\cdots+b_nz^n$$

has only real roots. Therefore, if it is precisely of degree n, two *adjacent* coefficients in it cannot vanish [**49**]. $G(z)$ is transcendental. Determine $n>m+1$ such that $b_n\neq 0$.

167. By assumption $\beta_\nu < 0$, $\nu = 1, 2, 3, \ldots$. Let $\sqrt{k} > n\sqrt{\alpha}$ and set

$$\left(1-\frac{\alpha x^2}{k}\right)^k \left(1-\frac{x}{\beta_1}\right)\left(1-\frac{x}{\beta_2}\right)\cdots\left(1-\frac{x}{\beta_k}\right) = Q(x),$$

$$B = \beta + \beta_1^{-1} + \beta_2^{-1} + \cdots + \beta_k^{-1}.$$

Passing from

$$a_0 + a_1 z + a_2 z^2 + \cdots + a_n z^n = 0$$

to the equation

$$a_0 + a_1\, e^B z + a_2\, e^{2B} z^2 + \cdots + a_n\, e^{nB} z^n = 0$$

leaves the number of non-real zeros unaltered, and passing then to the equation

$$a_0 Q(0) + a_1 Q(1)\, e^B z + a_2 Q(2)\, e^{2B} z^2 + \cdots + a_n Q(n)\, e^{nB} z^n = 0$$

does not increase it [**67**]. Now let $k \to \infty$ [III **201**].

168.

$$\frac{1}{\Gamma(z+1)} = e^{Cz}\left(1+\frac{z}{1}\right)e^{-\frac{z}{1}}\left(1+\frac{z}{2}\right)e^{-\frac{z}{2}}\cdots\left(1+\frac{z}{n}\right)e^{-\frac{z}{n}}\cdots$$

is of the form **165** and its zeros are negative. The polynomial

$$\left(1-\frac{z^2}{4n}\right)^n = 1 - \frac{1}{1!}\left(\frac{z}{2}\right)^2 + \frac{1}{2!}\left(\frac{z}{2}\right)^4\left(1-\frac{1}{n}\right) - \frac{1}{3!}\left(\frac{z}{2}\right)^6\left(1-\frac{1}{n}\right)\left(1-\frac{2}{n}\right) + \cdots$$

has only real zeros and therefore [**167**, $G(z) = \Gamma((z/2)+1)^{-1}$] so has the following polynomial

$$\frac{1}{\Gamma(1)} - \frac{1}{\Gamma(2)1!}\left(\frac{z}{2}\right)^2 + \frac{1}{\Gamma(3)2!}\left(\frac{z}{2}\right)^4\left(1-\frac{1}{n}\right) - \cdots \to J_0(z)$$

as $n \to \infty$ [III **203**]. Clearly **65** is also a special case of **167**.

169. [Op. cit. **160**.] If q is a positive integer, then the function

$$\frac{\Gamma(z+1)}{\Gamma(qz+1)} = G(z)$$

is entire, for the poles of the numerator $-q/q, -2q/q, \ldots$ are contained among those of the denominator $-1/q, -2/q, \ldots$, and in fact it is of the form **165** [solution **168**]. Apply **167** to this function $G(z)$ and to the polynomial

$$\left(1+\frac{z}{n}\right)^n = 1 + \frac{z}{1!} + \frac{z^2}{2!}\left(1-\frac{1}{n}\right) + \cdots$$

Subsequently let $n \to \infty$ [III **203**].

170. [G. Pólya: Messenger, **52**, pp. 185–188 (1923).]

$$\alpha F_\alpha(z) = \sum_{k=0}^{\infty} (-1)^k \frac{z^{2k}}{(2k)!}\, \alpha \int_0^{\infty} e^{-t^\alpha} t^{2k}\, dt = \sum_{k=0}^{\infty} (-1)^k \frac{z^{2k}}{k!} \frac{\Gamma(k+1)\Gamma\left(\frac{2k+1}{\alpha}\right)}{\Gamma(2k+1)}\,.$$

Apply **167** to the polynomial

$$\left(1-\frac{z^2}{n}\right)^n=\sum_{k=0}^{n}(-1)^k\frac{z^{2k}}{k!}\left(1-\frac{1}{n}\right)\cdots\left(1-\frac{k-1}{n}\right)$$

and to the entire function

$$\frac{\Gamma\left(\frac{z}{2}+1\right)\Gamma\left(\frac{z+1}{2q}\right)}{\Gamma(z+1)}=G(z)$$

where $\alpha=2q$, q an integer. Subsequently a limiting process [III **203**]. The function $G(z)$ is entire, for the poles of the numerator

$$-2,\ -4,\ -6,\ \ldots,\qquad -1,\ -(1+2q),\ -(1+4q),\ \ldots$$

are absorbed by those of the denominator $-1, -2, -3, \ldots$ $G(z)$ is also of the form **165** [solution **168**].—The function $F_2(z)=(\sqrt{\pi}/2)\,e^{-\frac{z^2}{4}}$ has no zeros at all.

171. If $\alpha\neq 2,4,6,\ldots$ and x tends to ∞ through *real* values, $\lim_{x\to\infty}x^{\alpha+1}F_\alpha(x)\neq 0$ [III **154**]. It is instructive that the function $F_\alpha(x)$ for $\alpha\neq 2$ has infinitely many zeros [op. cit. **170**]. It thus has infinitely many real and no non-real zeros, if $\alpha=4,6,8,\ldots$, and finitely many real and infinitely many non-real zeros in all other cases. Cf. III **201**.

171.1. [G. Pólya: Jber. deutsch. Math. Verein. **38**, pp. 161–168 (1929).] Let $\alpha_1, \alpha_2, \ldots, \alpha_s$ denote the zeros of $G(z)$ in $[0, m]$,

$$0\leqq\alpha_1<\alpha_2<\cdots<\alpha_s\leqq m,$$

and set

$$G(z)=(z-\alpha_1)(z-\alpha_2)\ldots(z-\alpha_s)G^*(z).$$

Then argue as follows:

(1)

$$a_0G^*(0)+a_1G^*(1)z+\cdots+a_nG^*(n)z^n$$

has precisely m changes of sign (by **47** and the definition of $G^*(z)$) and precisely m positive zeros (by a slight modification of **167**—make use of **67** more completely).

(2)

$$a_0G^*(0)(0-\alpha_1)+a_1G^*(1)(1-\alpha_1)z+\cdots+a_mG^*(m)(m-\alpha_1)z^m$$

has precisely $m-1$ changes of sign (attention to the various possible cases) and so by **36** at most $m-1$ positive zeros. Yet it has (by a slight modification of solution **66**) at least $m-1$, and so finally exactly $m-1$, positive zeros;

(3) Introduce successively $\alpha_2, \alpha_3, \ldots, \alpha_s$ and repeat appropriately the arguments of (2) (attention at each step to the changes wrought by the preceding steps) to arrive at the desired proposition.

171.2. Use **171.1** and **162** [details loc. cit. **171.1**].

171.3. Apply **171.2** to

$$g(z)=e^{-z}=\lim_{n\to\infty}\left(1-\frac{z}{n}\right)^n=\sum_{n=0}^{\infty}\frac{(-z)^n}{n!},$$

$$G(z)=\frac{1}{\Gamma(z+1-\nu)},$$

$$s=[\nu].$$

171.4. Apply **171.1** to the polynomial with only positive zeros

$$b_0-b_1z+b_2z^2-\cdots+(-1)^m b_m z^m$$

and the entire function

$$G(z)=\frac{(-1)^k \sin \pi(z-k)}{\mathfrak{l} \sin \dfrac{\pi(z-k)}{\mathfrak{l}}}.$$

Each of the s zeros of $G(z)$ that falls in the interval $[0, m]$ destroys just one change of sign and, by **171.1**, just one zero of the initial polynomial. Hence the resulting polynomial

$$b_k z^k - b_{k+\mathfrak{l}} z^{k+\mathfrak{l}} + b_{k+2\mathfrak{l}} z^{k+2\mathfrak{l}} - \cdots + (-1)^q b_{k+q\mathfrak{l}} z^{k+q\mathfrak{l}}$$

has just as many positive zeros as changes of sign.

171.5. Similar to **171.4**, but now

$$G(z)=\frac{\Gamma\left(-\dfrac{z-k}{\mathfrak{l}}\right)}{\mathfrak{l}\Gamma(-z)}=\frac{(-1)^k \sin \pi(z-k)}{\mathfrak{l} \sin \dfrac{\pi(z-k)}{\mathfrak{l}}} \frac{\Gamma(1+z)}{\Gamma\left(1+\dfrac{z-k}{\mathfrak{l}}\right)}.$$

172. [Cauchy: Exercices de mathématiques (anciens exercices) 1826. Oeuvres, Series 2, **6**, pp. 354–400. Paris: Gauthiers-Villars 1887.]

(a) [A. Hurwitz, op. cit. **159**.] With the exception of the root $z=0$, we are concerned with the zeros of the meromorphic function

$$\cot z-\frac{1}{z}=\lim_{n\to\infty}\left(\frac{1}{z+n\pi}+\cdots+\frac{1}{z+2\pi}+\frac{1}{z+\pi}+\frac{1}{z-\pi}\right.$$
$$\left.+\frac{1}{z-2\pi}+\cdots+\frac{1}{z-n\pi}\right).$$

Under the limit sign we have a rational function whose numerator is of degree $\leqq 2n-1$. Between the $2n$ zeros of the denominator there lie $2n-1$ intervals, and in the interior of each interval there lies one zero of the numerator [**26**]. Consequently the numerator is of precise degree $2n-1$ and has no non-real zeros. Take the limit [III **203**].

(b) $$z^{-2}\cos z(\tan z-z)=\int_0^1 t \sin zt\, dt.$$

Obtain an analogue for sine polynomials to theorem III **185** for cosine polynomials (proof the same) and deduce from this, by taking the limit, an analogue to III **205** that may be applied to the present integral.

(c) $$2z^{-3}\cos z(\tan z-z)=\int_0^1 (1-t^2)\cos zt\, dt.$$

173 yields apart from the reality also the existence of infinitely many zeros.

173.

$$z^2F(z)=zf(1)\sin z-f'(0)(1-\cos z)$$
$$+\int_0^1 f''(t)(\cos z-\cos zt)\,dt$$

(integrate by parts twice), whence we deduce

$$F[(2m-1)\pi]>0, \qquad F(2m\pi)<0, \qquad m=1,2,3,\ldots,$$

and thus the existence of infinitely many zeros. We have F(0) > 0. Hence the rational function

$$(-1)^n\frac{F(-n\pi)}{z+n\pi}+\cdots+\frac{F(-2\pi)}{z+2\pi}-\frac{F(-\pi)}{z+\pi}+\frac{F(0)}{z}-\frac{F(\pi)}{z-\pi}$$
$$+\frac{F(2\pi)}{z-2\pi}-\cdots+(-1)^n\frac{F(n\pi)}{z-n\pi}$$

has $2n-2$ real and at most 2 non-real zeros [cf. **26**]. It tends to $F(z)/\sin z$ [cf. III **165**] and hence [III **201**] $F(z)$ has either 0 or 2 non-real zeros. $F(z)$ is an even function. If it had exactly two non-real zeros, then these would be purely imaginary. Now if y is real we have

$$F(iy)=\int_0^1 f(t)\frac{e^{yt}+e^{-yt}}{2}\,dt>0.$$

174. It is sufficient to consider the case $\int_0^1|\varphi(t)|\,dt<1$ [III **203**]. Then we have for sufficiently large integers n

$$\frac{1}{n}\left|\varphi\left(\frac{1}{n}\right)\right|+\frac{1}{n}\left|\varphi\left(\frac{2}{n}\right)\right|+\cdots+\frac{1}{n}\left|\varphi\left(\frac{n-1}{n}\right)\right|<1$$

and hence

$$\sin\frac{nz}{n}-\frac{1}{n}\varphi\left(\frac{1}{n}\right)\sin\frac{z}{n}-\frac{1}{n}\varphi\left(\frac{2}{n}\right)\sin\frac{2z}{n}-\cdots-\frac{1}{n}\varphi\left(\frac{n-1}{n}\right)\sin\frac{(n-1)z}{n}$$

has no complex zeros [method of proof in **27**]; $n\to\infty$ [III **203**].

175. Excluding the trivial case $f(1)=0$, we have

$$\frac{z}{f(1)}\int_0^1 f(t)\cos zt\,dt=\sin z-\int_0^1\frac{f'(t)}{f(1)}\sin zt\,dt \qquad [\mathbf{174}].$$

176. The absolute values of the terms steadily increase from the first term 1 to the maximum term and then steadily decrease from the maximum term on to infinity [I **117**]. For this reason, if n denotes the central index, i.e. if $(-x)^na^{-n^2}$ is the maximum term for $z=-x$, $x>0$, we have

$$(-1)^nF(-x)=x^na^{-n^2}-x^{n-1}a^{-(n-1)^2}+x^{n-2}a^{-(n-2)^2}-\cdots$$
$$-x^{n+1}a^{-(n+1)^2}+x^{n+2}a^{-(n+2)^2}-\cdots$$
$$>x^na^{-n^2}-x^{n-1}a^{-(n-1)^2}-x^{n+1}a^{-(n+1)^2}.$$

If $x=a^{2n}$, then the central index is n [III **200**]. It follows that

$$(-1)^nF(-a^{2n})>a^{n^2}-2a^{n^2-1}\geqq 0.$$

These considerations may be applied with appropriate changes also to the partial sum

$$F_n(-x)=1-\frac{x}{a}+\frac{x^2}{a^4}+\cdots+(-1)^n\frac{x^n}{a^{n^2}}.$$

We find that

$$F_n(0)>0, \qquad F_n(-a^2)<0, \qquad F_n(-a^4)>0,\ldots, \qquad (-1)^n F_n(-a^{2n})>0,$$

(in the second inequality we have $=$ instead of $<$ for $n=2$, $a=2$), from which it follows that $F(x)$ has only real simple zeros, and in fact in the interior of each of the intervals $(-a^2, 0)$, $(-a^4, -a^2)$, $(-a^6, -a^4), \ldots, (-a^{2n}, -a^{2n-2})$ precisely one. By passing to the limit [III **201**] we deduce the result for $F(z)$. (Cf. III **200**.)

177. [Cf. G. Pólya: Math. Z. **2**, pp. 355–358 (1918).] The transformation of $p_0+p_1z+\cdots+p_nz^n$ in the proof of III **22** is essentially a partial summation.—The function $F(z)$ has no real zeros. Let $f'(t)>0$, $z=x+iy$, $x\geqq 0$, $y>0$, $e^{xt}f(t)=f_1(t)$, thus also $f_1(t)>0, f_1'(t)>0$. Let $0<\tau<1$. By integration by parts we obtain

$$iyF(z)-iy\int_\tau^1 f_1(t)\,e^{iyt}\,dt=f_1(\tau)\,e^{iy\tau}-f_1(0)-\int_0^\tau f_1'(t)\,e^{iyt}\,dt,$$

$$y|F(z)|+y\left|\int_\tau^1 f_1(t)\,e^{iyt}\,dt\right|\geqq f_1(\tau)-f_1(0)-\left|\int_0^\tau f_1'(t)\,e^{iyt}\,dt\right|.$$

The right-hand side is equal to $\int_0^\tau f_1'(t)\,dt-\left|\int_0^\tau f_1'(t)\,e^{iyt}\,dt\right|$ and increases monotonically as τ increases [III **14**]. The second term on the left-hand side becomes equal to 0 as $\tau\to 1$. Hence we have $y|F(z)|>0$.

178.

$$\frac{1}{z}\int_0^z e^{-u^2}\,du=\frac{1}{2}\int_0^1 e^{-z^2t}\,t^{-1/2}\,dt.$$

Now $t^{-1/2}$ is a decreasing function, hence [**177**] no zeros lie in the region $\Re(-z^2)\leqq 0$.

179.

$$\frac{1}{\mu+1}+\frac{z}{(\mu+1)(\mu+2)}+\frac{z^2}{(\mu+1)(\mu+2)(\mu+3)}+\cdots$$

$$=\sum_{n=0}^{\infty}\frac{z^n}{n!}\frac{\Gamma(n+1)\Gamma(\mu+1)}{\Gamma(n+\mu+2)}=\sum_{n=0}^{\infty}\frac{z^n}{n!}\int_0^1 t^n(1-t)^\mu\,dt$$

$$=\int_0^1 e^{zt}(1-t)^\mu\,dt=\int_0^1 e^{(z-\mu)t}[e^t(1-t)]^\mu\,dt; \qquad \mu>-1.$$

Now $e^t(1-t)$ for $0<t<1$ is a monotonically decreasing function (differentiate). Hence [**177**] the zeros of the series

for	$\mu>0$	all lie in the half-plane	$\Re z>\mu$,
for	$-1<\mu<0$	all lie in the half-plane	$\Re z<\mu$,
for	$\mu=0$	all lie on the straight line	$\Re z=0$.

In the problem μ is an integer, $\mu=n$.

180. (a) [G. Pólya: Ens. math. **21**, p. 217 (1920).] Make use of the following important theorem [I. Schur: Math. Ann. **66**, pp. 489–501 (1909)]: Denoting by $\omega_1, \omega_2, \ldots, \omega_n$ the roots of the equation

$$\begin{vmatrix} z-a_{11} & -a_{12} & \ldots & -a_{1n} \\ -a_{21} & z-a_{22} & \ldots & -a_{2n} \\ \ldots & \ldots & \ldots & \ldots \\ -a_{n1} & -a_{n2} & \ldots & z-a_{nn} \end{vmatrix} = 0,$$

we have

$$|\omega_1|^2 + |\omega_2|^2 + \cdots + |\omega_n|^2 \leqq \sum_{\lambda=1}^{n} \sum_{\mu=1}^{n} |a_{\lambda\mu}|^2.$$

From this it follows [VII **11**] that, if we denote the zeros of the polynomial

$$a_0 + a_1 z + a_2 z^2 + \cdots + a_n z^n$$

by $z_{1n}, z_{2n}, \ldots, z_{nn}$, we have

$$\frac{1}{|z_{1n}|^2} + \frac{1}{|z_{2n}|^2} + \cdots + \frac{1}{|z_{nn}|^2} \leqq 2\left(\left|\frac{a_1}{a_0}\right|^2 + \left|\frac{a_2}{a_1}\right|^2 + \cdots + \left|\frac{a_{n-1}}{a_{n-2}}\right|^2\right) + \left|\frac{a_n}{a_{n-1}}\right|^2.$$

Take limits [III **201**].

(b) [G. Valiron: Darboux Bull. Series 2, **45**, p. 269 (1921).] Set $|a_{n-1}a_n^{-1}| = l_n$, $n = 1, 2, 3, \ldots$. Since $\lim_{n\to\infty} l_n = \infty$, we can by rearrangement always obtain from $l_1, l_2, l_3, \ldots, l_p, \ldots$ a monotonically increasing sequence $l_{n_1}, l_{n_2}, l_{n_3}, \ldots, l_{n_p}, \ldots$

$$l_{n_1} \leqq l_{n_2} \leqq l_{n_3} \leqq \cdots \leqq l_{n_p} \leqq \cdots.$$

We have

$$|a_p| = \frac{|a_0|}{l_1 l_2 \ldots l_p} \leqq \frac{|a_0|}{l_{n_1} l_{n_2} \ldots l_{n_p}}, \qquad p = 1, 2, 3, \ldots.$$

Denote by $M(r)$ the maximum of $|F(z)|$ for $|z| \leqq r$ and by $\mu(r)$ the maximum term of the everywhere convergent series [IV **29**]

$$|a_0| + \frac{|a_0|}{l_{n_1}} z + \frac{|a_0|}{l_{n_1} l_{n_2}} z^2 + \cdots = \Phi(z).$$

From the assumption and from IV **37**, IV **54**, IV **38**, respectively, there follows successively the convergence of

$$\sum_{p=1}^{\infty} l_{n_p}^{-2}, \quad \int_1^{\infty} \log \mu(r) r^{-3}\, dr, \quad \int_1^{\infty} \log \Phi(r) r^{-3}\, dr, \quad \int_1^{\infty} \log M(r) r^{-3}\, dr, \quad \sum_{p=1}^{\infty} |\alpha_p|^{-2}.$$

This deduction, in contrast to (a), is not tailored to the exponent 2.

181. No. Example: $f(z) = (z^2 - 4)e^{\frac{z^2}{3}}$, $f'(z) = \frac{2}{3}z(z^2 - 1)\, e^{\frac{z^2}{3}}$. H. M. Macdonald: Proc. Lond. Math. Soc., **29**, p. 578 (1898), asserted the contrary.

182. [G. Pólya: Arch. Math. Phys. Ser. 3, **25**, p. 337 (1917). Solution by H. Prüfer: Arch. Math. Phys. Ser. 3, **27**, pp. 92–94 (1918).] Setting $H'(x) = g(x)$, the theorem is contained in the following more exact theorem: If the polynomial

$g(x)$ of degree n has only real zeros, then $G(x)=[g(x)]^2+g'(x)$ has at most $n+1$ real zeros ($2n>n+1$ if $n\geqq 2$). Let $x_1, x_2, \ldots, x_k$ be the distinct zeros of $g(x)$,

$$g(x)=a(x-x_1)^{m_1}(x-x_2)^{m_2}\cdots(x-x_k)^{m_k}.$$

(a) There are $n-k$ zeros of $G(x)$ that are at the same time zeros of $g(x)$.

(b) A real zero of $G(x)$ that is not a zero of $g(x)$ satisfies the equation

$$\frac{g'}{g^2}=-1 \quad \text{or} \quad \left(\frac{a}{g}\right)^2\frac{g'}{a}=-a.$$

Hence, if at least one such zero exists, a is real. Now we have

$$g\frac{d}{dx}\left(\frac{g'}{g^2}\right)=\left(\frac{g'}{g}\right)'-\left(\frac{g'}{g}\right)^2=-\frac{m_1}{(x-x_1)^2}-\frac{m_2}{(x-x_2)^2}-\cdots-\frac{m_k}{(x-x_k)^2}-\left(\frac{g'}{g}\right)^2<0.$$

The zeros of $g(x)$ divide the real axis into $k+1$ intervals. In each interval the curve $y=g'(x)[g(x)]^{-2}$ is monotonic and intersects the straight line $y=-1$ at most once. The total number of zeros of $G(x)$ specified in (a) and (b) is thus $\leqq(n-k)+k+1$. [Starting point for later developments in various directions. See e.g. G. Pólya: Math. Z. **12**, pp. 36–60 (1922), Bull. Amer. Math. Soc. **49**, pp. 178–191 (1943); W. Saxer: Math. Z. **17**, pp. 206–227 (1923).]

183.

$$A_n^{(k)}=P\left(\frac{n}{k},\frac{1}{k}\right),\qquad B_n^{(k)}=Q\left(\frac{n}{k},\frac{1}{k}\right),\qquad C_n^{(k)}=R\left(\frac{n}{k},\frac{1}{k}\right).$$

The expression for $B_n^{(k)}$ is also valid if $k>m-1$ and k is not an integer.

184.

$$\frac{Q(z,\omega)}{\omega^m}=\frac{1}{\Gamma\left(\frac{1+z}{\omega}-\overline{m-1}\right)\Gamma\left(-\frac{z}{\omega}\right)}\sum_{\nu=0}^{m}(-1)^\nu a_\nu\Gamma\left(\frac{1+z}{\omega}-\overline{\nu-1}\right)\Gamma\left(-\frac{z}{\omega}+\nu\right)$$

$$=\frac{\Gamma\left(\frac{1}{\omega}+1\right)}{\Gamma\left(\frac{1+z}{\omega}-\overline{m-1}\right)\Gamma\left(-\frac{z}{\omega}\right)}\sum_{\nu=0}^{m}(-1)^\nu a_\nu\int_0^1(1-t)^{\frac{1+z}{\omega}-\nu}t^{-\frac{z}{\omega}-1+\nu}\,dt;$$

$$\frac{R(z,\omega)}{\omega^m}=\frac{\Gamma\left(m+\frac{1-z}{\omega}\right)}{\Gamma\left(\frac{1}{\omega}\right)\Gamma\left(-\frac{z}{\omega}\right)}\int_0^1 t^{-\frac{z}{\omega}-1}(1-t)^{\frac{1}{\omega}-1}f(-t)\,dt,$$

for $\omega>0$, $\Re z<0$;

$$\frac{R(z,\omega)}{(-\omega)^m}=\frac{\Gamma\left(\frac{\omega+z}{\omega}\right)}{\Gamma\left(\frac{(1-m)\omega+z-1}{\omega}\right)\Gamma\left(\frac{1}{\omega}\right)}\int_0^1 t^{\frac{z-1}{\omega}}(1-t)^{\frac{1}{\omega}-1}f\left(-\frac{1}{t}\right)dt,$$

for $\omega>0$, $\Re z>1+(m-1)\omega$;

$$P(z,\omega)=\frac{1}{\Gamma\left(-\frac{z}{\omega}\right)}\int_0^\infty e^{-t}t^{-\frac{z}{\omega}-1}f(-\omega t)\,dt,$$

for $\omega>0$, $\Re z<0$.

185. [Cf. Laguerre, op. cit. **1**, p. 23; G. Pólya, problem: Arch. Math. Phys. Ser. 3, **24**, p. 84 (1916).] The three relations in the first column follow from **183**, **38**, **8**, the remaining four from **184**, **80** (change of variable!), **24**.

186. We have already assumed that $a_m \gtrless 0$. Let the sign $\|$ placed between two real numbers α and β (viz. $\alpha\|\beta$) denote that α and β have the same sign. Then we have

$$\begin{array}{lll} f(-\infty)\|P(-\infty), & f(0)\|P(0), & f(\infty)\ \|P(\infty), \\ f(-\infty)\|Q(-1+\overline{m-1}\omega), & f(0)\|Q(0), & f(1)\ \|Q(+\infty), \\ f(-1)\ \|R(-\infty), & f(0)\|R(0), & f(+\infty)\|R(1+\overline{m-1}\omega), \end{array}$$

$$f(-\infty)\|(-1)^m R(1+\overline{m-1}\omega), \qquad f(-1)\|(-1)^m R(+\infty).$$

187. [Laguerre, op. cit. **1**, pp. 13–25; M. Fekete and G. Pólya: Palermo Rend. **34**, pp. 89–120 (1912); E. Bálint: Diss. Budapest, 1916; D. R. Curtiss: Annals of Math. Ser. 2, **19**, pp. 251–278 (1918).] (1) from **38**, (2) from **34**, **33**, **32**, (3) from **183**, **24**, III **201** noting that $P(z, \omega)$, $Q(z, \omega)$, $R(z, \omega)$ depend continuously on ω and that

$$P(z, 0)=f(z), \qquad Q(z, 0)=(1+z)^m f\left(\frac{z}{1+z}\right), \qquad R(z, 0)=(1-z)^m f\left(\frac{z}{1-z}\right).$$

The second case considered theoretically permits the exact determination of the number of zeros of $f(z)$ in the interval $0<x<1$. That the method is also frequently useful in practice is shown by **45**, **46**. [For the heuristic background of the proof cf. MPR, **2**, p. 44, ex. 12.]

188. [G. Pólya, op. cit. **80**.] From the formula

$$\frac{m!\,f(z)}{\dfrac{z}{\omega}\left(\dfrac{z}{\omega}-1\right)\cdots\left(\dfrac{z}{\omega}-m\right)} = \sum_{\nu=0}^{m}(-1)^{m-\nu}\binom{m}{\nu}\frac{f(\nu\omega)}{\dfrac{z}{\omega}-\nu}$$

$$=\omega\int_0^\infty J(e^{-\lambda\omega}, \omega)\, e^{-\lambda(z-m\omega)}\, d\lambda,$$

valid for $\Re z>m\omega$ [VI § 9], it follows that $\mathfrak{F}_{m\omega}^{\infty} \leqq \mathfrak{Z}_0^1$ [**80**, **24**]; from the analogous formula

$$\frac{(-1)^m m!\,f(-z)}{\dfrac{z}{\omega}\left(\dfrac{z}{\omega}+1\right)\cdots\left(\dfrac{z}{\omega}+m\right)}=\omega\int_0^\infty e^{-\lambda m\omega}J(e^{\lambda\omega}, \omega)\, e^{-\lambda z}\, d\lambda$$

valid for $\Re z>0$, it follows in the same way that $\mathfrak{F}_{-\infty}^{0} \leqq \mathfrak{Z}_1^{\infty}$, and finally from the definition of $J(z, \omega)$ it follows [**36**] that $\mathfrak{Z}_{-\infty}^{0} \leqq$ number of changes of sign of the sequence $f(0), f(\omega), \ldots, f(n\omega)$. Finally observe (cf. solution **186**) that

$$\begin{array}{ll} f(-\infty)\|(-1)^m J(1, \omega), & f(0)\|(-1)^m J(+\infty, \omega)\|J(-\infty, \omega), \\ f(m\omega)\|J(0, \omega), & f(+\infty)\|J(1, \omega). \end{array}$$

189. We find that [**188**]

$$J(1, \omega)=\Delta^m f(0), \qquad J'(1, \omega)=-m\Delta^{m-1}f(0), \qquad \ldots.$$

Hence, the polynomial in question is equal to $J(1-z, \omega)$.

189.1. From **80** and

$$\int_0^\infty e^{-xt} S(e^{-\frac{1}{t}}) t^{-1/2}\, dt = (\pi/x)^{1/2} f(e^{-2\sqrt{x}}),$$

which is based on the integral formula

$$\int_0^\infty e^{-at-\frac{b}{t}} t^{-1/2}\, dt = \sqrt{\frac{\pi}{a}}\, e^{-2\sqrt{ab}},$$

where $a>0$, $b>0$; see e.g. C. Jordan: Cours d'Analyse, 3rd ed. (1913), **2**, p. 198.

189.2. Essentially from **80** and

$$(1/\Gamma(x)) \int_0^\infty e^{x \log t} [f(e^{-t}) - a_0] t^{-1}\, dt = D(x).$$

With appropriate definitions, extension to Dirichlet series (see VIII p. 119) is easy.

189.3. [G. Pólya: Vierteljschr. Naturforsch. Gesellsch. Zürich, **73**, pp. 141–145 (1928) or G. H. Hardy, J. E. Littlewood and G. Pólya: Inequalities, Cambridge University Press, 1952, pp. 57–60.]

190. [H. Poincaré: C.R. Acad. Sci. (Paris), **97**, p. 1418 (1883); E. Meissner: Math. Ann. **70**, pp. 223–225 (1911).]

$$f(z) = \frac{f(z)(1+z)^k}{(1+z)^k}.$$

Choose k sufficiently large [**187** (3)].

191. It is sufficient that $f(x) = a(\alpha_1 - x)(\alpha_2 - x)\ldots(\alpha_m - x)$, where $\alpha_1, \alpha_2, \ldots, \alpha_m$ are real and positive [**30**]. Moreover it is also necessary: choose $P(x) = (1+x)^k$, k sufficiently large [**187** (3)].

192. Sufficient: For $e^{-ax+\frac{1}{2}bx^2}(d/dx)P(x)\, e^{ax-\frac{1}{2}bx^2}$ has not more non-real zeros than $P(x)$ (proof as in **58**). Necessary: Choose $P(x) = 1 + \varepsilon x$; it follows that $(1+\varepsilon x)f(x) + \varepsilon$ and thus $[\varepsilon \to 0] f(x)$ has only real zeros. Choose $P(x) = f(x)$. It follows [**182**] that the degree of $f(x)$ is either $=0$ or $=1$. In the latter case, where accordingly $f(x) = a - bx$, again set $P(x) = f(x)$. Observe that $(a-bx)(a-bx) - b > 0$, if $b < 0$.

193. [G. Pólya, problem: Arch. Math. Phys. Ser. 3, **21**, p. 289 (1913). Solution by G. Szegö: Arch. Math. Phys. Ser. 3, **23**, pp. 81–82 (1915).]

194. The expression considered is the resultant of $f(x)$ and $g(x)$, i.e. $= b_0^2 f(\beta_1) f(\beta_2)$, where β_1, β_2 denote the two zeros of $g(x)$. If β_1 is not real, and thus $\beta_2 = \overline{\beta_1}$, then the resultant $\geqq 0$. If β_1, β_2 are real and the resultant <0, then we must have that $\operatorname{sgn} f(\beta_1) = -\operatorname{sgn} f(\beta_2) \neq 0$ [**8**].

195. The case $n=1$ is clear. If $n=2$ and we denote the zeros of $P(x)$ by x_1, x_2, $x_1 \leqq x_2$, we determine a such that $a \leqq x_1$, $\int_a^{\frac{x_1+x_2}{2}} P(x)\, dx = 0$. It is sufficient to consider the case $n \geqq 3$. We first assume that the first three of the zeros $x_1, x_2, \ldots, x_n$ of $P(x)$, $x_1 \leqq x_2 \leqq \cdots \leqq x_n$, are distinct, $x_1 < x_2 < x_3$. Determine ξ and $\delta > 0$ such that $x_1 < \xi - \delta < x_2 < \xi + \delta < x_3$ and $\int_{\xi-\delta}^{\xi+\delta} P(x)\, dx = 0$, and then a such that $a < x_1$ and $\int_a^{\xi-\delta} P(x)\, dx = 0$. The polynomial $Q(x)$ then has at least the zeros $x = a$, $x = \xi - \delta$, $x = \xi + \delta$. If in addition n is odd, then the existence of a fourth real zero of $Q(x)$

is assured since the non-real zeros of $Q(x)$ occur in pairs.—If $P(x)$ has a multiple zero of multiplicity $m \geqq 2$, choose it as the value of a and $Q(x)$ will have at least $m+1 \geqq 3$ zeros.—The equality sign holds, e.g. for

$$P(x)=x(x-1)^2(x-2)^2(x-3)^2\dots\left(x-\frac{n-1}{2}\right)^2, \qquad n \text{ odd},$$

$$P(x)=(x-1)^2(x-2)^2(x-3)^2\dots\left(x-\frac{n}{2}\right)^2, \qquad n \text{ even}.$$

196. [I. Schur; cf. C. Siegel: Math. Zeitschr. **10**, p. 175 (1921).] By multiplication of the inequalities

$$1+|z_\nu| \leqq 2 \operatorname{Max}(1, |z_\nu|) = 2\, e^{\frac{1}{2\pi}\int_0^{2\pi} \log|e^{i\vartheta}-z_\nu|\, d\vartheta}, \qquad \nu=1, 2, \dots, n \quad [\text{II } \mathbf{52}]$$

it follows that

$$\prod_{\nu=1}^{n}(1+|z_\nu|) \leqq 2^n\, e^{\frac{1}{2\pi}\int_0^{2\pi}\log|f(e^{i\vartheta})|\, d\vartheta} \leqq 2^n \sqrt{\frac{1}{2\pi}\int_0^{2\pi} |f(e^{i\vartheta})|^2\, d\vartheta}$$

$$= 2^n\sqrt{1+|a_1|^2+|a_2|^2+\cdots+|a_n|^2}$$

[II **69**, III **122**].

196.1. [G. Pólya, problem: Jber. deutsch. Math. Verein, Section 2, **35**, p. 48 (1926). Solution by N. Obreschkov: Jber. deutsch. Math. Verein., Section 2, **36**, pp. 43–45 (1917).]

Part Six. Polynomials and Trigonometric Polynomials

[For a fuller account of various topics studied in Part Six, especially for a generalization of **1–7** and **84–102**, see G. Szegö: Orthogonal Polynomials, Amer. Math. Soc. Colloq. Publ., **23**, 1939.]

1. The zeros of $T_n(x)$ are $\cos(2\nu-1)[\pi/(2n)]$, those of $U_n(x)$ are $\cos\nu[\pi/(n+1)]$, $\nu=1,2,\ldots,n$.

2. Identical with

$$\cos(n+1)\vartheta=\cos\vartheta\cos n\vartheta-\sin\vartheta\sin n\vartheta,$$
$$\sin(n+1)\vartheta=\sin\vartheta\cos n\vartheta+\cos\vartheta\sin n\vartheta.$$

3. In the differential equations

$$\frac{d^2\cos n\vartheta}{d\vartheta^2}=-n^2\cos n\vartheta,\qquad \frac{d^2\sin n\vartheta}{d\vartheta^2}=-n^2\sin n\vartheta,$$

introduce the new independent variable $\cos\vartheta=x$. A special case of **98**; cf. solution (g).

4. If we set $x=\cos\vartheta$, then the integrals in question become

$$t_{mn}=\int_0^\pi\cos m\vartheta\cos n\vartheta\,d\vartheta,\qquad u_{mn}=\int_0^\pi\sin(m+1)\vartheta\sin(n+1)\vartheta\,d\vartheta.$$

Hence $t_{mn}=u_{mn}=0$, if $m\gtrless n$, and $t_{nn}=u_{nn}=\pi/2$ for $n>0$, $t_{00}=\pi$, $u_{00}=\pi/2$.

The orthogonality property mentioned in the problem determines the polynomials $T_n(x)$ and $U_n(x)$ up to a constant factor. It is equivalent to the conditions

$$\int_{-1}^{1}\frac{T_n(x)}{\sqrt{1-x^2}}K(x)\,dx=0\quad\text{and}\quad\int_{-1}^{1}\sqrt{1-x^2}\,U_n(x)K(x)\,dx=0,\quad\text{respectively,}$$

for all polynomials $K(x)$ of degree $n-1$ (cf. introduction to **84**).

5. A special case of **98**; cf. solution (a).

6. [C. G. J. Jacobi: J. reine angew. Math., **15**, p. 3 (1836).] From **5** we deduce by integration by parts, since all the derivatives of $(1-x^2)^{n-1/2}$ which are of lower order than the nth vanish at $x=-1$ and $x=1$, that

$$\int_{-1}^{1}f(x)\frac{T_n(x)}{\sqrt{1-x^2}}\,dx=\frac{1}{1\cdot3\cdot5\ldots(2n-1)}\int_{-1}^{1}f^{(n)}(x)(1-x^2)^{n-1/2}\,dx.$$

7. $|\cos n\vartheta|\leqq1$, $n=1,2,3,\ldots$; the equality sign holds only if $n\vartheta$ is a multiple of π [**1**]. From the identity

$$\frac{\sin(n+1)\vartheta}{\sin\vartheta}=\cos n\vartheta+\cos\vartheta\frac{\sin n\vartheta}{\sin\vartheta}$$

one can deduce the second inequality by means of mathematical induction. There we have equality only in the case $|\cos\vartheta|=1$.

8. $\cos\nu\vartheta$ is a polynomial of degree ν in $\cos\vartheta$; $\cos^\nu\vartheta$ is a cosine polynomial of order ν.

9. $(\sin(\nu+1)\vartheta)/(\sin\vartheta)$ is a polynomial of degree ν in $\cos\vartheta$; $\sin\vartheta\cos^\nu\vartheta$ is a sine polynomial of order $\nu+1$.

10. $\cos p\vartheta\cos q\vartheta$, $\cos p\vartheta\sin q\vartheta$, $\sin p\vartheta\sin q\vartheta$, for p, q integers $p\geqq 0$, $q\geqq 0$, are trigonometric polynomials of order $p+q$.

11. We have

$$\cos\nu\vartheta=\frac{z^\nu+z^{-\nu}}{2},\qquad \sin\nu\vartheta=\frac{z^\nu-z^{-\nu}}{2i},\qquad z=e^{i\vartheta},\qquad \nu=0,1,2,\ldots,n.$$

Substituting these expressions in $g(\vartheta)$ and multiplying by $e^{in\vartheta}$ we obtain

$$G(z)=\lambda_0 z^n+\sum_{\nu=1}^{n}\left(\frac{1}{2}\lambda_\nu(z^{n+\nu}+z^{n-\nu})+\frac{1}{2i}\mu_\nu(z^{n+\nu}-z^{n-\nu})\right).$$

We have

$$u_\nu=\bar{u}_{2n-\nu}=\frac{\lambda_{n-\nu}+i\mu_{n-\nu}}{2},\qquad \nu=0,1,2,\ldots,n-1,$$

and $u_n=\lambda_0$. $G(z)$ is exactly of degree $2n$ (and does not vanish for $z=0$), if $g(\vartheta)$ is exactly of order n. The converse is also true.

12. Let $G(z)=u_0+u_1z+u_2z^2+\cdots+u_{2n}z^{2n}$. Then

$$u_\nu=\bar{u}_{2n-\nu},\qquad \nu=0,1,2,\ldots,2n,$$

so that u_n is real and

$$g(\vartheta)-u_n=e^{-in\vartheta}\sum_{\nu=0}^{n-1}(u_\nu e^{i\nu\vartheta}+u_{2n-\nu}e^{i(2n-\nu)\vartheta})=2\Re\sum_{\nu=0}^{n-1}u_\nu e^{i(\nu-n)\vartheta}$$

represents a trigonometric polynomial of order n with real coefficients.

13. Let z_0 be a zero of $G(z)$ different from 0. Then we have $z_0^{2n}\bar{G}(z_0^{-1})=0$, i.e. $G(\bar{z}_0^{-1})=0$, so that $\bar{z}_0^{-1}$ (the inverse of z_0 with respect to the unit circle) is also a zero. By means of differentiation one also shows that if z_0 is a k-fold zero, then $\bar{z}_0^{-1}$ is also a k-fold zero. If in addition $G(z)$ has the k-fold zero $z=0$, then the degree of $G(z)$ is decreased exactly by the number k on transition to $z^{2n}\bar{G}(z^{-1})$ [i.e. $z^{2n}\bar{G}(z^{-1})$ considered as a polynomial of degree $2n$ has k zeros at infinity].

Thus the zeros of $G(z)$ are situated symmetrically with respect to the unit circle.

14. If $z_1, z_2, \ldots, z_{2n}$ denote the zeros of the polynomials $G(z)$ defined in **11**, $z_\nu\neq 0$, $\nu=1,2,\ldots,2n$, then the zeros of $g(\vartheta)$ are

$$\vartheta_\nu=\frac{1}{i}\log z_\nu,\qquad 0\leqq\Re\vartheta_\nu<2\pi,\qquad \nu=1,2,\ldots,n.$$

15. With the notation of **11** we require the following identity in α and β to hold:

$$\sum_{k,\,l=0}^{2n}u_ku_l\sum_{\nu=0}^{n}e^{i(k-n)\left(\alpha-\frac{\nu\pi}{n+1}\right)+i(l-n)\left(\frac{\nu\pi}{n+1}-\beta\right)}=\sum_{k=0}^{2n}u_k\,e^{i(k-n)(\alpha-\beta)}.$$

We have

$$\sum_{\nu=0}^{n} e^{i(l-k)\frac{\nu\pi}{n+1}} = \begin{cases} n+1 & \text{for } k=l, \\ \dfrac{1-(-1)^{l-k}}{1-e^{i\frac{(l-k)\pi}{n+1}}} = \gamma_{kl} & \text{for } k \neq l, \end{cases}$$

and hence

$$(n+1)\sum_{k=0}^{2n} u_k^2\, e^{i(k-n)(\alpha-\beta)} + \sum_{\substack{k,\,l=0\\ k\neq l}}^{2n} \gamma_{kl} u_k u_l\, e^{i(k-n)\alpha + i(n-l)\beta} = \sum_{k=0}^{2n} u_k\, e^{i(k-n)(\alpha-\beta)}.$$

Obviously we have $\gamma_{kl}=0$ for even $l-k$, and $\gamma_{kl}\neq 0$ for odd $l-k$. From this we deduce, since

$$\sum_{\substack{k,\,l=0\\ k\neq l}}^{2n} \gamma_{kl} u_k u_l\, e^{i(k\alpha - l\beta)}$$

must be a polynomial in $e^{i(\alpha-\beta)}$, that $u_k u_l=0$ for $l-k$ odd, and further that

$$(n+1)u_k^2 = u_k, \qquad u_k=0 \quad \text{or} \quad u_k=\frac{1}{n+1}.$$

The trigonometric polynomials in question are characterized by the following conditions:

(1) They are cosine polynomials and contain terms with either only odd or only even multiples of ϑ.

(2) The terms which they actually contain all have the same coefficient $2/(n+1)$, with the exception of the constant term which, in case it is present, has the value $1/(n+1)$.

The number of such trigonometric polynomials is

$$2^{\left[\frac{n+1}{2}\right]} + 2^{\left[\frac{n+2}{2}\right]} - 1.$$

An example is given by

$$g(\vartheta) = \frac{1}{n+1}\,\frac{\sin (n+1)\vartheta}{\sin \vartheta}. \qquad \textbf{[16.]}$$

16. If we set $z=e^{i\vartheta}$, the first expression becomes

$$= \Re\left(\tfrac{1}{2} + z + z^2 + \cdots + z^n\right) = \Re\,\frac{1+z-2z^{n+1}}{2(1-z)} = \Re\,\frac{e^{-\frac{i\vartheta}{2}} + e^{\frac{i\vartheta}{2}} - 2e^{i\left(n+\frac{1}{2}\right)\vartheta}}{2\left(e^{-\frac{i\vartheta}{2}} - e^{\frac{i\vartheta}{2}}\right)}.$$

We can calculate similarly the three other expressions.—The proof may also be carried out by mathematical induction.

17. According to the last formula of **16** the expression in question is equal to

$$\frac{1}{\sin\vartheta}\left(\frac{\sin n\vartheta \sin \dfrac{2n+1}{2}\vartheta}{\sin \dfrac{\vartheta}{2}} - \frac{\sin n\vartheta \sin (n+1)\vartheta}{\sin\vartheta}\right).$$

(Alternatively this may be obtained by direct calculation.) For $\vartheta=0$ we obtain that the sum of the first n odd numbers is equal to the square of n.

18. By combining **16** and **17** we obtain:

$$\tfrac{1}{2}+\sum_{\nu=1}^{n}(\tfrac{1}{2}+\cos\vartheta+\cos 2\vartheta+\cdots+\cos\nu\vartheta)=\sum_{\nu=0}^{n}\frac{\sin\frac{2\nu+1}{2}\vartheta}{2\sin\frac{\vartheta}{2}}=\frac{1}{2}\left(\frac{\sin(n+1)\frac{\vartheta}{2}}{\sin\frac{\vartheta}{2}}\right)^2.$$

19. From **16–18** we find the zeros: (1) $\nu[2\pi/(2n+1)]$, $\nu=1,2,\ldots,2n$; (2) $\nu(2\pi/n)$, $\nu=1,2,\ldots,n-1$ and $(2\nu-1)[\pi/(n+1)]$, $\nu=1,2,\ldots,n+1$; (3) $\nu(\pi/(2n)$, $\nu=1,2,\ldots,2n-1,2n+1,\ldots,4n-1$; (4) $\nu(2\pi/n)$, $\nu[2\pi/(n+1)]$, $\nu=1,2,\ldots,n$; (5) $\nu(\pi/n)$, $\nu=1,2,\ldots,2n$, all double zeros with the exception of the cases $\nu=n$ and $\nu=2n$; (6) $\nu[2\pi/(n+1)]$, $\nu=1,2,\ldots,n$, all double zeros.

20. From **16** (2) or (1).

21. The sum in question is equal to

$$\frac{\sum_{\nu=1}^{n}\sin\nu\vartheta+\sum_{\nu=1}^{n+1}\sin\nu\vartheta}{2}=\sin^2(n+1)\frac{\vartheta}{2}\cot\frac{\vartheta}{2}. \qquad [\mathbf{16}.]$$

22. [L. Fejér: Math. Ann., **58**, p. 53 (1903.] From **18**.

23. [Cf. T. H. Gronwall: Math. Ann., **72**, pp. 229–230 (1912).] According to **16** we have

$$\frac{dA(n,\vartheta)}{d\vartheta}=\frac{\sin\frac{n}{2}\vartheta\cos\frac{n+1}{2}\vartheta}{\sin\frac{\vartheta}{2}}.$$

This expression vanishes only at the points mentioned in the problem in the interval $0\leqq\vartheta<\pi$; at these points it changes alternately from positive to negative and from negative to positive. If n is even, then it vanishes also for $\vartheta=\pi$. The point $x=\pi$, $y=0$ is however a center of symmetry [**29**] for the curve $y=A(n,x)$, and hence certainly not a maximum or a minimum.

24. [D. Jackson: Palermo Rend. **32**, pp. 257–258 (1911); T. H. Gronwall, loc. cit. **23**, p. 231; the equation

$$A(n,\vartheta)=A\left(n,\frac{\pi}{n+1}\right)$$

is due to L. Fejér; cf. D. Jackson, loc. cit.] According to **20** we have

$$A\left(n,(2\nu+1)\frac{\pi}{n+1}\right)-A\left(n,(2\nu-1)\frac{\pi}{n+1}\right)$$

$$=\int_{(2\nu-1)\frac{\pi}{n+1}}^{(2\nu+1)\frac{\pi}{n+1}}\frac{dA(n,\vartheta)}{d\vartheta}\,d\vartheta\leqq\frac{1}{2}\int_{(2\nu-1)\frac{\pi}{n+1}}^{(2\nu+1)\frac{\pi}{n+1}}\sin(n+1)\vartheta\cot\frac{\vartheta}{2}\,d\vartheta$$

$$=\frac{1}{2}\int_{(2\nu-1)\frac{\pi}{n+1}}^{2\nu\frac{\pi}{n+1}}\sin(n+1)\vartheta\left(\cot\frac{\vartheta}{2}-\cot\frac{\vartheta+\frac{\pi}{n+1}}{2}\right)d\vartheta,$$

$$\nu=1,2,\ldots,q-1;\quad n\geqq 3.$$

In the last integral $\sin(n+1)\vartheta \leqq 0$, the expression in brackets is positive, since $\cot x$ decreases monotonically for $0<x<\pi/2$.

25. [L. Fejér; cf. D. Jackson, loc. cit. **24**, p. 259; T. H. Gronwall, loc. cit. **23**, p. 253.] According to **24** we have

$$A\left(n, \frac{\pi}{n+1}\right) > A\left(n, \frac{\pi}{n}\right) = A\left(n-1, \frac{\pi}{n}\right).$$

Regarding the limit $\lim_{n\to\infty} A(n, \pi/(n+1))$, see II **6**.

26. [W. H. Young: Proc. Lond. Math. Soc. Series 2, **11**, p. 359 (1913).] According to **16** we have:

$$\frac{dB(n, \vartheta)}{d\vartheta} = -\frac{\sin\frac{n}{2}\vartheta \sin\frac{n+1}{2}\vartheta}{\sin\frac{\vartheta}{2}}. \qquad \text{[19, 23.]}$$

27. [W. H. Young, loc. cit. **26**.] Let $n \geqq 3$, κ, λ integers, $1 \leqq \kappa < \lambda \leqq [(n+1)/2]$. Then we have [**21**]

$$B\left(n, \lambda\frac{2\pi}{n+1}\right) - B\left(n, \kappa\frac{2\pi}{n+1}\right) = \int_{\kappa\frac{2\pi}{n+1}}^{\lambda\frac{2\pi}{n+1}} \frac{dB(n, \vartheta)}{d\vartheta}\, d\vartheta$$

$$= -\int_{\kappa\frac{2\pi}{n+1}}^{\lambda\frac{2\pi}{n+1}} \left(\sin\vartheta + \sin 2\vartheta + \cdots + \sin n\vartheta + \frac{\sin(n+1)\vartheta}{2}\right) d\vartheta < 0.$$

28. [W. H. Young, loc. cit. **26**.] According to **27** we have

$$\min B(n, \vartheta) = \begin{cases} B(n, \pi), & \text{for } n \text{ odd,} \\ B\left(n, n\frac{\pi}{n+1}\right), & \text{for } n \text{ even.} \end{cases}$$

In the first case we have

$$B(n, \pi) = -1 + \frac{1}{2} - \frac{1}{3} + \cdots + \frac{1}{n-1} - \frac{1}{n} \geqq -1.$$

For $n \geqq 5$, n odd, we have in fact

$$B(n, \vartheta) \geqq -1 + \frac{13}{60} > -1 + \frac{1}{n}.$$

In the second case we have

$$B\left(n, n\frac{\pi}{n+1}\right) = B\left(n+1, n\frac{\pi}{n+1}\right) - \frac{1}{n+1},$$

i.e. for $n \geqq 4$, n even, we have

$$B\left(n, n\frac{\pi}{n+1}\right) > -1 + \frac{1}{n+1} - \frac{1}{n+1} = -1.$$

In the case $n=2$ we have

$$B\left(2, \frac{2\pi}{3}\right) = \cos\frac{2\pi}{3} + \frac{1}{2}\cos\frac{4\pi}{3} = -\frac{1}{2} - \frac{1}{4} = -\frac{3}{4} > -1.$$

29. In the sequel, we set $n=0, \pm 1, \pm 2, \pm 3, \ldots$; such symmetry elements as cannot be derived from each other by means of motions of the plane of the graph are listed separately.

(1) The vertical lines $x=2n\pi$ and $x=(2n-1)\pi$ are axes of symmetry; $b_1 = b_2 = b_3 = \cdots = 0$.

(2) The points $x=2n\pi$ and $x=(2n-1)\pi$ of the x-axis $y=0$ are centers of symmetry; $a_0=a_1=a_2=a_3=\cdots=0$.

(3) The graph of the curve is mapped onto itself by a horizontal translation of length π along the x-axis and a subsequent reflection in the x-axis (the x-axis, $y=0$, is a "gliding symmetry axis"); $a_0=a_2=b_2=a_4=b_4=a_6=b_6=\cdots=0$.

(4a) The vertical lines $x=n\pi$ are axes of symmetry; the points $x=(n+\frac{1}{2})\pi$, $y=0$ are centers of symmetry; $b_1=b_2=b_3=\cdots=0$, $a_0=a_2=a_4=\cdots=0$.

(4b) The vertical lines $x=(n+\frac{1}{2})\pi$ are axes of symmetry, the points $x=n\pi$, $y=0$ are centers of symmetry; $a_0=a_1=a_2=\cdots=0$, $b_2=b_4=b_6=\cdots=0$. (This kind of symmetry is geometrically not distinct from (4a).)

(5) Here the period is in fact π; $a_1=b_1=a_3=b_3=a_5=b_5=\cdots=0$. The symmetry consists merely in the invariance under horizontal translation by multiples of π. (This kind of symmetry is geometrically not distinct from the symmetry of the general Fourier series.)

(There are apart from the five above-mentioned types of symmetry two additional ones (obtained by reflection in the x-axis), i.e. a total of seven different groups of motions which leave invariant an ornamental band, a "frieze", in fact an infinite number of equidistant points and a plane passing through them. Cf. MPR, **1**, p. 88.)

30. The first $2n+1$ Fourier coefficients of $f(\vartheta)$ are equal to the corresponding coefficients of $f(\vartheta)$ (with the notation of **11**: $a_0=\lambda_0$, $2a_\nu=\lambda_\nu$, $2b_\nu=\mu_\nu$, $\nu=1, 2, 3, \ldots, n$), all the others are equal to 0.

31.

$$\left(2\cos\frac{\vartheta}{2}\right)^n = \left(e^{\frac{i\vartheta}{2}} + e^{-\frac{i\vartheta}{2}}\right)^n = \sum_{\nu=0}^{n} \binom{n}{\nu} \left(e^{\frac{i\vartheta}{2}}\right)^{n-\nu} \left(e^{-\frac{i\vartheta}{2}}\right)^{\nu}$$

$$= \Re \sum_{\nu=0}^{n} \binom{n}{\nu} e^{i\left(\frac{n}{2}-\nu\right)\vartheta} = \sum_{\nu=0}^{n} \binom{n}{\nu} \cos\left(\frac{n}{2}-\nu\right)\vartheta.$$

[III **117**.]

32. By multiplication by $\cos n\vartheta$, or $\sin n\vartheta$, $n=0, 1, 2, \ldots$, and integration from 0 to 2π, we obtain that the Fourier coefficients of $f(\vartheta)$ are identical with $a_0, a_1, b_1, a_2, b_2, \ldots$, i.e. that the Fourier Series is identical with the given trigonometric series.

33. Expand the function $f(\vartheta)$ defined by

$$f(\vartheta) = \begin{cases} \dfrac{\pi-\vartheta}{2}, & \text{for} \quad 0<\vartheta<2\pi, \\ 0, & \text{for} \quad \vartheta=0, \end{cases}$$

in a Fourier series.

34. By direct calculation of the Fourier coefficients we obtain the first series. If we set in this $\vartheta=0$ and subtract the series obtained, we obtain the second series.

35. [G. Szegö: Math. Z. **9**, p. 163 (1921).] According to **34** we have

$$\rho_m=\frac{16}{\pi^2}\sum_{n=1}^{\infty}\frac{1}{4n^2-1}\int_0^{\frac{\pi}{2}}\frac{\sin^2 nm\vartheta}{\sin\vartheta}\,d\vartheta. \qquad [\mathbf{17}.]$$

36. From **29** (1) and (5) we deduce that the sine terms as well as the cosine terms of odd order vanish. We have

$$-c_n=\frac{1}{\pi}\int_0^{2\pi}f(\vartheta)\cos 2n\vartheta\,d\vartheta=\frac{2}{\pi}\int_0^{\pi}f(\vartheta)\cos 2n\vartheta\,d\vartheta$$

$$=\frac{2}{\pi}\sum_{\nu=1}^{n}\int_0^{\frac{\pi}{4n}}\left[f\left(\frac{\nu\pi}{n}+\vartheta\right)-f\left(\frac{\nu\pi}{n}+\frac{\pi}{2n}-\vartheta\right)\right.$$

$$\left.-f\left(\frac{\nu\pi}{n}+\frac{\pi}{2n}+\vartheta\right)+f\left(\frac{\nu\pi}{n}+\frac{\pi}{n}-\vartheta\right)\right]\cos 2n\vartheta\,d\vartheta.$$

In the interval $[0, \pi/(4n)]$ the function $\cos 2n\vartheta$ is positive; moreover [II **74**] for an arbitrary function $f(x)$ which is concave in the interval (a, b), if $a<x-v<x-u<x+u<x+v<b$, we have

$$f(x-u)\geqq\frac{(v+u)f(x-v)+(v-u)f(x+v)}{2v},$$

$$f(x+u)\geqq\frac{(v-u)f(x-v)+(v+u)f(x+v)}{2v},$$

and thus the following inequality holds:

$$f(x-v)-f(x-u)-f(x+u)+f(x+v)\leqq 0,$$

which is also evident geometrically.

37. The proof is obtained, as in the special case **35**, using **36**.

38. With the notation of **26–28**, we have [**34**]

$$\Gamma(n,\vartheta)=\frac{2}{\pi}\sum_{\nu=1}^{n}\frac{1}{\nu}-\frac{4}{\pi}\sum_{\nu=1}^{\infty}\frac{B(n,2\nu\vartheta)}{4\nu^2-1}\leqq\frac{2}{\pi}\sum_{\nu=1}^{n}\frac{1}{\nu}+\frac{4}{\pi}\sum_{\nu=1}^{\infty}\frac{1}{4\nu^2-1}$$

$$=\frac{2}{\pi}\sum_{\nu=1}^{n}\frac{1}{\nu}+\frac{2}{\pi}.$$

Moreover we have

$$M_n>\frac{1}{\pi}\int_0^{\pi}\Gamma(n,\vartheta)\,d\vartheta=\frac{2}{\pi}\sum_{\nu=1}^{n}\frac{1}{\nu}.$$

39. If we set $z=e^{i\vartheta}$ and

$$|x_0+x_1z+x_2z^2+\cdots+x_nz^n|^2$$
$$=\lambda_0+\lambda_1\cos\vartheta+\mu_1\sin\vartheta+\lambda_2\cos 2\vartheta+\mu_2\sin 2\vartheta+\cdots+\lambda_n\cos n\vartheta+\mu_n\sin n\vartheta,$$

then we have

$$\lambda_0=|x_0|^2+|x_1|^2+|x_2|^2+\cdots+|x_n|^2,$$
$$\tfrac{1}{2}(\lambda_\nu+i\mu_\nu)=x_0\bar{x}_\nu+x_1\bar{x}_{\nu+1}+\cdots+x_{n-\nu}\bar{x}_n,\qquad \nu=1,2,\ldots,n.$$

We have $\lambda_n + i\mu_n = 2x_0\bar{x}_n \neq 0$. (For $x_0 = x_1 = x_2 = \cdots = x_n$ we obtain from this a new proof of **18**.)

40. [F. Riesz; cf. L. Fejér: J. reine angew. Math. **146**, pp. 53–82 (1916).] The polynomial $G(z)$ defined in **11** consists of the following three factors, of which one or even two may be absent [**13**]:

$$\prod_{\mu=1}^{k} (z-\zeta_\mu), \qquad \prod_{\nu=1}^{l} (z-z_\nu)\left(z-\frac{1}{\bar{z}_\nu}\right), \qquad cz^r,$$
$$k+2l+2r=2n,$$

where k, l, r are integers, $k \geqq 0$, $l \geqq 0$, $r \geqq 0$, $|\zeta_\mu|=1$, $0<|z_\nu|<1$, $\mu=1,2,\ldots,k$, $\nu=1,2,\ldots,l$ and c is an arbitrary complex number. The zeros $\zeta_\mu = e^{i\vartheta_\mu}$ of $G(z)$ on the unit circle correspond to the zeros ϑ_μ of $g(\vartheta)$ on the real interval $0 \leqq \vartheta < 2\pi$. In fact, as one can see by differentiating the identity $g(\vartheta)=e^{-in\vartheta}G(e^{i\vartheta})$, a zero $\zeta_\mu = e^{i\vartheta_\mu}$ of $G(z)$ occurs with the same multiplicity as the corresponding zero ϑ_μ of $g(\vartheta)$. Since $g(\vartheta) \geqq 0$, every zero must be of even multiplicity. Let

$$\prod_{\mu=1}^{k} (z-\zeta_\mu) = \prod_{\mu=1}^{\frac{k}{2}} (z-\zeta_\mu)^2.$$

Then we obtain

$$g(\vartheta)=|g(\vartheta)|=|G(e^{i\vartheta})|=|c| \prod_{\mu=1}^{\frac{k}{2}} |e^{i\vartheta}-\zeta_\mu|^2 \prod_{\nu=1}^{l} \frac{|e^{i\vartheta}-z_\nu|^2}{|z_\nu|}.$$

41. [**40**.] The polynomial $G(z)$ in this case has only real coefficients and hence the complex zeros ζ_μ and z_ν occur in complex conjugate pairs.

42. One may replace any linear factor $z-z_0$ of $h(z)$ by $1-\bar{z}_0 z$ since on the unit circle we have $|z-z_0|=|1-\bar{z}_0 z|$ [III **5**]. Here we may also have $z_0=0$.

43. In the manner indicated in solution **42** all the zeros of $h(z)$ may be removed which have absolute value less than 1. Condition (2) can be satisfied by multiplication by a suitably chosen constant γ, $|\gamma|=1$. From III **274** it follows further that this polynomial $h(z)$ is uniquely determined.

44. The polynomial in question can be decomposed into factors of the form $(x-x_0)^2+y_0^2$, x_0, y_0 real. Apply repeatedly the identity

$$(p_1^2+q_1^2)(p_2^2+q_2^2)=(p_1p_2+q_1q_2)^2+(p_1q_2-p_2q_1)^2.$$

45. The polynomial in question can be decomposed into factors of the form

$$(x-x_0)^2+y_0^2, \quad x_0, y_0 \text{ real}; \quad x+x_1, \quad x_1 \geqq 0.$$

Apply the identity

$$\begin{aligned}
&[p_1^2+q_1^2+x(r_1^2+s_1^2)][p_2^2+q_2^2+x(r_2^2+s_2^2)] \\
&\quad =[(p_1^2+q_1^2)(p_2^2+q_2^2)+x^2(r_1^2+s_1^2)(r_2^2+s_2^2)] \\
&\qquad +x[(p_1^2+q_1^2)(r_2^2+s_2^2)+(r_1^2+s_1^2)(p_2^2+q_2^2)].
\end{aligned}$$

Furthermore, **44** must be used twice.

46. [M. Fekete.] Writing $P(x)$ for the polynomial in question, apply **41** to $P(\cos\vartheta)$ and make use of **8**, **9**. We have

$$P(\cos\vartheta)=|A(\cos\vartheta)+i\sin\vartheta B(\cos\vartheta)|^2,$$

where $A(x)$ and $B(x)$ denote polynomials with real coefficients.

47. [F. Lukács.] Let $P(x)$ be of degree $2m$. Then by **41** we have

$$P(\cos\vartheta)=|h(e^{i\vartheta})|^2=|e^{-im\vartheta}h(e^{i\vartheta})|^2,$$
$$e^{-im\vartheta}h(e^{i\vartheta})=A(\cos\vartheta)+i\sin\vartheta D(\cos\vartheta),$$

where $h(z)$ is a polynomial of degree $2m$ with real coefficients, $A(x)$ is of degree m and $D(x)$ is of degree $m-1$. If $P(x)$ is of degree $2m+1$, then we have

$$P(x)=(x-\alpha)P_1(x)=(x+1)P_1(x)+(-\alpha-1)P_1(x),$$

or

$$P(x)=(\beta-x)P_1(x)=(\beta-1)P_1(x)+(1-x)P_1(x),$$

where $\alpha\leqq-1$, $\beta\geqq 1$, respectively, and we apply the preceding considerations to $P_1(x)$.

48. [Ch. Hermite, problem: Interméd. des math. **1**, p. 65 (1894). Solution by J. Franel, E. Goursat, J. Sadier: Interméd. des. math. **1**, p. 251 (1894).] No. For if the representation were possible we could write

$$x^2+\varepsilon=\Sigma\, A(\varepsilon)(1-x)^\alpha(1+x)^\beta,$$

where $\varepsilon>0$, $A(\varepsilon)\geqq 0$. Here $\alpha+\beta\leqq 2$, and α, β are non-negative integers, so that the number of terms is exactly 6 for all ε. From this it follows, setting $x=0$, that $A(\varepsilon)$ is bounded for $0<\varepsilon\leqq 1$. Now, letting ε converge to 0, in such a manner that $\lim_{\varepsilon\to 0} A(\varepsilon)=A$ exists in all six terms, we obtain

$$x^2=\Sigma\, A(1-x)^\alpha(1+x)^\beta.$$

For $x=0$ this leads to a contradiction.

49. [F. Hausdorff: Math. Z. **9**, pp. 98–99 (1921).]

First solution: It is sufficient to consider the following two special cases:

(1) $P(x)$ linear:

$$P(x)=\frac{P(-1)}{2}(1-x)+\frac{P(1)}{2}(1+x).$$

(2) $P(x)$ quadratic, $P(x)=a+2b(1-x)+c(1-x)^2$, $c>0$, $ac-b^2>0$. Denoting by p an integer, $p\geqq 2$, to be determined later, set

$$\begin{aligned}2^pP(x)&=a\sum_{\nu=0}^{p}\binom{p}{\nu}(1-x)^\nu(1+x)^{p-\nu}\\&\quad+2b(1-x)2\sum_{\nu=1}^{p}\binom{p-1}{\nu-1}(1-x)^{\nu-1}(1+x)^{p-\nu}\\&\quad+c(1-x)^24\sum_{\nu=2}^{p}\binom{p-2}{\nu-2}(1-x)^{\nu-2}(1+x)^{p-\nu}\\&=\sum_{\nu=0}^{p}\frac{(p-2)!}{\nu!(p-\nu)!}f(\nu)(1-x)^\nu(1+x)^{p-\nu}.\end{aligned}$$

Here

$$\begin{aligned}f(\nu)&=ap(p-1)+4b(p-1)\nu+4c\nu(\nu-1)\\&=4c\nu^2+2(2bp-2b-2c)\nu+(ap^2-ap)\end{aligned}$$

is a polynomial of second degree in ν, which becomes positive definite if p is chosen sufficiently large, namely such that

$$4c(ap^2-ap)-(2bp-2b-2c)^2=4(ac-b^2)p^2+\cdots>0.$$

Second solution: Introduce a new variable z by means of the equation

$$(1+x)(1+z)=2.$$

Let $P(x)$ be of degree n. Then

$$P\left(\frac{1-z}{1+z}\right)(1+z)^n=f(z)$$

is a polynomial of degree n, $f(z)>0$ for $z>0$.

Hence for a sufficiently large integer p we have the representation

$$f(z)(1+z)^{p-n}=\sum_{\alpha=0}^{p} A_\alpha z^\alpha,$$

$A_\alpha \geqq 0$ for $\alpha=0, 1, 2, \ldots, p$ [solution V **187**]. From this it follows that

$$P(x)=f\left(\frac{1-x}{1+x}\right)\left(\frac{1+x}{2}\right)^n=2^{-p}\sum_{\alpha=0}^{p} A_\alpha(1-x)^\alpha(1+x)^{p-\alpha}.$$

50. *First proof.* [L. Fejér: C.R. Acad. Sci. (Paris) **157**, pp. 506–509 (1913); the case of equality is not discussed.] If we set

$$Q(z)=n+1-\sum_{\nu=1}^{n}\frac{1+ze^{i\vartheta_\nu}}{1-ze^{i\vartheta_\nu}}, \qquad \vartheta_\nu=\nu\frac{2\pi}{n+1}, \qquad \nu=1, 2, \ldots, n,$$

then we have [cf. III **6**]

$$\Re Q(z)<n+1, \qquad |z|<1.$$

Furthermore [cf. second proof]

$$Q(z)=1+2z+2z^2+\cdots+2z^n+q_{n+1}z^{n+1}+q_{n+2}z^{n+2}+\cdots.$$

We have for $0\leqq r<1$

$$\begin{aligned}&\lambda_0+\lambda_1 r+\lambda_2 r^2+\cdots+\lambda_n r^n\\ &=\frac{1}{2\pi}\int_0^{2\pi} g(\vartheta)(1+2r\cos\vartheta+\cdots+2r^n\cos n\vartheta)\,d\vartheta\\ &=\frac{1}{2\pi}\int_0^{2\pi} g(\vartheta)[\Re Q(re^{i\vartheta})]\,d\vartheta<(n+1)\frac{1}{2\pi}\int_0^{2\pi} g(\vartheta)\,d\vartheta=n+1.\end{aligned}$$

Let r converge to 1 and we obtain the desired estimate for $g(0)$. Then consider $g(\vartheta+\vartheta_0)$, ϑ_0 fixed.

Second proof. [L. Fejér: C.R. Acad. Sci. (Paris) **157**, pp. 571–572 (1913).] It is sufficient to prove that $g(0)\leqq n+1$. Setting

$$\vartheta_\nu=\nu\frac{2\pi}{n+1}, \qquad \nu=0, 1, 2, \ldots, n,$$

we have

$$\sum_{\nu=0}^{n} \cos k\vartheta_\nu = \sum_{\nu=0}^{n} \sin k\vartheta_\nu = 0, \qquad k=1, 2, \ldots, n,$$

since

$$\sum_{\nu=0}^{n} e^{ik\vartheta_\nu} = \frac{1-e^{i(n+1)k\frac{2\pi}{n+1}}}{1-e^{ik\frac{2\pi}{n+1}}} = 0.$$

Hence we have

$$g(0)+g(\vartheta_1)+g(\vartheta_2)+\cdots+g(\vartheta_n)=n+1, \qquad \text{i.e.} \quad g(0) \leqq n+1.$$

If the case of equality is attained, then we must have $g(\vartheta_1)=g(\vartheta_2)=\cdots=g(\vartheta_n)=0$, i.e. $g(\vartheta)$ then has, because of $g(\vartheta) \geqq 0$, the double zeros

$$\vartheta_\nu = \nu \frac{2\pi}{n+1}, \qquad \nu=1, 2, \ldots, n.$$

This property, together with the condition $\lambda_0=1$, determines $g(\vartheta)$ uniquely. [**18**.]

Third proof [L. Fejér, loc. cit. **40**, pp. 65–66]. By **40** the most general trigonometric polynomial $g(\vartheta)$ with the required properties has the form

$$g(\vartheta)=|x_0+x_1z+x_2z^2+\cdots+x_nz^n|^2, \qquad z=e^{i\vartheta},$$

where

$$\lambda_0=|x_0|^2+|x_1|^2+|x_2|^2+\cdots+|x_n|^2=1 \qquad [\mathbf{39}.]$$

Using II **80** we obtain that for any value of $\vartheta=\vartheta_0$ we have

$$g(\vartheta_0) \leqq (n+1)(|x_0|^2+|x_1|^2+|x_2|^2+\cdots+|x_n|^2)=n+1.$$

Here equality holds only for $x_\nu=\gamma e^{-i\nu\vartheta_0}$, $\nu=0, 1, 2, \ldots, n$; $|\gamma|=1/\sqrt{n+1}$.

51. [L. Fejér, loc. cit. **40**, p. 73.] We have $\lambda_n+i\mu_n=2x_0\bar{x}_n$ and

$$2|x_0\bar{x}_n| \leqq |x_0|^2+|x_n|^2 \leqq |x_0|^2+|x_1|^2+\cdots+|x_n|^2=1.$$

The equality sign holds only if $x_1=x_2=\cdots=x_{n-1}=0$, $|x_0|=|x_n|=1/\sqrt{2}$, i.e. for $g(\vartheta)=\frac{1}{2}|1+\gamma e^{in\vartheta}|^2$, $|\gamma|=1$.

52. [L. Fejér, loc. cit. **40**, p. 79.] By **39**, **40** we are concerned with the maximum of

$$4|x_0\bar{x}_1+x_1\bar{x}_2+\cdots+x_{n-1}\bar{x}_n|^2,$$

subject to the condition $|x_0|^2+|x_1|^2+|x_2|^2+\cdots+|x_n|^2=1$. We may confine ourselves to the consideration of real $x_0, x_1, x_2, \ldots, x_n$. The maximum of

$$2|x_0x_1+x_1x_2+\cdots+x_{n-1}x_n|$$

is equal to the absolute value of the zero of greatest absolute value of the determinant $D_n(f-h)$ of VII **70**, $f(\vartheta)=2\cos\vartheta$, i.e. equal to $h_{n,n}=2\cos[\pi/(n+2)]$ [cf. Kowalewski, p. 275].

53. Let $h(z)=c(1+z_1z)(1+z_2z)\ldots(1+z_nz)$, $|z_\nu| \leqq 1$, $\nu=1, 2, \ldots, n$, c real, $c>0$. By II **52** we have

$$\frac{1}{2\pi}\int_0^{2\pi} \log|1+z_\nu e^{i\vartheta}|^2\, d\vartheta=0,$$

and hence

$$\frac{1}{2\pi}\int_0^{2\pi} \log g(\vartheta)\, d\vartheta = \log c^2.$$

54. We obtain all trigonometric polynomials $g(\vartheta)$ of the required kind by considering $g(\vartheta)=|h(e^{i\vartheta})|^2$, where

$$h(z)=(1+z_1z)(1+z_2z)\ldots(1+z_nz)=x_0+x_1z+x_2z^2+\cdots+x_nz^n$$

and $z_1, z_2, \ldots, z_n$ are arbitrary complex numbers of absolute value $\leqq 1$. Hence for $|z|\leqq 1$ we have

$$|h(z)|\leqq(1+1)^n=2^n.$$

The equality sign holds only if all the z_ν are equal and of absolute value 1 and if $z=\bar{z}_\nu$.

55. We have

$$h(z)\ll(1+z)^n,$$

and hence

$$\lambda_0=|x_0|^2+|x_1|^2+|x_2|^2+\cdots+|x_n|^2\leqq 1+\binom{n}{1}^2+\binom{n}{2}^2+\cdots+\binom{n}{n}^2$$

[I **32**], equality as in **54**.

56. As in **55**

$$|\lambda_\nu+i\mu_\nu|=2|x_0\bar{x}_\nu+x_1\bar{x}_{\nu+1}+\cdots+x_{n-\nu}\bar{x}_n|$$
$$\leqq 2\left[\binom{n}{0}\binom{n}{\nu}+\binom{n}{1}\binom{n}{\nu+1}+\cdots+\binom{n}{n-\nu}\binom{n}{n}\right]=2\binom{2n}{n+\nu}.$$

57. *First proof.* By **40** and **39** from $g(\vartheta)\geqq 0$, $\lambda_0=0$, it follows that

$$g(\vartheta)=|x_0+x_1e^{i\vartheta}+x_2e^{2i\vartheta}+\cdots+x_ne^{ni\vartheta}|^2,$$
$$|x_0|^2+|x_1|^2+|x_2|^2+\cdots+|x_n|^2=0,$$

i.e. $x_0=x_1=x_2=\cdots=x_n=0$.

Second proof [L. Fejér, loc. cit. **50**, second proof]. We deduce as in solution **50**, second proof, that, setting

$$\vartheta_\nu=\nu\frac{2\pi}{n+1}, \qquad \nu=1,2,\ldots,n,$$

we have

$$g(0)+g(\vartheta_1)+g(\vartheta_2)+\cdots+g(\vartheta_n)=0,$$

i.e.

$$g(0)=g(\vartheta_1)=g(\vartheta_2)=\cdots=g(\vartheta_n)=0.$$

Thus $g(\vartheta)$ has zeros $0, \vartheta_1, \vartheta_2, \ldots, \vartheta_n$ and, since $g(\vartheta)\geqq 0$, they are in fact all double zeros. By **14** we deduce from this that $g(\vartheta)\equiv 0$.

Third proof. Mathematical induction. For $n=1$ we have

$$\lambda_1\cos\vartheta+\mu_1\sin\vartheta=\sqrt{\lambda_1^2+\mu_1^2}\cos(\vartheta-\vartheta_0), \qquad \tan\vartheta_0=\frac{\mu_1}{\lambda_1},$$

and thus the assertion is obvious. For $n>1$, if $g(\vartheta)\geqq 0$,

$$g^*(\vartheta)=\frac{g(\vartheta)+g(\vartheta+\pi)}{2}$$
$$=\lambda_2 \cos 2\vartheta+\mu_2 \sin 2\vartheta+\lambda_4 \cos 4\vartheta+\mu_4 \sin 4\vartheta+\cdots$$
$$+\lambda_{2p} \cos 2p\vartheta+\mu_{2p} \sin 2p\vartheta,$$

$p=[n/2]$, is a non-negative trigonometric polynomial of order p in 2ϑ and without a constant term. From $g^*(\vartheta)\equiv 0$, $g(\vartheta)\geqq 0$ it follows that also $g(\vartheta)\equiv 0$.

58. [L. Fejér, loc. cit. **40**, pp. 67–68; loc. cit. **50**, second proof, pp. 573–574.] If $g(\vartheta)\not\equiv 0$, then $m>0$, $M>0$ [**57**]. Apply **50** to $[g(\vartheta)+m]/m$ and again to $[M-g(\vartheta)]/M$.

59. [L. Fejér, loc. cit. **40**, p. 69.] Apply **58** to

$$g(\vartheta)-\frac{1}{2\pi}\int_0^{2\pi} g(\vartheta)\, d\vartheta.$$

60. [L. Fejér, loc. cit. **40**, pp. 80–81.] Let $g(\vartheta)\not\equiv$constant. Applying **51** to $[g(\vartheta)+m]/(\lambda_0+m)$ and again to $[M-g(\vartheta)]/(M-\lambda_0)$ we obtain

$$\sqrt{\lambda_n^2+\mu_n^2}\leqq\lambda_0+m, \qquad \sqrt{\lambda_n^2+\mu_n^2}\leqq M-\lambda_0,$$
$$\sqrt{\lambda_n^2+\mu_n^2}\leqq\frac{m+M}{2}\leqq\max(m, M).$$

61. [O. Szász: Münch. Ber. pp. 307–320 (1917).] By **40** we may set

$$M-\sum_{\nu=1}^{n}(\lambda_\nu \cos \nu\vartheta+\mu_\nu \sin \nu\vartheta)=|x_0+x_1z+x_2z^2+\cdots+x_nz^n|^2,$$

$z=e^{i\vartheta}$, from which we deduce [**39**] that

$$\sqrt{\lambda_\nu^2+\mu_\nu^2}\leqq 2(|x_0||x_\nu|+|x_1||x_{\nu+1}|+\cdots+|x_{n-\nu}||x_n|),$$
$$\nu=1, 2, \ldots, n.$$

Furthermore we have

$$M=|x_0|^2+|x_1|^2+\cdots+|x_n|^2,$$

and hence

$$\sum_{\nu=1}^{n}\sqrt{\lambda_\nu^2+\mu_\nu^2}\leqq(|x_0|+|x_1|+\cdots+|x_n|)^2-(|x_0|^2+|x_1|^2+\cdots+|x_n|^2)$$
$$\leqq(n+1)M-M=nM.$$

[II **80**.]

62. [P. Tchebychev: Oeuvres, **1**, pp. 387–469. St. Pétersbourg, 1899; cf. L. Fejér, loc. cit. **40**, pp. 81–82.] If $P(x)$ has real coefficients apply **60** to the trigonometric polynomial $P(\cos\vartheta)=2^{1-n}\cos n\vartheta+\cdots$. If $P(x)$ has arbitrary complex coefficients, then it may be separated into polynomials with real coefficients, $P(x)=P_1(x)+iP_2(x)$, where the coefficient of x^n in $P_1(x)$ is again 1 and $P_2(x)$ is of degree $n-1$. From $|P(x)|^2=[P_1(x)]^2+[P_2(x)]^2$ it follows that

$$\max|P(x)|\geqq\max|P_1(x)|\geqq\frac{1}{2^{n-1}}, \qquad -1\leqq x\leqq 1.$$

If here equality holds, then we must have $P_1(x)=2^{1-n}T_n(x)$ and at every point where $|P_1(x)|$ attains its maximum (at $n+1$ points) we must have $P_2(x)=0$, i.e. $P_2(x)\equiv 0$. This result is somewhat stronger than the extreme case of III **270**.

63. Under the linear transformation $[2/(\beta-\alpha)](x-\alpha)-1=y$, the interval $\alpha\leqq x\leqq\beta$ is mapped onto the interval $-1\leqq y\leqq 1$, and the polynomial $P(x)$ is transformed into $[(\beta-\alpha)/2]^n Q(y)$, where $Q(y)$ has the same form in y as $P(x)$ in x.

64. We may assume that $\beta>0$, $\alpha=-\beta$, $d<2\beta$. If $P(x)$ is any admissible polynomial, then

$$\frac{P(x)+(-1)^nP(-x)}{2}=\begin{cases} Q(x^2) & \text{for even } n,\\ xQ(x^2) & \text{for odd } n,\end{cases}$$

where $Q(\xi)$ denotes a polynomial of degree $[n/2]$ in $\xi=x^2$ with leading coefficient 1. The variable x runs through the two intervals $-\beta\leqq x\leqq -d/2$, $d/2\leqq x\leqq\beta$, and the variable ξ runs through the interval $(d/2)^2\leqq\xi\leqq\beta^2$. Hence, setting

$$2\left(\frac{\beta^2-\left(\frac{d}{2}\right)^2}{4}\right)^{\left[\frac{n}{2}\right]}=\mu,$$

we have [**63**]

$$(*)\qquad \max|P(x)|\geqq\max\left|\frac{P(x)+(-1)^nP(-x)}{2}\right|\begin{cases}=\max|Q(\xi)|\geqq\mu,\\ \geqq\frac{d}{2}\max|Q(\xi)|\geqq\frac{d}{2}\mu,\end{cases}$$

for n even or odd, respectively. The same estimate holds therefore for μ_n. On the other hand, let $Q_0(\xi)$ be the polynomial of degree $[n/2]$ with leading coefficient 1 for which $\max|Q_0(\xi)|=\mu$ [**62, 63**]. Set $P_0(x)=Q_0(x^2)$ or $P_0(x)=xQ_0(x^2)$ for n even or odd, respectively. We have

$$\mu_n\leqq\max|P_0(x)|\begin{cases}=\mu,\\ \leqq\beta\mu,\end{cases}$$

for n even or odd, respectively.

In the first case μ_n is attained, and this occurs *only* for $P(x)=P_0(x)$. For if $\max|P(x)|=\mu$, then by (*) we have also

$$\max\left|\frac{P(x)+P(-x)}{2}\right|=\mu,\quad\text{and hence}\quad\frac{P(x)+P(-x)}{2}=Q_0(x^2).$$

Furthermore we have

$$\left|\frac{P(x_\nu)+P(-x_\nu)}{2}\right|=\mu$$

at $(n/2)+1$ distinct points x_ν in the interval $d/2\leqq x\leqq\beta$. But now we have

$$\mu=\left|\frac{P(x_\nu)+P(-x_\nu)}{2}\right|\leqq\frac{|P(x_\nu)|+|P(-x_\nu)|}{2}\leqq\mu,$$

i.e. $P(x_\nu)=P(-x_\nu)$. The polynomial $P(x)-P(-x)$ of degree $n-1$ hence vanishes at $n+2$ points, $P(x)=P(-x)=Q_0(x^2)=P_0(x)$.

65. Set $M=\max|Q(z)|$ for $|z|=1$. The trigonometric polynomial of order $n+1$

$$\Re\left(1+\frac{1}{M}e^{i\vartheta}Q(e^{i\vartheta})\right)$$

is non-negative, has constant term equal to 1 and leading term $(1/M)\cos(n+1)\vartheta$ [**51**]. Cf. III **269**.

66. If the point P ranges over an arbitrary segment of the straight line L or over a disc D, and P_0 is an arbitrary point in space, then we have $\overline{PP_0}\geqq\overline{PP_0'}$, where P_0' denotes the projection of P_0 onto L or onto the plane of D, respectively. The "minimum maximorum" in question can thus be attained in both cases only for systems of points $P_1, P_2, \ldots, P_n$ which lie in the same plane as the given straight line segment or the given circle, respectively.

67. From the Lagrange interpolation formula, we obtain that

$$x^k=\frac{x_1^k}{f'(x_1)}\frac{f(x)}{x-x_1}+\frac{x_2^k}{f'(x_2)}\frac{f(x)}{x-x_2}+\cdots+\frac{x_n^k}{f'(x_n)}\frac{f(x)}{x-x_n},$$
$$k=0, 1, 2, \ldots, n-1.$$

Comparing the coefficients of x^{n-1} we obtain σ_k. For $k=0$ we obtain furthermore, by expanding in descending powers of x,

$$1=f(x)(\sigma_0x^{-1}+\sigma_1x^{-2}+\sigma_2x^{-3}+\cdots).$$

The coefficient of x^{-h-1} yields the recursion formula

$$a_0\sigma_{n+h}+a_1\sigma_{n+h-1}+\cdots+a_{n-1}\sigma_{h+1}+a_n\sigma_h=0, \qquad h\geqq 0,$$

so that

$$\begin{aligned}
&a_0\sigma_n \quad +a_1\sigma_{n-1} &&=0,\\
&a_0\sigma_{n+1} +a_1\sigma_n \quad +a_2\sigma_{n-1} &&=0,\\
&\cdots\cdots\cdots\cdots\cdots\cdots\cdots\cdots\cdots\cdots\\
&a_0\sigma_{2n-2}+a_1\sigma_{2n-3}+\ldots+a_{n-1}\sigma_{n-1} &&=0,\\
&a_0\sigma_{2n-1}+a_1\sigma_{2n-2}+\cdots+a_{n-1}\sigma_{n-2}+a_n\sigma_{n-1}&&=0.
\end{aligned}$$

68. [I. Schur.] Let the zeros of the polynomial $f(x)+\varepsilon$ be $x_\nu=x_\nu(\varepsilon)$, $\nu=1, 2, \ldots, n$. For sufficiently small ε these are differentiable functions of ε. We have $x_\nu(0)=x_\nu$. From $f[x_\nu(\varepsilon)]+\varepsilon=0$ we obtain by differentiation $f'(x_\nu)x_\nu'(0)=-1$. Apply **67** to $f(x)+\varepsilon$, differentiate with respect to ε and then set $\varepsilon=0$.

69. In (*) § 9 p. 82, set $P(x)=x^nf(x^{-1})+(-1)^{n-1}x_1x_2\ldots x_nf(x)$ and $x=-1$. We have

$$\sum_{\nu=1}^{n}\frac{P(x_\nu)}{f'(x_\nu)(1+x_\nu)}=-\frac{P(-1)}{f(-1)}=(-1)^{n-1}+(-1)^nx_1x_2\ldots x_n.$$

70. Let $P(x)$ be the polynomial considered, and

$$f(x)=(x-x_0)(x-x_1)(x-x_2)\ldots(x-x_n).$$

From

$$P(x)=\sum_{\nu=0}^{n}\frac{P(x_\nu)}{f'(x_\nu)}\frac{f(x)}{x-x_\nu}$$

we obtain, by comparing coefficients of x^n,

$$1=\sum_{\nu=0}^{n}\frac{P(x_\nu)}{f'(x_\nu)}, \quad \text{and hence} \quad 1\leqq M\sum_{\nu=0}^{n}\frac{1}{|f'(x_\nu)|},$$

where M denotes the greatest of the absolute values $|P(x_\nu)|$, $\nu=0, 1, 2, \ldots, n$. We have

$$\begin{aligned}|f'(x_\nu)| &= |(x_\nu-x_0)(x_\nu-x_1)\ldots(x_\nu-x_{\nu-1})(x_\nu-x_{\nu+1})\ldots(x_\nu-x_n)\\ &\geqq \nu!(n-\nu)!,\end{aligned}$$

$$\sum_{\nu=0}^{n}\frac{1}{|f'(x_\nu)|}\leqq\sum_{\nu=0}^{n}\frac{1}{\nu!(n-\nu)!}=\frac{2^n}{n!}.$$

71.

$$T_n'(x_\nu)=(-1)^{\nu-1}\frac{n}{\sqrt{1-x_\nu^2}}, \qquad \nu=1, 2, \ldots, n.$$

72.

$$U_n'(x_\nu)=(-1)^{\nu-1}\frac{n+1}{1-x_\nu^2}, \qquad \nu=1, 2, \ldots, n.$$

73. [Cf. I. Schur: Math. Z. **4**, pp. 273–274 (1919).] For $\nu=1, 2, \ldots, n-1$ we have

$$\left[\frac{d}{dx}[U_{n-1}(x)(x^2-1)]\right]_{x=x_\nu}=U_{n-1}'(x_\nu)(x_\nu^2-1)=(-1)^\nu n.$$

Furthermore

$$\left[\frac{d}{dx}[U_{n-1}(x)(x^2-1)]\right]_{x=\pm 1}=\pm 2U_{n-1}(\pm 1)=2n \quad \text{or} \quad (-1)^n 2n, \text{ respectively.}$$

74.

$$\left[\frac{d}{dx}(x^n-1)\right]_{x=\varepsilon_\nu}=n\varepsilon_\nu^{n-1}=n\varepsilon_\nu^{-1}, \qquad \nu=1, 2, \ldots, n.$$

75. The last equation states that the leading terms of the two polynomials $P(x)$ and $Q(x)$ are equal. Then we conclude from the second last equation that the terms following the leading terms are also equal, and so on. In other words, if we set

$$P(x)=a_0x^n+a_1x^{n-1}+\cdots+a_{n-1}x+a_n,$$

then, given the numbers $c_0, c_1, \ldots, c_n$, the coefficients $a_0, a_1, \ldots, a_n$ are successively and *uniquely* determined by the equations

$$\begin{aligned}
P^{(n)}(x_n) &= n!a_0 &&= c_n,\\
P^{(n-1)}(x_{n-1}) &= n!a_0x_{n-1}+(n-1)!a_1 &&= c_{n-1},\\
&\ldots\ldots\ldots\ldots\ldots\ldots\ldots\ldots\ldots\ldots\\
P'(x_1) &= na_0x_1^{n-1}+(n-1)a_1x_1^{n-2}+\cdots+a_{n-1} &&= c_1,\\
P(x_0) &= a_0x_0^n+a_1x_0^{n-1}+\cdots+a_{n-1}x_0+a_n &&= c_0.
\end{aligned}$$

76. [G. H. Halphen: Oeuvres, **2**, p. 520. Paris: Gauthier Villars, 1918.] $A_n(x)$ is uniquely determined by the conditions

$$A_n(0)=A_n'(1)=A_n''(2)=\cdots=A_n^{(n-1)}(n-1)=0, \qquad A_n^{(n)}(n)=1.$$

Clearly $A_0(x)=1$, $A_1(x)=x$ and in general $A_n'(1+x)$ satisfies the conditions imposed by the polynomial $A_{n-1}(x)$. Hence we have

$$A_n'(x)=A_{n-1}(x-1).$$

From this recurrence relation it follows by successive integration that

$$A_n(x)=\frac{x(x-n)^{n-1}}{n!}, \qquad n=1,2,3,4,\ldots.$$

Subsequent verification by differentiation is simpler.—A different proof is given in III **221**.

77. It follows from **71**, using the notation given there, that

$$a_0=\frac{2^{n-1}}{n}\sum_{\nu=1}^{n}(-1)^{\nu-1}\sqrt{1-x_\nu^2}P(x_\nu),$$

and hence $|a_0|\leqq(2^{n-1}/n)n=2^{n-1}$. Here equality holds if and only if

$$\sqrt{1-x_\nu^2}P(x_\nu)=(-1)^{\nu-1}\gamma, \qquad |\gamma|=1, \qquad \nu=1,2,\ldots,n.$$

By these n conditions the polynomial $P(x)$ of degree $n-1$ is determined uniquely. Since $\gamma U_{n-1}(x)$ satisfies these conditions we must have $P(x)=\gamma U_{n-1}(x)$.

78. Comparing the leading terms in the interpolation formula **73**, we obtain the following representation for the leading coefficient a_0 of $P(x)$

$$a_0=\frac{2^{n-2}}{n}[P(1)+(-1)^nP(-1)]+\frac{2^{n-1}}{n}\sum_{\nu=1}^{n=1}(-1)^\nu P(x_\nu).$$

Thus, if $|P(x)|\leqq 1$, $-1\leqq x\leqq 1$, we have

$$|a_0|\leqq\frac{2^{n-2}}{n}\cdot 2+\frac{2^{n-1}}{n}\cdot(n-1)=2^{n-1}.$$

The equality sign holds if and only if

$$P(1)=\gamma, \qquad P(-1)=(-1)^n\gamma, \qquad P(x_\nu)=(-1)^\nu\gamma,$$
$$\nu=1,2,\ldots,n-1, \qquad |\gamma|=1.$$

By these $n+1$ conditions $P(x)$ is determined uniquely. But $\gamma T_n(x)$ satisfies these conditions and hence we must have $P(x)=\gamma T_n(x)$.

79. If $P(x)$ is a polynomial of degree $n-1$ with leading coefficient a_0, then we have [**74**]

$$a_0=\frac{1}{n}\sum_{\nu=1}^{n}\varepsilon_\nu P(\varepsilon_\nu).$$

Thus if $|P(z)|\leqq 1$ for $|z|=1$, we have

$$|a_0|\leqq\frac{1}{n}\cdot n=1.$$

The equality sign holds only if $P(\varepsilon_\nu)=\gamma\bar{\varepsilon}_\nu$, $\nu=1,2,\ldots,n$, $|\gamma|=1$, i.e. $P(z)=\gamma z^{n-1}$.

80. From **71** it follows that for $x_1=\cos[\pi/(2n)]\leqq x\leqq 1$

$$|P(x)|\leqq\frac{1}{n}\sum_{\nu=1}^{n}\frac{T_n(x)}{x-x_\nu}=\frac{T_n'(x)}{n}=U_{n-1}(x),$$

i.e. by **7**: $|P(x)|\leqq n$. The equality sign holds only for $P(x)=\gamma U_{n-1}(x)$, $|\gamma|=1$, $x=1$. A similar result holds for $-1\leqq x\leqq x_n=-x_1$. For $x_n\leqq x\leqq x_1$, $n>1$ we have

$$\sqrt{1-x^2}\geqq \sin\frac{\pi}{2n}>\frac{2}{\pi}\cdot\frac{\pi}{2n}=\frac{1}{n}.$$

81. [Cf. M. Riesz: Jber. deutsch. Math. Verein. **23**, p. 354 (1914). Apply **80** to

$$P(x)=P(\cos\vartheta)=\frac{S(\vartheta)}{\sin\vartheta}. \qquad [\mathbf{9}.]$$

82. [S. Bernstein: Belg. Mém. 1912, p. 19, cf. M. Riesz, loc. cit. **81**. The idea used here is due to Fejér. Cf. M. Fekete: J. reine angew. Math. **146**, pp. 88–94, (1915).] By **81** we have

$$|S'(0)|=|g'(\vartheta_0)|\leqq n.$$

Cf. IV **201**.

83. [A. Markov: Abh. Akad. Wiss. St. Petersburg, **62**, pp. 1–24 (1889).] Apply **82** to $P(\cos\vartheta)$ and **80** to $n^{-1}P'(x)$. The equality sign holds only if $n^{-1}P'(x)=\gamma U_{n-1}(x)$, $|\gamma|=1$, i.e. $P(x)=c+\gamma T_n(x)$ where c is a constant. Since $|c\pm\gamma|\leqq 1$, we have $c=0$.

84. By integration by parts we obtain

$$\int_{-1}^{1}\left(\frac{1}{2^n n!}\frac{d^n}{dx^n}(x^2-1)^n\right)x^\nu\,dx=\frac{(-1)^n}{2^n n!}\int_{-1}^{1}(x^2-1)^n\frac{d^n x^\nu}{dx^n}\,dx,$$

since all the derivatives of $(x^2-1)^n$ of order less than n vanish at $x=1$ and $x=-1$. From this it follows that (1) of § 11 p. 85 is satisfied and (2) of § 11 p. 85 is also satisfied since by (1) we have

$$\int_{-1}^{1}\left(\frac{1}{2^n n!}\frac{d^n}{dx^n}(x^2-1)^n\right)^2 dx=\frac{(2n)!}{2^n n!^2}\int_{-1}^{1}\left(\frac{1}{2^n n!}\frac{d^n}{dx^n}(x^2-1)^n\right)x^n\,dx$$
$$=\frac{(2n)!}{2^{2n}n!^2}\int_{-1}^{1}(1-x^2)^n\,dx.$$

(3) is obvious. The coefficient of x^n in $P_n(x)$ is equal to

$$k_n=\frac{(2n)!}{2^n n!^2}.$$

85. From **84** we obtain by Leibniz's rule

$$P_n(x)=\frac{1}{2^n n!}\sum_{\nu=0}^{n}\binom{n}{\nu}\frac{n!}{(n-\nu)!}(x+1)^{n-\nu}\frac{n!}{\nu!}(x-1)^\nu.$$

86. We have [III **117**]

$$\frac{1}{2\pi}\int_0^{2\pi}\left(\sqrt{\frac{x+1}{2}}+\sqrt{\frac{x-1}{2}}\,e^{i\varphi}\right)^n\left(\sqrt{\frac{x+1}{2}}+\sqrt{\frac{x-1}{2}}\,e^{-i\varphi}\right)^n d\varphi$$
$$=\frac{1}{2\pi}\int_0^{2\pi}\left(\sum_{k=0}^{n}\binom{n}{k}\left(\frac{x+1}{2}\right)^{\frac{n-k}{2}}\left(\frac{x-1}{2}\right)^{\frac{k}{2}}e^{ik\varphi}\right)\left(\sum_{l=0}^{n}\binom{n}{l}\left(\frac{x+1}{2}\right)^{\frac{n-l}{2}}\left(\frac{x-1}{2}\right)^{\frac{l}{2}}e^{-il\varphi}\right)d\varphi$$
$$=\sum_{\nu=0}^{n}\binom{n}{\nu}^2\left(\frac{x+1}{2}\right)^{n-\nu}\left(\frac{x-1}{2}\right)^{\nu}=P_n(x) \qquad [\mathbf{85}].$$

87. Let k_n be the coefficient of x^n in $P_n(x)$, $k_n=[(2n)!/2^n n!^2]$ [**84**]. Express the polynomial of degree $n-1$, $P_n(x)-(k_n/k_{n-1})xP_{n-1}(x)$, as a linear combination of Legendre polynomials

$$P_n(x)-\frac{k_n}{k_{n-1}}xP_{n-1}(x)=c_0P_0(x)+c_1P_1(x)+\cdots+c_{n-1}P_{n-1}(x).$$

From (1) § 11, p. 85 we obtain $c_0=c_1=\cdots=c_{n-3}=0$. From $P_n(1)=1$, $P_n(-1)=(-1)^n$ [**85**] we then obtain c_{n-2} and c_{n-1}.

88. If there existed another polynomial $S_n^*(x)$ with the given properties we would have

$$\int_{-1}^{1}[S_n(x)-S_n^*(x)]^2\,dx=\int_{-1}^{1}S_n(x)[S_n(x)-S_n^*(x)]\,dx-\int_{-1}^{1}S_n^*(x)[S_n(x)-S_n^*(x)]\,dx$$
$$=S_n(1)-S_n^*(1)-[S_n(1)-S_n^*(1)]=0,$$

i.e. $S_n(x)=S_n^*(x)$. Now set

$$S_n(x)=\sum_{\nu=0}^{n}s_\nu P_\nu(x),\qquad K(x)=\sum_{\nu=0}^{n}t_\nu P_\nu(x).$$

If the equation

$$\int_{-1}^{1}S_n(x)K(x)\,dx=\sum_{\nu=0}^{n}\frac{2}{2\nu+1}s_\nu t_\nu=K(1)=\sum_{\nu=0}^{n}t_\nu$$

is to hold identically for $t_0, t_1, t_2, \ldots, t_n$, then we must have $s_\nu=(2\nu+1)/2$, i.e.

$$S_n(x)=\frac{1}{2}P_0(x)+\frac{3}{2}P_1(x)+\frac{5}{2}P_2(x)+\cdots+\frac{2n+1}{2}P_n(x).$$

Now let

$$(1-x)S_n(x)=u_0P_0(x)+u_1P_1(x)+\cdots+u_nP_n(x)+u_{n+1}P_{n+1}(x).$$

From

$$\int_{-1}^{1}(1-x)S_n(x)x^\nu\,dx=0,\qquad \nu=0, 1, 2, \ldots, n-1;\quad n\geqq 1$$

we deduce that $u_0=u_1=\cdots=u_{n-1}=0$. Setting now $x=1$ and $x=-1$, and observing that

$$P_n(1)=1,\qquad P_n(-1)=(-1)^n,\qquad S_n(1)=\frac{(n+1)^2}{2},\qquad S_n(-1)=(-1)^n\frac{n+1}{2},$$

we obtain

$$S_n(x)=\sum_{\nu=0}^{n}\frac{2\nu+1}{2}P_\nu(x)=\frac{n+1}{2}\,\frac{P_n(x)-P_{n+1}(x)}{1-x}$$

(Christoffel formula).

89. In **88** set $K(x)=(1-x)x^\nu$. Then we obtain

$$\int_{-1}^{1}(1-x)S_n(x)x^\nu\,dx=0,\qquad \nu=0, 1, 2, \ldots, n-1;\quad n\geqq 1.$$

We have [solution **88**]

$$\int_{-1}^{1}(1-x)[S_n(x)]^2\,dx=\frac{n+1}{2}\int_{-1}^{1}[P_n(x)-P_{n+1}(x)]\left(\sum_{\nu=0}^{n}\frac{2\nu+1}{2}P_\nu(x)\right)dx$$

$$=\frac{n+1}{2}\cdot\frac{2n+1}{2}\int_{-1}^{1}[P_n(x)]^2\,dx=\frac{n+1}{2}.$$

90. Integration by parts yields

$$\int_{-1}^{1}\left(\frac{d}{dx}(1-x^2)P_n'(x)\right)x^\nu\,dx=-\int_{-1}^{1}(1-x^2)P_n'(x)\cdot\nu x^{\nu-1}\,dx$$

$$=\int_{-1}^{1}P_n(x)\frac{d}{dx}((1-x^2)\nu x^{\nu-1})\,dx=0,$$

$$\nu=0,1,2,\ldots,n-1,$$

i.e. $(d/dx)(1-x^2)P_n'(x)=cP_n(x)$, $c=$constant. The constant c can be determined for example by comparing the terms of nth degree.

91. The coefficient of w^n in the expansion of $(1-2xw+w^2)^{-1/2}$ in ascending powers of w is evidently a polynomial of nth degree in x with a positive leading coefficient [§ 11 p. 85, condition (3)]. If we denote it by $P_n(x)$, we have identically in u and v

$$\sum_{k=0}^{\infty}\sum_{l=0}^{\infty}\int_{-1}^{1}P_k(x)P_l(x)\,dx\cdot u^kv^l=\int_{-1}^{1}\frac{dx}{\sqrt{1-2xu+u^2}\,\sqrt{1-2xv+v^2}}$$

$$=\frac{1}{\sqrt{uv}}\log\frac{1+\sqrt{uv}}{1-\sqrt{uv}}=\sum_{n=0}^{\infty}\frac{2}{2n+1}u^nv^n$$

[§ 11 p. 85, conditions (1), (2)]. The generating series may also be deduced directly from **84** [III **219**].

92. (a) Read III **219** in the reverse direction.

(b) [III **157**.] In the reverse direction the calculation proceeds as follows:

$$\sum_{n=0}^{\infty}\frac{1}{\pi}\int_{0}^{\pi}(x+\sqrt{x^2-1}\cos\varphi)^n\,d\varphi\cdot w^n=\frac{1}{\pi}\int_{0}^{\pi}\frac{d\varphi}{1-(x+\sqrt{x^2-1}\cos\varphi)w}.$$

This integral is equal to $(1-2xw+w^2)^{-1/2}$ for w sufficiently small and positive and for $x>1$ [III **149**, $n=0$].

(c) Set $F(x,w)=\sum_{n=0}^{\infty}P_n(x)w^n$. Then we have

$$\sum_{n=2}^{\infty}[nP_n(x)-(2n-1)xP_{n-1}(x)+(n-1)P_{n-2}(x)]w^{n-1}$$

$$=(1-2xw+w^2)\frac{\partial F}{\partial w}+(w-x)F\equiv 0.$$

(d)

$$\sum_{n=2}^{\infty}[(1-x^2)P_n''(x)-2xP_n'(x)+n(n+1)P_n(x)]w^n$$

$$=(1-x^2)\frac{\partial^2 F}{\partial x^2}-2x\frac{\partial F}{dx}+w\frac{\partial^2(wF)}{\partial w^2}\equiv 0.$$

93. From **91** we obtain

$$\frac{1}{\sqrt{1-2\cos\vartheta\cdot w+w^2}}=\frac{1}{\sqrt{1-e^{i\vartheta}w}}\,\frac{1}{\sqrt{1-e^{-i\vartheta}w}}$$

$$=\left(\sum_{k=0}^{\infty}\frac{1\cdot 3\cdots(2k-1)}{2\cdot 4\cdots 2k}e^{ik\vartheta}w^k\right)\left(\sum_{l=0}^{\infty}\frac{1\cdot 3\cdots(2l-1)}{2\cdot 4\cdots 2l}e^{-il\vartheta}w^l\right).$$

Hence, performing the multiplication [I **34**] and comparing coefficients, we obtain

$$P_n(\cos\vartheta)=g_0g_n\cos n\vartheta+g_1g_{n-1}\cos(n-2)\vartheta+g_2g_{n-2}\cos(n-4)\vartheta+\cdots$$
$$+g_ng_0\cos n\,\vartheta,$$

where we have used the abbreviations

$$g_0=1,\qquad g_n=\frac{1\cdot 3\cdots(2n-1)}{2\cdot 4\cdots 2n},\qquad n=1,2,3,\ldots.$$

We deduce that

$$|P_n(\cos\vartheta)|\leqq g_0g_n+g_1g_{n-1}+g_2g_{n-2}+\cdots+g_ng_0=P_n(1)=1.$$

The equality sign can hold only if $n\vartheta$, $(n-2)\vartheta,\ldots$ are simultaneously even or odd multiples of π, i.e. only if $\vartheta=k\pi$, where k is an integer.

94. If we set $x=1+\xi$, $\xi>0$, we have

$$1+\sum_{n=1}^{\infty}[P_n(x)-P_{n-1}(x)]w^n=\frac{1-w}{\sqrt{(1-w)^2-2\xi w}}=\frac{1}{\sqrt{1-2\xi\dfrac{w}{(1-w)^2}}}$$

$$=1+\sum_{n=1}^{\infty}\frac{1\cdot 3\cdots(2n-1)}{2\cdot 4\cdots 2n}\left(\frac{2\xi w}{(1-w)^2}\right)^n.$$

The power series $w(1-w)^{-2}=w+2w^2+3w^3+\cdots$ has all its coefficients positive.

95. [L. Fejér: Math. Ann. **67**, p. 83 (1909).] From **17** and III **157** we have

$$P_0(\cos\vartheta)+P_1(\cos\vartheta)+P_2(\cos\vartheta)+\cdots+P_n(\cos\vartheta)$$

$$=\frac{2}{\pi}\int_{\vartheta}^{\pi}\frac{\left(\sin(n+1)\dfrac{t}{2}\right)^2}{\sin\dfrac{t}{2}\sqrt{2(\cos\vartheta-\cos t)}}\,dt.$$

96. For $x\geqq 1$ we have $P_n(x)>0$. For $x<-1$, we have $\operatorname{sgn}P_n(x)=(-1)^n$, and $|P_n(x)|=|P_n(-x)|$ increases monotonically with n [**94**]. We have $P_n(-1)=(-1)^n$ [**85**].

97. A special case of II **140**: $a=-1$, $b=+1$, $f(x)=P_n(x)$.—Alternatively from **84** and V **58**, or from **85** and V **65**, or from **90** and III **34**, or from **90** and V **120**.

98. (a)

$$(1-x)^\alpha(1+x)^\beta P_n^{(\alpha,\beta)}(x)=\frac{(-1)^n}{2^nn!}\frac{d^n}{dx^n}(1-x)^{\alpha+n}(1+x)^{\beta+n}.$$

The coefficient of x^n in $P_n^{(\alpha,\beta)}$ is given by

$$k_n = \frac{1}{2^n}\binom{2n+\alpha+\beta}{n}.$$

(b)

$$(t-1)^n P_n^{(\alpha,\beta)}\left(\frac{t+1}{t-1}\right) = \sum_{\nu=0}^{n}\binom{n+\alpha}{\nu}\binom{n+\beta}{n-\nu}t^\nu.$$

It follows that

$$(-1)^n P_n^{(\alpha,\beta)}(-1) = \binom{n+\beta}{n}, \qquad P_n^{(\alpha,\beta)}(1) = \binom{n+\alpha}{n}.$$

(c)

$$P_n^{(\alpha,\beta)}(x) = \frac{1}{2\pi}\int_0^{2\pi}(x+\sqrt{x^2-1}\cos\varphi)^n\left(1+\sqrt{\frac{x+1}{x-1}}\,e^{i\varphi}\right)^\alpha\left(1+\sqrt{\frac{x-1}{x+1}}\,e^{i\varphi}\right)^\beta d\varphi,$$

if x lies in the plane slit along the interval -1, $+1$. Here those determinations of the square root or of the αth or the βth power respectively are to be chosen that are positive for $x>1$ and $\varphi=0$. For $\Re x \geqq 0$ we must have α integral (β arbitrary), for $\Re x \leqq 0$ we must have β integral (α arbitrary); for pure imaginary x the formula is true without any restrictions.

(d)

$$P_n^{(\alpha,\beta)}(x) = (A_n^{(\alpha,\beta)}x + B_n^{(\alpha,\beta)})P_{n-1}^{(\alpha,\beta)}(x) - C_n^{(\alpha,\beta)}P_{n-2}^{(\alpha,\beta)}(x),$$

$$A_n^{(\alpha,\beta)} = \frac{(2n+\alpha+\beta)(2n+\alpha+\beta-1)}{2n(n+\alpha+\beta)},$$

$$B_n^{(\alpha,\beta)} = \frac{\alpha^2-\beta^2}{2n}\,\frac{2n+\alpha+\beta-1}{(n+\alpha+\beta)(2n+\alpha+\beta-2)},$$

$$C_n^{(\alpha,\beta)} = \frac{(n+\alpha-1)(n+\beta-1)(2n+\alpha+\beta)}{n(n+\alpha+\beta)(2n+\alpha+\beta-2)}, \qquad n=2,3,4,\ldots.$$

(e) Let $S_n^{(\alpha,\beta)}$ be the polynomial of nth degree for which

$$\int_{-1}^{1}(1-x)^\alpha(1+x)^\beta S_n^{(\alpha,\beta)}(x)K(x)\,dx = K(1),$$

where $K(x)$ is an arbitrary polynomial of nth degree. We have

$$S_n^{(\alpha,\beta)}(x) = \sum_{\nu=0}^{n}\frac{2\nu+\alpha+\beta+1}{2^{\alpha+\beta+1}}\,\frac{\Gamma(\nu+\alpha+\beta+1)}{\Gamma(\alpha+1)\Gamma(\nu+\beta+1)}\,P_\nu^{(\alpha,\beta)}(x)$$

$$= \frac{1}{2^{\alpha+\beta}}\,\frac{n+\alpha+1}{2n+\alpha+\beta+2}\,\frac{\Gamma(n+\alpha+\beta+2)}{\Gamma(\alpha+1)\Gamma(n+\beta+1)}\,\frac{P_n^{(\alpha,\beta)}(x) - \dfrac{n+1}{n+\alpha+1}P_{n+1}^{(\alpha,\beta)}(x)}{1-x}$$

(Christoffel formula).

(f)

$$\int_{-1}^{1}(1-x)^{\alpha+1}(1+x)^\beta S_m^{(\alpha,\beta)}(x)S_n^{(\alpha,\beta)}(x)\,dx$$

$$= \begin{cases} 0, & \text{for } m \gtrless n, \\ \dfrac{2^{-\alpha-\beta}}{[\Gamma(\alpha+1)]^2(2n+\alpha+\beta+2)}\,\dfrac{\Gamma(n+\alpha+2)\Gamma(n+\alpha+\beta+2)}{\Gamma(n+1)\Gamma(n+\beta+1)}, & \text{for } m=n; \quad m,n=0,1,2,\ldots. \end{cases}$$

(g)

$$(1-x^2)P_n^{(\alpha,\beta)''}(x)+[\beta-\alpha-(\alpha+\beta+2)x]P_n^{(\alpha,\beta)'}(x)$$
$$+n(n+\alpha+\beta+1)P_n^{(\alpha,\beta)}(x)=0.$$

(h)

$$\frac{2^{\alpha+\beta}}{\sqrt{1-2xw+w^2}}(1-w+\sqrt{1-2xw+w^2})^{-\alpha}(1+w+\sqrt{1-2xw+w^2})^{-\beta}$$
$$=P_0^{(\alpha,\beta)}(x)+P_1^{(\alpha,\beta)}(x)w+P_2^{(\alpha,\beta)}(x)w^2+\cdots+P_n^{(\alpha,\beta)}(x)w^n+\cdots.$$

(i) The zeros of the Jacobi polynomials, are real and simple and lie in the interior of the interval $-1, 1$.

Prove similarly as in **84–91**, **97**. In the proof of (c) first write (b) in the form

$$P_n^{(\alpha,\beta)}(x)=\sum_{\nu=0}^{n}\binom{n+\alpha}{\nu}\binom{n+\beta}{n-\nu}\left(\frac{x-1}{2}\right)^{n-\nu}\left(\frac{x+1}{2}\right)^{\nu}$$
$$=\frac{1}{2\pi i}\oint\left(1+\frac{x+1}{2}z\right)^{n+\alpha}\left(1+\frac{x-1}{2}z\right)^{n+\beta}\frac{dz}{z^{n+1}}.$$

Integrate along the circular arc $|z|=2|x^2-1|^{-1/2}$. The integrand is continuous on this circular arc ($n\geqq 1$).

99. (a)

$$e^{-x}x^{\alpha}L_n^{(\alpha)}(x)=\frac{1}{n!}\frac{d^n}{dx^n}e^{-x}x^{n+\alpha}.$$

The coefficient of x^n in $L_n^{(\alpha)}(x)$ is given by

$$k_n=\frac{(-1)^n}{n!}.$$

(b)

$$L_n^{(\alpha)}(x)=\sum_{\nu=0}^{n}\binom{n+\alpha}{n-\nu}\frac{(-x)^{\nu}}{\nu!}.$$

It follows that

(c)

$$L_n^{(\alpha)}(0)=\binom{n+\alpha}{n}.$$

(d)

$$nL_n^{(\alpha)}(x)=(-x+2n+\alpha-1)L_{n-1}^{(\alpha)}(x)-(n+\alpha-1)L_{n-2}^{(\alpha)}(x),$$
$$n=2, 3, 4, \ldots.$$

(e) Let $S_n^{(\alpha)}(x)$ be the polynomial of nth degree for which

$$\int_0^{\infty} e^{-x}x^{\alpha}S_n^{(\alpha)}(x)K(x)\,dx=K(0),$$

$K(x)$ an arbitrary polynomial of nth degree. We have

$$S_n^{(\alpha)}(x)=\frac{1}{\Gamma(\alpha+1)}\sum_{\nu=0}^{n}L_\nu^{(\alpha)}(x)=\frac{n+\alpha+1}{\Gamma(\alpha+1)}\,\frac{L_n^{(\alpha)}(x)-\dfrac{n+1}{n+\alpha+1}L_{n+1}^{(\alpha)}(x)}{x}$$

(Christoffel formula).

(f)

$$\int_0^\infty e^{-x}x^{\alpha+1}S_m^{(\alpha)}(x)S_n^{(\alpha)}(x)\,dx=\begin{cases}0, & \text{if } m\gtrless n,\\ \dfrac{\alpha+1}{\Gamma(\alpha+1)}\dbinom{n+\alpha+1}{n}, & \text{if } m=n,\end{cases}$$

$$m, n=0, 1, 2, \ldots.$$

(g)

$$xL_n^{(\alpha)''}(x)+(\alpha+1-x)L_n^{(\alpha)'}(x)+nL_n^{(\alpha)}(x)=0.$$

(h)

$$\frac{e^{-\frac{xw}{1-w}}}{(1-w)^{\alpha+1}}=L_0^{(\alpha)}(x)+L_1^{(\alpha)}(x)w+L_2^{(\alpha)}(x)w^2+\cdots+L_n^{(\alpha)}(x)w^n\cdots.$$

(i) The zeros of the (generalized) Laguerre polynomials are real, positive and simple.

Proofs similar to those in **84, 85, 87–91, 97**.

100. (a)

$$e^{-\frac{x^2}{2}}H_n(x)=\frac{1}{n!}\frac{d^n}{dx^n}e^{-\frac{x^2}{2}}.$$

It follows from this by differentiation that

$$H_n'(x)-xH_n(x)=(n+1)H_{n+1}(x),$$

and we deduce that the coefficient of x^n in $H_n(x)$ is given by

$$k_n=\frac{(-1)^n}{n!}.$$

(d)

$$nH_n(x)=-xH_{n-1}(x)-H_{n-2}(x), \qquad n=2, 3, 4, \ldots.$$

(e) Let $S_n(x)$ be the polynomial of nth degree for which

$$\int_{-\infty}^{\infty} e^{-\frac{x^2}{2}}S_n(x)K(x)\,dx=K(0),$$

$K(x)$ an arbitrary polynomial of nth degree. We have

$$S_n(x)=\frac{1}{\sqrt{2\pi}}\sum_{\nu=0}^{n}(-1)^\nu\frac{(2\nu)!}{2^\nu\nu!}H_{2\nu}(x)$$

$$=\frac{(-1)^{p+1}}{\sqrt{2\pi}}\frac{(2p+1)!}{2^p p!}\frac{H_{2p+1}(x)}{x}, \qquad p=\left[\frac{n}{2}\right]$$

(Christoffel formula).

(g)

$$H_n''(x)-xH_n'(x)+nH_n(x)=0.$$

(h)

$$e^{-xw-\frac{w^2}{2}}=H_0(x)+H_1(x)w+H_2(x)w^2+\cdots+H_n(x)w^n+\cdots.$$

(i) The zeros of the Hermite polynomials are real and simple. The proofs of (a), (d), (e), (h), (i) are similar to those in **84**, **87**, **88**, **91**, **97**; deduce (g) from (a).

101. Follows from solution **98**(b). For, setting $(t+1)/(t-1)=1-\varepsilon$, $t=1-(2/\varepsilon)$ we have

$$\lim_{\beta\to+\infty}\binom{n+\beta}{n-\nu}\frac{t^{\nu}}{(t-1)^{n}}=\frac{(-x)^{n-\nu}}{(n-\nu)!}$$

[solution **99**(b)].

102. We have

$$\int_{-\infty}^{\infty} e^{-\frac{x^2}{2}}L_q^{(-1/2)}\left(\frac{x^2}{2}\right)x^{2k+1}\,dx=0, \qquad k=0,1,2,\ldots,q-1,$$

since the integrand is an odd function; also

$$\int_{-\infty}^{\infty} e^{-\frac{x^2}{2}}L_q^{(-1/2)}\left(\frac{x^2}{2}\right)x^{2k}\,dx=2^{k+1/2}\int_0^{\infty} e^{-y}L_q^{(-1/2)}(y)\,y^{k-1/2}\,dy=0,$$

$$k=0,1,2,\ldots,q-1,$$

by the definition of $L_q^{(-1/2)}(y)$. We thus have $L_q^{(-1/2)}(x^2/2)=\text{const. } H_{2q}(x)$. Similarly we show that $xL_q^{(1/2)}(x^2/2)=\text{const. } H_{2q+1}(x)$. We obtain the constant factor by comparing the coefficients of x^{2q} and x^{2q+1}, respectively [solution **99**(a), **100**(a)].

103. Setting

$$P(x)=\sum_{\nu=0}^{n} t_\nu\sqrt{\frac{2\nu+1}{2}}\,P_\nu(x),$$

we have

$$\int_{-1}^{1}[P(x)]^2\,dx=t_0^2+t_1^2+t_2^2+\cdots+t_n^2=1.$$

Furthermore from II **80** we obtain

$$[P(x)]^2\leqq\sum_{\nu=0}^{n}t_\nu^2\sum_{\nu=0}^{n}\frac{2\nu+1}{2}[P_\nu(x)]^2=\sum_{\nu=0}^{n}\frac{2\nu+1}{2}[P_\nu(x)]^2.$$

In this inequality the equality sign holds for an arbitrary $x=x_0$ only if $t_\nu=t\sqrt{(2\nu+1)/2}P_\nu(x_0)$, $\nu=0,1,2,\ldots,n$, where t is to be chosen such that the condition $t_0^2+t_1^2+t_2^2+\cdots+t_n^2=1$ is satisfied. Cf. also **93**.

104. Setting

$$P(x)=\sum_{\nu=0}^{n} t_\nu\sqrt{\frac{2}{\nu+1}}\,S_\nu(x)$$

we have

$$\int_{-1}^{1}(1-x)[P(x)]^2\,dx=t_0^2+t_1^2+t_2^2+\cdots+t_n^2=1 \qquad [\mathbf{89}].$$

Furthermore from II **80** we obtain

$$[P(x)]^2\leqq\sum_{\nu=0}^{n}t_\nu^2\sum_{\nu=0}^{n}\frac{2}{\nu+1}[S_\nu(x)]^2=\sum_{\nu=0}^{n}\frac{2}{\nu+1}[S_\nu(x)]^2.$$

We have [solution **88**]

$$S_n(1)=\frac{(n+1)^2}{2},\qquad S_n(-1)=(-1)^n\frac{n+1}{2}.$$

The bounds are attained for $t_\nu=t\sqrt{2/(\nu+1)}S_\nu(\pm 1)$, $\nu=0,1,2,\ldots,n$, where t is to be chosen in both cases such that the condition $t_0^2+t_1^2+t_2^2+\cdots+t_n^2=1$ is satisfied.

105. We have [**103**, solution **98**(e)]

$$\begin{aligned}\max\,[P(1)]^2&=S_n^{(\alpha,\beta)}(1)\\&=\frac{1}{2^{\alpha+\beta}}\frac{n+\alpha+1}{2n+\alpha+\beta+2}\frac{\Gamma(n+\alpha+\beta+2)}{\Gamma(\alpha+1)\Gamma(n+\beta+1)}\\&\quad\times\left(\frac{n+1}{n+\alpha+1}P_{n+1}^{(\alpha,\beta)\prime}(1)-P_n^{(\alpha,\beta)\prime}(1)\right).\end{aligned}$$

From solution **98**(g) and (b) we obtain

$$P_n^{(\alpha,\beta)\prime}(1)=\frac{n(n+\alpha+\beta+1)}{2(\alpha+1)}P_n^{(\alpha,\beta)}(1)=\frac{n(n+\alpha+\beta+1)}{2(\alpha+1)}\binom{n+\alpha}{n},$$

and hence

$$\begin{aligned}\max\,[P(1)]^2=S_n^{(\alpha,\beta)}(1)&=\frac{1}{2^{\alpha+\beta+1}}\frac{\Gamma(n+\alpha+2)\Gamma(n+\alpha+\beta+2)}{\Gamma(\alpha+1)\Gamma(\alpha+2)\Gamma(n+1)\Gamma(n+\beta+1)}\\&\sim\frac{1}{2^{\alpha+\beta+1}}\frac{n^{2\alpha+2}}{\Gamma(\alpha+1)\Gamma(\alpha+2)}.\end{aligned}$$

We have also

$$\begin{aligned}\max\,[P(-1)]^2=S_n^{(\beta,\alpha)}(1)&=\frac{1}{2^{\alpha+\beta+1}}\frac{\Gamma(n+\beta+2)\Gamma(n+\alpha+\beta+2)}{\Gamma(\beta+1)\Gamma(\beta+2)\Gamma(n+1)\Gamma(n+\alpha+1)}\\&\sim\frac{1}{2^{\alpha+\beta+1}}\frac{n^{2\beta+2}}{\Gamma(\beta+1)\Gamma(\beta+2)}.\end{aligned}$$

For $\alpha=1$, $\beta=0$ we obtain **104**.

106. We obtain as in **103** [**99**]

$$\max\,[P(0)]^2=S_n^{(\alpha)}(0)=\frac{\Gamma(n+\alpha+2)}{\Gamma(\alpha+1)\Gamma(\alpha+2)\Gamma(n+1)}\sim\frac{n^{\alpha+1}}{\Gamma(\alpha+1)\Gamma(\alpha+2)}.$$

107. We have [**103**, **100**]

$$\begin{aligned}\max\,[P(0)]^2=S_n(0)&=\frac{(-1)^{p+1}}{\sqrt{2\pi}}\frac{(2p+1)!}{2^p p!}H'_{2p+1}(0)\\&=\frac{1}{\sqrt{2\pi}}\frac{1\cdot 3\cdots(2p+1)}{2\cdot 4\cdots 2p},\qquad p=\left[\frac{n}{2}\right].\end{aligned}$$

By II **202** we have

$$\max\,[P(0)]^2=S_n(0)\sim\frac{1}{\pi}\sqrt{n}.$$

108. [F. Lukács: Math. Z. **2**, p. 299, 304 (1918).] Special case of **110**: $\alpha=\beta=0$. In the proof use **103, 104** but not **105**.

109. [F. Lukács, loc. cit. **108**.] It is sufficient to prove one inequality; then replace $P(x)$ by $-P(x)$. We may also assume that $m=0$. If $a<\xi<b$, then we have [**108**]

$$P(\xi)\leqq\frac{\alpha_n}{\xi-a}\int_a^\xi P(x)\,dx,\qquad P(\xi)\leqq\frac{\alpha_n}{b-\xi}\int_\xi^b P(x)\,dx.$$

Hence it follows that

$$P(\xi)\leqq\frac{\alpha_n}{b-a}\int_a^b P(x)\,dx,\qquad M\leqq\frac{\alpha_n}{b-a}\int_a^b P(x)\,dx.$$

110. We set [**47**]

$$P(x)=[A(x)]^2+(1-x)[B(x)]^2+(1+x)[C(x)]^2+(1-x^2)[D(x)]^2,$$

where $A(x)$, $B(x)$, $C(x)$, $D(x)$ are polynomials of degree $[n/2]=p$, $[(n-1)/2]=q-1$, $[(n-1)/2]=q-1$, $[n/2]-1=p-1$, respectively. By **105** we thus have, setting $S_n^{(\alpha,\beta)}(1)=S_n^{(\alpha,\beta)}$,

$$\begin{aligned}P(1)&=[A(1)]^2+2[C(1)]^2\\ &\leqq S_p^{(\alpha,\beta)}\int_{-1}^{1}(1-x)^\alpha(1+x)^\beta[A(x)]^2\,dx+2S_{q-1}^{(\alpha,\beta+1)}\int_{-1}^{1}(1-x)^\alpha(1+x)^{\beta+1}[C(x)]^2\,dx\\ &\leqq \text{Max}\,[S_p^{(\alpha,\beta)},2S_{q-1}^{(\alpha,\beta+1)}]\int_{-1}^{1}(1-x)^\alpha(1+x)^\beta P(x)\,dx\\ &\leqq \max\,[S_p^{(\alpha,\beta)},2S_{q-1}^{(\alpha,\beta+1)}].\end{aligned}$$

111. We set [**45**]

$$P(x)=[A(x)]^2+[B(x)]^2+x\{[C(x)]^2+[D(x)]^2\},$$

where $A(x)$ and $B(x)$ are polynomials of degree $[n/2]=p$, and $C(x)$ and $D(x)$ are polynomials of degree $[(n-1)/2]$. By **106** we thus have, setting $S_p^{(\alpha)}(0)=S_p^{(\alpha)}$,

$$\begin{aligned}P(0)=[A(0)]^2+[B(0)]^2&\leqq S_p^{(\alpha)}\int_0^\infty e^{-x}x^\alpha\{[A(x)]^2+[B(x)]^2\}\,dx\\ &\leqq S_p^{(\alpha)}\int_0^\infty e^{-x}x^\alpha P(x)\,dx=S_p^{(\alpha)}.\end{aligned}$$

112. Special case of **111**: $\alpha=0$.

113. Apply **112** to

$$\frac{P(x+\xi)}{\int_0^\infty e^{-x}P(x+\xi)\,dx}$$

We have

$$\begin{aligned}P(\xi)&\leqq\left(\left[\frac{n}{2}\right]+1\right)\int_0^\infty e^{-x}P(x+\xi)\,dx=\left(\left[\frac{n}{2}\right]+1\right)e^\xi\int_\xi^\infty e^{-x}P(x)\,dx\\ &\leqq\left(\left[\frac{n}{2}\right]+1\right)e^\xi.\end{aligned}$$

Part Seven. Determinants and Quadratic Forms

1. Interchanging two numbers in the numbering of the vertices is equivalent to the simultaneous interchange of two rows and two columns. If we give opposite vertices of the octahedron numbers that differ by 3, then the determinant

$$\begin{vmatrix} 0 & 1 & 1 & 0 & 1 & 1 \\ 1 & 0 & 1 & 1 & 0 & 1 \\ 1 & 1 & 0 & 1 & 1 & 0 \\ 0 & 1 & 1 & 0 & 1 & 1 \\ 1 & 0 & 1 & 1 & 0 & 1 \\ 1 & 1 & 0 & 1 & 1 & 0 \end{vmatrix} = 0.$$

The determinant for the tetrahedron is $= -3$, and for the hexahedron is $= 9$. [For further developments see F. Harary: (1) SIAM Review **4**, pp. 202–210 (1962) and (2) Graph Theory . Reading: Addison Wesley 1969.]

1.1. Add to the first column all the other columns.

2. Add the first row, multiplied by $-a_1$, to the second. Mathematical induction yields

$$(a_1-b_1)(a_2-b_2)\dots(a_n-b_n).$$

3. [Cauchy: Exercices d'analyse et de phys. math. **2**, second edition, pp. 151–159. Paris: Bachelier, 1841.] Subtract the last row from the $n-1$ preceding rows. Then from the columns we can take out the factors

$$\frac{1}{a_n+b_1}, \quad \frac{1}{a_n+b_2}, \quad \dots, \quad \frac{1}{a_n+b_{n-1}}, \quad \frac{1}{a_n+b_n},$$

and from the rows the factors

$$a_n-a_1, \quad a_n-a_2, \quad \dots, \quad a_n-a_{n-1}, \quad 1,$$

respectively, in front of the determinant. In the determinant remaining, subtract the last column from all the preceding columns and take out from the columns and rows the factors

$$b_n-b_1, \quad b_n-b_2, \quad \dots, \quad b_n-b_{n-1}, \quad 1,$$

and

$$\frac{1}{a_1+b_n}, \quad \frac{1}{a_2+b_n}, \quad \dots, \quad \frac{1}{a_{n-1}+b_n}, \quad 1$$

respectively. There remains an $(n-1)$-rowed corner minor of the given determinant. Use mathematical induction.

4. A special case of **3**: $a_\lambda=\lambda$, $b_\mu=\mu+\alpha$. We have

$$D_n(\alpha)=[1!2!\ldots(n-1)!]^2\,\frac{\Gamma(2+\alpha)\Gamma(3+\alpha)\ldots\Gamma(n+1+\alpha)}{\Gamma(n+2+\alpha)\Gamma(n+3+\alpha)\ldots\Gamma(2n+1+\alpha)}.$$

5. Row-by-row multiplication [cf. also II **51**, VIII **2**] of

$$\left|a_\lambda^{n-1}, -\binom{n-1}{1}a_\lambda^{n-2}, \binom{n-1}{2}a_\lambda^{n-3},\ldots,(-1)^{n-1}\right|\cdot\left|1, b_\mu, b_\mu^2,\ldots,b_\mu^{n-1}\right|.$$

6.
$$\begin{Vmatrix} p_0 & p_1\alpha_1 & p_2\alpha_1^2 & \ldots \\ p_0 & p_1\alpha_2 & p_2\alpha_2^2 & \ldots \\ \ldots & \ldots & \ldots & \ldots \\ p_0 & p_1\alpha_n & p_2\alpha_n^2 & \ldots \end{Vmatrix}\cdot\begin{Vmatrix} 1 & \beta_1 & \beta_1^2 & \ldots \\ 1 & \beta_2 & \beta_2^2 & \ldots \\ \ldots & \ldots & \ldots & \ldots \\ 1 & \beta_n & \beta_n^2 & \ldots \end{Vmatrix}$$

$$=\sum\cdots\sum p_{\nu_1}p_{\nu_2}\ldots p_{\nu_n}\begin{vmatrix} \alpha_1^{\nu_1} & \alpha_1^{\nu_2} & \ldots & \alpha_1^{\nu_n} \\ \alpha_2^{\nu_1} & \alpha_2^{\nu_2} & \ldots & \alpha_2^{\nu_n} \\ \ldots & \ldots & \ldots & \ldots \\ \alpha_n^{\nu_1} & \alpha_n^{\nu_2} & \ldots & \alpha_n^{\nu_n} \end{vmatrix}\begin{vmatrix} \beta_1^{\nu_1} & \beta_1^{\nu_2} & \ldots & \beta_1^{\nu_n} \\ \beta_2^{\nu_1} & \beta_2^{\nu_2} & \ldots & \beta_2^{\nu_n} \\ \ldots & \ldots & \ldots & \ldots \\ \beta_n^{\nu_1} & \beta_n^{\nu_2} & \ldots & \beta_n^{\nu_n} \end{vmatrix}.$$

The summation here is extended over all sets $\nu_1, \nu_2, \ldots, \nu_n$ of non-negative integers where $0\leqq\nu_1<\nu_2<\cdots<\nu_n$. All the terms in the last sum are non-negative. There are also positive terms among them if at least n of the numbers $p_0, p_1, p_2, \ldots$ do not vanish.

7. [A. Hurwitz; cf. O. Hölder: Leipz. Ber. **65**, pp. 110–120 (1913).] If in the determinant $D(x)$ obtained by introducing x in the specified manner we subtract the first column from the $n-1$ succeeding columns, then x remains only in the first column; thus $D(x)$ is linear in x, $D(x)=D+x\Delta$. For $x=-a$ and $x=-b$, $D(x)$ reduces to the product of the elements in the principal diagonal:

$$D-\Delta a=(r_1-a)(r_2-a)\ldots(r_n-a)=f(a),$$
$$D-\Delta b=(r_1-b)(r_2-b)\ldots(r_n-b)=f(b).$$

In the case $b=a$ [M. Roberts, problem: Nouv. Annls. Math. Series 2, **3**, p. 139 (1864)] the determinant $=f(a)-af'(a)$, as can also easily be seen otherwise.

8. [T. Muir: Amer. Math. Monthly, **29**, p. 12 (1922).] We have in general

$$\frac{\partial(\varphi f_1, \varphi f_2,\ldots,\varphi f_n)}{\partial(x_1, x_2,\ldots,x_n)}=\varphi^{n-1}\begin{vmatrix} \varphi & f_1 & f_2 & \ldots & f_n \\ -\dfrac{\partial\varphi}{\partial x_1} & \dfrac{\partial f_1}{\partial x_1} & \dfrac{\partial f_2}{\partial x_1} & \ldots & \dfrac{\partial f_n}{\partial x_1} \\ -\dfrac{\partial\varphi}{\partial x_2} & \dfrac{\partial f_1}{\partial x_2} & \dfrac{\partial f_2}{\partial x_2} & \ldots & \dfrac{\partial f_n}{\partial x_2} \\ \ldots & \ldots & \ldots & \ldots & \ldots \\ -\dfrac{\partial\varphi}{\partial x_n} & \dfrac{\partial f_1}{\partial x_n} & \dfrac{\partial f_2}{\partial x_n} & \ldots & \dfrac{\partial f_n}{\partial x_n} \end{vmatrix}.$$

Applying this identity, the given determinant is $=\Delta^3$, multiplied by

$$\begin{vmatrix} ad-bc & a & b & c & d \\ -d & 1 & 0 & 0 & 0 \\ c & 0 & 1 & 0 & 0 \\ b & 0 & 0 & 1 & 0 \\ -a & 0 & 0 & 0 & 1 \end{vmatrix}=3\Delta.$$

9. [Problem, Collège d'Aberystwyth: Mathesis, Series 2, **3**, p. 79 (1893). Solution by Retali et al: Mathesis, Series 2, **3**, p. 172 (1893).] The determinant in question for real l, m, n is associated with the quadratic form

$$(\rho-2)(x^2+y^2+z^2)+2\left(\frac{x}{l}+\frac{y}{m}+\frac{z}{n}\right)(lx+my+nz),$$

which for $\rho=2$ is equal to

$$2\left(\frac{x}{l}+\frac{y}{m}+\frac{z}{n}\right)(lx+my+nz)=\frac{1}{2}\left[\left(\frac{1}{l}+l\right)x+\left(\frac{1}{m}+m\right)y+\left(\frac{1}{n}+n\right)z\right]^2$$
$$-\frac{1}{2}\left[\left(\frac{1}{l}-l\right)x+\left(\frac{1}{m}-m\right)y+\left(\frac{1}{n}-n\right)z\right]^2.$$

Its rank for $\rho=2$ is thus less than 3. (Its rank $=2$, except if $l^2=m^2=n^2$; it is then $=1$.) Now set $\rho=(\rho-2)+2$ and expand in powers of $\rho-2$. We obtain

$$(\rho-2)^3+(2+2+2)(\rho-2)^2+\left(\begin{vmatrix} 2 & \frac{l}{m}+\frac{m}{l} \\ \frac{l}{m}+\frac{m}{l} & 2 \end{vmatrix}+\cdots\right)(\rho-2)+0$$
$$=(\rho-2)^3+6(\rho-2)^2+(9-P)(\rho-2),$$

where $P=(1/l^2+1/m^2+1/n^2)(l^2+m^2+n^2)$. The other two factors are

$$\rho+1-\sqrt{P}, \qquad \rho+1+\sqrt{P}.$$

10. From the system of equations

$$(-1)^{n-1}S_n+x_\nu(-1)^{n-2}S_{n-1}+x_\nu^2(-1)^{n-3}S_{n-2}+\cdots+x_\nu^{n-1}S_1=x_\nu^n,$$
$$(\nu=1,2,\ldots,n)$$

obtain the value of the unknown $(-1)^{q-1}S_q$.

11. From considerations of continuity, we can restrict ourselves to the case in which $a_0z^n+a_1z^{n-1}+\cdots+a_n$ has n distinct zeros. Denoting any such zero by z and setting

$$a_0z^{n-1}=x_0,\quad a_1z^{n-2}=x_1,\quad a_2z^{n-3}=x_2,\quad \ldots,\quad a_{n-2}z=x_{n-2},\quad a_{n-1}=x_{n-1},$$

we see that $x_0, x_1, \ldots, x_{n-1}$ satisfy the homogeneous relations

$$\left(z+\frac{a_1}{a_0}\right)x_0+\frac{a_2}{a_1}x_1+\frac{a_3}{a_2}x_2+\cdots+\frac{a_{n-1}}{a_{n-2}}x_{n-2}+\frac{a_n}{a_{n-1}}x_{n-1}=0,$$

$$-\frac{a_1}{a_0}x_0+zx_1=0,\qquad -\frac{a_2}{a_1}x_1+zx_2=0,\qquad \ldots,\qquad -\frac{a_{n-1}}{a_{n-2}}x_{n-2}+zx_{n-1}=0,$$

whose determinant must be $=0$. The given determinant is thus a polynomial of nth degree in z, with leading coefficient 1, that has the same zeros as $a_0z^n+a_1z^{n-1}+\cdots+a_n$.

11.1. By considering the coefficient of z^k for $k=0, 1, 2, \ldots, n$, we obtain a system of $n+1$ linear equations for the unknowns $b_0, b_1, b_2, \ldots, b_n$:

$$\begin{aligned}
&a_0b_0 &&=1,\\
&a_1b_0+a_0b_1 &&=0,\\
&a_2b_0+a_1b_1+a_0b_2 &&=0,\\
&\cdots\cdots\cdots\cdots\cdots\cdots\cdots\cdots\cdots\cdots\\
&a_nb_0+a_{n-1}b_1+a_{n-2}b_2+\cdots+a_0b_n &&=0.
\end{aligned}$$

Solve for b_n.

11.2. [G. Pólya, problem: Amer. Math. Monthly, **54**, p. 107 (1947).] Compute the coefficient of z^n in the expansion of $[(1+z)^a-xz]^{-1}$ first from **11.1** then directly as follows:

$$\begin{aligned}
\frac{1}{(1+z)^a-xz}&=\frac{(1+z)^{-a}}{1-xz(1+z)^{-a}}\\
&=\sum_{k=0}^{\infty} x^k z^k (1+z)^{-(k+1)a}\\
&=\sum_{k=0}^{\infty}\sum_{n=k}^{\infty} x^k z^n \binom{-(k+1)a}{n-k}\\
&=\sum_{n=0}^{\infty}\sum_{k=0}^{\infty} z^n \binom{-(k+1)a}{n-k} x^k\\
&=\sum_{n=0}^{\infty} z^n \sum_{k=0}^{n} (-1)^{n-k}\binom{(k+1)a+n-k-1}{n-k} x^k
\end{aligned}$$

11.3. Analogously to **11.2**,

$$\begin{aligned}
\frac{1}{e^z-xz}&=\frac{e^{-z}}{1-xze^{-z}}\\
&=\sum_{k=0}^{\infty} x^k z^k e^{-(k+1)z}\\
&=\sum_{k=0}^{\infty}\sum_{n=k}^{\infty} x^k z^n \frac{(-1)^{n-k}(k+1)^{n-k}}{(n-k)!}\\
&=\sum_{n=0}^{\infty} (-1)^n z^n \sum_{k=0}^{n} \frac{(-1)^k (k+1)^{n-k}}{(n-k)!} x^k.
\end{aligned}$$

Also limiting case of **11.2** for $a\to\infty$.

12. We are concerned with the rank of the two matrices

$$\begin{pmatrix} 0 & -a_3 & a_2\\ a_3 & 0 & -a_1\\ -a_2 & a_1 & 0\\ a_1 & a_2 & a_3 \end{pmatrix}, \qquad \begin{pmatrix} 0 & -a_3 & a_2 & b_1\\ a_3 & 0 & -a_1 & b_2\\ -a_2 & a_1 & 0 & b_3\\ a_1 & a_2 & a_3 & c \end{pmatrix}.$$

The matrix with three columns contains four determinants of third order equal to

$$a_1(a_1^2+a_2^2+a_3^2), \quad -a_2(a_1^2+a_2^2+a_3^2), \quad a_3(a_1^2+a_2^2+a_3^2), \quad 0,$$

respectively. Hence it has either rank 0 or rank 3. In the former case the compatibility condition is $b_1=b_2=b_3=c=0$, in the latter case it is $a_1b_1+a_2b_2+a_3b_3=0$.

If the four equations are compatible, then in the former case x_1, x_2, x_3 are completely undetermined, in the latter case they are completely determined. Viewed in terms of vector multiplication the result is evident.

13. We have identically in z

$$1+u_1^{(n)}z+u_2^{(n)}z^2+\cdots+u_n^{(n)}z^n=\left(1-\frac{z}{a_1}\right)\left(1-\frac{z}{a_2}\right)\cdots\left(1-\frac{z}{a_n}\right).$$

Since the infinite product $\prod_{n=1}^{\infty}[1-(z/a_n)]$ is uniformly convergent in every finite domain, the same holds for the sequence of polynomials

$$1+u_1^{(n)}z+u_2^{(n)}z^2+\cdots+u_n^{(n)}z^n, \qquad n=1,2,3,\ldots.$$

From this it follows that all the limits

$$\lim_{n\to\infty} u_k^{(n)}=u_k, \qquad k=1,2,3,\ldots$$

exist, and we have

$$1+u_1z+u_2z^2+u_3z^3+\cdots=\prod_{n=1}^{\infty}\left(1-\frac{z}{a_n}\right). \qquad \text{[I } \mathbf{179}\text{].}$$

In general this function differs from the given function by an exponential factor $e^{g(z)}$, where $g(z)$ is an entire function.

14. The following determinants must be interpreted as determinants of $2n$th order (and not for instance as determinants of second order whose elements are $|a_{\lambda\mu}|$, $|-b_{\lambda\mu}|$, $|b_{\lambda\mu}|$, $|a_{\lambda\mu}|$!) We have

$$\begin{vmatrix}(a_{\lambda\mu}) & (-b_{\lambda\mu})\\ (b_{\lambda\mu}) & (a_{\lambda\mu})\end{vmatrix}=\begin{vmatrix}(a_{\lambda\mu}+ib_{\lambda\mu}) & (-b_{\lambda\mu}+ia_{\lambda\mu})\\ (b_{\lambda\mu}) & (a_{\lambda\mu})\end{vmatrix}$$

$$=\begin{vmatrix}(a_{\lambda\mu}+ib_{\lambda\mu}) & (0)\\ (b_{\lambda\mu}) & (a_{\lambda\mu}-ib_{\lambda\mu})\end{vmatrix}$$

$$=|a_{\lambda\mu}+ib_{\lambda\mu}|\cdot|a_{\lambda\mu}-ib_{\lambda\mu}|=A^2+B^2,$$

if we set $|a_{\lambda\mu}+ib_{\lambda\mu}|=A+iB$, where A, B are real. The vanishing of A^2+B^2 is equivalent to the simultaneous vanishing of A and B.

15. [G. Rados, problem: Math. és phys. lapok, **15**, p. 389 (1906). Solution by M. Fekete, et al: Math. és phys. lapok, **16**, p. 310 (1907).] The product of the three terms $a_{11}a_{22}a_{33}$, $a_{12}a_{23}a_{31}$, $a_{13}a_{21}a_{32}$ is equal in absolute value but opposite in sign to the product of the remaining three terms.

16. [G. Pólya, problem: Arch. Math. Phys. Series 3, **20**, p. 271 (1913). Solution by G. Szegö: Arch. Matn. Phys. Series 3, **21**, p. 291 (1913).] If the signs in question were $\varepsilon_{\lambda\mu}$, then according to the supposed law we would have that the determinant $|\varepsilon_{\lambda\mu}|_{\lambda,\mu=1,2,\ldots,n}=n!$, in contradiction to Hadamard's theorem for determinants $|\varepsilon_{\mu\lambda}|_{\lambda,\mu=1,2,\ldots,n}\leqq n^{\frac{n}{2}}$ [Kowalewski, p. 460], since $n^n<n!^2$ for $n\geqq 3$. Note that it is sufficient to carry out the proof only for the case of the determinant of third order [**15**].

17. If the denominator is $z^q-c_1z^{q-1}-c_2z^{q-2}-\cdots-c_q$, then the product

$$(a_0+a_1z+a_2z^2+\cdots+a_nz^n+\cdots)(z^q-c_1z^{q-1}-\cdots-c_q)$$

reduces to a polynomial of degree $p-1$. Hence in the product series the coefficient of z^{q+n}

$$(*) \qquad a_n - a_{n+1}c_1 - a_{n+2}c_2 - \cdots - a_{n+q}c_q = 0$$

for $n=p-q$, $p-q+1, \ldots$, if $p \geqq q$ and for $n=0, 1, 2, \ldots$, if $p \leqq q$. The compatibility of the $q+1$ successive linear equations which we obtain from (*) for $n=k$, $k+1, k+2, \ldots, k+q$, implies that $A_k^{(q+1)}=0$.

18. From $A_n^{(q+1)}=0$, $A_n^{(q)} \neq 0$, it follows that $L_{n+q}(x)$ is linearly dependent on $L_n(x), L_{n+1}(x), \ldots, L_{n+q-1}(x)$ [Kowalewski, p. 53]. Hence $L_n(x)$ for $n \geqq d$ is linearly dependent on $L_d(x), L_{d+1}(x), \ldots, L_{d+q-1}(x)$, and the equation $L_n(x)=0$ is satisfied by a solution of the q simultaneous equations

$$L_d(x)=0, \qquad L_{d+1}(x)=0, \quad \ldots, \qquad L_{d+q-1}(x)=0.$$

But these equations, since $A_{d+1}^{(q)} \neq 0$, have a solution of the form

$$x_0=1, \qquad x_1=-c_1, \qquad x_2=-c_2, \quad \ldots, \qquad x_q=-c_q.$$

This implies the vanishing of the coefficients of z^{q+d}, $z^{q+d+1}, \ldots$ in the product

$$(a_0+a_1z+a_2z^2+\cdots)(z^q-c_1z^{q-1}-c_2z^{q-2}-\cdots-c_q).$$

19. [Kowalewski, p. 80, 109.]

20. From the assumptions it follows by **19** that

$$(A_{m+1}^{(q)})^2 = A_m^{(q)}A_{m+2}^{(q)}, \qquad (A_{m+2}^{(q)})^2 = A_{m+1}^{(q)}A_{m+3}^{(q)}, \ldots,$$
$$(A_{m+t-1}^{(q)})^2 = A_{m+t-2}^{(q)}A_{m+t}^{(q)}, \qquad (A_{m+t}^{(q)})^2 = A_{m+t-1}^{(q)}A_{m+t+1}^{(q)}.$$

Thus the vanishing of one of the t determinants of order q entails the vanishing of the two neighbouring determinants (or the neighbouring determinant). More easily seen from **22**.

21. Obvious. Example: $A_0^{(3)}$.

22. [A. Stoll.] (1) from **19**, (2) and (3) from **19** and (1). Observe the "cross-shaped" arrangement of the five determinants occurring in **19** in the array **21**. To see the validity of the statement on the boundary, insert a_{-1} before a_0, a_1, $a_2, \ldots$ and a corresponding slanting line before the array **21**.

23. [Cf. É. Borel: Darboux Bull, Series 2, **18**, pp. 22–25 (1894).] If in the first row of the array **21** there are only finitely many elements different from 0, then the power series $a_0+a_1z+\cdots$ reduces to a rational *entire* function. Otherwise by repeatedly using the result **20** we obtain that there exist numbers d and q, $1 \leqq q \leqq k$, for which we have the situation discussed in **18**.

24. [L. Kronecker: Monatsber. Akad. Berlin, pp. 566–567 (1881).] By repeated application of **22** (1), reduce to **23**.

25. [G. Pólya: Math. Ann. **77**, p. 507 (1916).] By repeated application of **22** (1), reduce to **23**. Represent the determinantal conditions **23**, **24**, **25** by means of "paths" in the array **21**.

26. Cf. **27**.

27. If the rank is finite, then all the determinants mentioned in **23** vanish for sufficiently large k, and hence the power series is rational. If the power series is rational, employ the notation of **17**, and consider the linear forms

$$\Lambda_n(x) = a_nx_0 + a_{n+1}x_1 + a_{n+2}x_2 + \cdots.$$

The number of variables is arbitrarily (not infinitely!) large. By the equations (*) in solution **17** we have

$$\Lambda_n = c_1\Lambda_{n+1} + c_2\Lambda_{n+2} + \cdots + c_q\Lambda_{n+q}$$

for $n=d, d+1, d+2, \ldots$. Hence the forms $\Lambda_d, \Lambda_{d+1}, \ldots, \Lambda_{d+\nu}, \nu \geqq q$, depend on the last q of them. As thus any $q+1$ of these forms are linearly dependent, all determinants of order $q+1$ contained in $\mathfrak{H}_d$ must vanish.

28. Cf. **29**.

29. In accordance with the statement of **28**, let

$$A_0^{(p)} \neq 0, \qquad A_0^{(p+1)} = A_0^{(p+2)} = A_0^{(p+3)} = \cdots = 0,$$
$$A_1^{(p)} = A_2^{(p-1)} = A_3^{(p-2)} = \cdots = A_{p-q}^{(q+1)} = 0, \qquad A_{p-q+1}^{(q)} \neq 0.$$

Moreover assume for definiteness that

$$0 < q < p.$$

From these assumptions, if we observe the position of the above determinants in the array **21**, it follows by **22** (1), that

$$(**) \qquad \begin{cases} A_{p-q}^{(q)} \neq 0, & A_{p-q+1}^{(q)} \neq 0, & A_{p-q+2}^{(q)} \neq 0, & \ldots, \\ A_{p-q-1}^{(q+1)} \neq 0, & A_{p-q}^{(q+1)} = 0, & A_{p-q+1}^{(q+1)} = 0, & \ldots. \end{cases}$$

I.e., the border of an infinite trapezium-shaped region in the array **21** consisting entirely of zeros must consist of determinants that are all different from zero.

By **23** the degree of the denominator $\leqq q$; cf. second line of (**).

By **17** the degree of the denominator $\geqq q$; cf. first line of (**).

The net rank is $\leqq$ rank of $\mathfrak{H}_{p-q}$, thus $\leqq q$, by solution **27**.

The net rank is $\geqq q$; cf. first line of (**).

By **18** the degree of the numerator $\leqq q+(p-q)-1$; cf. (**).

By **17** the degree of the numerator $> q+(p-q-1)-1$, since $A_{p-q-1}^{(q+1)} \neq 0$.

The rank of $\mathfrak{H}_{p-q}$ is precisely $=q$, thus that of $\mathfrak{H}_0$ is certainly $\leqq q+(p-q)$.

The gross rank, $=$rank of $\mathfrak{H}_0$, is $\geqq p$, since $A_0^{(p)} \neq 0$.

30. [E. Beke: Math. és term. ért. **34**, p. 25 (1916).] The property of the power series $\sum_{n=0}^{\infty} (a_n/n!)z^n$ in question holds if and only if the power series $\sum_{n=0}^{\infty} a_n z^n$ represents a rational function. [**24**, **26**.] In both cases we are concerned with the same recurrence formula for the coefficients, $a_0, a_1, a_2, \ldots$.

31. [G. Pólya: Proc. Lond. Math. Soc. Series 2, **21**, pp. 25–26 (1922).] By addition of rows and columns. For the case $Q_n(z) = (1-z)^n$, $n=0, 1, 2, \ldots$, cf. Kowalewski, p. 112.

32. The numerator is of degree $\leqq q-1$, hence we have

$$a_n c_q + a_{n+1}c_{q-1} + a_{n+2}c_{q-2} + \cdots + a_{n+q-1}c_1 = a_{n+q} \qquad \text{for} \quad n=0, 1, 2, \ldots.$$

If therefore the λth row (column) of $\mathfrak{A}_m$ is multiplied by the μth column of $\mathfrak{C}$, we obtain $a_{m+\lambda+\mu-1}$, $\lambda, \mu = 1, 2, \ldots, q$, that is $\mathfrak{A}_m\mathfrak{C} = \mathfrak{A}_{m+1}$.

33. Let the rank be r and the determinant

$$\begin{vmatrix} a_{1\nu_1} & a_{1\nu_2} & \cdots & a_{1\nu_r} \\ a_{2\nu_1} & a_{2\nu_2} & \cdots & a_{2\nu_r} \\ \cdots & \cdots & \cdots & \cdots \\ a_{r\nu_1} & a_{r\nu_2} & \cdots & a_{r\nu_r} \end{vmatrix} \neq 0,$$

$0 \leqq \nu_1 < \nu_2 < \cdots < \nu_r$. If $c_1 f_1(z) + c_2 f_2(z) + \cdots + c_r f_r(z) \equiv 0$, then in particular the coefficients of $z^{\nu_1}, z^{\nu_2}, \ldots, z^{\nu_r}$ on the left-hand side $=0$. From the resulting homogeneous linear equations with *non-vanishing* determinant it follows that $c_1 = c_2 = \cdots = c_r = 0$. If $m > r$, set $a_{1\nu}x_1 + a_{2\nu}x_2 + \cdots + a_{r\nu}x_r + a_{r+1,\nu}x_{r+1} \equiv L_\nu(x)$. Any number of linear forms $L_1, L_2, \ldots, L_n$ is by assumption linearly dependent on $L_{\nu_1}, L_{\nu_2}, \ldots, L_{\nu_r}$ [Kowalewski, p. 53]. There exists a set of values $x_1 = c_1, x_2 = c_2, \ldots, x_r = c_r, x_{r+1} = -1$, satisfying the simultaneous equations

$$(*) \qquad L_{\nu_1}(x) = 0, \qquad L_{\nu_2}(x) = 0, \qquad \ldots, \qquad L_{\nu_r}(x) = 0.$$

This set of values makes any form $L_\nu(x)$ vanish, i.e. we have identically in z

$$c_1 f_1(z) + c_2 f_2(z) + \cdots + c_r f_r(z) = f_{r+1}(z).$$

34. From **33** and the definitions. Observe in particular that there exists an N, such that the matrices which arise from $\mathfrak{M}$ (defined p. 99, preceding **33**) by omitting $N, N+1, N+2, \ldots$ columns respectively, all have the same rank, $=$net rank of $\mathfrak{M}$.

35. [I. Schur: J. reine angew. Math. **140**, p. 14 (1911).] Proof under the more stringent assumptions. There exists an orthogonal matrix $(l_{\lambda\mu})$ such that the first form can be decomposed as follows:

$$\sum_{\lambda=1}^{n} \sum_{\mu=1}^{n} a_{\lambda\mu} x_\lambda x_\mu = \sum_{\nu=1}^{n} h_\nu (l_{\nu 1} x_1 + l_{\nu 2} x_2 + \cdots + l_{\nu n} x_n)^2,$$

where $h_1, h_2, \ldots, h_n$ are positive numbers, uniquely determined by the form, its *eigenvalues* [Kowalewski, p. 275]. From

$$a_{\lambda\mu} = \sum_{\nu=1}^{n} h_\nu l_{\nu\lambda} l_{\nu\mu}, \qquad \lambda, \mu = 1, 2, \ldots, n,$$

setting $h = \min(h_1, h_2, \ldots, h_n)$, it follows that

$$\begin{aligned} \sum_{\lambda=1}^{n} \sum_{\mu=1}^{n} a_{\lambda\mu} b_{\lambda\mu} x_\lambda x_\mu &= \sum_{\nu=1}^{n} h_\nu \left\{ \sum_{\lambda=1}^{n} \sum_{\mu=1}^{n} b_{\lambda\mu} (l_{\nu\lambda} x_\lambda)(l_{\nu\mu} x_\mu) \right\} \\ &\geqq h \sum_{\lambda=1}^{n} \sum_{\mu=1}^{n} b_{\lambda\mu} x_\lambda x_\mu \left(\sum_{\nu=1}^{n} l_{\nu\lambda} l_{\nu\mu} \right) \\ &= h(b_{11} x_1^2 + b_{22} x_2^2 + \cdots + b_{nn} x_n^2). \end{aligned}$$

36.

$$\sum_{\lambda=1}^{n} \sum_{\mu=1}^{n} e^{a_{\lambda\mu}} x_\lambda x_\mu = \sum_{k=0}^{\infty} \frac{1}{k!} \sum_{\lambda=1}^{n} \sum_{\mu=1}^{n} a_{\lambda\mu}^k x_\lambda x_\mu,$$

whence the first part of the theorem follows by **35**. Because of the positivity we have

$$\sum_{\lambda=1}^{n} \sum_{\mu=1}^{n} a_{\lambda\mu} x_\lambda x_\mu = \xi_1^2 + \xi_2^2 + \cdots + \xi_l^2,$$

where

$$\begin{aligned} \xi_1 &= \alpha_1' x_1 + \alpha_2' x_2 + \cdots + \alpha_n' x_n, \\ \xi_2 &= \alpha_1'' x_1 + \alpha_2'' x_2 + \cdots + \alpha_n'' x_n, \\ &\cdots\cdots\cdots\cdots\cdots\cdots\cdots\cdots \\ \xi_l &= \alpha_1^{(l)} x_1 + \alpha_2^{(l)} x_2 + \cdots + \alpha_n^{(l)} x_n. \end{aligned}$$

If in the matrix $(a_{\lambda\mu})$ no two rows are identical, then no two of the n l-dimensional vectors $(\alpha_\nu', \alpha_\nu'', \ldots, \alpha_\nu^{(l)})$, $\nu=1, 2, \ldots, n$ are equal. There thus exist l numbers β', $\beta'', \ldots, \beta^{(l)}$, such that, setting $\beta'^2+\beta''^2+\cdots+\beta^{(l)2}=1$ and $\alpha_\nu'\beta'+\alpha_\nu''\beta''+\cdots+\alpha_\nu^{(l)}\beta^{(l)}=\gamma_\nu$, no two of the n numbers $\gamma_1, \gamma_2, \ldots, \gamma_n$ are equal; i.e., geometrically, the projections of the n different vectors onto the unit vector $(\beta', \beta'', \ldots, \beta^{(l)})$ are all different; or $(\beta', \beta'', \ldots, \beta^{(l)})$ lies outside certain $\frac{1}{2}n(n-1)$ planes. We obtain by a suitably chosen orthogonal transformation of the $\xi_1, \xi_2, \ldots, \xi_l$ into $\eta_1, \eta_2, \ldots, \eta_l$, where

$$\eta_1=\beta'\xi_1+\beta''\xi_2+\cdots+\beta^{(l)}\xi_l=\gamma_1x_1+\gamma_2x_2+\cdots+\gamma_nx_n,$$

$$\sum_{\lambda=1}^{n}\sum_{\mu=1}^{n} a_{\lambda\mu}x_\lambda x_\mu=\eta_1^2+\eta_2^2+\cdots+\eta_l^2=\sum_{\lambda=1}^{n}\sum_{\mu=1}^{n}\gamma_\lambda\gamma_\mu x_\lambda x_\mu+\sum_{\lambda=1}^{n}\sum_{\mu=1}^{n}\alpha_{\lambda\mu}'x_\lambda x_\mu;$$

the two forms on the right-hand side are $\geqq 0$. The difference $\exp(\gamma_\lambda\gamma_\mu+a_{\lambda\mu}')-\exp(\gamma_\lambda\gamma_\mu)$ is made up of products of powers of $\gamma_\lambda\gamma_\mu$ and $a_{\lambda\mu}'$ with positive coefficients. From this it follows by **35**, similarly as in the case of the first part of the assertion that we have already proved, that, setting

$$\sum_{\lambda=1}^{n}\sum_{\mu=1}^{n} e^{a_{\lambda\mu}}x_\lambda x_\mu=\sum_{\lambda=1}^{n}\sum_{\mu=1}^{n} e^{\gamma_\lambda\gamma_\mu}x_\lambda x_\mu+\sum_{\lambda=1}^{n}\sum_{\mu=1}^{n} b_{\lambda\mu}x_\lambda x_\mu,$$

the second form on the right-hand side is positive; the first is in fact definite [V **76**].

37. Similarly as in **36** [**35**, V **86**]. That the restrictions are not superfluous we see from the example of the matrix

$$\begin{pmatrix} 1 & -1 \\ -1 & 1 \end{pmatrix}.$$

38. [R. Remak: Math. Ann. **72**, p. 153 (1912); cf. also A. Hurwitz: Math. Ann. 73, p. 173 (1913).] Since we can consider $f(x+x_0)$, x_0 arbitrary, instead of $f(x)$, the condition given in the problem is equivalent to

$$A(f)=a_0f(0)+\frac{a_1}{1!}f'(0)+\frac{a_2}{2!}f''(0)+\cdots+\frac{a_n}{n!}f^{(n)}(0)\geqq 0$$

(or >0, respectively).

By VI **44** it suffices to set

$$f(x)=(t_0+t_1x+\cdots+t_nx^n)^2=\sum_{\lambda=0}^{n}\sum_{\mu=0}^{n} t_\lambda t_\mu x^{\lambda+\mu}$$

with arbitrary $t_0, t_1, t_2, \ldots, t_n$. We thus have

$$A(f)=\sum_{\lambda=0}^{n}\sum_{\mu=0}^{n} A(x^{\lambda+\mu})t_\lambda t_\mu=\sum_{\lambda=0}^{n}\sum_{\mu=0}^{n} a_{\lambda+\mu}t_\lambda t_\mu.$$

39. [Cf. G. Pólya: Math. és. term. ért. **32**, pp. 662–665 [1914).] Since we may consider $f(x+x_0)$, $x_0\geqq 0$, instead of $f(x)$, the condition in the problem is equivalent to

$$A(f)=a_0f(0)+\frac{a_1}{1!}f'(0)+\frac{a_2}{2!}f''(0)+\cdots+\frac{a_n}{n!}f^{(n)}(0)\geqq 0,$$

(or >0, respectively).

Set [VI **45**]

$$f(x)=(t_0+t_1x+\cdots+t_px^p)^2+x(u_0+u_1x+\cdots+u_{q-1}x^{q-1})^2,$$

$$p=\left[\frac{n}{2}\right],\qquad q=\left[\frac{n+1}{2}\right].$$

We then have

$$A(f)=\sum_{\lambda=0}^{p}\sum_{\mu=0}^{p}a_{\lambda+\mu}t_\lambda t_\mu+\sum_{\lambda=0}^{q-1}\sum_{\mu=0}^{q-1}a_{\lambda+\mu+1}u_\lambda u_\mu.$$

40. If we set firstly $f(x)=(1-x)\sum_{\lambda=0}^{q-1}\sum_{\mu=0}^{q-1}t_\lambda t_\mu(x-1)^{\lambda+\mu}$, $x=1$, and secondly $f(x)=(1+x)\sum_{\lambda=0}^{q-1}\sum_{\mu=0}^{q-1}t_\lambda t_\mu(x+1)^{\lambda+\mu}$, $x=-1$, where $q=[(n+1)/2]$, then we obtain from the assumption

$$-\sum_{\lambda=0}^{q-1}\sum_{\mu=0}^{q-1}a_{\lambda+\mu+1}t_\lambda t_\mu\geqq 0,\qquad \sum_{\lambda=0}^{q-1}\sum_{\mu=0}^{q-1}a_{\lambda+\mu+1}t_\lambda t_\mu\geqq 0,$$

and thus $a_1=a_2=\cdots=a_{2q-1}=0$. For odd n the assertion clearly follows. For even n we have in addition

$$a_0(t_0+t_1x+\cdots+t_px^p)^2+a_nt_p^2\geqq 0,\tag{1}$$

$$a_0(1-x^2)(u_0+u_1x+\cdots+u_{p-1}x^{p-1})^2-a_nu_{p-1}^2\geqq 0\tag{2}$$

for all values of $t_0, t_1, \ldots, t_p, u_0, u_1, \ldots, u_{p-1}$, $p=n/2$ and for $-1\leqq x\leqq 1$. From (1) for $t_0=t_1=\cdots=t_{p-1}=0$, $t_p=1$, $x=0$, it follows that $a_n\geqq 0$. From (2) for $x=1$, $u_{p-1}=1$, it follows that $a_n\leqq 0$.

41. [Cf. loc. cit. **39**, pp. 665–667. Cf. also M. Fujiwara: Tôhoku Math. J. **6**, pp. 20–26 (1914–1915).] The two sets of numbers $a_0, a_1, \ldots, a_{2n}$ and $b_0, b_1, b_2, \ldots, b_{2n}$ have the property defined in **38**. If $f(x)$ is admissible in the sense of **38**, then

$$f^*(x)=a_0f(x)+\frac{a_1}{1!}f'(x)+\frac{a_2}{2!}f''(x)+\cdots+\frac{a_{2n}}{(2n)!}f^{(2n)}(x)$$

is also admissible. Thus we have

$$b_0f^*(x)+\frac{b_1}{1!}f^{*\prime}(x)+\frac{b_2}{2!}f^{*\prime\prime}(x)+\cdots+\frac{b_{2n}}{(2n)!}f^{*(2n)}(x)$$

$$=c_0f(x)+\frac{c_1}{1!}f'(x)+\frac{c_2}{2!}f''(x)+\cdots+\frac{c_{2n}}{(2n)!}f^{(2n)}(x)\geqq 0\quad(\text{or }>0\text{ respectively})$$

for all values of x. Thus the system $c_0, c_1, c_2, \ldots, c_{2n}$ also has the same property.

42. [Cf. loc. cit. **39**, pp. 667–668.] Follows in a similar fashion from **39** as does **41** from **38**. Set $n=2m$ or $n=2m+1$ respectively.

43. [Cf. O. Szász: Math. Z. **1**, pp. 150–152 (1918).] Since we may consider $g(\vartheta+\vartheta_0)$, ϑ_0 arbitrary, instead of $g(\vartheta)$, the condition is equivalent to

$$A(g)=\sum_{\nu=-n}^{n}c_\nu\gamma_\nu\geqq 0\qquad(\text{or }>0\text{, respectively}).$$

Setting [VI **40**]

$$g(\vartheta)=|x_0+x_1e^{i\vartheta}+x_2e^{2i\vartheta}+\cdots+x_ne^{in\vartheta}|^2,$$

we obtain

$$A(g)=\sum_{\lambda=0}^{n}\sum_{\mu=0}^{n} A(e^{i(\lambda-\mu)\vartheta})x_\lambda\bar{x}_\mu=\sum_{\lambda=0}^{n}\sum_{\mu=0}^{n} c_{\mu-\lambda}x_\lambda\bar{x}_\mu.$$

43.1.

$$-\frac{G'(z)}{G(z)}=2\alpha z-\beta+\sum_{k=1}^{\infty}\left(\frac{1}{\beta_k-z}-\frac{1}{\beta_k}\right)$$
$$=-\beta+2\alpha z+\sum_{k=1}^{\infty}\left(\frac{z}{\beta_k^2}+\frac{z^2}{\beta_k^3}+\frac{z^3}{\beta_k^4}+\cdots\right)$$
$$=s_1+s_2z+s_3z^2+\cdots.$$

Hence

$$s_2=2\alpha+\sum_{k=1}^{\infty}\frac{1}{\beta_k^2},$$

$$s_m=\sum_{k=1}^{\infty}\frac{1}{\beta_k^m}\qquad \text{for}\quad m=3,4,5,\ldots.$$

The quadratic form

$$\sum_{\lambda=1}^{n}\sum_{\mu=1}^{n} s_{\lambda+\mu}x_\lambda x_\mu=2\alpha x_1^2+\sum_{\lambda=1}^{n}\sum_{\mu=1}^{n}\sum_{k=1}^{\infty}\frac{x_\lambda}{\beta_k^\lambda}\frac{x_\mu}{\beta_k^\mu}$$
$$=2\alpha x_1^2+\sum_{k=1}^{\infty}\left(\frac{x_1}{\beta_k}+\frac{x_2}{\beta_k^2}+\cdots+\frac{x_n}{\beta_k^n}\right)^2$$

is not only positive, but positive definite (unless $x_1=x_2=\cdots=x_n=0$ only a finite number of squares included under the last summation sign can vanish). The positive definiteness of the quadratic form is equivalent to the positivity of the determinants.

(The theorem proved here was discovered by A. Hurwitz with a view to investigating the Riemann hypothesis. The converse theorem was proved by J. Grommer: J. reine angew. Math. **144**, pp. 114–166 (1914), as a particular case of a more general fact. For another proof see N. Kritikos: Math. Ann. **81**, pp. 97–118 (1920).)

43.2. [G. Pólya: J. reine angew. Math. **145**, pp. 224–249 (1915), especially p. 235, theorem IV.] By refining the demonstration of V **65.1**, we find: If the polynomial $P(x)\geqq 0$ for all real values of x but does not vanish identically, then the polynomial

$$Q(x)=c_0P(x)+\frac{c_1}{1!}P'(x)+\frac{c_2}{2!}P''(x)+\cdots>0$$

for all real values of x (for details see op. cit.). Apply **38** and pass to the determinants as in **43.1**.

44. [I. Schur, problem: Arch. Math. Phys. Series 3, **19**, p. 276, (1912). Solution by G. Pólya: Arch. Math. Phys. Series 3, **24**, p. 369 (1916).]

45. [I. Schur, problem: Arch. Math. Phys. Series 3, **27**, p. 162 (1918).] Cf. **46**.

46. [I. Schur, problem: Arch. Math. Phys. Series 3, **27**, p. 163 (1918).] Consider four cases: (a) an arbitrary determinant; (b) elements in the principal diagonal $=0$, the remaining elements arbitrary; (c) a symmetric determinant [**45**];

(d) a symmetric determinant with elements in the principal diagonal $=0$ [**46**]. The notation for the numbers is summarized in the following table:

	Independent elements	Different terms in the expansion Positive	Negative		
(a)	n^2	S'_n	S''_n	$S'_n+S''_n=S_n,$	$S'_n-S''_n=D_n,$
(b)	n^2-n	Σ'_n	Σ''_n	$\Sigma'_n+\Sigma''_n=\Sigma_n,$	$\Sigma'_n-\Sigma''_n=\Delta_n,$
(c)	$\frac{n^2+n}{2}$	s'_n	s''_n	$s'_n+s''_n=s_n,$	$s'_n-s''_n=d_n,$
(d)	$\frac{n^2-n}{2}$	σ'_n	σ''_n	$\sigma'_n+\sigma''_n=\sigma_n,$	$\sigma'_n-\sigma''_n=\delta_n.$

We associate with the term $a_{1k_1}a_{2k_2}\ldots a_{nk_n}$ in the expansion the permutation

$$\begin{pmatrix}1 & 2 & \ldots & n\\ k_1 & k_2 & \ldots & k_n\end{pmatrix}.$$

If the term in the expansion contains the product $a_{\alpha\beta}a_{\beta\gamma}\ldots a_{\chi\lambda}a_{\lambda\alpha}$, then the associated permutation contains the cycle $(\alpha\beta\gamma\ldots\chi\lambda)$; if the above term in the expansion belongs to a symmetric determinant, then the permutation associated with another term of equal value contains *either* the cycle $(\alpha\beta\gamma\ldots\chi\lambda)$ *or* the inverse cycle $(\lambda\chi\ldots\gamma\beta\alpha)$.

The number of permutations of n elements, that contain k_1 one-term cycles, k_2 two-term cycles, k_3 three-term cycles, ... is known to be [cf. Riordan, Introduction to Combinatorial Analysis, p. 67; New York: Wiley 1958]

$$\frac{n!}{k_1!1^{k_1}\cdot k_2!2^{k_2}\cdot k_3!3^{k_3}\ldots}=Z_{k_1k_2k_3\ldots},$$

where $1k_1+2k_2+3k_3+\cdots=n$. If this number is multiplied by $+1$ or -1 according as the permutation in question is even or odd, then we obtain

$$(-1)^{k_2+k_4+k_6+\cdots}Z_{k_1k_2k_3\ldots}=\frac{n!}{k_1!1^{k_1}\cdot k_2!(-2)^{k_2}\cdot k_3!3^{k_3}\cdot k_4!(-4)^{k_4}\ldots}$$

[Kowalewski, p. 16] (applicable in the calculation of D_n, Δ_n). If permutations that arise from each other by replacing certain cycles by the *inverse cycles* are not regarded as different, then the number of different ones is given by

$$2^{-k_3-k_4-k_5-\cdots}Z_{k_1k_2k_3\ldots}=\frac{n!}{k_1!1^{k_1}\cdot k_2!2^{k_2}\cdot k_3!6^{k_3}\cdot k_4!8^{k_4}\ldots}$$

(applicable in the calculation of s_n, d_n, σ_n, δ_n). We have clearly

$$S_n=\Sigma Z_{k_1k_2k_3\ldots}, \qquad D_n=\Sigma(-1)^{k_2+k_4+k_6+\cdots}Z_{k_1k_2k_3\ldots},$$

$$\Sigma_n=\Sigma^* Z_{0k_2k_3\ldots}, \qquad \Delta_n=\Sigma^*(-1)^{k_2+k_4+k_6+\cdots}Z_{0k_2k_3\ldots},$$

$$s_n=\Sigma 2^{-k_3-k_4-\cdots}Z_{k_1k_2k_3\ldots}, \qquad d_n=\Sigma(-1)^{k_2+k_4+\cdots}2^{-k_3-k_4-\cdots}Z_{k_1k_2k_3\ldots},$$

$$\sigma_n=\Sigma^* 2^{-k_3-k_4-\cdots}Z_{0k_2k_3\ldots}, \qquad \delta_n=\Sigma^*(-1)^{k_2+k_4+\cdots}2^{-k_3-k_4-\cdots}Z_{0k_2k_3\ldots}.$$

Σ is extended over those non-negative sets of number $k_1, k_2, k_3, \ldots$ for which

$k_1+2k_2+3k_3+\cdots=n$, Σ^* over those for which $k_1=0$, $2k_2+3k_3+\cdots=n$. We have

$$\frac{S_n x^n}{n!}=\sum\frac{x^{k_1}}{k_1!1^{k_1}}\cdot\frac{x^{2k_2}}{k_2!2^{k_2}}\cdot\frac{x^{3k_3}}{k_3!3^{k_3}}\cdots,$$

$$\sum_{n=0}^{\infty}\frac{S_n x^n}{n!}=\sum_{k_1=0}^{\infty}\frac{x^{k_1}}{k_1!1^{k_1}}\cdot\sum_{k_2=0}^{\infty}\frac{x^{2k_2}}{k_2!2^{k_2}}\cdot\sum_{k_3=0}^{\infty}\frac{x^{3k_3}}{k_3!3^{k_3}}\cdots$$

$$=e^x\cdot e^{\frac{x^2}{2}}\cdot e^{\frac{x^3}{3}}\cdots=e^{x+\frac{x^2}{2}+\frac{x^3}{3}+\cdots}$$

$$=\frac{1}{1-x}=1+x+x^2+\cdots,$$

from which it follows that $S_n=n!$, which serves as a check on our deductions. In the same way we obtain

$$\sum_{n=0}^{\infty}\frac{D_n x^n}{n!}=e^{x-\frac{x^2}{2}+\frac{x^3}{3}-\cdots}=1+x,\qquad D_n=0\quad\text{for}\quad n\geqq 2,$$

which yields a further check.

$$\sum_{n=0}^{\infty}\frac{\Sigma_n x^n}{n!}=e^{\frac{x^2}{2}+\frac{x^3}{3}+\cdots}=\frac{e^{-x}}{1-x},\qquad \frac{\Sigma_n}{n!}=1-\frac{1}{1!}+\frac{1}{2!}-\cdots+\frac{(-1)^n}{n!}$$

[VIII **23**].

$$\sum_{n=0}^{\infty}\frac{\Delta_n x^n}{n!}=e^{-\frac{x^2}{2}+\frac{x^3}{3}-\frac{x^4}{4}+\cdots}=(1+x)e^{-x},\qquad \Delta_n=(-1)^n(1-n);$$

special case of **7** for $r_1=r_2=\cdots=r_n=0$, $a=b=1$.

$$\sum_{n=0}^{\infty}\frac{s_n x^n}{n!}=e^{x+\frac{x^2}{2}+\frac{1}{2}\left(\frac{x^3}{3}+\frac{x^4}{4}+\cdots\right)}=\frac{1}{\sqrt{1-x}}\,e^{\frac{x}{2}+\frac{x^2}{4}},$$

$$\sum_{n=0}^{\infty}\frac{d_n x^n}{n!}=e^{x-\frac{x^2}{2}+\frac{1}{2}\left(\frac{x^3}{3}-\frac{x^4}{4}+\cdots\right)}=\sqrt{1+x}\,e^{\frac{x}{2}-\frac{x^2}{4}},$$

$$\sum_{n=0}^{\infty}\frac{\sigma_n x^n}{n!}=e^{\frac{x^2}{2}+\frac{1}{2}\left(\frac{x^3}{3}+\frac{x^4}{4}+\cdots\right)}=\frac{1}{\sqrt{1-x}}\,e^{-\frac{x}{2}+\frac{x^2}{4}},$$

$$\sum_{n=0}^{\infty}\frac{\delta_n x^n}{n!}=e^{-\frac{x^2}{2}+\frac{1}{2}\left(\frac{x^3}{3}-\frac{x^4}{4}+\cdots\right)}=\sqrt{1+x}\,e^{-\frac{x}{2}-\frac{x^2}{4}}.$$

The last four lines satisfy the differential equations

$$2(1-x)y'=(2-x^2)y,\qquad 2(1+x)y'=(2-x^2)y,$$
$$2(1-x)y'=x(2-x)y,\qquad 2(1+x)y'=-x(2+x)y,$$

respectively, from which we deduce the recursion formulae which were to be proved. The formulae for the limits are obtained from I **178** for

$$f(x)=e^{\frac{x}{2}+\frac{x^2}{4}},\quad e^{\frac{x}{2}-\frac{x^2}{4}},\quad e^{-\frac{x}{2}+\frac{x^2}{4}},\quad\text{and}\quad e^{-\frac{x}{2}-\frac{x^2}{4}},\qquad\text{respectively,}$$

$$b_n=\frac{1\cdot 3\ldots(2n-1)}{2\cdot 4\ldots 2n}\sim\frac{1}{\sqrt{\pi n}}\quad\text{and}\quad(-1)^{n-1}\frac{1\cdot 3\ldots(2n-3)}{2\cdot 4\ldots 2n}\sim\frac{(-1)^{n-1}}{2\sqrt{\pi}\cdot n^{\frac{3}{2}}},$$

respectively [II **202**], and

$$q=1 \quad \text{and} \quad -1, \quad \text{respectively.}$$

46.1. Definition of the product of two matrices.

46.2. [Result found in answering a question of Sami Beraha.] Let a_{ij} and b_{ij} be the entries of A and B, respectively. For $j=1, 2, \ldots, n$ there exist i, $1 \leqq i \leqq n$ and $i \neq j$, and such that

$$a_{i1}b_{1j}+a_{i2}b_{2j}+\cdots+a_{in}b_{nj}=0.$$

All entries b_{kj} in the jth column cannot be 0, since the determinant of B is not 0, and so there exists in this column at least one positive, and at least one negative, entry, $j=1, 2, \ldots, n$.

If $a>0$ and $A=I+aZ$, then $B=I+bZ$ with

$$b=-\frac{a}{1+na}>-1$$

and it has exactly n positive entries.

If $a>1$ and $A=-I+aZ$ then $B=-I+bZ$ with

$$b=\frac{a}{na-1}<1$$

and it has exactly n^2-n positive entries.

47. [I. Schur, problem: Arch. Math. Phys. Series 3, **27**, p. 163 (1918).] Setting

$$\varphi=\frac{1}{1-x_1x_2\ldots x_n}\bar{\varphi},$$

we have

$$\Sigma \pm \varphi(x_{\lambda_1}, x_{\lambda_2}, \ldots, x_{\lambda_n})=\frac{1}{1-x_1x_2\ldots x_n}\sum_{\nu=1}^{n}(-1)^{n-\nu}[\Sigma_\nu \pm \bar{\varphi}(x_{\mu_1}, x_{\mu_2}, \ldots, x_{\mu_{n-1}})],$$

where Σ_ν is extended over the permutations of $x_1, x_2, \ldots, x_{\nu-1}, x_{\nu+1}, \ldots, x_n$, and the determination of signs is carried out analogously to that in the sum on the left-hand side.

Assuming then that the assertion is correct if n is replaced by $n-1$, the identity to be proved has the form

$$\Delta\Phi=\frac{1}{1-x_1x_2\ldots x_n}\sum_{\nu=1}^{n}(-1)^{n-\nu}\Delta_\nu\Phi_\nu,$$

where Δ_ν and Φ_ν are formed from $x_1, x_2, \ldots, x_{\nu-1}, x_{\nu+1}, \ldots, x_n$ in an analogous manner as Δ and Φ are from $x_1, x_2, \ldots, x_n$; Φ is to be considered in the form of a product as given at the end of the problem while the Φ_ν contain fewer factors. Now apply VI **69**, noting the equations (notation and assumptions as given there, $a_0=1$)

$$\frac{\Delta_\nu}{\Delta}=\frac{1}{(x-x_\nu)\ldots(x_{\nu-1}-x_\nu)(x_\nu-x_{\nu+1})\ldots(x_\nu-x_n)}=\frac{(-1)^{\nu-1}}{f'(x_\nu)},$$

$$\frac{\Phi_\nu}{\Phi}=(1-x_\nu)(1-x_\nu x_1)\ldots(1-x_\nu x_{\nu-1})(1-x_{\nu+1}x_\nu)\ldots(1-x_nx_\nu)=\frac{x_\nu^n f(x_\nu^{-1})}{1+x_\nu}.$$

48. [P. Epstein, problem: Arch. Math. Phys. Series 3, **8**, p. 262 (1905). From $\alpha_\rho x_\lambda = \sum_{\mu=1}^{n} a_{\lambda\mu} x_\mu$, $\lambda=1, 2, \ldots, n$, observing that

$$\sum_{\lambda=1}^{n} a_{\lambda\mu}\chi_{\lambda\sigma}(z) = z\chi_{\mu\sigma}(z) + \varepsilon_{\mu\sigma}\chi(z),$$

$\varepsilon_{\mu\sigma}=1$ for $\mu=\sigma$, $\varepsilon_{\mu\sigma}=0$ for $\mu\gtrless\sigma$, we deduce the equations

$$(\alpha_\rho - z)\sum_{\lambda=1}^{n} x_\lambda \chi_{\lambda\sigma}(z) = x_\sigma \chi(z), \qquad \sigma=1, 2, \ldots, n.$$

49. [M. Riesz.] Cf. **50**.

50. The property required in the problem is equivalent to the relations

$$\sum_{\nu=0}^{n} g\left(\frac{\pi}{n+1}(p-\nu+\alpha)\right) g\left(\frac{\pi}{n+1}(\nu-q+\beta)\right) = g\left(\frac{\pi}{n+1}(p-q+\alpha+\beta)\right).$$

Cf. VI **15**. In **49** (if we replace x_p by $(-1)^p x_p$ and y_p by $(-1)^p y_p$) we have

$$g(\vartheta) = \frac{1}{n+1}\frac{\sin(n+1)\vartheta}{\sin\vartheta}.$$

51. The determinant $D(\alpha)$ of S_α satisfies the functional equation

$$D(\alpha)D(\beta) = D(\alpha+\beta).$$

Moreover $D(\alpha)$ is real, continuous and periodic (with period $2n+2$), thus $D(\alpha)=0$ or 1. The first case holds if and only if the system of linear equations

$$\sum_{q=0}^{n} g\left(\frac{\pi}{n+1}(p-q)\right) x_q = 0, \qquad p=0, 1, \ldots, n,$$

has solutions that are different from $x_q=0$, $q=0, 1, \ldots, n$. This is certainly the case if $x_0, x_1, \ldots, x_n$ satisfy the following equations

$$(*) \qquad \sum_{q=0}^{n} x_q e^{ikq\frac{\pi}{n+1}} = 0, \qquad \sum_{q=0}^{n} x_q e^{-ikq\frac{\pi}{n+1}} = 0,$$

where k runs through all the values for which the coefficient of $\cos k\vartheta$ in $g(\vartheta)$ is different from 0. (For $k=0$ only one of the two equations is to be retained.) By solution VI **15**, k can take either only even or only odd values; the number of the coefficients above is thus $\leqq 1+[n/k]$ and $\leqq [(n+1)/2]$ respectively. The number of equations (*) is in the first case $\leqq 1+2[n/2] \leqq n+1$, in the second case $\leqq 2[(n+1)/2] \leqq n+1$. The equality sign in these inequalities holds if and only if

$$g(\vartheta) = \frac{1}{n+1}\frac{\sin(n+1)\vartheta}{\sin\vartheta} \qquad [49]$$

and n is even or odd respectively. Apart from this case the number of equations (*) is always less than $n+1$, i.e. less than the number of unknowns. There thus exist solutions of the stated kind.

In **49** we obtain the constant value of the determinant if we let $\alpha \to 0$.

52. [M. Riesz.] Setting

$$g(\vartheta)=\frac{1}{n+1}\frac{\sin(n+1)\vartheta}{\sin\vartheta},$$

we are concerned with the sum

$$\sum_{p=0}^{n}\sum_{k,l=0}^{n}(-1)^{k+l}x_kx_l g\left(\frac{\pi}{n+1}(p-k+\alpha)\right)g\left(\frac{\pi}{n+1}(p-l+\alpha)\right)$$
$$=\sum_{k,l=0}^{n}(-1)^{k+l}x_kx_l\sum_{p=0}^{n}g\left(\frac{\pi}{n+1}(k-\alpha+p)\right)g\left(\frac{\pi}{n+1}(p+\alpha-l)\right).$$

By VI **15** this is

$$=\sum_{k,l=0}^{n}(-1)^{k+l}x_kx_l g\left(\frac{\pi}{n+1}(k-l)\right),$$

i.e. $\sum_{p=0}^{n} y_p^2$ is independent of α. Set $\alpha=0$.

53. [I. Schur.] The equations to be proved are:

$$u_1^2+u_2^2+\cdots+u_n^2+u_n^2=u_nu_{n+2},$$
$$u_1^2+u_2^2+\cdots+u_n^2 \qquad =u_nu_{n+1},$$
$$\frac{u_{n-1}}{u_{n+1}}+u_n^2\left(\frac{1}{u_nu_{n+2}}+\frac{1}{u_{n+1}u_{n+3}}+\frac{1}{u_{n+2}u_{n+4}}+\cdots\right)=1,$$
$$-\frac{1}{u_{n+2}}+u_{n+1}\left(\frac{1}{u_{n+1}u_{n+3}}+\frac{1}{u_{n+2}u_{n+4}}+\frac{1}{u_{n+3}u_{n+5}}+\cdots\right)=0, \qquad n=1,2,3,\ldots.$$

From the definition of the u_n it follows that

$$\sum_{\nu=1}^{n}u_\nu^2=\sum_{\nu=1}^{n}u_\nu(u_{\nu+1}-u_{\nu-1})=\sum_{\nu=1}^{n}u_\nu u_{\nu+1}-\sum_{\nu=1}^{n-1}u_\nu u_{\nu+1}=u_nu_{n+1},$$

i.e. the second equation. The first results if we set in the second $u_{n+1}=u_{n+2}-u_n$. Observe moreover that

$$\frac{1}{u_nu_{n+2}}+\frac{1}{u_{n+1}u_{n+3}}+\frac{1}{u_{n+2}u_{n+4}}+\cdots$$
$$=\frac{1}{u_{n+1}}\left(\frac{1}{u_n}-\frac{1}{u_{n+2}}\right)+\frac{1}{u_{n+2}}\left(\frac{1}{u_{n+1}}-\frac{1}{u_{n+3}}\right)+\frac{1}{u_{n+3}}\left(\frac{1}{u_{n+2}}-\frac{1}{u_{n+4}}\right)+\cdots$$
$$=\frac{1}{u_{n+1}u_n}.$$

54. [E. C. Titchmarsh: Proc. Lond. Math. Soc. Series 2, **22**, No. 5, III (1924).] Clearly we have

$$\sum_{n=-\infty}^{+\infty}\left(\frac{1}{\lambda+n-\alpha}-\frac{1}{\mu+n-\alpha}\right)=0$$

for $\lambda\lessgtr\mu$; $\lambda,\mu=\cdots,-1,0,1,\ldots$. As is well known, we have

$$\sum_{n=-\infty}^{+\infty}\frac{1}{(m+n-\alpha)^2}=\sum_{n=-\infty}^{+\infty}\frac{1}{(n-\alpha)^2}=\left(\frac{\pi}{\sin\alpha\pi}\right)^2.$$

54.1. The system of equations

$$\begin{aligned} &1 &&=1\\ &1+x &&=y\\ &1+2x+x^2 &&=y^2\\ &1+3x+3x^2+x^3 &&=y^3\\ &\cdots\cdots\cdots\cdots \end{aligned}$$

implies the system

$$\begin{aligned} &1 &&=1\\ &-1+y &&=x\\ &1-2y+y^2 &&=x^2\\ &-1+3y-3y^2+y^3 &&=x^3\\ &\cdots\cdots\cdots\cdots \end{aligned}$$

Replace

$$x^0,\ x^1,\ x^2,\ \ldots,\ x^n;\quad y^0,\ y^1,\ y^2,\ \ldots,\ y^n$$

by the indeterminates

$$x_0,\ x_1,\ x_2,\ \ldots,\ x_n;\quad y_0,\ y_1,\ y_2,\ \ldots,\ y_n.$$

The two systems still imply each other and so we obtain two linear transformations that are inverse to each other. See also I **34.1**.

54.2. From I **191** and I **199** by using the idea of solution **54.1**. For the numerical values involved see I **186** and I **197**.

54.3. (1) Immediate from I **34.1**. (2) Denote the three matrices arising in **54.1** by A, B and I, respectively, so that

$$AB=I.$$

Define

$$\tilde{I}=\begin{bmatrix} 1 & & & & \\ 0 & -1 & & & \\ 0 & 0 & 1 & & \\ 0 & 0 & 0 & -1 & \\ \cdots & \cdots & \cdots & \cdots & \end{bmatrix}.$$

Then

$$\tilde{I}^2=I,\qquad A\tilde{I}=\tilde{I}B$$

and the desired relation is obtained from

$$(A\tilde{I})(\tilde{I}B)=I.$$

54.4. By the required condition

$$y_0=a_0x_0,\qquad x_0=a_0y_0$$

and thus

$$a_0^2=1,\qquad a_0=1\quad\text{or}\quad -1.$$

Guided by the solution of I **34.1**, we introduce the power series

$$a_0+\frac{a_1z}{1!}+\frac{a_2z^2}{2!}+\cdots=a_0F(z)$$

and start from

$$\sum_{n=0}^{\infty} \frac{y_n z^n}{n!} = \sum_{k=0}^{\infty} \frac{a_k z^k}{k!} \sum_{l=0}^{\infty} \frac{(-1)^l x_l z^l}{l!} = a_0 F(z) \sum_{n=0}^{\infty} \frac{(-1)^n x_n z^n}{n!},$$

from which it follows (substitute $-z$ for z) that

$$\sum_{n=0}^{\infty} \frac{x_n z^n}{n!} = [a_0 F(-z)]^{-1} \sum_{n=0}^{\infty} \frac{(-1)^n y_n z^n}{n!}.$$

The required condition demands that

$$[F(-z)]^{-1} = F(z)$$

or

$$-\log F(-z) = \log F(z).$$

That is, $\log F(z)$ is an *odd function*. Hence the sequence $a_0, a_1, a_2, \ldots$ is obtained from

$$a_0 + \frac{a_1 z}{1!} + \frac{a_2 z^2}{2!} + \cdots = \pm \exp(c_1 z + c_3 z^3 + c_5 z^5 + \cdots),$$

where $c_1, c_3, c_5, \ldots$ are arbitrary. The example given in **54.3**, I **34.1** results from $c_1 = 1$, $c_n = 0$ for $n > 1$.

The infinite power series are used in a purely formal way (which can be explicitly described); convergence is not required.

55. Multiplication theorem.

56. Addition of columns.

57. Leibniz formula for $(uv)^{(n)}$, addition of columns.

58. In **57** set $\varphi = f_1^{-1}$.

59. After performing the differentiation, the numerator is a determinant of second order, whose elements are minors of order $n-1$ of $W(f_1, f_2, \ldots, f_n)$. [Kowalewski, pp. 80, 109.]

60. Determine the $n-1$ functions $\varphi_1(x), \varphi_2(x), \ldots, \varphi_{n-1}(x)$ from the $n-1$ equations

$$\begin{aligned}
&\varphi_1 f_1 + \varphi_2 f_2 + \cdots + \varphi_{n-1} f_{n-1} = f_n,\\
&\varphi_1 f_1' + \varphi_2 f_2' + \cdots + \varphi_{n-1} f_{n-1}' = f_n',\\
&\cdots\cdots\cdots\cdots\cdots\cdots\cdots\cdots\\
&\varphi_1 f_1^{(n-2)} + \varphi_2 f_2^{(n-2)} + \cdots + \varphi_{n-1} f_{n-1}^{(n-2)} = f_n^{(n-2)},
\end{aligned}$$

whose determinant $\neq 0$. By **59** and the assumptions we have

$$\frac{d\varphi_{n-1}}{dx} = \frac{d}{dx} \frac{W(f_1, \ldots, f_{n-2}, f_n)}{W(f_1, \ldots, f_{n-2}, f_{n-1})} = 0$$

and similarly

$$\varphi_1' = \varphi_2' = \cdots \varphi_{n-2}' = 0.$$

61. Eliminate $\varphi_n, \varphi_{n-1}, \ldots, \varphi_1, 1$ from the $n+1$ homogeneous equations

$$\begin{aligned}
&\varphi_n f_1 + \varphi_{n-1} f_1' + \cdots + \varphi_1 f_1^{(n-1)} + 1 \cdot f_1^{(n)} = 0,\\
&\cdots\cdots\cdots\cdots\cdots\cdots\cdots\cdots\\
&\varphi_n f_n + \varphi_{n-1} f_n' + \cdots + \varphi_1 f_n^{(n-1)} + 1 \cdot f_n^{(n)} = 0,\\
&\varphi_n y + \varphi_{n-1} y' + \cdots + \varphi_1 y^{(n-1)} + 1(y^{(n)} - L) = 0.
\end{aligned}$$

and calculate L by setting the determinant equal to zero.

62. Setting

$$y=Y_0,\quad W(f_1,y)=Y_1,\quad W(f_1,f_2,y)=Y_2,\quad \ldots,\quad W(f_1,f_2,\ldots,f_n,y)=Y_n$$

we find by **59**

$$\frac{d}{dx}\frac{Y_0}{W_1}=\frac{W_0Y_1}{W_1^2},\qquad \frac{d}{dx}\frac{Y_1}{W_2}=\frac{W_1Y_2}{W_2^2},\qquad \ldots,\qquad \frac{d}{dx}\frac{Y_{n-1}}{W_n}=\frac{W_{n-1}Y_n}{W_n^2}.$$

From these relations calculate Y_n/W_n. [**61**.]

63. [J. P. Gram: J. reine angew. Math. **94**, pp. 41–73 (1883).] The corresponding quadratic form is:

$$\int_a^b [t_1f_1(x)+t_2f_2(x)+\cdots+t_nf_n(x)]^2\,dx.$$

This integral is not negative, and $=0$ if and only if in the interval $a\leqq x\leqq b$ we have identically

$$t_1f_1(x)+t_2f_2(x)+\cdots+t_nf_n(x)=0.$$

Cf. also II **68**.

64. [G. Pólya, problem: Arch. Math. Phys. Series 3, **20**, p. 271 (1913).] The corresponding quadratic form

$$\int_a^b \{(t_1^2+t_2^2+\cdots+t_n^2)[(f_1(x))^2+(f_2(x))^2+\cdots+(f_n(x))^2]$$

$$-[t_1f_1(x)+t_2f_2(x)+\cdots+t_nf_n(x)]^2\}\,dx=\int_a^b \sum_{j>k}^{1,2,\ldots,n}[t_jf_k(x)-t_kf_j(x)]^2\,dx$$

is positive. It vanishes for a set of numbers $t_1, t_2, \ldots, t_n$ with $t_1^2+t_2^2+\cdots+t_n^2>0$, if and only if $f_\nu(x)=t_\nu\varphi(x)$, $\nu=1,2,\ldots,n$, where $\varphi(x)$ does not depend on ν.

65. [E. H. Moore, problem: Amer. Math. Monthly, **24**, p. 293 (1916). Solution by C. F. Gummer: Amer. Math. Monthly, **24**, pp. 293, 333–334 (1916) (the first solution given on p. 293 is incorrect as is noted on p. 333).] [**36**, **63**.] If, e.g., the first two rows are identical, we have in particular that

$$\frac{a_{11}a_{22}}{a_{21}a_{12}}=e^{\int_a^b [f_1(x)-f_2(x)]^2\,dx}=1,\qquad f_1(x)=f_2(x).$$

66. [G. Pólya, problem: Arch. Math. Phys. Series 3, **28**, p. 174 (1920).] In view of II **112**, II **113** we have

$$(a_\lambda+a_\mu+1)\int_0^\infty x^{a_\lambda+a_\mu}f(x)\,dx=x^{a_\lambda+a_\mu+1}f(x)\Big|_0^\infty-\int_0^\infty x^{a_\lambda+a_\mu+1}\,df(x),$$

$$=-\int_0^\infty x^{a_\lambda+a_\mu+1}\,df(x),$$

where the last integral is interpreted as a Stieltjes integral. Hence we have

$$\sum_{\lambda=1}^n\sum_{\mu=1}^n(a_\lambda+a_\mu+1)\int_0^\infty x^{a_\lambda+a_\mu}f(x)\,dx\cdot t_\lambda t_\mu$$

$$=-\int_0^\infty (t_1x^{a_1}+t_2x^{a_2}+\cdots+t_nx^{a_n})^2x\,df(x)\geqq 0.$$

If $f(x)$ is piece-wise constant, and if there exist at least n points with a negative jump, then the last integral can vanish only if $t_1 = t_2 = \cdots = t_n = 0$ [V **76**].

67. [For **67, 68**, cf. O. Toeplitz: Gött, Nachr., pp. 110–115 (1907); pp. 489–506 (1910).]

For $f(\vartheta) = a_0 + 2(a_1 \cos \vartheta + b_1 \sin \vartheta)$ we obtain

$$T_n(f) = a_0(|x_0|^2 + |x_1|^2 + \cdots + |x_n|^2) + 2\Re(a_1 + ib_1)(x_0\bar{x}_1 + x_1\bar{x}_2 + \cdots + x_{n-1}\bar{x}_n).$$

For

$$f(\vartheta) = \frac{1 - r^2}{1 - 2r \cos \vartheta + r^2} = 1 + 2r \cos \vartheta + 2r^2 \cos 2\vartheta + \cdots + 2r^n \cos n\vartheta + \cdots$$

we obtain

$$T_n(f) = \sum_{\lambda=0}^{n} \sum_{\mu=0}^{n} r^{|\mu-\lambda|} x_\lambda \bar{x}_\mu.$$

68. We have

$$c_n = \frac{1}{2\pi} \int_0^{2\pi} f(\vartheta) e^{in\vartheta} \, d\vartheta, \qquad n = 0, \pm 1, \pm 2, \pm 3, \ldots;$$

hence we have for the Toeplitz form $T_n(f)$ the representation

$$T_n(f) = \frac{1}{2\pi} \int_0^{2\pi} f(\vartheta) |x_0 + x_1 e^{-i\vartheta} + x_2 e^{-2i\vartheta} + \cdots + x_n e^{-in\vartheta}|^2 \, d\vartheta.$$

From this the assertion follows, if we note the equation

$$T_n(1) = \frac{1}{2\pi} \int_0^{2\pi} |x_0 + x_1 e^{-i\vartheta} + x_2 e^{-2i\vartheta} + \cdots + x_n e^{-in\vartheta}|^2 \, d\vartheta$$
$$= |x_0|^2 + |x_1|^2 + |x_2|^2 + \cdots + |x_n|^2.$$

69. [For **69–71**, cf. G. Szegö: Math. Ann. **76**, pp. 490–503 (1915); Math. es. term. ért. **35**, pp. 185–222 (1917); Math. Z. **6**, pp. 167–202 (1920). For heuristic considerations see MPR, **2**, pp. 45–46. For later developments see U. Grenander and G. Szegö, Toeplitz forms and their applications, Berkeley and Los Angeles: University of California Press, 1958.]

For $f(\vartheta) = a_0 + 2(a_1 \cos \vartheta + b_1 \sin \vartheta)$ we have

$$D_n(f) = \begin{vmatrix} a_0 & a_1 + ib_1 & 0 & \ldots & 0 \\ a_1 - ib_1 & a_0 & a_1 + ib_1 & \ldots & 0 \\ 0 & a_1 - ib_1 & a_0 & \ldots & 0 \\ \ldots & \ldots & \ldots & \ldots & \ldots \\ 0 & 0 & 0 & \ldots & a_0 \end{vmatrix}$$

By expanding the determinant in terms of the last row we obtain the recurrence relations

$$D_n(f) = a_0 D_{n-1}(f) - (a_1^2 + b_1^2) D_{n-2}(f), \qquad n = 2, 3, 4, \ldots;$$
$$D_0(f) = a_0, \qquad D_1(f) = a_0^2 - (a_1^2 + b_1^2).$$

Let α and β be the two roots of the quadratic equation:

$$x^2 = a_0 x - (a_1^2 + b_1^2).$$

Mathematical induction yields

$$D_n(f)=\begin{cases}\dfrac{[D_1(f)-D_0(f)\beta]\alpha^n-[D_1(f)-D_0(f)\alpha]\beta^n}{\alpha-\beta}, & \text{if } \alpha\neq\beta,\\[2ex] (n+2)\alpha^{n+1}=(n+2)\beta^{n+1}=(n+2)\left(\dfrac{a_0}{2}\right)^{n+1}, & \text{if } \alpha=\beta.\end{cases}$$

For $f(\vartheta)=\dfrac{1-r^2}{1-2r\cos\vartheta+r^2}$ we have

$$D_n(f)=\begin{vmatrix}1 & r & r^2 & \dots & r^n\\ r & 1 & r & \dots & r^{n-1}\\ r^2 & r & 1 & \dots & r^{n-2}\\ \dots & \dots & \dots & \dots & \dots\\ r^n & r^{n-1} & r^{n-2} & \dots & 1\end{vmatrix}.$$

If we subtract r times the second row from the first row, and then r times the third row from the second row, and so on, then we obtain that $D_n(f)=(1-r^2)^n$.

If $f(\vartheta)=a_0+2(a_1\cos\vartheta+b_1\sin\vartheta)$ is positive, then α and β are positive, $\alpha\neq\beta$. Let $\alpha>\beta>0$, then we have $\lim_{n\to\infty}\sqrt[n+1]{D_n(f)}=\alpha$. In the second case this limit $=1-r^2$. The geometric mean [II **48**] $\mathfrak{G}(f)$ of $a_0+2(a_1\cos\vartheta+b_1\sin\vartheta)$ in the interval $0, 2\pi$ is identical with that of $a_0-2\sqrt{a_1^2+b_1^2}\cos\vartheta=\alpha(1+\rho^2-2\rho\cos\vartheta)$, if ρ is determined in accordance with $\sqrt{a_1^2+b_1^2}=\alpha\rho$; $0\leqq\rho<1$, $a_0=\alpha(1+\rho^2)$. For $\mathfrak{G}(1+\rho^2-2\rho\cos\vartheta)$ cf. II **52**.

70. Because $D_n(f)$ is a Hermitian determinant, all the zeros of $D_n(f-h)$ are real and lie between the minimum and maximum of the form $T_n(f)$ for $|x_0|^2+|x_1|^2+\cdots+|x_n|^2=1$. [Kowalewski, pp. 130, 283.] By **68** they thus have the form

$$h=a_0-2\sqrt{a_1^2+b_1^2}\cos\varphi,\qquad 0<\varphi<\pi.$$

It then follows, by everywhere replacing a_0 by a_0-h in solution **69**, that

$$D_n(f-h)=(a_1^2+b_1^2)^{\frac{n+1}{2}}\frac{\sin(n+2)\varphi}{\sin\varphi}.$$

From this we obtain the zeros of $D_n(f-h)$:

$$h_{\nu n}=a_0-2\sqrt{a_1^2+b_1^2}\cos\frac{\nu+1}{n+2}\pi,\qquad \nu=0,1,2,\dots,n.$$

71. Cf. **70**.

$$\frac{F(h_{0n})+F(h_{1n})+\cdots+F(h_{nn})}{n+1}$$

$$=\frac{1}{n+1}\sum_{\nu=0}^{n}F\left(a_0-2\sqrt{a_1^2+b_1^2}\cos\frac{\nu+1}{n+2}\pi\right)\to\frac{1}{\pi}\int_0^{\pi}F(a_0-2\sqrt{a_1^2+b_1^2}\cos\vartheta)\,d\vartheta$$

$$=\frac{1}{2\pi}\int_0^{2\pi}F[f(\vartheta)]\,d\vartheta.$$

The property of the numbers $h_{\nu n}$ formulated in the problem holds not only in the special case when $f(\vartheta)$ is a trigonometric polynomial of first order, but in general for an arbitrary properly integrable function $f(\vartheta)$. [Cf. G. Szegö, op. cit. **69**.]

72. *First solution:* We use the following symbolic notation [cf. M. Riesz: Ark. för Mat., Astron. och Fys. **17**, No. 16, p. 1 (1923).] Let

$$f(z)=\sum_{\lambda=0}^{n}\sum_{\mu=0}^{n} a_{\lambda\mu}z^{\lambda}\bar{z}^{\mu},$$

where $a_{\lambda\mu}$ are arbitrary constants. We set

$$f(c)=\sum_{\lambda=0}^{n}\sum_{\mu=0}^{n} a_{\lambda\mu}c_{\lambda-\mu}.$$

The assumptions of the problem imply for example that

$$|x_0+x_1c+x_2c^2+\cdots+x_nc^n|^2>0$$

if the polynomial $x_0+x_1z+x_2z^2+\cdots+x_nz^n$ does not vanish identically. Let $P(z)$ be the polynomial in question, z_0 one of its zeros and $P(z)=(z-z_0)Q(z)$. Denoting by $z(\bar{Q})$ the polynomial with conjugate complex coefficients, we set

$$Q(c)\bar{Q}(\bar{c})=C_0, \qquad Q(c)\bar{Q}(\bar{c})c=C_1, \qquad Q(c)\bar{Q}(\bar{c})\bar{c}=C_{-1}.$$

Then we have, with x_0, x_1 arbitrary,

$$\begin{aligned} C_0(|x_0|^2+|x_1|^2)+C_{-1}x_0\bar{x}_1+C_1\bar{x}_0x_1 &= Q(c)\bar{Q}(\bar{c})(|x_0|^2+|x_1|^2+\bar{c}x_0\bar{x}_1+c\bar{x}_0x_1) \\ &= Q(c)(x_0+x_1c)\bar{Q}(\bar{c})(\bar{x}_0+\bar{x}_1\bar{c})>0, \end{aligned}$$

if $|x_0|^2+|x_1|^2>0$. Hence we have $|C_1|<C_0$.

We now have

$$P(c)\bar{c}^{\nu}=0, \qquad \nu=0, 1, \ldots, n-1,$$

and hence

$$(c-z_0)Q(c)\bar{Q}(\bar{c})=0, \qquad z_0=\frac{Q(c)\bar{Q}(\bar{c})c}{Q(c)\bar{Q}(\bar{c})}=\frac{C_1}{C_0}.$$

Second solution: By a theorem of C. Carathéodory [Palermo Rend. **32**, p. 205 (1911)], there exist $2n$ numbers

$$\begin{matrix} \rho_1, & \rho_2, & \ldots, & \rho_n, \\ \varepsilon_1, & \varepsilon_2, & \ldots, & \varepsilon_n, \end{matrix}$$

such that $\rho_\nu \geqq 0$, $|\varepsilon_\nu|=1$, $\nu=1, 2, \ldots, n$, and in addition

$$c_k=\rho_1\varepsilon_1^k+\rho_2\varepsilon_2^k+\cdots+\rho_n\varepsilon_n^k, \qquad k=1, 2, \ldots, n.$$

We have

$$c_0-h=\rho_1+\rho_2+\cdots+\rho_n,$$

where h denotes the minimum of the form mentioned in the problem with the subsidiary condition $|x_0|^2+|x_1|^2+\cdots+|x_n|^2=1$. In our own case we have $h>0$.

From this we deduce

$$\sum_{\lambda=0}^{n-1}\sum_{\mu=0}^{n-1} c_{\lambda-\mu}x_\lambda\bar{x}_\mu = h(|x_0|^2+|x_1|^2+\cdots+|x_{n-1}|^2) + \sum_{\nu=1}^{n} \rho_\nu |x_0+x_1\varepsilon_\nu+\cdots+x_{n-1}\varepsilon_\nu^{n-1}|^2,$$

$$\sum_{\lambda=0}^{n-1}\sum_{\mu=0}^{n-1} c_{\lambda-\mu+1}x_\lambda\bar{x}_\mu = h(x_0\bar{x}_1+x_1\bar{x}_2+\cdots+x_{n-2}\bar{x}_{n-1}) + \sum_{\nu=1}^{n} \rho_\nu\varepsilon_\nu |x_0+x_1\varepsilon_\nu+\cdots+x_{n-1}\varepsilon_\nu^{n-1}|^2,$$

and hence

$$\left|\sum_{\lambda=0}^{n-1}\sum_{\mu=0}^{n-1} c_{\lambda-\mu+1}x_\lambda\bar{x}_\mu\right| < \sum_{\lambda=0}^{n-1}\sum_{\mu=0}^{n-1} c_{\lambda-\mu}x_\lambda\bar{x}_\mu,$$

assuming that the numbers $x_0, x_1, \ldots, x_{n-1}$ are not all equal to zero.

Now we can write our equation in the form

$$|c_{\lambda-\mu+1} - zc_{\lambda-\mu}|_{\lambda,\,\mu=0,\,1,\ldots,\,n-1} = 0.$$

Thus if z_0 denotes a root, then there exist numbers $x_0, x_1, \ldots, x_{n-1}$ not all vanishing that satisfy the system of equations

$$\sum_{\mu=0}^{n-1} (c_{\lambda-\mu+1} - z_0c_{\lambda-\mu})\bar{x}_\mu = 0, \qquad \lambda=0, 1, \ldots, n-1.$$

We deduce that

$$z_0 = \frac{\sum_{\lambda=0}^{n-1}\sum_{\mu=0}^{n-1} c_{\lambda-\mu+1}x_\lambda\bar{x}_\mu}{\sum_{\lambda=0}^{n-1}\sum_{\mu=0}^{n-1} c_{\lambda-\mu}x_\lambda\bar{x}_\mu}, \qquad |z_0| < 1.$$

Part Eight. Number Theory

1.

$$x+n-1<[x+n]\leqq x+n, \qquad x-1<[x+n]-n\leqq x.$$

Hence, since $[x+n]-n$ is an integer, we have

$$[x+n]-n=[x].$$

2. We must show that

$$(n-1)+(n-2)+\cdots+1=\frac{n(n-1)}{2}\equiv\left[\frac{n}{2}\right] \qquad (\text{mod } 2).$$

We show this for $n=0, 1, 2, 3$ by calculation. Both sides increase by an even number if we substitute $n+4$ for n.

3. If $x-[x]<\frac{1}{2}$, then $0\leqq 2x-2[x]<1$. If $x-[x]\geqq\frac{1}{2}$, then $1\leqq 2x-2[x]<2$. By **1**, we have

$$[2x-2[x]]=[2x]-2[x].$$

4. From **1** in a similar fashion as **3**.

5. From **3** we have

$$[2x]-2[x]+[x]=[2x]-[x].$$

6. The required integer n satisfies:

$$n-1<x\leqq n, \quad \text{therefore} \quad -n\leqq -x<-n+1,$$

and hence $n=-[-x]$.

7.

$$0\leqq\alpha+\beta-[\alpha]-[\beta]=\alpha-[\alpha]+\beta-[\beta]<2,$$

$$-1<\alpha-\beta-[\alpha]+[\beta]=\alpha-[\alpha]-(\beta-[\beta])<1.$$

8. From **1** we see that both sides change by the same quantity if either α or β changes by an integer. It is thus sufficient to prove the theorem only for the case $0\leqq\alpha<1$, $0\leqq\beta<1$. It then reads as follows

$$[2\alpha]+[2\beta]\geqq[\alpha+\beta].$$

If $[\alpha+\beta]=0$, we have nothing to prove. If $[\alpha+\beta]=1$ then $\alpha+\beta\geqq 1$, and hence at least one of the two numbers, say α, is $\geqq\frac{1}{2}$, and thus $[2\alpha]+[2\beta]\geqq[2\alpha]\geqq 1$.

9. [Ch. Hermite: Acta Math. **5**, p. 315 (1884).] It suffices to consider the case $0 \leqq x < 1$ [solution **8**]. Determine k such that

$$x + \frac{k-1}{n} < 1 \leqq x + \frac{k}{n}, \qquad \text{i.e.} \quad -k = [nx - n] = [nx] - n.$$

Both sides are $= n - k$.

10. We may assume that $0 \leqq x < 1$ [solution **8**, solution **9**]. For $0 \leqq x < 1$ the right-hand side is $= 0$, and we have also

$$[nx] \leqq nx, \qquad \frac{[nx]}{n} \leqq x < 1, \qquad \left[\frac{[nx]}{n}\right] = 0.$$

11. [J. J. Sylvester, problem: Nouv. Annls Math. Sér. 1, **16**, p. 125 (1857). Solution by E. Prouhet, Lebesgue: Nouv. Annls Math. Sér. 1, **16**, p. 184, p. 262 (1857).]

$$[(1+\sqrt{3})^{2m+1}] = (1+\sqrt{3})^{2m+1} + (1-\sqrt{3})^{2m+1},$$

since $-1 < (1-\sqrt{3})^{2m+1} < 0$ and the right-hand side is an integer. We obtain further that

$$\begin{aligned}&(1+\sqrt{3})(4+2\sqrt{3})^m + (1-\sqrt{3})(4-2\sqrt{3})^m \\ &\quad = 2^m\{(1+\sqrt{3})(2+\sqrt{3})^m + (1-\sqrt{3})(2-\sqrt{3})^m\}.\end{aligned}$$

The expression in curly brackets is of the form

$$2(a+b\sqrt{3}) + (1+\sqrt{3})(\sqrt{3})^m + 2(a-b\sqrt{3}) + (1-\sqrt{3})(-\sqrt{3})^m,$$

where a, b are rational and integral, and hence is divisible only by the first power of 2.

12. The numbers in question are $a, 2a, 3a, \ldots, ka$, where $ka \leqq n < (k+1)a$, and thus $k = [n/a]$. We can also pose the question in the following way: How many of the fractions $1/a, 2/a, \ldots, n/a$ are integers, i.e. how many integers are contained in the interval $0 < x \leqq n/a$?

13. The number of integers in the interval $a/\pi < x \leqq b/\pi$ is equal to [solution **12**]

$$\left[\frac{b}{\pi}\right] - \left[\frac{a}{\pi}\right].$$

The number of integers in the interval $-b/\pi < x \leqq -a/\pi$ is equal to

$$\left[-\frac{a}{\pi}\right] - \left[-\frac{b}{\pi}\right].$$

14. [J. König: Math. Ann. **9**, p. 530 (1876). Also Nouv. Corresp. Math. **5**, p. 222 (1879); solution by Radicke: ibid. **6**, p. 82 (1880).] For $\alpha = 0$ and $\alpha = \pi$ the assertion is obvious; thus assume $0 < \alpha < \pi$. Then no two successive terms of the sequence can vanish simultaneously. If two *adjacent* terms $\cos \nu\alpha$, and $\cos(\nu+1)\alpha$ constitute a change of sign, then between $\nu\alpha$ and $(\nu+1)\alpha$ there lies just one zero of $\cos x$. Two *non-adjacent* terms $\cos(\nu-1)\alpha$ and $\cos(\nu+1)\alpha$ can also constitute a change of sign, namely in the case that $\cos \nu\alpha = 0$. Thus $V_n(\alpha)$ is equal to the number of zeros of $\cos x$ in the interval $0 \leqq x < n\alpha$, i.e. to the number of zeros of $\sin x$ in the interval $(\pi/2) - n\alpha < x \leqq \pi/2$. This number is equal to

$$-\left[\frac{1}{2} - n\frac{\alpha}{\pi}\right] \qquad [\mathbf{13}].$$

15.

$$g_n = [n\vartheta] - [(n-1)\vartheta].$$

16. $N(r, a, \alpha) = 0$ for $r < |\log a|$; $N(r, a, \alpha) = l - k$ for $r \geqq |\log a|$, where

$$\alpha + 2l\pi \leqq \sqrt{r^2 - (\log a)^2} < \alpha + 2(l+1)\pi,$$
$$\alpha + 2k\pi < -\sqrt{r^2 - (\log a)^2} \leqq \alpha + 2(k+1)\pi,$$

i.e. we have in this case

$$N(r, a, \alpha) = 1 + \left[\frac{\sqrt{r^2 - (\log a)^2} - \alpha}{2\pi}\right] + \left[\frac{\sqrt{r^2 - (\log a)^2} + \alpha}{2\pi}\right].$$

Cf. III **73**.

17.

$$[f(a)] + [f(a+1)] + [f(a+2)] + \cdots + [f(b-1)] + [f(b)].$$

18. The left-hand side represents [**17**] the number of lattice points in the region $1 \leqq x \leqq p-1$, $0 < y \leqq (q/p)x$. In the rectangle

$$1 \leqq x \leqq p-1, \qquad 1 \leqq y \leqq q-1$$

there lie a total of $(p-1)(q-1)$ lattice points. They are distributed symmetrically with respect to the point $x = p/2$, $y = q/2$, and there are as many above as below the straight line $y = (q/p)x$ but none on the straight line itself, for a lattice point lies on this straight line if and only if x is an integral multiple of p.

19. [Gauss: Theorematis arithmetici demonstratio nova, 1808, Werke, **2**, pp. 1–8, Göttingen: Königl, Ges. der Wiss. (1863). G. Eisenstein, problem: J. reine angew. Math. **27**, p. 281 (1844).] We are concerned with the lattice points in the rectangle

$$1 \leqq x \leqq p', \qquad 1 \leqq y \leqq q'.$$

Their total number is $p'q'$. The first p' terms on the left-hand side give the number of lattice points below, the last q' terms the number of lattice points above, the straight line $y = (q/p)x$ [**17**].

20. [V. Bunyakovski: C.R. Acad. Sci. (Paris), **94**, pp. 1459–1461 (1882).] Let the variable x run through the numbers $1, 2, \ldots, n$, and the variable y for given x run through the numbers $[\sqrt{(x-1)p}] + 1$, $[\sqrt{(x-1)p}] + 2, \ldots, [\sqrt{xp}]$. Setting $r = r(x, y) = xp - y^2$, we have $0 < r < p$ and $-r$ is a quadratic residue modulo p. Since -1 is a quadratic residue, the same is true for r. The number of all the numbers r, since $4n = p - 1$, is

$$= \sum_{x=1}^{n} ([\sqrt{xp}] - [\sqrt{(x-1)p}]) = [\sqrt{np}] = \frac{p-1}{2},$$

i.e. as many as there are quadratic residues mod p. Moreover they are all distinct: From $x_1p-y_1^2=x_2p-y_2^2$ it follows that p divides $y_1^2-y_2^2=(y_1+y_2)(y_1-y_2)$, i.e. $y_1=y_2$, and hence also $x_1=x_2$. The sum of all the numbers r is given by

$$\sum_{x=1}^{n}\sum_{y=[\sqrt{(x-1)p}]+1}^{[\sqrt{xp}]} r(x,y)=p\sum_{x=1}^{n} x([\sqrt{xp}]-[\sqrt{(x-1)p}])-\sum_{y=1}^{2n} y^2$$

$$=-p\sum_{x=1}^{n}[\sqrt{xp}]+(n+1)p[\sqrt{np}]-\frac{n}{3}(2n+1)(4n+1),$$

and hence is equal to the sum of all the smallest positive quadratic residues mod p, i.e. $=[(p-1)/4]p$.

21. [J. J. Sylvester: C.R. Acad. Sci. (Paris), **96**, p. 463 (1883).] Let n be the number of properties $\alpha, \beta, \gamma, \ldots, \kappa, \lambda$. If an object has k of these properties $(1\leqq k\leqq n)$, then it contributes to the sum written out in the problem precisely

$$1-\binom{k}{1}+\binom{k}{2}-\binom{k}{3}+\cdots+(-1)^k\binom{k}{k}=0$$

units. If however an object has none at all of the properties $\alpha, \beta, \gamma, \ldots, \kappa, \lambda$, then the number it contributes is precisely 1, contributed in fact to the first summand N.—The theorem actually belongs to formal logic. For a proof based on mathematical induction cf. for example U. Yule, An Introduction to the Theory of Statistics, Chapter 2, London: Griffin, 1916.

21.1. (1) $1-A$,

(2) AB,

(3) $1-(1-A)(1-B)=A+B-AB$.

21.2. Let $A, B, C, \ldots$ be the subsets characterized by the properties $\alpha, \beta, \gamma, \ldots$, respectively. We are required to find the number of elements in the intersection of the complements of $A, B, C, \ldots$ whose characteristic function is, by **21.1**, (1) and (2),

$$\begin{aligned}&(1-A)(1-B)(1-C)\ldots\\&=1\\&-A-B-C-\cdots\\&+AB+AC+BC+\cdots\\&-ABC-\cdots\end{aligned}$$

Pass from the characteristic function to the number of elements by summation over all the elements of U.

22. $N=n$, $N_\alpha=N_\beta=\cdots=n-1$, $N_{\alpha\beta}=N_{\alpha\gamma}=\cdots=n-2, \ldots$. We have

$$n-\binom{n}{1}(n-1)+\binom{n}{2}(n-2)-\cdots+(-1)^n\binom{n}{n}(n-n)=0,$$

since there are no objects that have none of the properties $\alpha, \beta, \gamma, \ldots, \kappa, \lambda$. [I **37**.]

22.1. We are required to paint the n houses of a settlement with k different colors, using just one color for each house [I **192**]. We regard each color distribution as one of our "objects"; the number of all such objects is $N=k^n$. If the first color is not used, the object (color distribution) has the property α, if the second color is not used, it has the property β, and so on. Hence [**21**]

$$N_\alpha = N_\beta = \cdots = N_\kappa = N_\lambda = (k-1)^n,$$
$$N_{\alpha\beta} = N_{\alpha\gamma} = \cdots = N_{\kappa\lambda} = (k-2)^n,$$
$$N_{\alpha\beta\gamma} = \cdots = (k-3)^n,$$
$$\cdots\cdots\cdots\cdots\cdots\cdots$$

and the number of those color distributions in which every one of the k available colors is used at least once is

$$k^n - \binom{k}{1}(k-1)^n + \binom{k}{2}(k-2)^n - \cdots + (-1)^k 0^n$$

which equals $k!S_k^n$ [I **192**].

22.2. Apply **21**. There are altogether $T_n = N$ different partitions of a set S of n elements. The property α belongs to those $N_\alpha = T_{n-1}$ partitions that contain a class (subset of S) consisting of the first element of S alone. The property β is similarly linked to the second element of S, and so on. Let $\tilde{T}_n$ be the number of those partitions that have neither the property α, nor β, nor $\gamma, \ldots$.

22.3. Similarly to **22.2**. Can also be linked to I **204**.

23. [In a different guise as "jeu de recontre" by P. R. Montmort, A. de Moivre. Cf. L. Euler: Opera Omnia, Series 1, 7, p. 11. Leipzig and Berlin: B. G. Teubner, 1923.] The total number of terms in the expansion is $n!$. The number of those terms that contain $a_{\nu\nu}$ is $(n-1)!$, the number of those that contain $a_{\mu\mu}a_{\nu\nu}$ is $(n-2)!$ and so on. By **21** the required number is thus

$$n! - \binom{n}{1}(n-1)! + \binom{n}{2}(n-2)! - \binom{n}{3}(n-3)! + \cdots + (-1)^n\binom{n}{n}$$
$$= n!\left(1 - \frac{1}{1!} + \frac{1}{2!} - \frac{1}{3!} + \cdots + \frac{(-1)^n}{n!}\right).$$

(Cf. also solution VII **46**.)

24. The number of numbers less than or equal to n that are divisible by a (property α) is [**12**] equal to $[n/a]$. The number of those that are divisible simultaneously by a and b (properties α and β) is equal to the number of those that are divisible by ab, since a and b are relatively prime, and hence is equal to $[n/(ab)]$, and so on. The required number is thus [**21**] equal to

$$n - \left[\frac{n}{a}\right] - \left[\frac{n}{b}\right] - \left[\frac{n}{c}\right] - \cdots - \left[\frac{n}{k}\right] - \left[\frac{n}{l}\right]$$
$$+ \left[\frac{n}{ab}\right] + \left[\frac{n}{ac}\right] + \cdots + \left[\frac{n}{kl}\right]$$
$$- \left[\frac{n}{abc}\right] - \cdots$$
$$\cdots\cdots\cdots\cdots\cdots\cdots$$
$$\pm \left[\frac{n}{abc\ldots kl}\right].$$

25. [Euler.] In **24** set: $a=p, b=q, c=r, \ldots$. The required number is equal to

$$
\begin{aligned}
&n-\frac{n}{p}-\frac{n}{q}-\frac{n}{r}-\cdots\\
&\quad+\frac{n}{pq}+\frac{n}{pr}+\cdots\\
&\quad-\frac{n}{pqr}-\cdots\\
&\quad\cdots\cdots\cdots\cdots\cdots\cdots\cdots\cdots\\
&\quad\pm\frac{n}{pqr\ldots}=n\left(1-\frac{1}{p}\right)\left(1-\frac{1}{q}\right)\left(1-\frac{1}{r}\right)\cdots.
\end{aligned}
$$

26. By analogous considerations to those that prove the special case **21**, where the equal values 1 are associated with every object.

27. Apply **26**. Let the objects be the numbers $1, 2, 3, \ldots, n$, and let the property α be divisibility by p, the property β divisibility by q and so on. By the value of an object let us mean the square of the number in question. We have

$$
\begin{aligned}
W&=1^2+2^2+3^2+\cdots+n^2=\frac{n(n+1)(2n+1)}{6}\\
&=An^3+Bn^2+Cn+D, \qquad A=\tfrac{1}{3}, \qquad B=\tfrac{1}{2}, \qquad C=\tfrac{1}{6}, \qquad D=0,\\
W_\alpha&=p^2+(2p)^2+(3p)^2+\cdots+\left(\frac{n}{p}p\right)^2=p^2\left[A\left(\frac{n}{p}\right)^3+B\left(\frac{n}{p}\right)^2+C\frac{n}{p}+D\right],\ldots,\\
W_{\alpha\beta}&=p^2q^2\left[A\left(\frac{n}{pq}\right)^3+B\left(\frac{n}{pq}\right)^2+C\frac{n}{pq}+D\right],\ldots
\end{aligned}
$$

The sum in question is thus

$$
\begin{aligned}
&=An^3+Bn^2+Cn+D\\
&\quad-\sum p^2\left[A\left(\frac{n}{p}\right)^3+B\left(\frac{n}{p}\right)^2+C\frac{n}{p}+D\right)\\
&\quad+\sum p^2q^2\left[A\left(\frac{n}{pq}\right)^3+B\left(\frac{n}{pq}\right)^2+C\frac{n}{pq}+D\right]\\
&\quad\cdots\cdots\cdots\cdots\cdots\cdots\cdots\cdots\cdots\cdots\\
&\quad\pm p^2q^2r^2\ldots\left[A\left(\frac{n}{pqr\ldots}\right)^3+B\left(\frac{n}{pqr\ldots}\right)^2+C\frac{n}{pqr\ldots}+D\right]\\
&=an^3+bn^2+cn+d,
\end{aligned}
$$

where

$$
\begin{aligned}
a&=A\left(1-\sum\frac{1}{p}+\sum\frac{1}{pq}-\cdots\pm\frac{1}{pqr\ldots}\right)\\
&=\frac{1}{3}\left(1-\frac{1}{p}\right)\left(1-\frac{1}{q}\right)\left(1-\frac{1}{r}\right)\cdots=\frac{\varphi(n)}{3n},\\
b&=B(1-\sum 1+\sum 1-\cdots\pm 1)=\tfrac{1}{2}(1-1)(1-1)(1-1)\cdots=0,\\
c&=C(1-\sum p+\sum pq-\cdots\pm pqr\ldots)\\
&=\tfrac{1}{6}(1-p)(1-q)(1-r)\cdots=\frac{(-1)^\nu}{6}\frac{\varphi(n)}{n}pqr\ldots,\\
d&=0.
\end{aligned}
$$

27.1. Consider the partial sum S of the expression given in **21** that contains all the terms involving s or less properties, but contains none of the terms involving $s+1$ or more properties.

(1) An object that has none of the properties $\alpha, \beta, \ldots, \lambda$ contributes just one unit to S (as it is counted just once, in the initial term N).

(2) An object that has exactly k of the properties where $1 \leqq k \leqq s$ contributes to S

$$1-\binom{k}{1}+\binom{k}{2}-\cdots+(-1)^k\binom{k}{k}=0.$$

(3) An object that has exactly k of the properties where $k>s$ contributes to S

$$1-\binom{k}{1}+\binom{k}{2}-\cdots+(-1)^s\binom{k}{s}=(-1)^s\binom{k-1}{s}.$$

[I **31.2**]. All these contributions are of the same sign and cause S to deviate from the desired value given by (1) in the direction indicated by $(-1)^s$.

Another proof, that proceeds along the lines of **21.2**, may be based on a more general lemma, namely V **163**, applied appropriately to the polynomial in t defined by

$$g(t)=(1-At)(1-Bt)\ldots(1-Lt)$$

(which depends on the parameter x involved in $A=A(x)$, $B=B(x), \ldots$) at the point $t=1$.

We can also append to **26** a system of inequalities analogous to the one we have just proved. Both methods of proof apply.

27.2. [Cf. W. Feller: An Introduction to Probability Theory and its Applications. New York: J. Wiley and Sons, 1950, I, pp. 64–65.] From VII **54.1**; if two matrices are inverse to each other, then the transposed matrices are also inverse to each other. The same proof applies also to a parallel generalization of **26**.

28. Let $N=\max(a, b, c, \ldots, k, l)$. For $N=0$ there is nothing to prove. For $N>0$, apply **21** to the numbers $1, 2, \ldots, N$ considered as the objects. Let a number have property α if it is a part of a $(\leqq a)$, β if it is a part of b, and so on. The number of objects that have neither property α nor property $\beta, \ldots$ is obviously $=0$. Hence express N as required.

29. Let p be any prime number and $a=p^\alpha a'$, $b=p^\beta b'$, $c=p^\gamma c', \ldots, k=p^\kappa k'$, $l=p^\lambda l'$, where $a', b', c', \ldots, k', l'$ are not divisible by p. The exponent of p on the left-hand side is equal to $\max(\alpha, \beta, \gamma, \ldots, \kappa, \lambda)$, and the exponent of p on the right-hand side is equal to

$$\begin{aligned}\alpha+\beta+\gamma+\cdots+\kappa+\lambda-\min(\alpha,\beta)-\min(\alpha,\gamma)-\cdots-\min(\kappa,\lambda)\\ +\min(\alpha,\beta,\gamma)+\cdots\\ \cdots\cdots\cdots\cdots\cdots\\ \pm\min(\alpha,\beta,\gamma,\ldots,\kappa,\lambda).\end{aligned}$$

By **28** these two expressions are equal. Since this holds for every prime number p, the assertion follows. Cf. **46**.

30. By row-by-row multiplication.

31. The determinant in question is equal to the square of the determinant $|\eta_{\lambda\mu}|_{\lambda,\,\mu=1,\,2,\ldots n}$, in which $\eta_{\lambda\mu}=1$ if μ is a divisor of λ and $\eta_{\lambda\mu}=0$ otherwise. The square is formed by row-by-row multiplication.

32. Multiply row-by-row the two determinants

$$\begin{vmatrix} 1 & 0 & 0 & 0 & \ldots & 0 \\ 1 & 1 & 0 & 0 & \ldots & 0 \\ 1 & 1 & 1 & 0 & \ldots & 0 \\ 1 & 1 & 1 & 1 & \ldots & 0 \\ \ldots & \ldots & \ldots & \ldots & \ldots & \ldots \\ 1 & 1 & 1 & 1 & \ldots & 1 \end{vmatrix} \quad \begin{vmatrix} a_0 & 0 & 0 & 0 & \ldots & 0 \\ a_0 & a_1 & 0 & 0 & \ldots & 0 \\ a_0 & a_1 & a_2 & 0 & \ldots & 0 \\ a_0 & a_1 & a_2 & a_3 & \ldots & 0 \\ \ldots & \ldots & \ldots & \ldots & \ldots & \ldots \\ a_0 & a_1 & a_2 & a_3 & \ldots & a_n \end{vmatrix}.$$

33. Multiply row-by-row the two determinants

$$\begin{vmatrix} 1 & 0 & 0 & 0 & \ldots & 0 \\ 1 & 1 & 0 & 0 & \ldots & 0 \\ 1 & 0 & 1 & 0 & \ldots & 0 \\ 1 & 1 & 0 & 1 & \ldots & 0 \\ \ldots & \ldots & \ldots & \ldots & \ldots & \ldots \\ 1 & \eta_{n2} & \eta_{n3} & \eta_{n4} & \ldots & 1 \end{vmatrix} \cdot \begin{vmatrix} a_1 & 0 & 0 & 0 & \ldots & 0 \\ a_1 & a_2 & 0 & 0 & \ldots & 0 \\ a_1 & 0 & a_3 & 0 & \ldots & 0 \\ a_1 & a_2 & 0 & a_4 & \ldots & 0 \\ \ldots & \ldots & \ldots & \ldots & \ldots & \ldots \\ a_1 & . & . & . & \ldots & a_n \end{vmatrix},$$

where the $\eta_{\lambda\mu}$ have the same meaning as in **31**. The second determinant is formed from the first by replacing $\eta_{\lambda\mu}$ by $a_\mu\eta_{\lambda\mu}$.

34. The first part is obvious.—Let $p, q, r, \ldots$ be the prime factors of n. Apply **26** to the divisors t of n. Let the value of t be a_t. Let us say that a divisor has the property α if it divides not only n but also n/p, the property β if it divides not only n but also n/q, and so on. The total value of those divisors t that divide neither n/p, nor n/q, nor $n/r, \ldots$ is then equal to

$$\begin{aligned} &\sum_{t/n} a_t - \sum_{t/\frac{n}{p}} a_t - \sum_{t/\frac{n}{q}} a_t - \sum_{t/\frac{n}{r}} a_t - \cdots \\ &\quad + \sum_{t/\frac{n}{pq}} a_t + \sum_{t/\frac{n}{pr}} a_t + \cdots \\ &\quad - \sum_{t/\frac{n}{pqr}} a_t - \cdots \\ &\quad \ldots\ldots\ldots\ldots\ldots\ldots \\ &\quad \pm \sum_{t/\frac{n}{pqr\cdots}} a_t = \sum_{t/n} \mu(t) A_{\frac{n}{t}} \end{aligned}$$

by the definition of $\mu(n)$. The only divisor, however, that has neither property α, nor β, nor $\gamma, \ldots$, is n and its value is a_n.

35. [A. Hurwitz.] By **34** it suffices to prove that

$$g(n) = \sum_{t/n} f(t)$$

But this is evident. For if we reduce the fractions $1/n, 2/n, 3/n, \ldots, n/n$ to their lowest terms, then we obtain the fractions of the form r/t where $r \leqq t$, $(r, t)=1$, and t is a divisor of n, and each of these fractions we obtain precisely once.

36. In **35** set

$$\psi(y) = \log(x - e^{2\pi i y}).$$

We then have $g(n) = \log(x^n - 1)$, $f(n) = \log K_n(x)$, and thus

$$\log K_n(x) = \sum_{t/n} \mu(t) \log(x^{\frac{n}{t}} - 1).$$

37. In **35** set $\psi(y) = e^{2\pi i y}$. We then have

$$g(1) = 1, \qquad g(n) = 0 \qquad \text{for} \qquad n > 1, \qquad \text{hence} \qquad f(n) = \sum_{(r,n)=1} e^{\frac{2\pi i r}{n}} = \mu(n).$$

38.

n	$\varphi(n)$	$\tau(n)$	$\sigma(n)$	$\sigma_2(n)$	$\nu(n)$	$\mu(n)$	$\lambda(n)$	$e^{\Lambda(n)}$
1	1	1	1	1	0	1	1	1
2	1	2	3	5	1	−1	−1	2
3	2	2	4	10	1	−1	−1	3
4	2	3	7	21	1	0	1	2
5	4	2	6	26	1	−1	−1	5
6	2	4	12	50	2	1	1	1
7	6	2	8	50	1	−1	−1	7
8	4	4	15	85	1	0	−1	2
9	6	3	13	91	1	0	1	3
10	4	4	18	130	2	1	1	1

38.1. It is advantageous to make use of **43**.

(1) If p is a prime number and $m \geqq 1$, then

$$\varphi(p^m) = p^{m-1}(p-1)$$

is divisible by $p-1$ and thus an even number if $p > 2$. If n is divisible by p^m, then $\varphi(n)$ is divisible by $\varphi(p^m)$ and therefore by $p-1$, and so it is even if p is odd. Hence $\varphi(n)$ can be odd only if n is a power of 2, but

$$\varphi(2^m) = 2^{m-1}$$

is even if $m \geqq 2$.

(2) Proved by (1).

(3) If n is divisible by a prime $p > 3$, then $\varphi(n)$ has a factor $> 3-1$, and so $\varphi(n) > 2$. Hence we need to examine only the cases

$$\varphi(2^l) = 2^{l-1}, \qquad \varphi(3^m) = 3^{m-1} \cdot 2, \quad \varphi(2^l 3^m) = 2^l \cdot 3^{m-1}$$

for $l \geqq 1$, $m \geqq 1$.

(4) If n has the divisor d, it has also the "complementary" divisor n/d and these two divisors are different unless $n = d^2$.

(5) No. If n is not a perfect square multiplied by some power of 2, then there is an odd prime number p whose highest power that divides n is p^{2k-1}; k is a

positive integer. Hence $\sigma(n)$ is divisible by

$$\sigma(p^{2k-1})=1+p+p^2+\cdots+p^{2k-1},$$

which is an even number.

(6) Further exploration may suggest results similar to **45**.

39. By the Cauchy and the Dirichlet multiplication rules respectively, cf. introduction to **39** p. 119, $a_n=b_n=1$.

40. By the Cauchy and the Dirichlet multiplication rules respectively, cf. introduction to **39** p. 119, $b_n=1$.

41. Special case of **42**: $A_0=A_1=A_2=A_3=\cdots=1$.

42. By Cauchy and Dirichlet multiplication respectively, with the aid of **34**.

43. If m and n have no common divisors, then every divisor of mn is formed by multiplying one divisor of m by one divisor of n. We therefore have

$$\sum_{t_1/m} t_1^{\alpha}\cdot\sum_{t_2/n} t_2^{\alpha}=\sum_{t/mn} t^{\alpha}, \qquad \text{i.e.} \quad \sigma_\alpha(m)\sigma_\alpha(n)=\sigma_\alpha(mn).$$

For $\varphi(n)$ the assertion follows from **25.** For the other functions the assertion is obvious.—For $f(n)=n^\alpha$, $\lambda(n)$, the relation $f(mn)=f(m)f(n)$ is satisfied not only for relatively prime but in fact for arbitrary m and n.

44. By **43**, it is sufficient to consider the case $n=p^k$, p a prime number. The divisors are then $1, p, p^2, \ldots, p^k$, and hence

$$\sigma_\alpha(n)=1+p^\alpha+p^{2\alpha}+\cdots+p^{k\alpha}=\frac{1-p^{\alpha(k+1)}}{1-p^\alpha}.$$

45. The values of the quotient

$$\frac{\tau(p^k)}{\varphi(p^k)}=\frac{k+1}{p^{k-1}(p-1)}$$

can be arranged in a two-way classification table.

p \ k	1	2	3	4
2	$\mathbf{\frac{2}{1}}$	$\mathbf{\frac{3}{2}}$	$\mathbf{\frac{4}{4}}$	$\frac{5}{8}$
3	$\mathbf{\frac{2}{2}}$	$\frac{3}{6}$	$\frac{4}{18}$	$\frac{5}{54}$
5	$\frac{2}{4}$	$\frac{3}{20}$	$\frac{4}{100}$	$\frac{5}{500}$

In the above table entries $\geqq 1$ are printed in bold type. The entries decrease with increasing p as well as with increasing k (differentiate!) and tend to 0. Since $\tau(n)/\varphi(n)$ is a multiplicative function of n, we require to form all products of value $\geqq 1$ whose factors ($\geqq 1$ in number) are taken from *different* rows of the table. There exist only ten such products corresponding to the numbers $n=2$, 3, 4, 6, 8, 10, 12, 18, 24, 30, which together with $n=1$, comprise all the solutions of the inequality $\tau(n)\geqq\varphi(n)$.

46. It suffices to consider a single fixed prime number p and a specific multiplicative function that is associated with p in the following manner: $f(n)=g(k)$, if p^k is the highest power of p that divides n; $g(0)=1$ and $g(k)$ is a fixed but otherwise arbitrary function of the exponent k for $k\geqq 1$. All multiplicative functions may be constructed from such functions by multiplication. But for such a function the theorem reads as follows:

$$g(\max(\alpha,\beta,\gamma,\delta,\ldots,\kappa,\lambda))g(\min(\alpha,\beta))g(\min(\alpha,\gamma))\cdots$$
$$g(\min(\kappa,\lambda))g(\min(\alpha,\beta,\gamma,\delta))\cdots=g(\alpha)g(\beta)\cdots g(\lambda)g(\min(\alpha,\beta,\gamma))\cdots,$$

where $\alpha,\beta,\gamma,\delta,\ldots,\kappa,\lambda$ are the non-negative exponents of p in $a,b,c,d,\ldots,k,l$. This is now a generalization of **28**, that can be proved in an analogous fashion to **28** with the single difference that here we must use **26** instead of **21**, setting the "value" of a number m equal to $\log g(m)-\log g(m-1)$; $\log g(-1)=0$.

47. Every positive integer may be written in one and only one way as a product of powers of prime numbers. By the rules for the construction of the infinite product in question we thus obtain every term $f(n)n^{-s}$, where $n=p_1^{k_1}p_2^{k_2}\ldots p_\nu^{k_\nu}$, once and exactly once, namely in the form $f(p_1^{k_1})p_1^{-k_1s}f(p_2^{k_2})p_2^{-k_2s}\ldots f(p_\nu^{k_\nu})p_\nu^{-k_\nu s}$. The theorem is completely equivalent to the multiplicative property of $f(n)$.

48. [Euler: Introductio in analysin infinitorum, **1**, Opera Omnia, Series 1, **8**, p. 288. Leipzig and Berlin: B. G. Teubner, 1922.] $f(n)=1$, $n=1,2,3,4,\ldots$ is multiplicative [**47**].

49. Because of the multiplicative property we have

$$\sum_{n=1}^{\infty}\sigma_\alpha(n)n^{-s}=\prod_p\frac{(1-p^\alpha)1^{-s}+(1-p^{2\alpha})p^{-s}+(1-p^{3\alpha})p^{-2s}+\cdots}{1-p^\alpha}$$
$$=\prod_p\frac{1}{(1-p^{-s})(1-p^{\alpha-s})},$$
$$\sum_{n=1}^{\infty}2^{\nu(n)}n^{-s}=\prod_p(1+2p^{-s}+2p^{-2s}+2p^{-3s}+\cdots)=\prod_p\frac{1-p^{-2s}}{(1-p^{-s})^2},$$
$$\sum_{n=1}^{\infty}\lambda(n)n^{-s}=\prod_n(1-p^{-s}+p^{-2s}-p^{-3s}+\cdots)=\prod_p\frac{1-p^{-s}}{1-p^{-2s}},$$
$$\sum_{n=1}^{\infty}\varphi(n)n^{-s}=\prod_p\left[1+\left(1-\frac{1}{p}\right)p^{1-s}+\left(1-\frac{1}{p}\right)p^{2-2s}+\cdots\right]=\prod_p\frac{1-p^{-s}}{1-p^{1-s}}.$$

50. [E. Cesàro, problem: Mathesis, **6**, p. 192 (1886). Solution by Mantel: Mathesis, **8**, p. 208 (1888).] $a(n)$ is a multiplicative function, hence [**47**]

$$\sum_{n=1}^{\infty}a(n)n^{-s}=\prod_p[1+a(p)p^{-s}+a(p^2)p^{-2s}+\cdots]$$
$$=(1^{-s}+2^{-s}+2^{-2s}+\cdots)\prod_{p>2}(1+p^{1-s}+p^{2(1-s)}+\cdots)$$
$$=\frac{1}{1-2^{-s}}\prod_{p>2}\frac{1}{1-p^{1-s}}.$$

51. By **48** we have

$$-\frac{\zeta'(s)}{\zeta(s)}=\sum_p \frac{p^{-s}\log p}{1-p^{-s}}.$$

52.

$$\sum_{n=1}^{\infty} n^{-s} \sum_{n=1}^{\infty} \mu(n)n^{-s}=1.$$

Equivalent to the second part of **41**.—**52–56**, **58** may also be proved without using Dirichlet series.

53. By **49** we have

$$\sum_{n=1}^{\infty} (n^2)^{-s}=\sum_{n=1}^{\infty} n^{-s} \sum_{n=1}^{\infty} \lambda(n)n^{-s}.$$

54. By **49** we have

$$\sum_{n=1}^{\infty} n\cdot n^{-s}=\sum_{n=1}^{\infty} n^{-s} \sum_{n=1}^{\infty} \varphi(n)n^{-s}.$$

55. By **49** we have

$$\sum_{n=1}^{\infty} n\cdot n^{-s} \sum_{n=1}^{\infty} \mu(n)n^{-s}=\sum_{n=1}^{\infty} \varphi(n)n^{-s};$$

alternatively by applying **34** to **54**. Equivalent to **25**.

56.

$$-\zeta'(s)=\left(-\frac{\zeta'(s)}{\zeta(s)}\right)\zeta(s),$$

i.e. $\sum_{n=1}^{\infty} \log n\cdot n^{-s}=\sum_{n=1}^{\infty} \Lambda(n)n^{-s} \sum_{n=1}^{\infty} n^{-s}$.

57. From **33**, **54**. Further particular cases of **33** follow from **52–56**.

58. [Jacobi, problem: Nouv. Ann. **11**, p. 45 (1852). Solution by A. Dallot *et al.*: Nouv. Ann. **11**, p. 126, p. 186 (1852).]

$$\sum \alpha^{-s}=1^{-s}+3^{-s}+5^{-s}+\cdots=(1-2^{-s})\zeta(s),$$
$$\sum \beta^{-s}=2^{-s}+4^{-s}+6^{-s}+\cdots=2^{-s}\zeta(s);$$

hence we have

$$\begin{aligned}\sum\sum(\beta-\alpha)(\alpha\beta)^{-s}&=\sum\sum \alpha^{-s}\beta^{1-s}-\sum\sum \alpha^{1-s}\beta^{-s}\\ &=(1-2^{-s})\zeta(s)2^{1-s}\zeta(s-1)-(1-2^{1-s})\zeta(s-1)2^{-s}\zeta(s)\\ &=2^{-s}\zeta(s)\zeta(s-1)=\sum_{m=1}^{\infty} \sigma(m)(2m)^{-s}.\end{aligned}$$

58.1. [A. F. Möbius; see E. Netto: Lehrbuch der Kombinatorik, 2nd ed., Leipzig, Teubner, 1927, pp. 173–176, p. 277.]

$$\frac{1}{\zeta(s)}=1-(\zeta(s)-1)+(\zeta(s)-1)^2-(\zeta(s)-1)^3+\cdots,$$

$$\sum_{n=1}^{\infty} \mu(n)n^{-s}=1-\sum_{k=2}^{\infty} k^{-s}+\sum_{k=2}^{\infty}\sum_{l=2}^{\infty} (kl)^{-s}-\sum_{k=2}^{\infty}\sum_{l=2}^{\infty}\sum_{m=2}^{\infty} (klm)^{-s}+\cdots.$$

In fact, for $k=1, 2, 3, \ldots$,

$$(\zeta(s)-1)^k=\sum_{n=1}^{\infty} P(n, k)n^{-s}.$$

58.2. [Viggo Brun; see E. Netto, op. cit., **58.1**, p. 276.] By **48**

$$\begin{aligned}\log \zeta(s) &= -\sum_p \log(1-p^{-s}) \\ &= \sum_p (p^{-s}+\tfrac{1}{2}p^{-2s}+\tfrac{1}{3}p^{-3s}+\cdots) \\ &= (\zeta(s)-1)-\tfrac{1}{2}(\zeta(s)-1)^2+\tfrac{1}{3}(\zeta(s)-1)^3-\cdots;\end{aligned}$$

see **58.1**.

58.3. A decomposition of such an n into k factors (each >1) is, in fact, a partition of its m prime factors into k classes, if we disregard the order of the factors. Observe that the k factors must be different. For an explicit formula for S_k^m see **22.1**, I **189**.

58.4. Proof by following up the intuitive solution of I **21**; cf. MD, **2**, p. 189, problem 3.40.1.

58.2 is now also readily verified:

$$\frac{1}{1}\binom{m-1}{0}-\frac{1}{2}\binom{m-1}{1}+\cdots+\frac{(-1)^{m-1}}{m}\binom{m-1}{m-1}=\int_0^1 (1-x)^{m-1}\,dx=\frac{1}{m}.$$

58.5. See solution **58.3**.

58.6.

$$\begin{aligned}\sum_{n=1}^{\infty} Q(n)n^{-s} &= \prod_{j=2}^{\infty} (1+j^{-s}+j^{-2s}+j^{-3s}+\cdots) \\ &= \exp\left(-\sum_{j=2}^{\infty} \log(1-j^{-s})\right) \\ &= \exp\left(\sum_{j=2}^{\infty} j^{-s}+\tfrac{1}{2}\sum_{j=2}^{\infty} j^{-2s}+\tfrac{1}{3}\sum_{j=2}^{\infty} j^{-3s}+\cdots\right).\end{aligned}$$

58.7.

$$\begin{aligned}\sum_{n=1}^{\infty} \delta(n)n^{-s} &= (1+2^{-s}+2^{-2s}+\cdots) \\ &\quad \times\prod_p (1+2p^{-s}+3p^{-2s}+\cdots)\prod_q (1+q^{-2s}+q^{-4s}+\cdots) \\ &= \frac{1}{1-2^{-s}}\prod_p \frac{1}{(1-p^{-s})^2}\prod_q \frac{1}{1-q^{-2s}} \\ &= \zeta(s)\prod_p \frac{1}{(1-p^{-s})}\prod_q \frac{1}{1+q^{-s}} \\ &= \zeta(s)\prod_p (1+p^{-s}+p^{-2s}+\cdots)\prod_q (1-q^{-s}+q^{-2s}-\cdots) \\ &= \zeta(s)\left(1-\frac{1}{3^s}+\frac{1}{5^s}-\frac{1}{7^s}+\frac{1}{9^s}-\cdots\right).\end{aligned}$$

$\prod_p$ and $\prod_q$ are extended over the primes $\equiv 1$ and $\equiv 3 \pmod 4$, respectively. It can also be proved without the use of Dirichlet series; see Hardy-Wright, pp. 240–242. See also Gauss: Disquisitiones arithmeticae, Art. 182, annotation, Werke, **1**, p. 161, Göttingen, 1863. See also **237.2**.

58.8. Only $\delta(n)$. Thus, for $k \geqq 2$,

$$P(2^k, k)=1, \qquad P(3^k, k)=1, \qquad P(6^k, k)>1.$$

59. By **47**, a function $h(n)$ is multiplicative if

$$\prod_p (1^{-s}+h(p)p^{-s}+h(p^2)p^{-2s}+\cdots)=\sum_{n=1}^{\infty} h(n)n^{-s}.$$

In our case, however, we have

$$\begin{aligned}&1^{-s}+h(p)p^{-s}+h(p^2)p^{-2s}+\cdots\\&\quad=(1^{-s}+f(p)p^{-s}+f(p^2)p^{-2s}+\cdots)(1^{-s}+g(p)p^{-s}+g(p^2)p^{-2s}+\cdots),\end{aligned}$$

whence the assertion follows.

60. Let the corners of the regular *convex* n-gon, numbered consecutively, be $A_1, A_2, A_3, \ldots, A_n$. The line joining the corners A_k and A_l belongs to a regular (convex or self-intersecting) n/t-gon, where $t=(n, l-k)$; after tracing this line n/t times, we arrive back at the starting point. Thus from every corner, in accordance with $t=1$, there emanate $\varphi(n)$ connecting lines that belong to one of the n-gons; these have therefore a total of $[n\varphi(n)]/2$ sides.

61. If k/n is in its lowest terms, then so is $(n-k)/n$. The sum of the two $=1$.

62. [G. Frobenius; see A. Errera: Palermo Rend. **35**, p. 110 (1913).] Let $(b, c)=d$. If $(a, b, c)=1$, then only those fractions can be reduced whose numerators are divisible by prime numbers that divide c but do not divide b and thus do not divide d. Analogously to **25**, it therefore follows that the number of irreducible fractions is

$$N=c\prod_{\substack{p/c\\(p,d)=1}}\left(1-\frac{1}{p}\right)=\frac{c\prod_{q/c}\left(1-\frac{1}{q}\right)}{\prod_{r/d}\left(1-\frac{1}{r}\right)}=\frac{\varphi(c)}{\frac{\varphi(d)}{d}}.$$

But $\varphi(bc)=d\varphi(bc/d)$, since bc and bc/d contain the same prime numbers. Moreover by **46** we have $\varphi(bc/d)\varphi(d)=\varphi(b)\varphi(c)$.

63.
$$\varphi(1)+2[\varphi(2)+\varphi(3)+\cdots+\varphi(n)].$$

64. (1) The number of columns of the square array in which the imaginary parts of the numerator have a fixed greatest common divisor t with n is $\varphi(n/t)$. In each of these columns the number of irreducible fractions is equal to the number of numbers less than n that are relatively prime with t, i.e. $(n/t)\varphi(t)$. The number of all the irreducible fractions is therefore

$$\sum_{t/n}\varphi(t)\cdot\frac{n}{t}\,\varphi\left(\frac{n}{t}\right),$$

and thus [**43**, **59**] multiplicative.—For $n=p^k$ the number of reducible fractions is $p^{k-1} \cdot p^{k-1}$, and hence the number of irreducible fractions $p^{2k} \cdot \cdot p^{2k-2}$.—As a matter of fact, from $\Phi(n) = \sum_{t/n} \varphi(t)(n/t)\varphi(n/t)$, (4) also follows directly, for we obtain

$$\sum_{n=1}^{\infty} \Phi(n)n^{-s} = \sum_{k=1}^{\infty} \varphi(k)k^{-s} \cdot \sum_{l=1}^{\infty} \varphi(l)l^{1-s} = \frac{\zeta(s-1)}{\zeta(s)} \cdot \frac{\zeta(s-2)}{\zeta(s-1)} \qquad \textbf{[49]}.$$

(2) Follows from **21** in the same way as **25.**

(3) The number of fractions of the array that may be reduced precisely by the factor t is $\Phi(n/t)$. Summing over all n/t it follows that

$$n^2 = \sum_{t/n} \Phi\left(\frac{n}{t}\right) = \sum_{t/n} \Phi(t).$$

(1) follows from (4) by **47** and conversely.

(2) follows from (4) and conversely; for we have

$$\sum_{n=1}^{\infty} \Phi(n)n^{-s} = \frac{\zeta(s-2)}{\zeta(s)} = \sum_{n=1}^{\infty} \sum_{d/n} \mu(d) \frac{n^2}{d^2} n^{-s}$$

[41], and hence

$$\Phi(n) = \sum_{d/n} \mu(d) \frac{n^2}{d^2} = n^2\left(1 - \frac{1}{p^2} - \frac{1}{q^2} - \cdots + \frac{1}{p^2q^2} + \cdots\right)$$
$$= n^2\left(1 - \frac{1}{p^2}\right)\left(1 - \frac{1}{q^2}\right) \cdots.$$

(3) follows from (4) and conversely by Dirichlet multiplication.

65. From the first identity we obtain

$$A_n = \sum_{t/n} a_t,$$

and from the second identity

$$\sum_{n=1}^{\infty} \frac{a_n x^n}{1-x^n} = \sum_{m=1}^{\infty} a_m(x^m + x^{2m} + x^{3m} + \cdots) = \sum_{n=1}^{\infty} \left(\sum_{t/n} a_t\right) x^n.$$

66. We have

$$(1 - 2^{1-s})\zeta(s) = 1^{-s} - 2^{-s} + 3^{-s} - 4^{-s} + \cdots.$$

From the first identity it therefore follows that

$$B_n = \sum_{t/n} (-1)^{t-1} a_{\frac{n}{t}},$$

and from the second identity

$$\sum_{n=1}^{\infty} \frac{a_n x^n}{1+x^n} = \sum_{m=1}^{\infty} a_m(x^m - x^{2m} + x^{3m} - \cdots) = \sum_{n=1}^{\infty} \left(\sum_{t/n} (-1)^{t-1} a_{\frac{n}{t}}\right) x^n.$$

67. Writing A_n in the form $A_n = \sum_{t/n} a_t$ and collecting the factors with exponent a_m, we obtain on the right-hand side the product

$$\left(1 - \frac{x^2}{m^2\pi^2}\right)^{a_m} \left(1 - \frac{x^2}{(2m)^2\pi^2}\right)^{a_m} \left(1 - \frac{x^2}{(3m)^2\pi^2}\right)^{a_m} \cdots.$$

On the left-hand side we have correspondingly the a_mth power of

$$\frac{m}{x}\sin\frac{x}{m}=\left(1-\frac{x^2}{m^2\pi^2}\right)\left(1-\frac{x^2}{(2m)^2\pi^2}\right)\left(1-\frac{x^2}{(3m)^2\pi^2}\right)\cdots.$$

68. Cf. solution **66**:

$$\prod_{k=1}^{\infty}\left(1-\frac{x^2}{(km)^2\pi^2}\right)^{(-1)^{k-1}}=\frac{x}{2m}\cot\frac{x}{2m}.$$

68.1. The left-hand side of the required equation may be written in the form

$$\begin{aligned}a_1f(x)+a_1f(2x)+a_1f(3x)+a_1f(4x)+&\cdots\\ +a_2f(2x)\qquad\qquad +a_2f(4x)+&\cdots\\ +a_3f(3x)\qquad\qquad +&\cdots\\ +a_4f(4x)+&\cdots.\\ &\cdot\\ &\cdot\\ &\cdot\end{aligned}$$

Cf. a diagram due to Leibniz, MD, **2**, p. 168. (Questions of convergence are not considered in the present chapter although they may be important in concrete applications.)

If $f(x)=e^{-x}$, then $F(x)=\dfrac{e^{-x}}{1-e^{-x}}$.

If $f(x)=\log\left(1-\dfrac{1}{\pi^2x^2}\right)$, then $F(x)=\log\left(x\sin\dfrac{1}{x}\right)$. Hence **65** and **67** easily follow.

If $f(x)=x^{-s}$, then $F(x)=\zeta(s)f(x)$. How is this result related to **40**?

68.2. Analogous to **68.1**.

68.3. [Hardy-Wright, p. 237, theorem 270.] Assume the first equation and apply **68.1** to the particular case

$$a_n=\mu(n),\qquad A_1=1,\qquad A_n=0\qquad\text{for }n\geqq 2.$$

Assume the second equation. Imitate solution **68.1**, add the equations:

$$\begin{aligned}f(x)&=\mu(1)F(x)+\mu(2)F(2x)+\mu(3)F(3x)+\mu(4)F(4x)+\cdots\\ f(2x)&=\qquad\qquad \mu(1)F(2x)\qquad\qquad\qquad +\mu(2)F(4x)+\cdots\\ f(3x)&=\qquad\qquad\qquad\qquad \mu(1)F(3x)\qquad\qquad +\cdots\\ f(4x)&=\qquad\qquad\qquad\qquad\qquad\qquad\qquad \mu(1)F(4x)+\cdots\end{aligned}$$

and use **52** (which was in fact also used in the first case.)

(It deserves to be mentioned that the "inversion formula" just proved played a role in a problem of electrophysiology raised by B. W. Knight, and discussed at Rockefeller University, especially by J. Towber.—Communication by M. Schreiber.)

69. From **65**, observing **41** and **49**, we find that the desired functions are x and $x/(1-x)^2$.

70. [Baschwitz, problem: Mathesis, Series 2, **3**, p. 80 (1893). Solution by

E. Cesàro: Mathesis, Series 2, **3**, p. 205 (1893). Laguerre: Oeuvres, **1**, p. 216. Paris: Gauthier-Villars, 1898.] From **65** and **49**.

71. From **66**, observing **41**, **49**.

72. From **66**, observing **49**.—By I **93** the sum of the series

$$\sum_{n=1}^{\infty} \lambda(n) \frac{x^n}{1+x^n}$$

converges to $-\infty$ as $x \to 1$. An analogous result for the series $\sum_{n=1}^{\infty} \lambda(n)x^n$ would decide the Riemann hypothesis. [For a related conjecture, its heuristic background and its fate, see MPR, **2**, p. 49, ex. 16 and p. 209, ex. 13.20.]

72.1. [Cf. Richard Bellman, problem: Amer. Math. Monthly, **50**, pp. 124–125 (1943).] From **69** by integration.

73. Cf. **64**, **65**, **66**.

74. By **39**, **65** we have

$$\begin{aligned}\sum_{n=1}^{\infty} \tau(n)x^n = \sum_{n=1}^{\infty} \frac{x^n}{1-x^n} = {} & x + x^2 + x^3 + x^4 + \cdots \\ & + x^2 + x^4 + x^6 + x^8 + \cdots \\ & + x^3 + x^6 + x^9 + x^{12} + \cdots \\ & \cdots\cdots\cdots\cdots\cdots\cdots\end{aligned}$$

Sum successively the terms lying *to the right and below* each of the diagonal elements $x, x^4, x^9, \ldots$. Cf. also II **32**.

74.1. See **58.7**, **65**.

***75.** First combine **49** and **65**:

$$\begin{aligned}\sum_{n=1}^{\infty} \sigma(n)x^n = \sum_{n=1}^{\infty} n \frac{x^n}{1-x^n} = {} & x + x^2 + x^3 + x^4 + \cdots \\ & + 2x^2 + 2x^4 + 2x^6 + 2x^8 + \cdots \\ & + 3x^3 + 3x^6 + 3x^9 + 3x^{12} + \cdots \\ & \cdots\cdots\cdots\cdots\cdots\cdots\end{aligned}$$

The double series, obtained by rows, should then be summed by columns.

75.1. Use **75**. From I **54**; differentiate the logarithm of both sides.

75.2. Use **75**. From A I **54.1**; differentiate the logarithm of both sides.

75.3. [Euler: Opera Omnia, Series 1, **2**, Leipzig and Berlin: B. G. Teubner, 1915, especially pp. 241–253 (translated and annotated MPR, **1**, pp. 90–107) and p. 390–398 (proof).] From **75.1**.

75.4. From **75.2**.

75.5. From **75** and A I **20.2**.

$$\sigma(1)x + \sigma(2)x^2 + \sigma(3)x^3 + \cdots = \frac{\sum_1^{\infty} np(n)x^n}{\sum_0^{\infty} p(n)x^n}.$$

A common generalization of **75.3**, **75.4** and **75.5** is obvious, but to discover some interesting new particular case of this generalization is not obvious.

76.
$$\sum_{n=0}^{\infty} \frac{p^n}{1-qx^n} = \sum_{m=0}^{\infty} \sum_{n=0}^{\infty} p^n q^m x^{mn} = \sum_{m=0}^{\infty} \frac{q^m}{1-px^m}.$$

77. In **76** replace p by x, q by $-x^2$, x by x^2.

78. In the expansion of $x/(1-x^2)$ we have as exponents the odd numbers, in that of $x^2/(1-x^4)$ the numbers that are divisible by 2 but not by 4, in that of $x^4/(1-x^8)$ the numbers that are divisible by 4 but not by 8, and so on. Cf. solution I **19**, where the classification of all the numbers according to the highest power of 2 which they contain as a factor also plays a part. The required identity may also be obtained from I **164** by logarithmic differentiation.—The second identity follows from **66** or from solution I **14** by logarithmic differentiation.

79. [Cf. P. G. Lejeune-Dirichlet: Werke, **2**, p. 52, Berlin: G. Reimer, 1897.] The lattice points in question may be counted in two different ways. The number of lattice points on the hyperbola $xy=k$ is equal to $\tau(k)$. From this relation for $k=1, 2, \ldots, n$ we obtain the left-hand side of the equation. The number of lattice points on the straight line $x=k$ parallel to the y-axis is $[n/k]$. From this relation for $k=1, 2, \ldots, n$ we obtain the right-hand side. Cf. II **46**.

80. The lattice points considered in **79** may also be counted in the following way: Take the number of lattice points in the two strips

$$1 \leqq x \leqq \nu, \qquad xy \leqq n \quad \text{and} \quad 1 \leqq y \leqq \nu, \qquad xy \leqq n,$$

and subtract from this the number of lattice points in $1 \leqq x \leqq \nu$, $1 \leqq y \leqq \nu$, i.e. ν^2. [In the region $x>\nu$, $y>\nu$, $xy \leqq n$ there does not lie a single lattice point since $(\nu+1)^2>n$.]

80.1. If p is a prime number and the integer m is characterized by the two-sided inequality

$$p^m \leqq n < p^{m+1},$$

then the proposed sum has m terms equal to $\log p$, $U(n)$ clearly contains the factor p^m, and $m=[\log n/\log p]$. Hence, with the notation of **82**, the proposed sum

$$=\sum_{p \leqq n} \left[\frac{\log n}{\log p}\right] \log p = \log U(n).$$

80.2. Cf. last line of solution **80.1.** The ratio of the two sides of the required inequality tends to 1 as $n \to \infty$; see Hardy-Wright, p. 347, theorem 427.

81. With each lattice point (k, l) in the region $x>0$, $y>0$, $xy \leqq n$ associate the value $a_k b_l$. By methods of enumeration similar to those in **79** we obtain different representations of Γ_n, which is in fact the "total value" of all the above-mentioned lattice points.

82. In **81** set $a_n=\Lambda(n)$, $b_n=1$. Then $c_n=\log n$ and $B_n=n$, $\Gamma_n=\log n!$, and hence we obtain

$$\log n! = \sum_{r=1}^{n} a_r B_{\left[\frac{n}{r}\right]} = \sum_{r=1}^{n} \Lambda(r)\left[\frac{n}{r}\right] = \sum_{p \leqq n} \sum_{m=1}^{\infty} \log p \left[\frac{n}{p^m}\right].$$

83. [**80**.]

84. The binomial coefficients are integers. $\binom{x}{m}$ has integral values also for negative integers x, since

$$\binom{-x}{m} = (-1)^m \binom{x+m-1}{m}.$$

85. The functions $1, x, x^2, \ldots, x^n$ may be successively expressed as linear combinations with constant coefficients of

$$1, \frac{x}{1}, \frac{x(x-1)}{2!}, \ldots, \frac{x(x-1)\ldots(x-n+1)}{n!} \qquad \text{(cf. I } \mathbf{191}\text{)}.$$

The coefficients $b_0, b_1, \ldots, b_m$ may be determined from

$$P(0)=b_m,$$
$$P(1)=b_m+\binom{1}{1}b_{m-1},$$
$$P(2)=b_m+\binom{2}{1}b_{m-1}+\binom{2}{2}b_{m-2},$$
$$\cdots\cdots\cdots\cdots\cdots\cdots$$
$$P(m)=b_m+\binom{m}{1}b_{m+1}+\cdots+\binom{m}{m}b_0.$$

If $P(0), P(1), \ldots, P(m)$ are integers, we deduce that $b_m, b_{m-1}, \ldots, b_0$ are integers.

86. [**85**.]

87. We may assume that the points at which by hypothesis $P(x)$ assumes integral values are $0, 1, \ldots, m$. Then the assertion follows from solution **85**.

88. [G. Pólya: Palermo Rend. **40**, p. 5 (1915).]

$$\binom{x+m-1}{2m-1}=\frac{x(x^2-1^2)(x^2-2^2)\ldots[x^2-(m-1)^2]}{(2m-1)!}.$$

The coefficients $c_1, c_2, \ldots, c_m$ may be determined successively from

$$P(1)=c_1,$$
$$P(2)=c_1\binom{2}{1}+c_2,$$
$$P(3)=c_1\binom{3}{1}+c_2\binom{4}{3}+c_3,$$
$$\cdots\cdots\cdots\cdots\cdots\cdots$$

[solution **85**.]

89. The polynomial of degree $2m$

$$\frac{x}{m}\binom{x+m-1}{2m-1}= \quad 1, \quad 0, \quad \ldots, \quad 0, 0, 0, \ldots, 0, \quad 1,$$
$$\text{for } x=-m, -m+1, \ldots, -1, 0, 1, \ldots, m-1, m;$$

thus it is integral-valued [**87**]. Moreover

$$P(0)=d_0,$$
$$P(1)=d_0+d_1,$$
$$P(2)=d_0+d_1\frac{2}{1}\binom{2}{1}+d_2,$$
$$\cdots\cdots\cdots\cdots\cdots\cdots$$

[solution **85, 88**.]

90. [G. Pólya: Jber. deutsch. Math. Verein. **28**, pp. 31–40 (1919).] By VI **70**

a polynomial $P(x)=a_0x^m+a_1x^{m-1}+\cdots+a_m$, $a_0\neq 0$, with integral coefficients assumes at at least one of $m+1$ distinct integral points a value whose absolute value is $\geqq (m!/2^m)|a_0|\geqq m!/2^m$. For $m\geqq 4$, we have $m!/2^m>1$. With regard to $m\leqq 3$ cf. the examples:

$$\begin{array}{llll} m=1, & P(x)=x & \text{for} & x=-1,1, \\ m=2, & P(x)=x(x-1)-1 & \text{for} & x=-1,0,1, \\ m=3, & P(x)=(x+1)x(x-2)+1 & \text{for} & x=-1,0,1,2. \end{array}$$

91. From the Lagrange interpolation formula, cf. VI, introduction to § 9 p. 82.

92. *First solution.* Let the function in question be $R(x)=P(x)/Q(x)$, where $P(x)$, $Q(x)$ are relatively prime polynomials, r the sum of the degrees of $P(x)$ and $Q(x)$. For $r=0$ the theorem is obvious. Consider if necessary $R(x)^{-1}$ in place of $R(x)$ and assume that the degree of $P(x)$ is not less than that of $Q(x)$ and furthermore that a is a positive integer such that $Q(a)\neq 0$. Then $P(a)/Q(a)$ is rational and

$$\frac{1}{x-a}\left(R(x)-\frac{P(a)}{Q(a)}\right)=\frac{P^*(x)}{Q(x)}$$

is a rational function whose value is rational for integral x, $x>a$, and the degree of

$$P^*(x)=\frac{P(x)Q(a)-Q(x)P(a)}{Q(a)(x-a)}$$

is less than that of $P(x)$; thus the sum of the degrees of $P^*(x)$ and $Q(x)$ is less than that of $P(x)$ and $Q(x)$. The assertion follows from this by mathematical induction.

Second Solution. Let the function in question be $R(x)=P(x)/Q(x)$, where $P(x)$ and $Q(x)$ are relatively prime polynomials of degree m and n respectively. Let the values of the function for $x=0, 1, 2, \ldots, m+n$ be the rational numbers $r_0, r_1, r_2, \ldots, r_{m+n}$. Consider the system of $m+n-1$ homogeneous linear equations

$$u_0+u_1k+u_2k^2+\cdots+u_mk^m-v_0r_k-v_1r_kk-v_2r_kk^2-\cdots-v_nr_kk^n=0$$
$$(k=0,1,2,\ldots,m+n)$$

for the $m+n+2$ unknowns $u_0, u_1, \ldots, u_m, v_0, v_1, \ldots, v_n$. To two solutions of this system there correspond two pairs of polynomials $P(x)$, $Q(x)$ and $P^*(x)$, $Q^*(x)$ where $P(x)$, $P^*(x)$ are of degree $\leqq m$, and $Q(x)$, $Q^*(x)$ of degree $\leqq n$, such that

$$P(k)-r_kQ(k)=0, \qquad P^*(k)-r_kQ^*(k)=0, \qquad k=0,1,\ldots,m+n.$$

But now from the vanishing of the function $P(x)Q^*(x)-P^*(x)Q(x)$ of degree $\leqq m+n$, for $x=0, 1, \ldots, m+n$ there follows that it vanishes identically. Since $P(x)$ and $Q(x)$ are relatively prime, $P(x)$ must divide $P^*(x)$ and thus $P^*(x)=cP(x)$, $Q^*(x)=cQ(x)$, c constant. I.e. the system has, apart from proportionality factors only one solution. Its rank is therefore $m+n+1$. Consequently the matrix of the system contains a determinant of order $m+n+1$ that is $\neq 0$. All the elements of this matrix are rational numbers. Hence we may assume a solution $u_0, u_1, \ldots, u_m, v_0, v_1, \ldots, v_n$ that is integral.

It would be sufficient to assume that $R(x)$ is finite and rational for $m+n+1$ distinct rational values of x.

93. The function $f(x)$ considered is [solution **92**, cf. the final remark] $=P(x)/Q(x)$, where $P(x)$ and $Q(x)$ are polynomials with integral coefficients. We can find an integer q, such that $qf(x)=G(x)+r(x)$, where $G(x)$ denotes a polynomial *with integral coefficients* and $r(x)$ a rational function whose numerator is of lower degree than its denominator. The value of $r(x)$ is integral for infinitely many integral values of x. Since $\lim_{x\to\infty} r(x)=0$, from a certain point of this kind on we must have $|r(x)|<1$, thus $r(x)=0$ and hence $r(x)\equiv 0$, for a rational function that does not vanish identically can be $=0$ for only a finite number of values of x.

94. From

$$2^m \equiv -1 \qquad (\text{mod } p)$$

it follows that

$$2^{2km}\equiv 1, \qquad 2^{2km}+1\equiv 2 \qquad (\text{mod } p),$$

where m and k are positive integers, p an odd prime number. From this we deduce that the number of prime numbers up to the bound x is certainly $\geqq$ const. log log x.

95. [G. Pólya: Math. Z. **1**, p. 144 (1918).] Leaving aside the case $a=0$, we may assume that $(a, d)=1$, $d\geqq 1$, $a>d$. The numbers

$$n=\frac{a}{d}\left(a^{\varphi(d)\nu}-1\right),$$

$\nu=1, 2, 3, \ldots$, are integral, and the numbers

$$a+nd=a^{\varphi(d)\nu+1}$$

contain only prime factors of a.

96. A product of numbers of the form $6n+1$ is again of the same form.

97. [Goldbach; cf. Euler: Opera Omnia, Series 1, **3**, p. 4, p. 337. Leipzig: B. G. Teubner 1917.]

First solution. Let us suppose [**85**] that

$$P(x+n)=b_0\binom{x}{m}+b_1\binom{x}{m-1}+\cdots+b_{m-1}\binom{x}{1}+b_m, \qquad b_m=P(n)$$

and that b_0 is positive. Let n be chosen sufficiently large so that the prime number $P(n)=p$ is larger than m and $P(p+n)>p$. [$P(x)$ increases to infinity, and in fact monotonically for large x.] For $x=p$ all the terms then are divisible by p and hence p is a proper divisor of $P(p+n)$.

Second solution. If $P(x)$ is of the mth degree, then $m!P(x)$ has integral coefficients [**86**]. Let a and b be positive integers, $P(a)=p$ and $P(b)=q$, where p, q, are prime numbers, $q>p>m$. Choose c such that $c\equiv a$ (mod p), $c\equiv b$ (mod q). We then have $m!P(c)\equiv 0$ (mod pq), and thus $P(c)\equiv 0$ (mod pq).

98. The odd prime number p is a prime divisor of x^2+1 if and only if $\left(\frac{-1}{p}\right)=(-1)^{\frac{p-1}{2}}=1$.

99. The prime number p, $p\neq 2, 3, 5$ is a prime divisor of x^2+15 if and only if $\left(\frac{-15}{p}\right)=\left(\frac{-1}{p}\right)\left(\frac{3}{p}\right)\left(\frac{5}{p}\right)=1$. Since

$$\left(\frac{-1}{p}\right)=(-1)^{\frac{p-1}{2}},\qquad \left(\frac{3}{p}\right)=(-1)^{\frac{p-1}{2}}\left(\frac{p}{3}\right),\qquad \left(\frac{5}{p}\right)=\left(\frac{p}{5}\right),$$

$$\left(\frac{p}{3}\right)\equiv p \pmod 3,\qquad \left(\frac{p}{5}\right)\equiv p^2 \pmod 5,$$

$\left(\frac{-15}{p}\right)=\left(\frac{p}{3}\right)\left(\frac{p}{5}\right)=1$, if and only if p is contained either in both the sequences

$$4,\ 7,\ 10,\ 13,\ 16,\ 19,\ \ldots,\qquad 4,\ 6,\ 9,\ 11,\ 14,\ 16,\ 19,\ 21,\ 24,\ 26,\ \ldots$$

or else in neither of them. The required prime divisors are 3, 5, and all prime numbers of the form $15x+y$, $y=1, 2, 4, 8$.

100. The polynomial ax^2+bx+c is irreducible if and only if b^2-4ac is not a square. Let p be not a divisor of b^2-4ac. From

$$4a(ax^2+bx+c)=(2ax+b)^2+4ac-b^2\equiv 0 \pmod p$$

it follows that $((b^2-4ac)/p)=1$. By the theorem of reciprocity, applied as in solution **99**, and by the Dirichlet theorem concerning the prime numbers in an arithmetic progression [cf. the footnote to **110**], there are infinitely many prime numbers such that $((b^2-4ac)/p)=-1$. With the aid of more advanced results we can extend the theorem to irreducible polynomials of arbitrary degree. Cf. G. Frobenius: Berl. Ber. I, pp. 689–703 (1896).

101. After multiplication by an integer $\neq 0$, the polynomial considered is written in the form $(ax+b)Q(x)$, where a, b are integers, $a\neq 0$, $Q(x)$ integral-valued, $Q(x)\not\equiv 0$. If p is a prime number that does not divide a, then there exist infinitely many x, such that $ax+b\equiv 0 \pmod p$, i.e. $(ax+b)Q(x)\equiv 0 \pmod p$.

102. $$x^6-11x^4+36x^2-36=(x^2-2)(x^2-3)(x^2-6).$$

If $p>3$, we cannot have that simultaneously $\left(\frac{2}{p}\right)=\left(\frac{3}{p}\right)=\left(\frac{6}{p}\right)=-1$, since $\left(\frac{2}{p}\right)\left(\frac{3}{p}\right)\left(\frac{6}{p}\right)=1$.

103. From **104** it follows that m is a divisor of $p-1$.

104. From

$$x^m-1=K_m(x)\prod_{t/m,\ t<m} K_t(x)$$

[**36**] it follows that $a^m-1\equiv 0 \pmod p$, and thus that a is relatively prime to p. If a did not have order $m \pmod p$, then we would have that $a^t-1\equiv 0 \pmod p$, where t is a proper divisor of m. Hence in view of

$$x^t-1=\prod_{t'/t} K_{t'}(x)$$

at least one other factor of x^m-1 apart from $K_m(x)$ would be divisible by p. Thus we would have $a^m-1\equiv 0 \pmod{p^2}$ and at the same time $(a+p)^m-1\equiv 0 \pmod{p^2}$, which is impossible, since $(a+p)^m-1\equiv a^m-1+mpa^{m-1} \pmod{p^2}$.

105. $6P-1$ has a prime factor of the form $6n-1$ [**96**]; this factor is not a divisor of P and hence is distinct from all the already known prime numbers of the form $6n-1$.

106. Cf. solution **105**.

107. [G. Pólya: J. reine angew. Math. **151**, pp. 19–21 (1921).] Let $p_1, p_2, \dots, p_k, q_1, q_2, \dots, q_l$ be prime divisors of $ab^x + c$, where every q but no p is a divisor of b. Every q must therefore divide c. If the highest power of q_ν that divides c is $q_\nu^{\beta_\nu}$, then, for $x > \beta_\nu$, $q_\nu^{\beta_\nu}$ is also the highest power of q_ν that divides $ab^x + c$. Let x_0 be an integer for which $ab^{x_0} + c \gtrless 0$ and let $p_\mu^{\alpha_\mu}$ be the highest power of p_μ that divides $ab^{x_0} + c$. Setting

$$\varphi(p_1^{\alpha_1+1} p_2^{\alpha_2+1} \dots p_k^{\alpha_k+1}) = r,$$

we have for every integer $x \geqq 0$ and for $\mu = 1, 2, \dots, k$,

$$ab^{x_0+rx} + c \equiv ab^{x_0} + c \not\equiv 0 \qquad (\mathrm{mod}\ p_\mu^{\alpha_\mu+1}).$$

If now $p_1, p_2, \dots, p_k, q_1, q_2, \dots, q_l$ were *all* the prime divisors of $ab^x + c$, then for all sufficiently large values integral values of x we would have

$$ab^{x_0+rx} + c \not\equiv 0 \quad (\mathrm{mod}\ p_1^{\alpha_1+1}), \quad (\mathrm{mod}\ p_2^{\alpha_2+1}), \quad \dots, \quad (\mathrm{mod}\ q_l^{\beta_l+1}),$$

$$|ab^{x_0+rx} + c| \leqq p_1^{\alpha_1} p_2^{\alpha_2} \dots p_k^{\alpha_k} q_1^{\beta_1} q_2^{\beta_2} \dots q_l^{\beta_l},$$

and thus bounded, which is not the case.

108. Every integral-valued polynomial that is not a constant has at least one prime divisor, for it assumes the values 0, 1, -1 at only a finite number of points. Assume that $P(x)$ is a polynomial with integral coefficients [**86**], $P(a) = b \neq 0$ and let us assume that we know that $p_1, p_2, \dots, p_l$ are prime divisors of $P(x)$. The polynomial $b^{-1}P(a + bp_1 p_2, \dots, p_l x)$ is integral-valued, $\equiv 1 \ (\mathrm{mod}\ p_1 p_2 \dots p_l)$ for integral x, and thus has a prime divisor *distinct* from $p_1, p_2, \dots, p_l$. This prime divisor is also a prime divisor of $P(x)$. Also possible by the method of **107** [cf. op. cit. **107**].

109. Yes, if $P(x)$ is linear or a power of a linear function [**95**], but in no other case as can be shown with the aid of more advanced results. Cf. op. cit. **95** and C. Siegel, Math. Z. **10**, pp. 204–205 (1921).

110. [J. A. Serret and others; cf. E. Landau: Handbuch der Lehre von der Verteilung der Primzahlen, p. 436, p. 897. Leipzig and Berlin: B. G. Teubner, 1909.] [**108**, **103**.]

111. There exist two polynomials $p(x)$ and $q(x)$ with integral coefficients such that $p(x)P(x) + q(x)Q(x) = m \neq 0$, where m is an integer. Consequently $(P(n), Q(n))$ is a divisor of m, while $P(x)$, $Q(x)$ have prime divisors that do not divide m [**108**].

112. First let $P(x)$ have integral coefficients. $P(x)$ has no common divisors with $P'(x)$. There exist infinitely many prime divisors p of $P(x)$, such that $P(n) \equiv P(n+p) \equiv 0 \ (\mathrm{mod}\ p)$ and $P'(n) \not\equiv 0 \ (\mathrm{mod}\ p)$ [**111**]. Since $P(n+p) - P(n) \equiv pP'(n) \ (\mathrm{mod}\ p^2)$ [**130**], both of the numbers $P(n)$, $P(n+p)$ cannot be divisible by p^2. In general consider $m!P(x)$ where m is the degree of $P(x)$ [**86**].

113. Let $J(x), J_1(x), J_2(x), \dots$ be the distinct irreducible factors of $P(x)$, ordered according to multiplicity, so that

$$P(x) = [J(x)]^m [J_1(x)]^{m_1} [J_2(x)]^{m_2} \dots, \qquad m \leqq m_1 \leqq m_2 \leqq \cdots.$$

There exist [**111**] infinitely many prime divisors p of $J(x)$, such that from $J(n)\equiv 0$ $(\mathrm{mod}\, p)$ it will follow that $J'(n)\not\equiv 0, J_1(n)\not\equiv 0, J_2(n)\not\equiv 0, \ldots$ $(\mathrm{mod}\, p)$. One of the two numbers $J(n)$ and $J(n+p)$ is divisible only by p [**112**] and hence one of the two numbers $P(n)$ and $P(n+p)$ is divisible only by p^m and by no higher power of p.

114. [Ch. Brisse, problem: Interméd. des math. **1**, p. 10 (1894). R. Jentzsch, problem: Arch. Math. Phys. Ser. 3, **19**, p. 361 (1912). Solution by W. Grosch: Arch. Math. Phys. Ser. 3, **21**, p. 368 (1913). See also Ser. 3, **25**, p. 86 (1917).] If $P(x)$ is not an exact kth power, then there exists an integer a and two polynomials $Q(x)$, $R(x)$ with integral coefficients such that $a^kP(x)=[Q(x)]^kR(x)$, where $Q(x)$ may possibly be $=1$, but $R(x)$ in any case has a zero whose multiplicity is $<k$ (can be seen from the factorization of $P(x)$ into irreducible factors). There exist arbitrarily large integers n for which $R(n)$ is divisible by a prime number p but is not divisible by p^k [**113**]: consequently neither $R(n)$ nor $P(n)$ is an exact kth power. A different solution is given in **190**.

115. Refinement of the method of **107**. [Cf. op. cit. **107**, pp. 19–21.]

116. [Gauss. Cf. Hecke, p. 14, p. 78.]

117. [Gauss. Cf. Nouv. Annls Math. Ser. 1, **15**, p. 383 (1856). Solution by De Rochas, *et al.*: Nouv. Annls Math. Ser. 1, **16**, p. 9, p. 10, p. 71 (1857).] If $f(a)=0$, where a is an integer, then $f(x)=(x-a)\varphi(x)$, where $\varphi(x)$ has integral coefficients [**116**]. Of the two numbers $-a$, $1-a$ one is even and therefore of the two numbers $f(0)=-a\varphi(0)$, $f(1)\equiv(1-a)\varphi(1)$ at least one is even.

118. [Op. cit. **90**.] Let $f(x)=a_0x^m+a_1x^{m-1}+\cdots+a_m$, $a_0\neq 0$, be a factor of $P(x)$ with integral coefficients of the highest possible degree m. Then we have $n-[n/2]\leqq m<n$. By the assumptions $f(x)$ would have to assume for n and thus for at least $m+1$ values of x integral values that are of absolute value less than

$$\frac{\left(n-\left[\frac{n}{2}\right]\right)!}{2^{n-\left[\frac{n}{2}\right]}}\leqq\frac{m!}{2^m}.$$

From the Lagrange interpolation formula (cf. VI, introduction to § 9 p. 85) it then follows [VI **70**] that $|a_0|<1$, and hence $a_0=0$: Contradiction. Cf. **116**.

119. [Op. cit. **90**.] By modification of the argument in **118**.

120. [P. Stäckel: J. für Math. **148**, p. 104 (1918); G. Pólya, op. cit. **90**.] Denoting the polynomial in question by $P(x)$, we have for integral x

$$P(x)\equiv x(x-1)(x-2)\ldots(x-n+1)\equiv 0 \qquad (\mathrm{mod}\, n!) \qquad [\mathbf{130}].$$

The irreducibility of $P(x)$ follows from **119** since

$$n!<\left(\frac{n!+1}{2}\right)^{n-\left[\frac{n}{2}\right]}\left(n-\left[\frac{n}{2}\right]\right)!, \qquad n\geqq 3.$$

The given irreducible polynomial has the maximum constant numerical factor $n!$ that a polynomial of nth degree with integral coefficients with no common divisor can have [**86**].—Another counter-example [A. and R. Brauer]: The polynomial $x^n+105x+12$ has the constant numerical factor 2, and it is irreducible by the well-known *Eisenstein* criterion (O. Perron, Algebra, Vol. 1, p. 196, Satz 104). However, it cannot assume the values ± 2 since by the same criterion also the polynomials

$$x^n+105x+10, \qquad x^n+105x+14$$

are irreducible.

121. [I. Schur, problem: Arch. Math. Phys. Ser. 3, **13**, p. 367 (1908). Solution by W. Flügel: Arch. Math. Phys. Ser. 3, **15**, pp. 271–272 (1909).] If $\varphi(x)$ and $\psi(x)$ are polynomials with integral coefficients [**116**] and

$$(x-a_1)(x-a_2)\ldots(x-a_n)-1=\varphi(x)\psi(x),$$

then we must have $\varphi(a_\nu)=-\psi(a_\nu)=\pm 1$ for $\nu=1,2,\ldots,n$. If the polynomials $\varphi(x)$ and $\psi(x)$ and thus also $\phi(x)+\psi(x)$ were of degree $\leqq n-1$, then we would necessarily have $\varphi(x)+\psi(x)\equiv 0$, since it is $=0$ at n values of x; hence we would also have

$$(x-a_1)(x-a_2)\ldots(x-a_n)-1\equiv -[\varphi(x)]^2,$$

which is impossible because of the coefficient of x^n.

122. [Loc. cit. **121**.] If $\varphi(x)$ and $\psi(x)$ are polynomials of degree $\leqq n-1$ with integral coefficients and

$$F(x)\equiv x(x-a_1)(x-a_2)\ldots(x-a_{n-1})+1\equiv\varphi(x)\psi(x),$$

where $0<a_1<a_2<\cdots<a_{n-1}$, then by the argument used in **121** it follows that

$$\varphi(x)\equiv\psi(x), \qquad F(x)\equiv[\varphi(x)]^2, \qquad n=2m,$$

where m is an integer. Now we have

$$F(\tfrac{1}{2})=1-\frac{1}{2}\,\frac{2a_1-1}{2}\,\frac{2a_2-1}{2}\cdots\frac{2a_{n-1}-1}{2}$$

$$\leqq 1-\frac{1}{2}\cdot\frac{1}{2}\cdot\frac{3}{2}\cdots\frac{4m-3}{2}<0 \quad \text{for} \quad m\geqq 3,$$

and thus $F(x)$ is not a square. Only for $F(\frac{1}{2})>0$ is further discussion required. $F(x)$ is reducible (and thus a square) only in the following two cases:

$$m=2, \quad n=4, \quad a_1=1, \quad a_2=2, \quad a_3=3.$$
$$m=1, \quad n=2, \quad a_1=2.$$

123. [I. Schur, problem: Arch. Math. Phys. Ser. 3, **15**, p. 259 (1909).] If $\varphi(x)$, $\psi(x)$ are polynomials with integral coefficients and of degrees k and l respectively, and if further the leading coefficient of $\varphi(x)$ is positive and we have that

$$(x-a_1)^2(x-a_2)^2\ldots(x-a_n)^2+1\equiv\varphi(x)\psi(x),$$

then $\varphi(x)$ and $\psi(x)$ are always >0 for real values of x, $\varphi(a_\nu)=\psi(a_\nu)=1$, $\varphi(x)=x^k+\cdots$, $\psi(x)=x^l+\cdots$, $k+l=2n$. If $k<l$, then $\varphi(x)\equiv 1$, since the value of $\varphi(x)$ is $=1$ for $n\geqq k+1$ values of x. If $k=l=n$, then the polynomial of degree $n-1$

$\varphi(x)-\psi(x)$, vanishes for n values of x and hence vanishes identically. An identity of the form

$$1\equiv[\varphi(x)]^2-(x-a_1)^2(x-a_2)^2\dots(x-a_n)^2$$
$$\equiv[\varphi(x)+(x-a_1)(x-a_2)\dots(x-a_n)][\varphi(x)-(x-a_1)(x-a_2)\dots(x-a_n)]$$

is, however, impossible.

124. [I. Schur, problem: op. cit. **123**. Solution by A. and R. Brauer.] We set $p_0(x)=(x-a_1)(x-a_2)\dots(x-a_n)$. If $p_0^4(x)+1$ were reducible, then it could be factorized in the form

$$p_0^4(x)+1=[1-p_0(x)p_{-1}(x)][1-p_0(x)p_1(x)], \tag{1}$$

where $p_{-1}(x)$ and $p_1(x)$ are polynomials with integral coefficients and with leading coefficient -1. From (1) it follows that

$$p_0^4(x)=-[p_{-1}(x)+p_1(x)]\cdot p_0(x)+p_{-1}(x)\cdot p_0^2(x)\cdot p_1(x),$$
$$p_{-1}(x)+p_1(x)=-p_0(x)\cdot t(x), \tag{2}$$

where $t(x)$ is a polynomial with integral coefficients. Hence we have that

$$p_0^2(x)=t(x)+p_{-1}(x)\cdot p_1(x). \tag{3}$$

If now the degrees n_{-1} and n_1 of $p_{-1}(x)$ and $p_1(x)$ are equal, we deduce from (1): $n_{-1}=n_1=n$. Comparing the leading coefficients in (2), we conclude that $t(x)=2$. From (2) it follows moreover that $p_{-1}(a_\nu)=-p_1(a_\nu)$; hence we deduce by (3): $p_1^2(a_\nu)=2$, $\nu=1, 2, \dots, n$. Since $p_1(a_\nu)$ is rational and an integer, this is impossible. Let therefore $n_{-1}>n_1$. We have $1\equiv p_1^4(x)p_0^4(x)[\text{mod } 1-p_1(x)\cdot p_0(x)]$.
Hence it follows from (1) that

$$p_1^4(x)+1\equiv p_1^4(x)+p_1^4(x)\cdot p_0^4(x)\equiv 0[\text{mod } 1-p_1(x)\cdot p_0(x)],$$
$$p_1^4(x)+1=[1-p_1(x)\cdot p_0(x)][1-p_1(x)\cdot p_2(x)], \tag{4}$$

where $p_2(x)$, as also in the sequel $p_\lambda(x)$ and $t_\lambda(x)$ in general, denotes a polynomial with integral coefficients. In the deduction of (4) from (1) we did not use the properties of the roots of $p_0(x)$. Hence we obtain analogously to (4), (2) and (3)

$$p_\lambda^4(x)+1=[1-p_\lambda(x)\cdot p_{\lambda-1}(x)][1-p_\lambda(x)\cdot p_{\lambda+1}(x)], \tag{5}$$

$$p_{\lambda-1}(x)+p_{\lambda+1}(x)=-p_\lambda(x)t_\lambda(x), \tag{6}$$

$$p_\lambda^2(x)=t_\lambda(x)+p_{\lambda-1}(x)\cdot p_{\lambda+1}(x), \qquad \lambda=0, 1, 2, \dots. \tag{7}$$

By eliminating $p_{\lambda-1}(x)$ and $p_{\lambda+1}(x)$ from (6) and (7) respectively we obtain

$$\frac{p_\lambda^2(x)+p_{\lambda+1}^2(x)}{1-p_\lambda(x)\cdot p_{\lambda+1}(x)}=t_\lambda(x)=\frac{p_\lambda^2(x)+p_{\lambda-1}^2(x)}{1-p_\lambda(x)p_{\lambda-1}(x)}=t_{\lambda-1}(x)=t_{\lambda-2}(x)=\dots=t(x).$$

The degrees n_λ of the $p_\lambda(x)$ decrease by a constant amount, for from (5) it follows, since $n_{-1}>n_1$, that

$$2n_\lambda=n_{\lambda-1}+n_{\lambda+1}, \qquad n_{\lambda-1}-n_\lambda=n_\lambda-n_{\lambda+1}, \qquad n_{\lambda+1}<n_\lambda.$$

Hence there must exist a first polynomial $p_{\nu+1}(x)$ that vanishes identically. Set $p_\nu(x)=y$. From (7) we deduce for $\lambda=\nu$: $y^2=t(x)$, and thus from (6)

$$p_{\nu-1}(x)=-y^3, \qquad p_{\nu-2}(x)=-p_{\nu-1}(x)\cdot y^2-p_\nu(x)=y^5-y, \qquad \dots$$

From (6) it further follows that all the p_λ become polynomials in y, $p_\lambda(x)=q_\lambda(y)$; in every polynomial q_λ all the exponents are congruent (mod 4). Thus, together with α, $i\alpha$ is also a root of $q_\lambda(y)=0$. Apart from $q_\nu(y)$ and $q_{\nu-1}(y)$, all the $q_\lambda(y)$ have zeros different from zero and hence also non-real zeros. The same is true for $p_\lambda(x)$, since $y(x)$ has rational coefficients. Since $p_0(x)$ has only real zeros, we must have that either $\nu=1$ or $\nu=0$. The former alternative is not possible since then we would have $p_0(x)=-p_1^3(x)$, which is impossible since the zeros of $p_0(x)$ are all distinct.

125. [A. and R. Brauer.] Analogously to **124**, we obtain a series of polynomials with integral coefficients $p_{-1}(x), p_0(x), p_1(x), \dots$, that satisfy the following equations

$$(1) \qquad F[p_\lambda(x)]=\{1-p_\lambda(x)\cdot p_{\lambda-1}(x)\}\{1-p_\lambda(x)\cdot p_{\lambda+1}(x)\},$$

$$(2) \qquad p_{\lambda-1}(x)+p_{\lambda+1}(x)=-p_\lambda(x)\cdot t(x)-A,$$

$$(3) \qquad p_\lambda^2(x)+Ap_\lambda(x)+B=t(x)+p_{\lambda-1}(x)p_{\lambda+1}(x), \qquad \lambda=0, 1, 2, \dots,$$

where $t(x)$ denotes a polynomial with integral coefficients that is independent of λ. If $p_{-1}(x)$ and $p(x)$ are of the same degree, then analogously to **124** we will have

$$p_1(a_k)=\frac{-A}{2}\pm\frac{1}{2}\sqrt{A^2-4B+8}, \qquad k=1, 2, \dots, n.$$

Hence we must have $A^2-4B+8=C^2$. But if this is the case, then $F(z)$ is reducible,

$$(4) \qquad F(z)=\{z^2+\tfrac{1}{2}(A+C)z+1\}\{z^2+\tfrac{1}{2}(A-C)z+1\}.$$

If $n_{-1}>1$, let $p_{\nu+1}(x)$ again be the first identically vanishing polynomial, $p_\nu(x)=y$. Then we obtain $t(x)=u(y)=y^2+Ay+B$, and $p_\lambda(x)=q_\lambda(y)$ are polynomials in y with integral coefficients. Together with $p_0(x)$, $q_0(y)$ has only distinct integral zeros $b_1, b_2, \dots, b_m$. For $y=b_\mu$, $\mu=1, 2, \dots, m$, we obtain, setting $q_1(b_\mu)=b_\mu^*$,

$$(5) \qquad q_1^2(b_\mu)+Aq_1(b_\mu)+B=u(b_\mu)=b_\mu^2+Ab_\mu+B$$

[from (3) for $\lambda=1$], and hence

$$(6) \qquad u(b_\mu^*)=b_\mu^{*2}+Ab_\mu^*+B=u(b_\mu)=b_\mu^2+Ab_\mu+B.$$

Hence either $b_\mu^*=b_\mu$ or $b_\mu^*=-A-b_\mu$. Now we have $q_0(b_\mu)=q_{\nu+1}(b_\mu^*)=0$, $b_\mu^*=q_1(b_\mu)=q_\nu(b_\mu^*)$, since $q_\nu(y)=y$. From $q_\lambda(b_\mu)=q_{\nu-\lambda+1}(b_\mu^*)$ and $q_{\lambda+1}(b_\mu)=q_{\nu-\lambda}(b_\mu^*)$ there follows from (2):

$$\begin{aligned} q_{\lambda+2}(b_\mu)&=-A-q_{\lambda+1}(b_\mu)u(b_\mu)-q_\lambda(b_\mu)\\ &=-A-q_{\nu-\lambda}(b_\mu^*)u(b_\mu^*)-q_{\nu-\lambda+1}(b_\mu^*)=q_{\nu-\lambda-1}(b_\mu^*),\\ &\qquad \lambda=0, 1, \dots, \nu-1, \end{aligned}$$

and hence

$$(7) \qquad b_\mu = q_\nu(b_\mu) = q_1(b_\mu^*).$$

Those b_μ, for which $b_\mu^* = b_\mu$, satisfy the equation

$$(8) \qquad q_1(y) = y,$$

and those for which $b_\mu^* = -A - b_\mu$, the equation

$$(9) \qquad q_1(y) = -y - A.$$

(8) and (9) are of degree $m-2$ if $\nu > 1$. Hence at least 2 of the b_μ, b_1 and b_2 say, satisfy equation (8) and at least 2 satisfy equation (9), b_3 and b_4 say. Since $b_3 \neq b_4$, we may assume that $b_3 \neq -A/2$, and thus $b_3 \neq b_3^*$. Now by (7) we have that

$$\frac{q_1(b_3) - q_1(b_1)}{b_3 - b_1} = \frac{b_3^* - b_1}{b_3 - b_1}$$

and

$$\frac{q_1(b_3^*) - q_1(b_1)}{b_3^* - b_1} = \frac{b_3 - b_1}{b_3^* - b_1}$$

are integers. Hence $b_3^* - b_1 = \pm(b_3 - b_1)$. The + sign is not applicable since $b_3^* \neq b_3$. Thus we obtain $b_1 = (b_3^* + b_3)/2$ and analogously $b_2 = (b_3^* + b_3)/2$ and hence $b_1 = b_2$. This is impossible. Hence $\nu = 1$:

$$q_0(y) = -y(y^2 + Ay + B) - A = -(y^3 + Ay^2 + By + A).$$

From $b_1 + b_2 + b_3 = -A = b_1 b_2 b_3$, it follows with suitable numbering that $b_1 = 1$, $b_2 = 2$, $b_3 = 3$ or $b_1 = -1$, $b_2 = -2$, $b_3 = -3$, or $b_1 = -b_2 > 0$, $b_3 = 0$. In the first two cases we obtain that $F(z) = z^4 \pm 6z^3 + 11z^2 \pm 6z + 1$ is reducible by (4), in the last case $F(z) = z^4 - b_1^2 z^2 + 1$ is positive definite only for $b_1^2 = 1$. Hence we obtain $F(z) = z^4 - z^2 + 1$; $b_1 = 1$, $b_2 = -1$, $b_3 = 0$; $q_0(y) = -y(y-1)(y+1) = p_0(x) = \prod_{\nu=1}^n (x - a_\nu)$. $y(x)$ can have only a linear factor as otherwise $y(x)$ and $y(x) - 1$ would have not only integral roots a_ν. Hence we obtain that

$$p_0(x) = (x - \alpha)(x - \alpha - 1)(x - \alpha - 2).$$

In this case $F[p_0(x)]$ is in fact reducible.

126. [A. and R. Brauer.] Cf. solution **124, 125**. The recurrence relations here become

$$\left.\begin{aligned} &A p_\lambda^4(x) + 1 = [1 - p_\lambda(x) p_{\lambda-1}(x)][1 - p_\lambda(x) p_{\lambda+1}(x)] \\ &p_{\lambda-1}(x) + p_{\lambda+1}(x) = -p_\lambda(x) \cdot t(x) \\ &A p_\lambda^2(x) = t(x) + p_{\lambda-1}(x) p_{\lambda+1}(x) \end{aligned}\right\} \quad \text{for even } \lambda,$$

$$\left.\begin{aligned} &\frac{1}{A} p_\lambda^4(x) + 1 = [1 - p_\lambda(x) p_{\lambda-1}(x)][1 - p_\lambda(x) p_{\lambda+1}(x)] \\ &p_{\lambda-1}(x) + p_{\lambda+1}(x) = -p_\lambda(x) \cdot \frac{1}{A}\, t(x) \\ &\frac{1}{A} p_\lambda^2(x) = \frac{1}{A}\, t(x) + p_{\lambda-1}(x) \cdot p_{\lambda+1}(x) \end{aligned}\right\} \quad \text{for odd } \lambda.$$

$p_{-1}(x)$ and $p_1(x)$ here have integral coefficients; let their leading coefficients be A_{-1} and A_1; the remaining $p_\lambda(x)$ need not have integral coefficients. Nevertheless in the case $n_{-1} > n$, we can argue as in **124, 125**.

For $n_{-1} = n_1$, we deduce from the recurrence relations, since $A_{-1}A_1 = A > 0$, that

$$[p_{-1}(x) - p_1(x)]^2 = p_0^2(x)(A_{-1} - A_1)^2 - 4(A_{-1} + A_1),$$

$$\{p_{-1}(x) - p_1(x) + p_0(x)(A_{-1} - A_1)\}\{p_{-1}(x) - p_1(x) - p_0(x)(A_{-1} - A_1)\}$$
$$= -4(A_{-1} + A_1).$$

From $A_{-1}A_1 > 0$ it follows that $A_{-1} + A_1 \neq 0$; the two factors on the left-hand side are constants; the leading coefficient in the first factor is thus $2A_{-1} - 2A_1 = 0$. Hence $A_{-1} = A_1$, $[p_{-1}(x) - p_1(x)]^2 = -8A_1$, $2A_1 = -C^2$, $A_1 = -2D^2$, $A = 4D^4$. Conversely we have that $4D^4z^4 + 1 = (2D^2z^2 + 2Dz + 1)(2D^2z^2 - 2Dz + 1)$ is reducible.

127. [Generalization of a remark of O. Gmelin, cf. P. Stäckel, op. cit. **120**, pp. 109–110.] Let $\varphi(x)$ be a factor with integral coefficients of $P(x) = \varphi(x)\psi(x)$ [**116**]. Then all the zeros of $\varphi(x)$ also lie in the half-plane $\Re x < n - \frac{1}{2}$. We deduce that $|\varphi(n - \frac{1}{2} - t)| < |\varphi(n - \frac{1}{2} + t)|$, if $t > 0$. Since $\varphi(n-1) \neq 0$, $\varphi(n-1)$ integral, we have that $|\varphi(n-1)| \geqq 1$ and $\varphi(n)$ integral, $|\varphi(n)| > 1$. A similar statement is true for $\psi(x)$. Thus $P(n)$ would have the proper divisors $\varphi(n)$ and $\psi(n)$: Contradiction.

128. [A. Cohn.] Apply **127** with $n = 10$ [III **24**].

129. [D. Hilbert: Gött. Nachr. p. 53 (1897); proof by A. Hurwitz.] $\sqrt{r}$, $\sqrt{s}$, $\sqrt{rs}$, $\sqrt{r} + \sqrt{s}$, $\sqrt{r} - \sqrt{s}$ are irrational numbers. Hence in every linear factor the ratios of the coefficients are irrational and similarly in every quadratic factor that is obtained by combination of any two linear factors. Hence the given polynomial is irreducible in the absolute sense [introduction to § 3 p. 132]. If we have the function congruence [Hecke, pp. 11–12]

$$P(x) \equiv (x^2 + a_1x + a_2)(x^2 + a_3x + a_4) \pmod{a},$$
$$P(x) \equiv (x^2 + b_1x + b_2)(x^2 + b_3x + b_4) \pmod{b}$$

and if $(a, b) = 1$, then there exist numbers c_1, c_2, c_3, and c_4 such that

$$c_\nu \equiv a_\nu \pmod{a}, \qquad c_\nu \equiv b_\nu \pmod{b}, \qquad \nu = 1, 2, 3, 4,$$
$$P(x) \equiv (x^2 + c_1x + c_2)(x^2 + c_3x + c_4) \pmod{ab}.$$

Thus it is sufficient to demonstrate the factorization modulo powers of prime numbers. r is a quadratic residue mod 8 and thus mod 2^n, where n is arbitrary. $P(x)$ appears in (1) as the difference of two squares mod 2^n and hence decomposes into two factors of the second degree. $P(x)$ decomposes mod r^n by (2) and mod s^n by (1). If p is a prime number, $p \neq 2$, $p \neq r$, $p \neq s$, then certainly at least one of the three Legendre symbols $\left(\frac{r}{p}\right)$, $\left(\frac{s}{p}\right)$ and $\left(\frac{rs}{p}\right)$ is $= 1$, since $\left(\frac{r}{p}\right)\left(\frac{s}{p}\right)\left(\frac{rs}{p}\right) = 1$. According as $\left(\frac{r}{p}\right)$, $\left(\frac{s}{p}\right)$ or $\left(\frac{rs}{p}\right) = 1$, (1), (2) or (3) then shows the decomposition of $P(x)$ mod p^n. The basis for the possibility of such examples can be found in G. Frobenius, op. cit. **100**.

130. $$\frac{a(a+1)\dots(a+m-1)}{m!} = \binom{a+m-1}{n}.$$ [**84**, solution **136**.]

131. If the numbers in question are $a, a+d, a+2d, \ldots, a+(m-1)d$ and $(d, m!)=1$, there exists a number d' satisfying $(d', m!)=1$ such that $dd'\equiv 1 \pmod{m!}$. We have, setting $ad'=a'$,

$$d'^m a(a+d)(a+2d)\ldots(a+(m-1)d)$$
$$\equiv a'(a'+1)(a'+2)\ldots(a'+m-1) \pmod{m!} \qquad \textbf{[130.]}$$

132. [K. Hensel: J. reine angew. Math. **116**, p. 354 (1896).] [Solution V **96.**]

133. (a) $n=4$ or a prime number. For if n is a composite number, $n=ab$ and $a<b<n-1$, then $(n-1)!$ is divisible by n. If n is the square of a prime number, $n=p^2$ and $p>2$, then $n-1>2p$ and $(n-1)!$ is again divisible by p^2.

(b) $n=8, 9, p, 2p$ where p denotes a prime number. If n does *not* have the form p, $2p$, p^2 and $n\neq 8, 16$ then $n=ab$, where $3\leqq a<b$. Either the numbers a, $2a$, b, $2b$ or the numbers a, b, $3a$, $2b$ are all distinct. In either case from $2b<n$ we deduce the divisibility of $(n-1)!$ by a^2b^2. We now have

$$\left[\frac{p^2-1}{p}\right]=p-1\geqq 4 \qquad \text{for} \quad p\geqq 5,$$

$$\left[\frac{2^n-1}{2}\right]+\left[\frac{2^n-1}{4}\right]+\left[\frac{2^n-1}{8}\right]+\cdots=2^n-1-n>2n \qquad \text{for} \quad n\geqq 4,$$

and hence by **134** there remain only the enumerated exceptional cases.

134. By **82** the exponent in question is

$$\left[\frac{n}{p}\right]+\left[\frac{n}{p^2}\right]+\left[\frac{n}{p^3}\right]+\cdots=\sum\left[\frac{n}{p^\nu}\right].$$

The sum is extended over l terms where $p^l\leqq n<p^{l+1}$, or over infinitely many terms: $\nu=1, 2, 3, \ldots$.

135. [E. Lucas: Théorie des nombres, **1**, p. 363. Paris: Gauthier-Villars, 1891.] The highest power of 10 that divides 1000! evidently has the same exponent as the highest power of 5 that divides 1000! The exponent is [**134**]

$$\left[\frac{1000}{5}\right]+\left[\frac{1000}{25}\right]+\left[\frac{1000}{125}\right]+\left[\frac{1000}{625}\right]=200+40+8+1=249.$$

136. [E. Catalan: Nouv. Annls Math. Ser. 2, **13**, p. 207 (1874); E. Landau: Nouv. Annls Math. Ser. 3, **19**, pp. 344–362 (1900).] Let p be a prime number, ν a positive integer. Setting $ap^{-\nu}=a'$, $bp^{-\nu}=b'$ it suffices [**134**] to prove the following:

$$[2a']+[2b']\geqq[a']+[b']+[a'+b'].$$

Cf. **8**.

137. [F. G. Teixeira: C.R. Acad. Sci. (Paris) **92**, p. 1066 (1881); M. Weill: S. M. F. Bull. **9**, p. 172 (1881).] It suffices [**134**] to show that

$$\left[\frac{hn}{p}\right]+\left[\frac{hn}{p^2}\right]+\cdots\geqq n\left(\left[\frac{h}{p}\right]+\left[\frac{h}{p^2}\right]+\cdots\right)+\left[\frac{n}{p}\right]+\left[\frac{n}{p^2}\right]+\cdots$$

where p is a prime number.

(a) $(h, p)=1$: $h\geqq[h/p^\nu]p^\nu+1$, hence

$$\left[\frac{nh}{p^\nu}\right]\geqq n\left[\frac{h}{p^\nu}\right]+\left[\frac{n}{p^\nu}\right], \qquad \nu=1, 2, 3, \ldots.$$

(b) $h=p^\alpha h'$, $(h',p)=1$. Then on the left-hand side and on the right-hand side the first α terms cancel and the assertion becomes

$$\left[\frac{h'n}{p}\right]+\left[\frac{h'n}{p^2}\right]+\cdots\geqq n\left(\left[\frac{h'}{p}\right]+\left[\frac{h'}{p^2}\right]+\cdots\right)+\left[\frac{n}{p}\right]+\left[\frac{n}{p^2}\right]+\cdots.$$

Cf. (a).

138. Let $m!=\tau M$, where τ contains only prime factors of t and $(M,t)=1$. Let $tt'\equiv 1 \pmod{M}$. Then we have, similarly as in **131**,

$$t'^m s(s-t)(s-2t)\ldots[s-(m-1)t]\equiv t's(t's-1)\ldots[t's-(m-1)]\equiv 0 \qquad (\text{mod } M).$$

139. By **134** and **138** we have

$$\alpha_n=n\alpha+\left[\frac{n}{p}\right]+\left[\frac{n}{p^2}\right]+\left[\frac{n}{p^3}\right]+\cdots, \qquad \lim_{n\to\infty}\frac{\alpha_n}{n}=\alpha+\frac{1}{p-1}.$$

140. If $t=p^\alpha q^\beta r^\gamma\ldots$, where $p,q,r,\ldots$ are distinct prime numbers, then $T=p^{\alpha+1}q^{\beta+1}r^{\gamma+1}\ldots$ [**139**], since

$$n\alpha+\left[\frac{n}{p}\right]+\left[\frac{n}{p^2}\right]+\cdots<n\left(\alpha+\frac{1}{p-1}\right)\leqq n(\alpha+1),$$

$$n\beta+\left[\frac{n}{q}\right]+\left[\frac{n}{q^2}\right]+\cdots<n(\beta+1),\ldots, \qquad \text{and so on.}$$

141. Let $f(z)=P(z)/Q(z)$, where $Q(0)\neq 0$, and $P(z)$ and $Q(z)$ are polynomials with integral coefficients [**149**]. $Q(0)f[zQ(0)]$, after simplification by cancellation of $Q(0)$, may be represented as the quotient of two polynomials with integral coefficients, where the denominator assumes the value 1 for $z=0$. Cf. the last part of solution **142**.

142. If $f(z)$ and $g(z)$ satisfy the Eisenstein condition, then $f(z)-f(0)$ and $g(z)-g(0)$ may be *simultaneously* transformed into power series with integral coefficients by replacing z by Tz (T a suitably chosen positive integer).

If $F(z)$ and $G(z)$ are power series *with integral coefficients*, then so are $F(z)+G(z)$, $F(z)-G(z)$, $F(z)G(z)$, and also $F[G(z)]$, if we assume that $G(0)=0$, and $F(z)/G(z)$ if we assume that $G(0)=1$. For we have

$$G(z)=1-a_1z-a_2z^2-\cdots=1-H(z),$$

where $a_1, a_2, \ldots$ are integers, and we deduce that

$$[G(z)]^{-1}=[1-H(z)]^{-1}=1+H(z)+[H(z)]^2+[H(z)]^3+\cdots$$

arranged in ascending powers of z is a power series with integral coefficients.

143. From the definition.

144. [Cf. G. Pólya: Acta Math. **42**, p. 314 (1920).] Conditions (1) and (2) follow directly from the Eisenstein condition. That (1) and (2) imply the Eisenstein condition we can show as follows: Assume that

$$\frac{\log t_n}{n}<A, \qquad n=1,2,3,\ldots.$$

Then if a prime factor p occurs exactly ν_n times in t_n we have that

$$\frac{\nu_n}{n}\log p < A, \qquad \frac{\nu_n}{n} < \frac{A}{\log 2} = B.$$

If P denotes the product of those prime numbers that divide $t_1, t_2, t_3, \ldots$, and k is an integer, $k > B$, then we may set $T = P^k$. For every factor p of P occurs in T^n exactly kn times, and $kn > \nu_n$ for $n = 1, 2, 3, \ldots$.

145.
$$\sum_{n=0}^{\infty} \frac{2^{n+1}}{2^{n+1}-1} z^n, \qquad \sum_{n=1}^{\infty} \frac{u_n}{2^{n^2}} z^n,$$

where $u_n/2^{n^2} = \sum_{t/n} 1/2^{nt}$, and thus u_n is odd. The first series satisfies (2) but not (1) [**107**], the second series satisfies (1) but not (2), and neither of the two satisfies the complete Eisenstein condition, neither is algebraic.

146. Setting

$$(1-z)^{-\alpha} = \sum_{n=0}^{\infty} a_n z^n, \qquad (1-z)^{-\beta+\gamma-1} = \sum_{n=0}^{\infty} b_n z^n,$$

we have

$$\frac{\beta-\gamma+1}{1}\cdot\frac{\beta-\gamma+2}{2}\cdots\frac{\beta-1}{\gamma-1} F(\alpha, \beta, \gamma; z) = \sum_{n=0}^{\infty} a_n b_{n+\gamma-1} z^n$$

[**140, 143**]. For further investigations and literature see A. Errera: Palermo Rend. **35**, pp. 107–144 (1913).

147. The quotients of two successive coefficients is $[(\alpha+n)(\beta+n)]/[(1+n)(\gamma+n)]$. If $\alpha \neq \gamma$, $\beta \neq \gamma$ and $[(\alpha+x)(\beta+x)]/[(1+x)(\gamma+x)]$ is rational for $x = 0, 1, 2, \ldots$, then $\alpha\beta$, $\alpha+\beta$ and γ are rational numbers [**92**].

148. Choose $a \gtrless 0$ such that $a(\alpha+x)(\beta+x) = ax^2+bx+c$, where a, b, and c are integers. If $F(\alpha, \beta, \gamma; z)$ were algebraic, then [**144** (1)] every prime number p (with the exception of a finite number) would have to be a prime divisor of ax^2+bx+c. Cf. however **100**.

149. A special case of **150**.

150. A special case of **151**.

151. [Cf. E. Heine: J. reine angew. Math. **48**, pp. 269–271 (1854); Theorie der Kugelfunktionen, 2nd ed., pp. 52–53. Berlin: Reimer, 1878.]

First Proof. Consider the coefficients of R. If the function R does not vanish identically, then it may be written in the form $R_0 + \alpha_1 R_1 + \alpha_2 R_2 + \cdots + a_l R_l$ where $R_0, R_1, R_2, \ldots, R_l$ are rational entire functions in $z, y, y', \ldots, y^{(r)}$ with rational coefficients and where $1, \alpha_1, \alpha_2, \ldots, \alpha_l$ are *rationally independent*, i.e. from $n_0 + n_1\alpha_1 + n_2\alpha_2 + \cdots + n_l\alpha_l = 0$ with $n_0, n_1, n_2, \ldots, n_l$ rational it follows that $n_0 = n_1 = n_2 = \cdots = n_l = 0$. If in $R_0, R_1, R_2, \ldots, R_l$ we set for y the power series $a_0 + a_1 z + \cdots + a_n z^n + \cdots$, then we obtain the power series $\sum_{n=0}^{\infty} t_n^{(\nu)} z^n$, $\nu = 0, 1, \ldots, l$, where the coefficients $t_n^{(\nu)}$ are rational, respectively. By comparing the coefficients of z^n, it follows that $t_n^{(0)} + \alpha_1 t_n^{(1)} + \cdots + \alpha_l t_n^{(l)} = 0$, i.e. $t_n^{(0)} = t_n^{(1)} = \cdots = t_n^{(l)} = 0$, so that

$$\sum_{n=0}^{\infty} t_n^{(\nu)} z^n \equiv 0, \qquad \nu = 0, 1, \ldots, l.$$

Second Proof. Consider the coefficients of y. The existence of the relation $R=0$ implies that the power series of a certain finite number of functions of the form $z^{\mu}y^{\nu}(y')^{\nu_1}(y'')^{\nu_2}\ldots(y^{(r)})^{\nu_r}$ are linearly dependent [introduction to VII **33** p. 100]. Express one of these power series linearly in terms of a distinct set of linearly independent power series. The constant factors required for this representation are constructed by means of the system of equations (*) occurring in solution VII **33** rationally from the coefficients of the power series y, and thus in the present instance are rational numbers.

152. [E. Heine, op. cit. **151**, p. 50.] The discriminant of the equation of degree l in w

$$(*)\qquad F_0(z)w^l+F_1(z)w^{l-1}+\cdots+F_{l-1}(z)w+F_l(z)=0$$

is a rational entire expression in $F_0, F_1, \ldots, F_{l-1}, F_l$, and hence an analytic function that is regular in a neighbourhood of $z=0$. In case it should vanish identically, we can construct an equation by means of rational operations which is satisfied by $f(z)$ and whose discriminant does not vanish identically. Thus let us assume that the discriminant of (*) itself does not vanish identically. Then there exist l distinct functions $w_1=f(z), w_2, \ldots, w_l$, that are regular in some annular region $0<|z|<\rho$, $\rho>0$, are perhaps many-valued, and satisfy the equation (*). At the point $z=0$ they may have an algebraic singularity. If a function $\varphi(z)$ with the same properties as w has the property that

$$(\varphi(z)-c_0-c_1z-\cdots-c_{n-1}z^{n-1})z^{-n}$$

remains bounded in the neighbourhood of $z=0$ for arbitrarily large values of n, then we have that identically $\varphi(z)=f(z)$ [IV **166**]. Consider a number m such that the $l-1$ functions

$$(w_\lambda-c_0-c_1z-\cdots-c_{m-1}z^{m-1})z^{-m},\qquad \lambda=2, 3, \ldots, l$$

are all unbounded in the neighbourhood of $z=0$. If we set in (*)

$$w=c_0+c_1z+\cdots+c_{m-1}z^{m-1}+z^my,$$

then we obtain an equation for y of the same type as (*). Dividing the coefficients by a suitable power of z this equation may be expressed in the form

$$(**)\qquad G_0(z)y^l+G_1(z)y^{l-1}+\cdots+G_{l-1}(z)y+G_l(z)=0,$$

where not all the numbers $G_0(0), G_1(0), \ldots, G_{l-1}(0), G_l(0)$ vanish. If $G_0(0)=G_1(0)=\cdots=G_{h-1}(0)=0$, $G_h(0)\neq 0$, then (**) has (for instance by the Weierstrass preparation theorem) $l-h$ solutions bounded for $z=0$. (**) has in fact only one solution that is bounded for $z=0$, namely $c_m+c_{m+1}z+c_{m+2}z^2+\cdots$. Hence $l-h=1$, $h=l-1$, q.e.d.

153. [Op. cit. **151**.] We may assume that, if we set $f(z)=c_0+c_1z+c_2z^2+\cdots$, then $P_0(0)=P_1(0)=\cdots=P_{l-2}(0)=0$, $P_{l-1}(0)=a\neq 0$ [**152**], $P_0(z), P_1(z), \ldots, P_l(z)$ are power series with integral coefficients, $c_0\neq 0$ and c_0 is an integer. Setting $z=0$ we obtain $ac_0+P_l(0)=0$, and thus $P_l(0)$ is divisible by a. Hence $a^{-1}P_\lambda(az)$ is a power series with integral coefficients, $\lambda=0, 1, 2, \ldots, l$, and in particular $a^{-1}P_{l-1}(0)=1$. Hence we may set

(*) $$f(az)=Q_0(z)+Q_2(z)[f(az)]^2+\cdots+Q_l(z)[f(az)]^l,$$

where $Q_\lambda(z)=-[a^{-1}P_{l-\lambda}(az)]/[a^{-1}P_{l-1}(az)]$ is again a power series with integral coefficients [solution **142**], $\lambda=0, 2, 3, \ldots, l$, and $Q_\lambda(0)=0$ for $\lambda=2, 3, \ldots, l$. From (*) we can determine $a^n c_n$ by comparing coefficients as an entire function with integral coefficients of $c_0, ac_1, a^2c_2, \ldots, a^{n-1}c_{n-1}$, and hence (by recursion) we find that it is an integer.

154. Let $P(z), Q(z), Q_2(z), Q_3(z), \ldots$ be power series with rational coefficients, and set $y=c_0+c_1z+c_2z^2+\cdots$. If one of the following six equations holds

$$y=P(z)+Q(z), \qquad y=P(z)-Q(z), \qquad y=P(z)Q(z),$$

$$y=\frac{P(z)}{Q(z)}, \qquad \text{assuming } Q(0)=1 \qquad [\mathbf{142}],$$

$$y=P[Q(z)], \qquad \text{assuming } Q(0)=0,$$

$$y=Q(z)+Q_2(z)y^2+Q_3(z)y^3+\cdots+Q_l(z)y^l, \qquad \text{assuming}$$
$$Q_2(0)=Q_3(0)=\cdots=Q_l(0)=0 \qquad [\mathbf{152}],$$

then c_n is a uniquely determined rational number, given as a rational and *entire* function of the coefficients of $1, z, z^2, \ldots, z^n$ in the expansions of $P(z)$, $Q(z)$, $Q_2(z), \ldots$, which therefore can have only such prime numbers in its denominator as occur also in the denominators of the coefficients mentioned above. [Cf., op. cit., **144**.]

155. [A. Hurwitz, Cf. G. Pólya: Math. Ann. **77**, pp. 510–512 (1916).] Let p be a prime number, that divides all the $c_n=\sum_{\nu=0}^{n} a_\nu b_{n-\nu}$ but not all the a_n, $n=0, 1, 2, \ldots$. Let for example $a_0, a_1, \ldots, a_{k-1}\equiv 0$, $a_k \not\equiv 0 \pmod p$. From $c_k\equiv a_k b_0 \pmod p$ it then follows that $b_0\equiv 0 \pmod p$, from $c_{k+1}\equiv a_k b_1\equiv 0 \pmod p$ it follows that $b_1\equiv 0 \pmod p$, from $c_{k+2}\equiv a_k b_2\equiv 0 \pmod p$ it follows that $b_2\equiv 0 \pmod p$ and so on.

156. [P. Fatou: Acta Math. **30**, p. 369 (1906). The solution given here is due to A. Hurwitz; cf. G. Pólya, op. cit., **155**.] It is sufficient to prove the theorem for primitive power series $f(z)=a_0+a_1z+a_2z^2+\cdots$ [**155**]. By **149** we have $f(z)=P(z)/Q(z)$, where $P(z)$ and $Q(z)$ have integral coefficients and their coefficients are relatively prime. Then the (terminating) power series $Q(z)$ is primitive. For if its coefficients had a common divisor t, then, since $P(z)=t[Q(z)/t]f(z)$, t would also have to divide the coefficients of $P(z)$. Now determine two polynomials $p(z), q(z)$ with integral coefficients, such that $p(z)P(z)+q(z)Q(z)=m\neq 0$, where m is an integer. Setting $q(z)+p(z)f(z)=R(z)$, we have $m=Q(z)R(z)$. Here $R(z)$ has integral coefficients and is not primitive if $m\neq \pm 1$ (otherwise [**155**] m would also have to be primitive); its coefficients are in any case divisible by m. From $1=Q(0)[R(0)/m]$ it follows that $Q(0)=\pm 1$.

157. For rational ϑ the sequence $a_1, a_2, a_3, \ldots$ is periodic from a certain term on and hence $f(z)=a_1z+a_2z^2+\cdots+a_nz^n+\cdots$ is rational. For irrational ϑ, $f(z)$ cannot be rational since it would then [**149**] be a quotient of two polynomials with integral coefficients, and thus $f(1/10)$ would be rational.

158. [Cf. E. Landau: Nouv. Annls Math. Ser. 4, **3**, pp. 333–336 (1903). R. Jentzsch: Math. Ann. **78**, p. 277 (1918).] "If the sequence of coefficients is periodic, then the function represented by it is rational" is easily proved (geometric series). Conversely: Let the numbers $a_0, a_1, a_2, \ldots$ take only m different values, and let the denominator be $=1+l_1z+l_2z^2+\cdots+l_kz^k$. We have $a_n+l_1a_{n-1}+l_2a_{n-2}+\cdots+l_ka_{n-k}=0$ for sufficiently large n. Since there are only m^k different systems $a_{n-1}, a_{n-2}, \ldots, a_{n-k}$, there exist two numbers $\mu, \nu, \mu<\nu\leqq\mu+m^k$, such that

$$a_{\mu-1}=a_{\nu-1}, \qquad a_{\mu-2}=a_{\nu-2}, \qquad \ldots, \qquad a_{\mu-k}=a_{\nu-k}.$$

But then it follows that $a_\mu=a_\nu$, and hence also $a_{\mu+1}=a_{\nu+1}, a_{\mu+2}=a_{\nu+2}, \ldots$.

159. The coefficients of

$$(1-z)^{-l-1}=\sum_{n=0}^{\infty}\binom{l+n}{l}z^n$$

are periodic modulo any prime number p. If $p>l$, then $l!$ and p are relatively prime and

$$(n+1)(n+2)\ldots(n+l)\equiv(n+p+1)(n+p+2)\ldots(n+p+l) \pmod p.$$

If $p\leqq l!$, $l!=p^aL$, $(L,p)=1$, then $\binom{n+p^{a+1}}{l}\equiv\binom{n}{l} \pmod p$, since the symbolic numerators of these binomial coefficients are congruent $(\bmod\, p^{a+1})$.—The sequence of coefficients obtained after multiplication by $P(z)$ is periodic from a certain term on.

160.

$$\frac{(D-1)z}{(1-Dz)(1-z)}=\sum_{n=1}^{\infty}(D^n-1)z^n.$$

If k is the smallest number such that $D^{n-k}\equiv D^n \pmod p$, i.e. $D^k\equiv 1 \pmod p$, then k is a divisor of $p-1$.

161.

$$(1-Dz^2)^{-1}=\sum_{n=0}^{\infty}D^nz^{2n}.$$

The length of the period k is even, $k=2k'$; k' is the smallest positive integer such that $D^{k'}\equiv 1 \pmod p$. Since $D^{p-1}\equiv 1 \pmod p$, k' is a divisor of $p-1$. However, for odd p we have that $D^{\frac{p-1}{2}}\equiv 1 \pmod p$, if $\left(\frac{D}{p}\right)=1$, and thus k' divides $(p-1)/2$. Conversely $D^{\frac{p-1}{2}}\equiv 1 \pmod p$ follows from "$D^{k'}\equiv 1 \pmod p$, k' divides $\frac{1}{2}(p-1)$."

162.

p	2	3	5	7	11	13	17	19	23	29
length of period	3	8	20	16	10	28	36	18	48	14
$p-1$	—	—	—	—	10	—	—	18	—	28
p^2-1	3	8	—	48	—	168	288	—	528	—.

Of the two numbers $p-1$ and p^2-1, $p-1$ is given in the table if $p-1$ is a multiple of the length of the period and p^2-1 is given if $p-1$ is not a multiple but p^2-1 is a multiple of the length of the period. Whether the first or the second case occurs depends on whether p is or is not a quadratic residue (mod 5). The number 5 occupies an exceptional position. Cf. A. Speiser: Trans. Amer. Math. Soc. **23**, p. 177 (1922).

163. If a_n are the coefficients in the expansion considered, then $a_n+l_1a_{n-1}+l_2a_{n-2}+\cdots+l_ka_{n-k}=0$ for n sufficiently large, where $l_1, l_2, \ldots, l_k$ are integers

[**156**]. Since mod m there are only m^k different systems $a_{n-1}, a_{n-2}, \ldots, a_{n-k}$, there exist two numbers μ and ν such that $a_{\mu-1}\equiv a_{\nu-1}$, $a_{\mu-2}\equiv a_{\nu-2}, \ldots, a_{\mu-k}\equiv a_{\nu-k}$ (mod m), $\mu\neq\nu$. Hence $a_\mu\equiv a_\nu$, and $a_{\mu+1}\equiv a_{\nu+1}, \cdots$ (mod m) [**158**].

164. [G. Pólya: Tôhoku Math. J. **22**, p. 79 (1922).] We have $(1-4z)^{-1/2}=\sum_{m=0}^{\infty}\binom{2m}{m}z^m$. Let p be an odd prime number and p^{r-1} its highest power that divides $(2k-1)!$, $k\geqq 1$, $r\geqq 1$. We deduce from **134** that

$$\binom{2p^r}{p^r}=\frac{(2p^r)!}{p^r!p^r!}\not\equiv 0 \quad (\text{mod}\, p).$$

On the other hand we have

$$\binom{2p^r}{p^r}=\frac{2}{1}\cdot\frac{2p^r-1}{p^r}\cdot\frac{2p^r-2}{p^r-1}\cdot\frac{2p^r-3}{p^r-1}\cdots\frac{2p^r-2q+2}{p^r-q+1}\cdot\frac{2p^r-2q+1}{p^r-q+1}\binom{2(p^r-q)}{p^r-q},$$

$$-2(2q-1)!\binom{2(p^r-q)}{p^r-q}\equiv 0 \quad (\text{mod}\, p^r), \qquad \binom{2(p^r-q)}{p^r-q}\equiv 0 \quad (\text{mod}\, p)$$

for $q=1, 2, 3, \ldots, k$. Since k and thus r may be arbitrarily large, in the sequence of coefficients reduced mod p there occur arbitrarily long sequences of zeros, followed by a term that is $\not\equiv 0$.

165. For $z=1$ the general term does not tend to 0.

[For the background of **165–173** see G. Pólya: Math. Ann. **77**, pp. 497–513 (1916), also Proc. Lond. Math. Soc., Ser. 2, **21**, pp. 22–38 (1923), and F. Carlson: Math. Z. **9**, pp. 1–13 (1921). For heuristic considerations see also MPR, **2**, p. 46, ex. 14.]

166. [P. Fatou: Acta. Math. **30**, pp. 368–371 (1906).] Setting $f(z)=\sum_{n=0}^{\infty}a_nz^n$, the series

$$|a_0|^2+|a_1|^2+\cdots+|a_n|^2+\cdots$$

would have to converge [III **122**].

167. [P. Fatou, op. cit. **166**.] By **150** the algebraic function considered satisfies an equation of the form

$$P_0(z)[f(z)]^l+P_1(z)[f(z)]^{l-1}+\cdots+P_{l-1}(z)f(z)+P_l(z)=0,$$

where $P_0(z), P_1(z), \ldots, P_l(z)$ have rational integral coefficients. It follows from this that $y=P_0(z)f(z)$ satisfies the equation

$$y^l+P_1(z)y^{l-1}+P_2(z)P_0(z)y^{l-2}+\cdots+P_l(z)P_0(z)^{l-1}=0.$$

y cannot be infinite for any finite value of z, since, if y were infinite, y^l on the left-hand side would be of higher order than the other terms and so the equation could not be satisfied. However, $y=P_0(z)f(z)$ is by hypothesis a non-terminating power series with integral coefficients [**166**].

168. [F. Carlson: Math. Z., **9**, p. 1 (1921).]

$$\frac{1}{\sqrt{1-4z^l}}=\sum_{n=0}^{\infty}\binom{2n}{n}z^{ln},$$

where l is an arbitrarily large integer.

169. [G. Pólya, problem: Arch. Math. Phys. Ser. 3, **23**, p. 289 (1915).] In the first case the numbers $P(n)-[P(n)]$ are periodic [**158**]. In the second case let us assume that the power series $f(z)$ under consideration is rational. We may assume a_0 irrational; otherwise consider

$$\frac{f(z)+f\left(ze^{\frac{2\pi i}{k}}\right)+\cdots+f\left(ze^{(k-1)\frac{2\pi i}{k}}\right)}{k}=\sum_{n=0}^{\infty}[P(kn)]z^{kn}$$

and choose k such that a_0k^r is an integer, and accordingly $P(kn)$ is given by $[P(kn)]=a_0k^rn^r+[a_1k^{r-1}n^{r-1}+\cdots]$. Applying I **85** we obtain

$$\lim_{z\to 1-0}(1-z)^{r+1}f(z)=\lim_{n\to\infty}\frac{[P(n)]}{\binom{r+n}{r}}=r!a_0,$$

which is impossible since [**149**] the limit on the left-hand side must be rational.

170. [Cf. D. Hansen: Thèse, p. 65. Copenhagen 1904. G. Pólya, problem: Arch. Math. Phys. Ser. 3, **27**, pp. 161–162 (1918). Proc. Lond. Math. Soc. Ser. 2, **21**, pp. 36–38 (1922).] That the assumptions regarding the a_n cannot be weakened in any very obvious fashion is shown by **69**.

171. [Op. cit. **170**.] Cf. **71**.

172. We have $Q_n-Q_{n-1}=1-9A_n$, where A_n denotes the number of zeros at the end of the decimal representation of n. Hence we have

$$(1-z)\sum_{n=1}^{\infty}Q_nz^n=\sum_{n=1}^{\infty}(Q_n-Q_{n-1})z^n=\frac{z}{1-z}-9\left(\frac{z^{10}}{1-z^{10}}+\frac{z^{100}}{1-z^{100}}+\frac{z^{1000}}{1-z^{1000}}+\cdots\right).$$

The function

$$f(z)=\frac{z}{1-z}+\frac{z^{10}}{1-z^{10}}+\frac{z^{100}}{1-z^{100}}+\frac{z^{1000}}{1-z^{1000}}+\cdots$$

has $|z|=1$ as its natural boundary. In fact $z=1$ is a singular point $[\lim_{z\to 1-0}f(z)=+\infty]$, and also $z=e^{\frac{2\pi i\nu}{10^m}}$, $\nu=1,2,\ldots,10^m-1$, since the function obtained by deleting the first m terms is transformed into $f(z)$ if we replace z^{10^m} by z, whereas the m omitted terms behave as regular functions at those 10^mth roots of unity that are not 10^{m-1}th roots of unity.

173. [G. Pólya, problem: Arch. Math. Phys. Ser. 3, **25**, p. 85 (1917). Solution by R. Jentzsch: Arch. Math. Phys. Ser. 3, **27**, pp. 90–91 (1918).] Apply the *Hadamard gap theorem* to $(1-z)\sum_{n=1}^{\infty}d_nz^n$.

174. Differentiation replaces the sequence of coefficients $a_0, a_1, a_2, a_3, \ldots$ by $a_1, a_2, a_3, a_4, \ldots$, and the indicated integration replaces them by $0, a_0, a_1, a_2, \ldots$.

175. For the multiplication cf. I **34**. If

$$g(z)=1+\frac{b_1}{1!}z+\frac{b_2}{2!}z^2+\cdots+\frac{b_n}{n!}z^n+\cdots=1-h(z),$$

then $h(z)$ is a power series with H-integral coefficients and so is

$$\frac{1}{g(z)}=\frac{1}{1-h(z)}=1+h(z)+[h(z)]^2+[h(z)]^3+\cdots.$$

176. If we assume that $[f(z)]^m/m!$ is a power series with H-integral coefficients, then it follows [**174, 175**] that the same is true of

$$\int_0^z \frac{[f(z)]^m}{m!} f'(z)\,dz = \frac{[f(z)]^{m+1}}{(m+1)!}.$$

177.

$$g(z) = b_0 + \frac{b_1}{1!} z + \frac{b_2}{2!} z^2 + \cdots + \frac{b_m}{m!} z^m + \cdots,$$

$$g[f(z)] = b_0 + \frac{b_1}{1!} f(z) + \frac{b_2}{2!} [f(z)]^2 + \cdots + \frac{b_m}{m!} [f(z)]^m + \cdots$$

is a power series with H-integral coefficients in view of **176**.

178. In general: if y is determined by a differential equation of the form

$$\frac{d^n y}{dx^n} = P\left(x, y, \frac{dy}{dx}, \frac{d^2 y}{dx^2}, \ldots, \frac{d^{n-1} y}{dx^{n-1}}\right)$$

and by the initial conditions $y = m_0, y' = m_1, \ldots, y^{(n-1)} = m_{n-1}$ for $x = 0$, where P is a rational entire function with integral coefficients and $m_0, m_1, \ldots, m_{n-1}$ are integers, then all the derivatives of y at $x = 0$ have integral values, and thus y is a power-series with H-integral coefficients. In particular: we have $\varphi'' = -2\varphi^3$, $\varphi(0) = 0$, $\varphi'(0) = 1$.

179.

$$\begin{aligned}(e^z - 1)^3 &= e^{3z} - 3\,e^{2z} + 3\,e^z - 1\\ &= \sum_{n=2}^{\infty} \frac{3^n - 3\cdot 2^n + 3}{n!} z^n \equiv \sum_{n=2}^{\infty} \frac{(-1)^n - 1}{n!} z^n \pmod 4.\end{aligned}$$

180.

$$\begin{aligned}(e^z - 1)^{p-1} &= z^{p-1} + \cdots \equiv \frac{-1}{(p-1)!} z^{p-1} + \cdots \pmod p\\ &= e^{(p-1)z} - \binom{p-1}{1} e^{(p-2)z} + \binom{p-1}{2} e^{(p-3)z} - \cdots - \binom{p-1}{p-2} e^z + 1\\ &= \sum_{n=1}^{\infty} \frac{(p-1)^n - \binom{p-1}{1}(p-2)^n + \binom{p-1}{2}(p-3)^n - \cdots - \binom{p-1}{p-2}}{n!} z^n.\end{aligned}$$

In the first line we have used *Wilson's* theorem $(p-1)! \equiv -1 \pmod p$. From the third line (where incidentally we have assumed $p \geqq 3$) it follows that the coefficient of $z^n/n!$ is periodic mod p with period $p-1$, $n = 1, 2, 3, \ldots, p-1, p, \ldots$. *Now* again make use of the first line!

181. From **176** and **133** or from I **41** and **133**.

182. [Theorem of K. G. Ch. v. Staudt and Th. Clausen. For the proof see J. C. Kluyver: Math. Ann. **53**, pp. 591–592 (1900).] From

$$z = \log\,[1 + (e^z - 1)],$$

$$\frac{z}{e^z - 1} = 1 - \frac{e^z - 1}{2} + \frac{(e^z - 1)^2}{3} - \frac{(e^z - 1)^3}{4} + \cdots$$

it follows [**179–181**] that

$$\frac{z}{e^z-1}=g(z)-\frac{z}{2}-\frac{1}{2}\left(\frac{z^2}{2!}+\frac{z^4}{4!}+\frac{z^6}{6!}+\cdots\right)$$
$$-\sum_p \frac{1}{p}\left(\frac{z^{p-1}}{(p-1)!}+\frac{z^{2p-2}}{(2p-2)!}+\frac{z^{3p-3}}{(3p-3)!}+\cdots\right),$$

where $g(z)$ is a power series with H-integral coefficients and the summation $\sum_p$ is extended over the odd prime numbers 3, 5, 7, 11, ...

183.

$$\frac{dz}{d\varphi}=\frac{1}{\sqrt{1-\varphi^4}}=\sum_{n=0}^{\infty}\binom{2n}{n}\frac{\varphi^{4n}}{2^{2n}},\qquad \frac{z}{\varphi(z)}=\sum_{n=0}^{\infty}\binom{2n}{n}\frac{(4n)!}{2^{2n}(4n+1)}\frac{[\varphi(z)]^{4n}}{(4n)!};$$

$(4n)!$ is divisible by 2^{2n} [**134**] and also by $4n+1$, if $4n+1$ is not a prime number [**133**]; $[\varphi(z)]^{4n}/(4n)!$ is a power series with H-integral coefficients [**178, 176**].

184. From the differential equation **178** it follows that

$$\left[\frac{d}{dz}\left(\frac{1}{\varphi^2}\right)\right]^2=4\frac{1}{\varphi^6}-4\frac{1}{\varphi^2},$$

and hence $1/\varphi^2=\wp$, and moreover that

$$\frac{dz^2}{d\varphi}=\frac{2z}{\sqrt{1-\varphi^4}},\qquad (1-\varphi^4)\frac{d^2z^2}{d\varphi^2}-2\varphi^3\frac{dz^2}{d\varphi}=2.$$

We seek the solution z^2 that satisfies the initial condition $z^2=dz^2/d\varphi=0$ at $\varphi=0$ and find that

$$z^2=\varphi^2+\frac{3}{5}\frac{\varphi^6}{3}+\frac{3}{5}\cdot\frac{7}{9}\cdot\frac{\varphi^{10}}{5}+\frac{3}{5}\frac{7}{9}\frac{11}{13}\frac{\varphi^{14}}{7}+\cdots,$$
$$z^2\wp(z)=\left(\frac{z}{\varphi(z)}\right)^2=\sum_{n=0}^{\infty}\frac{G_nH_n}{(4n+1)(2n+1)}\frac{[\varphi(z)]^{4n}}{(4n)!},$$

where $G_n=3\cdot 7\cdot 11\ldots(4n-1)$, $H_n=2\cdot 3\cdot 4\cdot 6\cdot 7\cdot 8\ldots(4n-2)(4n-1)4n$. If $2n+1\equiv -1 \pmod 4$ then $2n+1$ divides G_n. Let $2n+1\equiv 1 \pmod 4$ and $2n+1=ab$, $a>1$, $b>1$. If $a\equiv b\equiv -1 \pmod 4$, then both the numbers a and b divide both the numbers G_n and H_n and thus ab divides G_nH_n. If however $a\equiv b\equiv 1 \pmod 4$, then among the factors of H_n we have $2a$ as well as $4b$. $2n+1$ thus does not divide G_nH_n in only one case, if it is a prime number and $\equiv 1 \pmod 4$, and we find that the same holds for $4n+1$. Note that $2n+1$ and $4n+1$ are relatively prime, and thus if each separately divides G_nH_n, then their product also divides G_nH_n. For deeper more exact results see A. Hurwitz, op. cit., p. 145.

185. [For more particulars cf. M. Fujiwara: Tôhoku Math. J. **2**, p. 57 (1912).] The power series

$$a_0+\frac{a_1}{1!}z+\frac{a_2}{1!}z^2+\cdots+\frac{a_n}{n!}z^n+\cdots$$

satisfies a differential equation of the required type if and only if $a_0+a_1z+a_2z^2+\cdots+a_nz^n+\cdots$ is a rational function. Both conditions require the same recurrence relation between $a_0, a_1, a_2, \ldots$. Cf. **163**.

186. [G. Pólya: Tôhoku Math. J. **22**, p. 79 (1922).] Cf. **164**; y satisfies [I **48**] the differential equation

$$xy''+(1-4x)y'-2y=0.$$

187. [S. Kakeya: Tôhoku Math. J. **10**, p. 70 (1916). G. Pólya: Tôhoku Math. J. **19**, p. 65 (1921).] If

$$g(z)=\sum_{n=0}^{\infty}\frac{a_n}{n!}z^n,$$

and if $a_n\neq 0$, then we have

$$|a_n|\geqq 1, \qquad M(n)\geqq\frac{|a_n|}{n!}n^n\geqq\frac{n^n}{n!}\sim\frac{e^n}{\sqrt{2\pi n}}.$$

The equality sign holds, e.g., for $g(z)=\sum_{n=0}^{\infty}[z^{2^n}/(2^n)!]$.

188. [Th. Skolem: Videnskapsselskapets Skrifter 1921, No. 17, Theorem 8.] Denoting by g the lowest common denominator of the rational numbers b_1, b_2, $\ldots$, b_m, we have that

$$gb_0+\frac{gb_{-1}}{z}+\frac{gb_{-2}}{z^2}+\cdots=gb_0+r(z)$$

is also integral for the infinitely many integral values in question. Since $\lim_{z\to\infty}r(z)=0$. gb_0 must approach arbitrarily closely to an integer, i.e. gb_0 must be an integer, Thus $r(z)$ also assumes integral values for infinitely many integral values of z, and hence, in view of $\lim_{z\to\infty}r(z)=0$, it is $=0$ for infinitely many integral values of z. It follows that $r(z)\equiv 0$. For $r(z^{-1})$ is regular in some closed circular disc with center $z=0$, it is $=0$ at infinitely many points of the disc, and hence it is identically $=0$.

189. The equation $y^2-2z^2=1$ has infinitely many solutions in the integers, as can be seen from

$$(3+2\sqrt{2})^n=y_n+z_n\sqrt{2}, \qquad (3-2\sqrt{2})^n=y_n-z_n\sqrt{2},$$

$$(9-8)^n=y_n^2-2z_n^2, \qquad y_n,\ z_n \text{ integers}, \qquad n=0,1,2,3,\ldots.$$

190. [J. Franel: Interméd. des math. **2**, p. 94 (1895).] If the integral-valued polynomial

$$P(x)=a_0x^n+a_1x^{n-1}+\cdots+a_n,$$

which accordingly has rational coefficients, represents for all sufficiently large integral values of x the kth power of an integer, but is itself *not* the kth power of a polynomial, then the polynomial

$$P(x+l_1)P(x+l_2)\ldots P(x+l_k)=a_0^kx^{nk}+b_1x^{nk-1}+\cdots$$

has the same two properties, provided that the integers $l_1, l_2, \ldots, l_k$ are chosen such that $P(x+l_1), P(x+l_2), \ldots, P(x+l_k)$ have no common zeros. Let $l_1<l_2<\cdots<l_k$ say, and assume further that $l_2-l_1, l_3-l_2, \ldots, l_k-l_{k-1}$ are sufficiently large.

Then the power series

$$\sqrt[k]{P(x+l_1)P(x+l_2)\ldots P(x+l_k)} = a_0 x^n \sqrt[k]{1+\frac{b_1}{a_0^k}\frac{1}{x}+\cdots}$$
$$= a_0 x^n + c_1 x^{n-1} + c_2 x^{n-2} + \cdots$$

with rational coefficients would represent a rational integer for all sufficiently large integral values of x, but would itself not be a rational entire function: Contradiction to **188**!

191. [Proof by H. Prüfer. For more particulars see op. cit. **188.**] If b_m, $b_{m-1}, \ldots, b_1$ are all rational, apply **188**. If not, we may assume that b_m is irrational. (If $b_m, b_{m-1}, \ldots, b_{m-k+1}$ were rational with lowest common denominator g, and b_{m-k} were irrational, $k \geqq 1$, consider

$$g[F(z) - b_m z^m - b_{m-1} z^{m-1} - \cdots - b_{m-k+1} z^{m-k+1}].)$$

The mth difference

$$F(z+m) - \binom{m}{1} F(z+m-1) + \binom{m}{2} F(z+m-2) - \cdots + (-1)^m F(z)$$

$$= m! b_m + \frac{b'_{-1}}{z} + \frac{b'_{-2}}{z^2} + \cdots$$

also has integral values for sufficiently large integral z, from which it follows by **188** that b_m is rational. Contradiction!

192. [G. Pólya, problem: Jber. deutsch. Math. Verein. **32**, Section 2, p. *16* (1923). Solution by T. Radó: Jber. deutsch. Math. Verein. **33**, Section 2, p. *30* (1924).] For a proof based on the method of **188**, **191** see **193**. Let the polynomials be denoted by $f(z)$ and $g(z)$. The two entire functions

$$e^{2\pi i f(z)} - 1, \qquad e^{2\pi i g(z)} - 1$$

have the same zeros but perhaps not with the same multiplicities. Their multiple zeros are contained among the zeros of the polynomials $f'(z)$ and $g'(z)$ respectively. Hence the function

$$g'(z)\,\frac{e^{2\pi i f(z)} - 1}{e^{2\pi i g(z)} - 1}$$

is entire and has only finitely many zeros; it is also of finite genus. Thus it is equal to $k(z)e^{h(z)}$, where $k(z)$ and $h(z)$ are polynomials. From the identity

$$g'(z)e^{2\pi i f(z)} - g'(z) = k(z)e^{h(z)+2\pi i g(z)} - k(z)e^{h(z)}$$

it follows that one of the two functions $h(z) + 2\pi i g(z)$ and $h(z)$ is equal to a constant and the other is equal to $2\pi i f(z)$ + a constant. [Cf. G. Pólya: Nyt Tidsskr. for Math. (B), **32**, p. 21 (1921).]

193. [G. Pick, problem: Jber. deutsch. Math. Verein. **32**, Section 2, p. *45* (1923). Solution by G. Szegö: Jber. deutsch. Math. Verein. **33**, Section 2, p. *31* (1924).] If m is the degree of $y = f(x)$ and n the degree of $z = g(x)$, then we have for sufficiently large $|y|$

$$x = a_1 y^{\frac{1}{m}} + a_0 + \frac{a_{-1}}{y^{\frac{1}{m}}} + \frac{a_{-2}}{y^{\frac{2}{m}}} + \cdots,$$

$$z = b_n y^{\frac{n}{m}} + b_{n-1} y^{\frac{n-1}{m}} + \cdots + b_0 + \frac{b_{-1}}{y^{\frac{1}{m}}} + \frac{b_{-2}}{y^{\frac{2}{m}}} + \cdots,$$

where all m determinations are admissible for $y^{\frac{1}{m}}$.

By assumption the *Laurent* series

$$\sum_{k=-\infty}^{n} b_k e^{i\frac{2\nu}{m}k\pi} Y_1^k, \qquad \nu = 0, 1, 2, \ldots, m-1, \tag{1}$$

as well as the *Laurent* series

$$\sum_{k=-\infty}^{\infty} b_k e^{i\frac{2\nu+1}{m}k\pi} Y_2^k, \qquad \nu = 0, 1, 2, \ldots, m-1, \tag{2}$$

have real values as Y_1 and Y_2 each run through some sequence of positive numbers tending to infinity. I.e. the numbers

$$b_k e^{i\frac{2\nu}{m}k\pi}, \qquad b_k e^{i\frac{2\nu+1}{m}k\pi}, \qquad \nu = 0, 1, 2, \ldots, m-1$$

are real and hence $b_k = 0$ if k is not divisible by m. [For odd m it is sufficient to consider simply the series (1).] We obtain that $g(x) = \varphi(y) + P(y^{-1})$, where $\varphi(y)$ is a polynomial in y and $P(y^{-1})$ a power series with no constant term, both with real coefficients. The polynomial $g(x) - \varphi[f(x)]$ accordingly converges to 0 as $x \to \infty$, i.e. $g(x) - \varphi[f(x)] \equiv 0$.—With the assumptions of **192**, the inverse function of the polynomial $\varphi(y)$ must also be a polynomial; thus $\varphi(y)$ is of the first degree. Moreover $\varphi(y)$ as well as the inverse function must be integral-valued.

194. From

$$\alpha^n + a_1\alpha^{n-1} + \cdots + a_{n-1}\alpha + a_n = 0$$

it follows that

$$(\sqrt{\alpha})^{2n} + a_1(\sqrt{\alpha})^{2n-2} + \cdots + a_{n-1}(\sqrt{\alpha})^2 + a_n = 0.$$

195. The case $s = 0$ is clear. Assume $s \neq 0$. If $r + s\sqrt{-1}$ satisfies an equation with rational integral coefficients and with leading coefficient 1, then the complex conjugate number $r - s\sqrt{-1}$ satisfies the same equation; it is thus also an integer. Hence the numbers

$$(r + s\sqrt{-1}) + (r - s\sqrt{-1}), \quad (r + s\sqrt{-1})(r - s\sqrt{-1})$$

are also integers; they are also rational. The coefficients of the polynomial

$$(x - r - s\sqrt{-1})(x - r + s\sqrt{-1}) = x^2 - 2rx + r^2 + s^2$$

are thus ordinary integers. $r^2 + s^2$ and $2r$ are integers and hence so are $4(r^2 + s^2)$ and $2s$. Let $2r = a$ and $2s = b$; $a^2 + b^2 = 4(r^2 + s^2) \equiv 0 \pmod 4$ *cannot* occur in the following three cases:

$$a \equiv 1, \quad b \equiv 1; \qquad a \equiv 1, \quad b \equiv 0; \qquad a \equiv 0, \quad b \equiv 1 \qquad \pmod 2.$$

Hence we must have $a \equiv b \equiv 0 \pmod 2$ and hence $r = a/2$ and $s = b/2$ are *integers*.

196. We deduce as in **195** that the coefficients of the polynomial $(x-r-s\sqrt{-5})(x-r+s\sqrt{-5})$, i.e. $2r$ and r^2+5s^2, and hence also $5(2s)^2=4(r^2+5s^2)-(2r)^2$ are *integers*. But if the rational number $2s$ were not an integer, then the *square* of a prime number would be a factor of the denominator of $(2s)^2$ and this could not be cancelled by 5 as required. Hence $2s=b$ is an integer and similarly $2r=a$. We now argue as in **195** from

$$a^2+b^2\equiv a^2+5b^2\equiv 4(r^2+5s^2)\equiv 0 \pmod 4.$$

197. We deduce as in **195**, **196** that $2r$ and r^2+3s^2 and hence also $2r=a$ and $2s=b$ are *integers*. That $a\equiv b \pmod 2$ follows from

$$(a+b)(a-b)\equiv a^2+3b^2\equiv 4(r^2+3s^2)\equiv 0 \pmod 4.$$

$\frac{1}{2}(-1+\sqrt{-3})$ is an integer, a zero of x^2+x+1.

197.1. [See G. Pólya: L'Enseignement Math., **15**, pp. 237–243 (1969).] Assume that ϑ/π is rational, $\vartheta=2\pi m/n$ where m and n are rational integers, $n\geqq 1$. Let $\zeta=e^{i\vartheta}$. Then

$$\zeta^n-1=0 \quad\text{and}\quad (\zeta^{-1})^n-1=0.$$

Therefore, ζ, ζ^{-1} and

$$2\cos\vartheta=\zeta+\zeta^{-1}$$

are algebraic integers. Assume now that also $\cos\vartheta$ is rational. Then $2\cos\vartheta$ is a rational integer, yet it is $\leqq 2$ in absolute value and so it must be equal to one of the five numbers $2, 1, 0, -1, -2$ which correspond to $\vartheta/\pi=0, \frac{1}{3}, \frac{1}{2}, \frac{2}{3}, 1$, respectively. See **227.2**.

197.2. Follows from **197.1** and the values of the cosines. To obtain any one of these cosines apply

$$\cos a=\cos b\cos c+\sin b\sin c\cos\alpha,$$

to an appropriate spherical triangle with sides a, b, c and angles α, β, γ on a sphere centered at a vertex of the solid concerned. We are required to find $\cos\alpha$. Each of the angles a, b, and c is of the form $[\pi(n-2)]/n$ since it is an angle in a regular polygon with n sides; n is 3 or 4 or 5, and $b=c$. See the table:

α	T	H	O	D	I
$\frac{a}{\pi}$	$\frac{1}{3}$	$\frac{1}{2}$	$\frac{1}{2}$	$\frac{3}{5}$	$\frac{3}{5}$
$\frac{b}{\pi}=\frac{c}{\pi}$	$\frac{1}{3}$	$\frac{1}{2}$	$\frac{1}{3}$	$\frac{3}{5}$	$\frac{1}{3}$
$\cos\alpha$	$\frac{1}{3}$	0	$-\frac{1}{3}$	$-\sqrt{5}/5$	$-\sqrt{5}/3$

$$\cos\frac{\pi}{3}=\frac{1}{2},\qquad \cos\frac{\pi}{2}=0,\qquad \cos\frac{3\pi}{5}=\frac{1-\sqrt{5}}{4}.$$

197.3. By virtue of the linear relation between T, H, and O, the rationality of O$/T$ would imply that of H/T which is contrary to **197.2**.

197.4. [H. Lebesgue, Ann. Soc. Polon. Math., **17**, pp. 193–226 (1938).]

(1) We assume first that there is a linear relation of the form

$$(*) \qquad lT+2mI=4hH+2kD,$$

where h, k, l and m are rational integers, k, l and m positive. The assumed relation (*) is equivalent to

$$(**) \qquad e^{ilT}\cdot e^{i2mI}=e^{i2kD}.$$

Now (cf. **197.2**)

$$e^{i2kD}=\left(\frac{-3-i4}{5}\right)^k=K+K'i$$

$$e^{ilT}=\left(\frac{1+2i\sqrt{2}}{3}\right)^l=L+L'i\sqrt{2},$$

$$e^{i2mI}=\left(\frac{1-4i\sqrt{5}}{9}\right)^m=M+M'i\sqrt{5},$$

where K, K', L, L', M, M' are rational.

Observe that $K'=0$ would imply $K=\pm 1$, and so D/π would be rational, contrary to **197.2**. Therefore $K'\neq 0$ and similarly $L'\neq 0$, $M'\neq 0$.

There follows from (**) that

$$L'M\sqrt{2}+LM'\sqrt{5}=K'.$$

Observe that $L=0$ would imply $M\neq 0$ and so would imply that $\sqrt{2}$ is rational. Therefore $L\neq 0$ and similarly $M\neq 0$.

Yet now we conclude that $LML'M'\neq 0$ and squaring the last equation we conclude further that

$$LML'M'\sqrt{10}$$

is rational—and this absurd conclusion shows that the equation (*) initially assumed is in fact impossible.

(2) In the foregoing we have dealt with the most crucial possibility of linear dependence between T, H, O, D, and I. The other possibilities can be reduced to the one we have considered, or can be treated either similarly or more simply. We briefly mention a few cases.

If there existed a linear relation containing O and essentially different from the one given in **197.3** we could eliminate O.

There is no restriction in assuming that in (*) one of the coefficients is a multiple of 4 and two others are even (multiplication by 4).

If in (*), $l<0$, we use

$$e^{ilT}=\left(\frac{1-2i\sqrt{2}}{3}\right)^{-l}.$$

If $l=0$ the argument is simpler.

197.5. From **197.1** and

$$\cos 2\vartheta=\frac{1-\tan^2\vartheta}{1+\tan^2\vartheta}.$$

197.6. From **197.5**; there are now only three exceptional values: 0, 1, and 1/0.

198. Let $\alpha_1, \alpha_2, \ldots, \alpha_n$ be conjugate integers, $|\alpha_\nu|<k$ for $\nu=1, 2, \ldots, n$, and

$$(x-\alpha_1)(x-\alpha_2)\ldots(x-\alpha_n)=x^n+a_1x^{n-1}+\cdots+a_n$$

identically. Then

$$|a_\nu|<\binom{n}{\nu}k^\nu, \qquad \nu=1, 2, \ldots, n.$$

For the rational integers $a_1, a_2, \ldots, a_n$ we thus have only finitely many admissible sets of values.

199. Notation as in solution **198.** According to the assumptions $|a_n|=|\alpha_1\alpha_2\ldots\alpha_n|<1$, and hence $a_n=0$. The only irreducible polynomial with vanishing constant term and leading coefficient 1 is x.

200. [L. Kronecker, Werke, **1**, p. 105. Leipzig; B. G. Teubner, 1895.] Let the polynomial in question be denoted by $F(x)$ and its zeros by $\alpha_1, \alpha_2, \ldots, \alpha_n$ (repeated according to multiplicity). Set

$$(x-\alpha_1^h)(x-\alpha_2^h)\ldots(x-\alpha_n^h)=F_h(x),$$

$h=1, 2, \ldots$; $F_1(x)=F(x)$; $F_h(x)$ has rational integral coefficients. The infinite sequence $F_1(x), F_2(x), \ldots, F_h(x), \ldots$ contains only finitely many different polynomials [**198**]. If $F_h(x)$ is identical with $F_k(x)$, $1\leqq h<k$, then the systems $\alpha_1^h, \alpha_2^h, \ldots, \alpha_n^h$ and $\alpha_1^k, \alpha_2^k, \ldots, \alpha_n^k$, coincide apart from their order. If $\alpha_1^h=\alpha_1^k$, then we already have the desired result. With suitable choice of numbering let

$$\alpha_1^h=\alpha_2^k, \qquad \alpha_2^h=\alpha_3^k, \qquad \ldots, \qquad \alpha_{l-1}^h=\alpha_l^k, \qquad \alpha_l^h=\alpha_1^k.$$

Then we have

$$\alpha_1^{h^l}=\alpha_2^{kh^{l-1}}=\alpha_3^{k^2h^{l-2}}=\cdots=\alpha_{l-1}^{k^{l-2}h^2}=\alpha_l^{k^{l-1}h}=\alpha_1^{k^l}.$$

201. The equation $x^2-\alpha x+1$ has by assumption two complex roots of absolute value 1. Let $\alpha_1, \alpha_2, \ldots, \alpha_n$ be α and its conjugates, $\alpha_1=\alpha$. The polynomial

$$(x^2-\alpha_1x+1)(x^2-\alpha_2x+1)\ldots(x^2-\alpha_nx+1)$$

has rational integral coefficients and its zeros are all of absolute value 1 and thus roots of unity [**200**]. Let the root of unity that makes the first factor vanish be $e^{\frac{2\pi ip}{q}}$. We have

$$e^{\frac{4\pi ip}{q}}-\alpha e^{\frac{2\pi ip}{q}}+1=0.$$

202. The numbers of the field that are generated by the number of the second degree ϑ may be expressed in the form $r+s\vartheta$, where r and s are rational [Hecke, Theorem 53, p. 68]. Cf. **195–197**. The equation

$$\sqrt{-3}=a+b\sqrt{-5},$$

where a and b are rational, is excluded since from it would follow that $a=0$, $b=\sqrt{3/5}$; the latter square root is known to be irrational. The same reasoning shows that all the three fields are distinct.

203. [G. Pólya: Proc. Lond. Math. Soc. Ser. 2, **21**, p. 27 (1921).] If for the moment we set aside the requirement with respect to $\zeta'\zeta''$ we have for $\mathfrak{S}$ as we know [Hurwitz-Courant, pp. 134–138] only the following two possibilities:

(a) $\mathfrak{S}$ consists of the numbers nA, $n=0, \pm 1, \pm 2, \ldots$;
(b) $\mathfrak{S}$ consists of the numbers $mA+nB$, $A\neq 0$, B/A not real, $m, n=0, \pm 1, \pm 2, \ldots$.

If we now apply the requirement with respect to $\zeta'\zeta''$ we deduce in case (a) that $A^2=nA$, i.e. if $A\neq 0$, $A=n$, where n is a rational integer. In case (b) we have

$$AB=lA+l'B, \qquad B^2=mA+m'B, \qquad AB^2=nA+n'B,$$

where l, l', m, m', n, n' are rational integers. Eliminating 1, B, B^2 from these three linear equations we obtain

$$mA^2+(lm'-l'm-n)A=ln'-l'n.$$

Thus $ln'-l'n$ belongs to $\mathfrak{S}$. If $ln'-l'n=0$, then $lm'-l'm-n$ belongs to $\mathfrak{S}$. If also this number $=0$, then $m=0$, $m'=B\neq 0$ and m' belongs to $\mathfrak{S}$. In any case $\mathfrak{S}$ contains rational integers other than 0. If $R=rA+r'B$ is the rational integer other than 0 of smallest absolute value that is contained in $\mathfrak{S}$ then r and r' are relatively prime. Let s and s' be rational integers, $rs'-r's=1$. Then $S=sA+s'B$ is a number of $\mathfrak{S}$ and every number of $\mathfrak{S}$ may be expressed in the form $pR+qS$, where $p, q=0, \pm 1, \pm 2, \ldots$. S is certainly not real. We have $S^2=pR+qS$.

204. If $3=(a+b\sqrt{-5})(c+d\sqrt{-5})$ then it follows that

$$9=(a^2+5b^2)(c^2+5d^2).$$

Thus the only possible cases are

$$a^2+5b^2=1, 3, 9$$

and consequently

$$a, b=\pm 1, 0; \qquad \pm 2, \pm 1; \qquad \pm 3, 0.$$

The case $a, b=\pm 2, \pm 1$ is excluded, for it would imply that $c^2+5d^2=1$ whereas $3\neq \pm 2\pm\sqrt{-5}$. 3 has only the divisors ± 1 and ± 3. Similarly we obtain that $1+2\sqrt{-5}$ has only the divisors ± 1 and $\pm(1+2\sqrt{-5})$. The greatest common divisor could thus only be 1. However

$$3(d+D\sqrt{-5})+(1+2\sqrt{-5})(c+C\sqrt{-5})=1$$

is impossible, since from

$$\begin{aligned} 3d+c-10C&=1,\\ 3D+2c+C&=0 \end{aligned}$$

it follows that $3(-2d+D+7C)=-2$. This is incompatible with the requirements that d, D and C are integers.

205. The divisors of 9 lying in the field are 1, 3, 9, $2+\sqrt{-5}$, $2-\sqrt{-5}$ [**204**]. We find using **196** that $-19+4\sqrt{-5}$ is not divisible only by 3, 9 and $2-\sqrt{-5}$ while

$-19+4\sqrt{-5}=(2+\sqrt{-5})(-2+3\sqrt{-5})$. We seek to determine $\xi=x+u\sqrt{-5}$ and $\eta=y+v\sqrt{-5}$ such that

$$9\xi+(-19+4\sqrt{-5})\eta=2+\sqrt{-5}.$$

Transforming this equation we find successively that

$$(2-\sqrt{-5})\xi+(-2+3\sqrt{-5})\eta=1,$$
$$2x+5u-2y-15v=1, \qquad -x+2u+3y-2v=0,$$
$$x=2u+3y-2v, \qquad 9u+4y-19v=1.$$

Since the g.c. divisor of 9, 4, 19 is in fact $=1$, a solution is possible. For example let $u=1$, $y=-2$, $v=0$, $x=-4$:

$$9(-4+\sqrt{-5})-(-19+4\sqrt{-5})2=2+\sqrt{-5},$$
$$9=(2+\sqrt{-5})(2-\sqrt{-5}), \qquad -19+4\sqrt{-5}=(2+\sqrt{-5})(-2+3\sqrt{-5}).$$

206. Observe **194** and the last three equations in solutions **205**. It follows that

$$3\sqrt{\frac{9}{2+\sqrt{-5}}}(-4+\sqrt{-5})-(1+2\sqrt{-5})\sqrt{\frac{-19+4\sqrt{-5}}{2+\sqrt{-5}}}\cdot 2=\sqrt{2+\sqrt{-5}},$$
$$3=\sqrt{2+\sqrt{-5}}\cdot\sqrt{2-\sqrt{-5}}, \qquad 1+2\sqrt{-5}=\sqrt{2+\sqrt{-5}}\sqrt{-2+3\sqrt{-5}}.$$

The g.c. divisor of 3 and $1+2\sqrt{-5}$ is thus $\sqrt{2+\sqrt{-5}}$.

207. From $\alpha_1=\gamma_1\delta$, $\alpha_2=\gamma_2\delta,\ldots,\alpha_m=\gamma_m\delta$ and from

$$\alpha_1\lambda_1+\alpha_2\lambda_2+\cdots+\alpha_m\lambda_m=\vartheta$$

(definition of the g.c. divisor) it follows that

$$\vartheta=(\gamma_1\lambda_1+\gamma_2\lambda_2+\cdots+\gamma_m\lambda_m)\delta.$$

208. The quotients ϑ/ϑ' and ϑ'/ϑ are both integers [**207**] and hence divisors of unity since $\vartheta/\vartheta'\cdot\vartheta'/\vartheta=1$.

209. Multiply the $m+1$ equations used in the definition of the g.c. divisor (introduction to § 2 p. 146) by γ.

210. The equation $\alpha\alpha'+\beta\gamma\beta'=1$ may also be interpreted as $\alpha\alpha'+\beta(\beta'\gamma)=1$ or as $\alpha\alpha'+\gamma(\beta\beta')=1$.

211. Special case of **212**.

212. The g.c. divisor δ of α and γ also divides $\beta\gamma$. From $\beta\beta'=1+\alpha\alpha'$, $\gamma\gamma'=\delta+\alpha\alpha''$ it follows that $\beta\gamma(\beta'\gamma')=\delta+\alpha(\alpha'\delta+\alpha''+\alpha\alpha'\alpha'')$.

213. α/δ and β/δ are relatively prime [definition!, cf. al o **209**], hence [**211**] so are $(\alpha/\delta)^n$, $(\beta/\delta)^n$ and hence δ^n is the g.c. divisor of α^n, β^n [**209**].

214. $\sqrt[n]{\alpha}$ is an integer, cf. **194**. From $\alpha=\alpha'\delta$, $\beta=\beta'\delta$, $\alpha\alpha''+\beta\beta''=\delta$ we deduce

$$\sqrt[n]{\alpha}=\sqrt[n]{\alpha'}\sqrt[n]{\delta}, \qquad \sqrt[n]{\beta}=\sqrt[n]{\beta'}\sqrt[n]{\delta}, \qquad \sqrt[n]{\alpha}(\sqrt[n]{\alpha'})^{n-1}\alpha''+\sqrt[n]{\beta}(\sqrt[n]{\beta'})^{n-1}\beta''=\sqrt[n]{\delta}.$$

215. Obvious for $m=2$. Mathematical induction from m to $m+1$. Assume that

$$\frac{\mu}{\alpha_1}\lambda_1+\frac{\mu}{\alpha_2}\lambda_2+\cdots+\frac{\mu}{\alpha_m}\lambda_m=1,$$

and let α_{m+1} be relatively prime to α_1, to $\alpha_2, \ldots$, and to α_m. Then α_{m+1} is also relatively prime to $\alpha_1\alpha_2\ldots\alpha_m$ [**211**]. In $\lambda\alpha_{m+1}+\lambda'\alpha_1\alpha_2\ldots\alpha_m=1$ substitute

$$\alpha_{m+1}=\frac{\mu\alpha_{m+1}}{\alpha_1}\lambda_1+\frac{\mu\alpha_{m+1}}{\alpha_2}\lambda_2+\cdots+\frac{\mu\alpha_{m+1}}{\alpha_m}\lambda_m.$$

216. If α is an integer and is $\neq 0$, it satisfies an equation of the form

$$a_n=-\alpha(a_{n-1}+a_{n-2}\alpha+\cdots+\alpha^{n-1}),$$

where $a_1, \ldots, a_{n-1}, a_n$ are rational integers, $a_n\neq 0$. Let p be a rational prime number that does *not* divide a_n. In

$$a_n u+pv=1$$

substitute for a_n the value given by the equation that defines α.

217. [D. Hilbert: Die Theorie der algebraischen Zahlkörper, Jber. deutsch. Math. Verein., **4**, p. 218 (1897).] From $a_\nu^N=d^\nu b_\nu^\nu$, where b_ν is a rational integer, it follows that $a_\nu=\delta^\nu\beta_\nu$ where β_ν is an integer. If α is any of the numbers $\alpha_1, \alpha_2, \ldots, \alpha_n$, then we have

$$\alpha^n+\delta\beta_1\alpha^{n-1}+\delta^2\beta_2\alpha^{n-2}+\cdots+\delta^n\beta_n=0,$$

$$\left(\frac{\alpha}{\delta}\right)^n+\beta_1\left(\frac{\alpha}{\delta}\right)^{n-1}+\beta_2\left(\frac{\alpha}{\delta}\right)^{n-2}+\cdots+\beta_n=0,$$

and hence α/δ is an integer [Hecke, p. 79, Theorem 62].—On the other hand, since $a_1, a_2, \ldots, a_n$ are *rational* numbers, there exist rational integers $c_1, c_2, \ldots, c_n$ such that

$$c_1a_1^N+c_2a_2^{\frac{N}{2}}+c_3a_3^{\frac{N}{3}}+\cdots+c_na_n^{\frac{N}{n}}=d.$$

Since $a_1, a_2, \ldots, a_n$ are homogeneous functions of $\alpha_1, \alpha_2, \ldots, \alpha_n$ we have

$$c_1a_1^N+c_2a_2^{\frac{N}{2}}+c_3a_3^{\frac{N}{3}}+\cdots+c_na_n^{\frac{N}{n}}=\sum C_{k_1k_2\ldots k_n}\alpha_1^{k_1}\alpha_2^{k_2}\ldots\alpha_n^{k_n},$$

where $C_{k_1k_2\ldots k_n}$ are rational integers, $k_1+k_2+\cdots+k_n=N$. From

$$\sum C_{k_1k_2\ldots k_n}\alpha_1^{k_1}\alpha_2^{k_2}\ldots\alpha_n^{k_n}=\delta^{N-1}\delta,$$

since δ divides $\alpha_1, \alpha_2, \cdots, \alpha_n$, it follows by dividing both sides by δ^{N-1} that

$$\alpha_1\gamma_1+\alpha_2\gamma_2+\cdots+\alpha_n\gamma_n=\delta,$$

where $\gamma_1, \gamma_2, \ldots, \gamma_n$ are integers. Cf. also **214**.

218. From $\beta=\alpha\gamma$ there follow analogous equations for the conjugates which on multiplication yield $N(\beta)=N(\alpha)N(\gamma)$.

219. [G. Rabinowitsch: J. reine angew. Math., **142**, pp. 153–164 (1913).] Necessity: If the g.c. divisor ϑ of α and β exists, i.e. if

$$\alpha=\alpha'\vartheta, \qquad \beta=\beta'\vartheta, \qquad \alpha\gamma+\beta\delta=\vartheta,$$

and neither α' nor β' is a unit, then we have $|N(\alpha')|>1$, $|N(\beta')|>1$, and consequently, since $N(\alpha)=N(\alpha')N(\vartheta)$, $N(\beta)=N(\beta')N(\vartheta)$, $0<|N(\vartheta)|=|N(\alpha\gamma+\beta\delta)|<|N(\alpha)|$ and also $<|N(\beta)|$.

Sufficiency: Let ξ and η run through all pairs of integers for which the linear form $\alpha\xi+\beta\eta$ assumes a value different from 0. Among all the positive rational integers $|N(\alpha\xi+\beta\eta)|$ thus obtained let $|N(\alpha\xi_0+\beta\eta_0)|$ be the smallest. Let $\alpha\xi_0+\beta\eta_0=\vartheta$, where accordingly $\vartheta\neq 0$. We must distinguish two cases: (1) α is a divisor of ϑ. Then, setting $\vartheta=\vartheta'\alpha$, we have $|N(\vartheta)|=|N(\vartheta')||N(\alpha)|\geqq|N(\alpha)|$. But we have on the other hand, by the choice of ϑ, $|N(\vartheta)|\leqq|N(1\cdot\alpha+0\cdot\beta)|=|N(\alpha)|$. Hence $|N(\vartheta)|=|N(\alpha)|$, $N(\vartheta')=\pm 1$, and thus ϑ' is a unit and ϑ is also a divisor of α. (2) α is not a divisor of ϑ. *If* also ϑ *were not* a divisor of α, we could by the assumptions determine ξ, η such that $0<|N(\alpha\xi+\vartheta\eta)|<|N(\vartheta)|$: a contradiction to the choice of ϑ! For $\alpha\xi+\vartheta\eta=\alpha(\xi+\eta\xi_0)+\beta\eta\eta_0$ is also a number of the set from which $\vartheta=\alpha\xi_0+\beta\eta_0$ was chosen. In any case ϑ is thus a divisor of α and by the same reasoning also a divisor of β and thus a g.c. divisor.

220. Let α, β be integers of the field, neither a divisor of the other, and $N(\beta)\leqq N(\alpha)$. α/β is a number of the field but not an integer and hence $\alpha/\beta=r+s\sqrt{-1}$, where r, s are rational numbers but not both integers [**202**]. Determine two rational integers R and S such that $|r-R|\leqq\frac{1}{2}$, $|s-S|\leqq\frac{1}{2}$, $r-R$, $s-S$ are not both $=0$. Hence, setting $\gamma=r-R+(s-S)\sqrt{-1}$, we have

$$0<N(\gamma)=(r-R)^2+(s-S)^2\leqq\tfrac{1}{4}+\tfrac{1}{4}.$$

Set $\alpha-\beta(R+S\sqrt{-1})=\delta$. By this definition δ is an integer. On the other hand we have $\delta=\beta(r+s\sqrt{-1}-(R+S\sqrt{-1}))=\beta\gamma$ and hence

$$N(\delta)=N(\gamma)N(\beta)\leqq\tfrac{1}{2}N(\beta)\leqq\tfrac{1}{2}N(\alpha).$$

If we set $\xi=1$, $\eta=-R-S\sqrt{-1}$, condition **219** is satisfied.

221. As in the theory of rational integers.

222. As in the theory of rational integers.

223. From $\alpha\alpha_1+\mu\mu_1=1$, $\alpha\equiv 0 \pmod{\mu}$ it follows that $1\equiv\alpha\alpha_1\equiv 0 \pmod{\mu}$, i.e. μ is a divisor of 1.

224. By repeated application of the congruence, obviously valid for any two integers β, γ,

$$(\beta+\gamma)^p=\beta^p+p\beta^{p-1}\gamma+\frac{p(p-1)}{2}\beta^{p-2}\gamma^2+\cdots+\gamma^p\equiv\beta^p+\gamma^p \pmod{p}.$$

225. [G. Pólya: J. reine angew. Math., **151**, p. 7 (1921).] Let the determinant $|\omega_l^k|_{k,\,l=1,\,2,\ldots m}$ be denoted by Δ. Let p be a rational prime number that is relatively prime to α_1 as well as to Δ [**216**].

Then the m congruences

$$\alpha_1\omega_1^{rp}+\alpha_2\omega_2^{rp}+\cdots+\alpha_m\omega_m^{rp}\equiv 0 \pmod{p}, \qquad r=1,2,\ldots,m$$

cannot be satisfied simultaneously. For if they were it would follow [**222, 224**] that

$$\alpha_1|\omega_l^{kp}|\equiv\alpha_1\Delta^p\equiv 0 \pmod{p},$$

whereas on the other hand $\alpha_1\Delta^p$ is relatively prime to p [**211**]: Contradiction [**223**].

226. [Cf. op. cit. **227**.] Solution follows problem **226** p. 148.

227. [F. Mertens: Wien. Ber. **117**, pp. 689–690 (1908). Cf. K. Grandjot: Math. Z., **19**, pp. 128–129 (1924).] Notation as in problem and solution **226** p. 148. Let P be the product of all primes $\leqq 2^h$, t the greatest divisor of P relatively prime to m, r an arbitrary number relatively prime to m. In addition let y denote a solution of the congruences

$$y \equiv r \pmod{m}, \qquad y \equiv 1 \pmod{t}.$$

y is relatively prime to m and t and hence to P and thus contains only prime numbers that are $>2^h$.

The method of proof given for problem **226** p. 148 yields $f(\alpha^p)=0$, where p denotes a prime factor of y. If $y>p$ the same reasoning is repeated for an arbitrary prime factor q of y/p; we obtain $f(\alpha^{pq})=0$. Finally we obtain

$$f(\alpha^y)=f(\alpha^r)=0,$$

i.e. $f(x)$ has all the numbers $\alpha^{r_1}, \alpha^{r_2}, \dots, \alpha^{r_h}$ as zeros, $f(x)\equiv K_m(x)$.

227.1. [Cf. D. H. Lehmer: Amer. Math. Monthly, **40**, pp. 165–166 (1933).] Since $n>2$, $\varphi(n)$ is even [**38.1** (1)]. We use the notation

$$\varphi(n)=2h, \qquad e^{2\pi i m/n}=\alpha$$

(different from **226**). The cyclotomic polynomial $K_n(x)$ has the property

$$x^{2h}K_n(x^{-1})\equiv K_n(x)$$

and so the equation $K_n(\alpha)=0$ has the form

$$(*) \qquad \alpha^h+\alpha^{-h}+a_1(\alpha^{h-1}+\alpha^{-h+1})+\cdots+a_{h-1}(\alpha+\alpha^{-1})+a_h=0,$$

where $a_1, a_2, \dots, a_h$ are rational integers. Since

$$\begin{aligned} &2\cos 2\pi m/n=\alpha+\alpha^{-1},\\ &(2\cos 2\pi m/n)^2=\alpha^2+\alpha^{-2}+2,\\ &(2\cos 2\pi m/n)^3=\alpha^3+\alpha^{-3}+3(\alpha+\alpha^{-1}),\\ &\dots\dots\dots\dots\dots\dots\dots\dots \end{aligned}$$

the equation (*) can be transformed into

$$(**) \qquad \left(2\cos\frac{2\pi m}{n}\right)^h+b_1\left(2\cos\frac{2\pi m}{n}\right)^{h-1}+\cdots+b_h=0,$$

where $b_1, b_2, \dots, b_h$ are rational integers. And (**) cannot be reducible, otherwise (*) would be reducible also, which is not the case [**226**].

227.2. An algebraic number is rational if, and only if, it is of degree 1. Now $\varphi(n)/2>1$ and so $2\cos((2\pi m)/n)$ is irrational except in the cases $n = 1, 2, 3, 4$ and 6 [**38.1** (2), (3)].

228. The rational function under consideration is the quotient of two polynomials with rational integral coefficients [**149**], and the first as well as the last non-vanishing coefficient of the denominator is $=\pm 1$ [**156**].

229. [P. Fatou: C.R. Acad. Sci. (Paris), **138**, pp. 342–344 (1904).] Because of the convergence of the series in the unit circle, the absolute values of all the poles are $\geqq 1$. By **156** the product of these absolute values is $\leqq 1$, and hence they are all $=1$ and the leading coefficient of the denominator is $=\pm 1$. Then the assertion follows from **200**. **157** illustrates a special case.

230. By **235**, or by direct considerations analogous to **149**, we see that the given function is the quotient of two polynomials whose coefficients are algebraic integers. The finite field from which they come contains also $\alpha_0, \alpha_1, \alpha_2, \ldots$. If we replace all the coefficients by the corresponding numbers of the fields conjugate to K and form the Cauchy product of the power series obtained in this way, then we obtain on the right-hand side a power series in z^{-1} with integral coefficients and on the left-hand side a rational function whose numerator and denominator are polynomials with integral coefficients, where the denominator has leading coefficient 1 [**156**]. The zeros of this denominator are hence algebraic integers.

231. Interpretation of **225**.

232. [G. Pólya, op. cit. **225**, pp. 3–9.] By carefully expressing the rational function $f'(z)$, without introducing irrationalities, in a form suitable for integration, we reduce the assertion to **231**. [For a conjecture that undertakes to generalize this proposition and awaits proof or disproof see MPR, **2**, p. 53, ex. 19.]

233. We may assume that the algebraic function $f(z)$ is entire; for $P_0(z)f(z)$ satisfies an equation of the stated form with leading coefficient 1. We may assume further that $z=\alpha$ is a regular point of the algebraic entire function $f(z)$. If the function $f(z)$ is expanded in ascending powers of $(z-\alpha)^{\frac{1}{m}}$ (m an ordinary integer), set $z=\alpha+\zeta^m$; then the branch point $z=\alpha$ is transformed into the regular point $\zeta=0$ and the new expansion in powers of ζ has the same coefficients as the original expansion in powers of $(z-\alpha)^{\frac{1}{m}}$; the given equation is transformed, when z is replaced by $a+\zeta^m$, into another equation of the given form. If

$$f(z)=a_0+a_1(z-\alpha)+\cdots+a_m(z-\alpha)^m+\cdots,$$

then

$$\begin{aligned}(z-\alpha)^{-m}[f(z)-a_0-a_1(z-\alpha)-\cdots-a_{m-1}(z-\alpha)^{m-1}]\\=a_m+a_{m+1}(z-\alpha)+\cdots=\varphi(z)\end{aligned}$$

is regular at the point $z=\alpha$ and satisfies an equation in which only algebraic numbers occur as coefficients of the polynomials involved, since $a_0, a_1, \ldots, a_{m-1}$ have already been demonstrated to be algebraic numbers. We must show that

$$a_m=\varphi(\alpha)$$

is algebraic. Thus the whole proposition is reduced to the special case to which we have drawn attention in the statement of the problem. But this case is obvious since we may assume from the outset that $P_0(z), P_1(z), \ldots, P_l(z)$ are not all divisible by $z-\alpha$. [Hecke, p. 66, Theorem 51.]

234. [H. Weyl.] Let $F(z, y)=0$ have a rational solution $f(z)$. We may assume without loss of generality that $f(z)$ is an entire function [solution **233**]. The coefficients of $f(z)$ are algebraic numbers [**233**], all of which belong to a finite field [Hecke, p. 67, Theorem 52], whose degree we denote by n. Replacing the coefficients in $f(z)$ by the conjugate numbers we obtain the additional polynomials $f_1(z), f_2(z), \ldots, f_{n-1}(z)$. The equation

$$(y-f(z))(y-f_1(z))\ldots(y-f_{n-1}(z))=0$$

has rational coefficients and has a root in common with the irreducible equation $F(z, y)=0$. Hence the roots of the latter are contained among $f(z), f_1(z), \ldots, f_{n-1}(z)$ and are thus rational functions of z.

235. If the power series expansion $y=\alpha_0+\alpha_1 z+\alpha_2 z^2+\cdots$ represents an algebraic function and $\alpha_0, \alpha_1, \alpha_2, \ldots$ are algebraic numbers, then the series satisfies an equation of the form $F(z, y)=0$, where $F(z, y)$ is a rational entire function, $F(z, y)\not\equiv 0$, with *algebraic coefficients* [**151**; both proofs can be adapted to this problem]. The coefficients $\alpha_0, \alpha_1, \alpha_2, \ldots$ are determined recursively from a certain point on by an equation of the form

$$(*) \qquad Y=\frac{P_1(z)}{Q(z)}+\frac{zP_2(z)}{Q(z)}Y^2+\frac{zP_3(z)}{Q(z)}Y^3+\cdots+\frac{zP_n(z)}{Q(z)}Y^n,$$

where $Y=\alpha_m+\alpha_{m+1}z+\cdots$ with suitable m, $P_1(z), P_2(z), \ldots, P_n(z)$, $Q(z)$ are polynomials with algebraic coefficients and $Q(0)\neq 0$ [**152**]. Hence all the coefficients depend rationally on finitely many algebraic numbers [Hecke, p. 67, Theorem 52].

236. Deduction from equation (*) in solution **235** similarly as in **153**.

237. [Th. Skolem: Videnskapsselskapets Skrifter 1921, No. 17, Theorem 41.]

237.1. [Cf. MPR, **1**, p. 9.] Set $l=x$, $m=y$ and regard x and y as rectangular coordinates. The three inequalities

$$x\leqq y, \qquad y\leqq n, \qquad x+y>n$$

($x>0$ follows from the last two) delimit a triangle which contains the lattice points (x, y) that we are required to count. We count by horizontal rows:

If $y=n$,	then $1\leqq x\leqq n$,	and there are	n	cases,
$y=n-1$,	$2\leqq x\leqq n-1$,		$n-2$	,
$y=n-2$,	$3\leqq x\leqq n-2$,		$n-4$	,
....	....			....

The total number of cases (lattice points, triangles) is the sum of non-negative terms

$$n+(n-2)+(n-4)+\cdots,$$

which equals

$$\left(\frac{n+1}{2}\right)^2 \quad \text{or} \quad \left(\frac{n+1}{2}\right)^2-\frac{1}{4},$$

according as n is odd or even. Cf. I **31**.

237.2. Draw the figure for some values of the variables, count the solutions in integers of the equation

$$x^2+y^2=n$$

in all cases. Interesting cases are

$$4\delta(25)=12, \qquad 4\delta(65)=16.$$

238. [E. Lucas: S. M. France Bull., **6**, p. 9 (1878).] See **238.1**.

238.1. We term a straight line that passes through two different lattice points a *lattice line*. The tangent of the angle included by a lattice line and the x-axis is obviously rational. Consider two lattice lines which include with the x-axis the angles α and β respectively; the angle between them is $\alpha-\beta$ and so

$$\tan(\alpha-\beta)=\frac{\tan\alpha-\tan\beta}{1+\tan\alpha\tan\beta}$$

is rational.

The exterior angle at any vertex of the polygon considered is $2\pi/n$ and is included by two lattice lines, and so its tangent must be rational; by **197.6**, $n=4$ or 8. The octagon with vertices (1, 0), (2, 0), (3, 1)(3, 2), (2, 3), (1, 3), (0, 2) and (0, 1) is equiangular.

238.2. [T. Gallai and P. Turán; cf. F. Kárteszi: Mat. és fizikai lapok, **50**, pp. 182–183 (1943) and W. Scherrer: Elemente der Math., **1**, pp. 97–98 (1946).] The regular octagon, admitted by **238.1**, is excluded by **244.1**.

239. [G. Pólya: Arch. Math. Phys. Ser. 3, **27**, p. 135 (1918); method of proof given here by A. Speiser.] We term a lattice point p, q *primitive* if it is visible from the origin, i.e. if p and q are relatively prime. If the relation $pv-qu=1$ holds, then the two lattice points p, q and u, v are primitive and connected by a parallelogram of area 1 (the two other corners are 0, 0 and $p+u$, $q+v$); we term u, v the *left neighbour* of p, q and p, q the *right neighbour* of u, v. We term the diagonal emanating from 0, 0 the *diagonal* of the connecting parallelogram. If the length of the diagonal of the connecting parallelogram of p, q and u, v is d, then p, q and u, v lie at the same distance $1/d$ from it. Every primitive lattice point has infinitely many left neighbours, all of which lie on a straight line and are in fact equally spaced.

(1) 1, 0 and $s-1$, 1 are neighbours. The diagonal of the connecting parallelogram is of length $\sqrt{s^2+1^2}$. If this diagonal extended to infinity is to be intercepted by a ρ-circle, then only the circles with center 1, 0 and $s-1$, 1 need be considered. Hence $\rho\geqq 1/\sqrt{s^2+1^2}$.

(2) For an arbitrary primitive lattice point p, q lying in the circle $x^2+y^2\leqq s^2$, determine the *furthest* left neighbouring point p', q' lying in the same circle, i.e. $p'+p$, $q'+q$ already lies outside the circle $x^2+y^2\leqq s^2$. In the same manner, let p'', q'' be the furthest left neighbour of p', q' and p''', q''' that of p'', q'' and so on. After a certain number n of steps we reach $p^{(n)}, q^{(n)}$ with the property that the connecting parallelograms of p, q and p', q', of p', q' and $p'', q'', \ldots$, and of $p^{(n-1)}q^{(n-1)}$ and $p^{(n)}, q^{(n)}$ completely cover the circle $x^2+y^2\leqq 1$. The diagonal of the connecting parallelogram of p, q and p', q' is $>s$ and the distance of the points p, q and p', q' from it is $<1/s$. Thus if we draw circles of radius $1/s$ about each of the points $p, q; p', q'; \ldots; p^{(n)}, q^{(n)}$, then every ray emanating from 0, 0 will be intersected by one of the circles and the diagonals in fact by two of the circles. Hence $\rho<1/s$.

240. [A. J. Kempner: Annals of Math. Series 2, **19**, pp. 127–136 (1917).] Let x be rational $=p/q$, where p, q denotes a primitive lattice point [solution **239**]. If the path in question has the point 0, 0 lying on its right boundary, then it is bounded on the right by the straight line connecting 0, 0 and p, q and on the left by the straight line containing the left neighbours of p, q [solution **239**]. The width of the path is equal to the height of a parallelogram of area 1 and base $\sqrt{p^2+q^2}$, that is, it is equal to

$$\frac{1}{\sqrt{p^2+q^2}}=\varphi\left(\frac{p}{q}\right), \quad \text{and thus} \quad f\left(\frac{p}{q}\right)=\varphi\left(\frac{p}{q}\right)\sqrt{1+\frac{p^2}{q^2}}=\frac{1}{q}.$$

We have $f(x)=\varphi(x)=0$ if x is irrational [II **166**]. (II **99**, II **169**.)

241. If we consider all the lattice points congruent mod n as a "residue class", then there are n^2 different residue classes. It is not possible to distribute kn^2+1 objects into n^2 boxes in such a way that there are not more than k objects in any one box.

242. [H. F. Blichfeldt: Trans. Amer. Math. Soc. **15**, pp. 227–235 (1914). W. Scherrer: Math. Ann. **86**, p. 99 (1922).] Consider the lattice with mesh size $1/N$, i.e. the set of all points x/N, y/N, where x, y, N are integers, x, y variable, $N>0$ given. If z_N of the points of this lattice fall into the domain of area F, then $\lim_{N\to\infty} z_N/N^2 \equiv F$. Let $F>[F]$. Then among the z_N points there are $[F]+1$ that are congruent mod N if N is sufficiently large [**241**], which is the required result. The case $F=[F]$ may be reduced to the preceding case by continuity considerations or else the proof may be suitably modified.

243. [For further details see R. Fueter and G. Pólya: Zürich. Naturf. Ges. **68**, p. 380 (1923).] Let N be an integer, $N>1$. In that part of the plane in which the three inequalities

$$(*) \qquad f(x,y)\leqq N, \qquad x\geqq 0, \qquad y\geqq 0$$

are simultaneously satisfied, there lie by conditions (1) and (2) exactly N lattice points, i.e. points for which x and y are integers. The first inequality (*) may be written in the following form:

$$\varphi_m\left(xN^{-\frac{1}{m}}, yN^{-\frac{1}{m}}\right)+N^{-\frac{1}{m}}\varphi_{m-1}\left(xN^{-\frac{1}{m}}, yN^{-\frac{1}{m}}\right)$$
$$+N^{-\frac{2}{m}}\varphi_{m-2}\left(xN^{-\frac{1}{m}}, yN^{-\frac{1}{m}}\right)+\cdots\leqq 1.$$

Denote by F the area of the domain defined by the inequalities

$$(**) \qquad \varphi_m(x,y)\leqq 1, \qquad x\geqq 0, \qquad y\geqq 0$$

(F is by supposition finite), and consider the points $xN^{-\frac{1}{m}}$, $yN^{-\frac{1}{m}}$, where x and y are integers (they form a fine-meshed lattice). Of the points of this fine-meshed lattice approximately $FN^{\frac{2}{m}}$ points lie in the region (**). We deduce that the number of points of the lattice of mesh width 1 that lie in the region (*) as N tends to infinity is asymptotically $=FN^{\frac{2}{m}}$. But now as we have stated above the exact number $=N$. From

$$FN^{\frac{2}{m}}\sim N$$

it follows that

$$m=2, \qquad F=1.$$

244. [Cf. W. Ahrens: Mathematische Unterhaltungen und Spiele, **2**, p. 364. Leipzig: B. G. Teubner, 1918.] The four systems of numbers

$$\begin{array}{llll} x_1, & x_2, & \ldots, & x_n, \\ y_1, & y_2, & \ldots, & y_n, \\ x_1-y_1, & x_2-y_2, & \ldots, & x_n-y_n, \\ x_1+y_1, & x_2+y_2, & \ldots, & x_n+y_n \end{array}$$

are required to be complete residue systems mod n. If we set $x_\mu - y_\mu = r_\mu$, $y_\mu = s_\mu$, then we are concerned with the simplest particular cases $p=2$, $p=3$ and $p \geqq 5$, respectively, of **247**.

244.1. [H. E. Chrestenson: Amer. Math. Monthly, **70**, pp. 447–448 (1963).] Three consecutive vertices of the polygon P considered are the vertices of an isosceles triangle with sides a, a, b. The sides of length a are consecutive sides of P, containing the angle $[\pi(n-2)]/n$. Therefore

$$b^2 = 2a^2 - 2a^2 \cos \frac{\pi(n-2)}{n}.$$

Yet a^2 and b^2 are integers. Therefore $\cos [\pi(n-2)]/n$ is rational, $=\frac{1}{2}$, 0, or $-\frac{1}{2}$ by **197.1**, and so $n=3$, 4, or 6.

244.2. The square is trivial. Triangle with vertices (1, 0, 0), (0, 1, 0), (0, 0, 1). Hexagon with vertices

$$(0, 1, -1) \quad (1, 0, -1) \quad (1, -1, 0) \quad (0, -1, 1) \quad (-1, 0, 1) \quad (-1, 1, 0).$$

244.3. A face of the dodecahedron is a regular pentagon. A given vertex of the icosahedron is joined by edges to five other vertices which form a regular pentagon. Thus these two solids are excluded by **244.1**.

244.4. The cube is trivial. Tetrahedron with vertices

$$(0, 0, 0), \quad (0, 1, 1), \quad (1, 0, 1), \quad (1, 1, 0).$$

Octahedron with vertices

$$(1, 0, 0) \quad (-1, 0, 0) \quad (0, 1, 0) \quad (0, -1, 0) \quad (0, 0, 1) \quad (0, 0, -1).$$

245. [A. Hurwitz, problem: Nouv. Annls Math., Ser. 3, **1**, p. 384 (1882).] $r_1 s_1, r_2 s_2, \ldots, r_q s_q$ certainly do not form a complete remainder system mod q if $r_\alpha \equiv s_\beta \equiv 0 \pmod q$ and $\alpha \neq \beta$. Thus let us assume that $r_q \equiv s_q \equiv 0 \pmod q$. Then we have

$$r_1 r_2 \ldots r_{q-1} \equiv s_1 s_2 \ldots s_{q-1} \equiv 1 \cdot 2 \ldots (q-1) \equiv -1 \pmod q,$$

and hence

$$r_1 s_1 \cdot r_2 s_2 \ldots r_{q-1} s_{q-1} \equiv 1 \not\equiv 1 \cdot 2 \ldots (q-1) \pmod q.$$

If we assume the reduced remainder system $1, 2, \ldots, q-1$ written in the form $g, g^2, \ldots, g^{q-1}$ (g a primitive root mod q), then the theorem is a special case of **247** for $n=q-1$, $p=2$.

246. [A. Hurwitz.] It is sufficient to consider the sum

$$S = 1^\lambda + 2^\lambda + \cdots + p^{\alpha\lambda}.$$

If λ is not a multiple of $p-1$ and g is a primitive root mod p, then $g^\lambda-1$ is not divisible by p. It then follows from

$$g^\lambda S \equiv g^\lambda+(g\cdot 2)^\lambda+(g\cdot 3)^\lambda+\cdots+(gp^\alpha)^\lambda \equiv S, \qquad S(g^\lambda-1)\equiv 0 \pmod{p^\alpha},$$

that $S\equiv 0 \pmod{p^\alpha}$. If λ is a multiple of $p-1$, then we have that

$$1^\lambda+2^\lambda+\cdots+(p^\alpha)^\lambda \equiv -p^{\alpha-1} \pmod{p^\alpha}$$

is true for $\alpha=1$. Assume that this congruence is true for a certain value of α. It then follows that

$$\begin{aligned}1^\lambda+2^\lambda+\cdots+(p^{\alpha+1})^\lambda &= \sum_{k=0}^{p-1}[(1+kp^\alpha)^\lambda+(2+kp^\alpha)^\lambda+\cdots+(p^\alpha+kp^\alpha)^\lambda]\\ &\equiv \sum_{k=0}^{p-1}(1^\lambda+2^\lambda+\cdots+p^{\alpha\lambda})\\ &\quad+\lambda p^\alpha\sum_{k=0}^{p-1}k(1^{\lambda-1}+2^{\lambda-1}+\cdots+p^{\alpha(\lambda-1)})\\ &\equiv p(1^\lambda+2^\lambda+\cdots+p^{\alpha\lambda})\\ &\quad+\lambda p^\alpha\cdot\frac{p(p-1)}{2}(1^{\lambda-1}+2^{\lambda-1}+\cdots+p^{\alpha(\lambda-1)})\\ &\equiv -p^\alpha \pmod{p^{\alpha+1}}.\end{aligned}$$

247. [A. Hurwitz, cf. op. cit. **244**.] (1) If we set $r_\mu=s_\mu$ then $(r+s)=(2r)$, $(r+2s)=(3r),\ldots,(r+(p-2)s)=((p-1)r)$ are complete remainder systems mod n, since $2, 3, \ldots, p-1$ are relatively prime to n.

(2) If $p=2$, and thus n is even, and if (r), (s), $(r+s)$ were simultaneously complete remainder systems, then it would follow that

$$\begin{aligned}r_1+r_2+\cdots+r_n&\equiv s_1+s_2+\cdots+s_n\equiv(r_1+s_1)+(r_2+s_2)+\cdots\\ &+(r_n+s_n)\equiv 1+2+\cdots+n\equiv\frac{n(n+1)}{2}\equiv\frac{n}{2} \pmod{n},\end{aligned}$$

and hence that

$$(r_1+s_1)+(r_2+s_2)+\cdots+(r_n+s_n)\equiv 0\equiv\frac{n}{2} \pmod{n}.$$

Contradiction!

(3) Let p be odd and p^α the highest power of p that divides n. Set

$$1^{p-1}+2^{p-1}+\cdots+n^{p-1}=S.$$

If (r), (s), $(r+s),\ldots,((r+(p-1)s)$ were simultaneously complete remainder systems, then it would follow that

$$\sum_{\nu=1}^{n}(r_\nu+ks_\nu)^{p-1}\equiv S \pmod{n}$$

and hence, setting

$$\binom{p-1}{\alpha}\sum_{\nu=1}^{n}r_\nu^{p-1-\alpha}s_\nu^\alpha=S_\alpha,$$

we would have

$$kS_1+k^2S_2+\cdots+k^{p-2}S_{p-2}+k^{p-1}S\equiv 0 \pmod n), \qquad k=1,2,\ldots,p-1.$$

Now the determinant of this system of $p-1$ linear congruences is formed from factors that are all >0 and $<p$ and is thus relatively prime to n. Thus it would finally follow that

$$S_1\equiv S_2\equiv\cdots\equiv S_{p-2}\equiv S\equiv 0 \pmod n.$$

Contradiction, since S is not divisible by p^α [**246**].

247.1. In the sequel, H-W refers to Hardy-Wright.

For $\binom{p}{k}$ see H-W, p. 64, Theorem 75.

For s_k^n see I* **199** and H-W, p. 85, Theorem 112.

From I **187**, I **191**, and H-W, p. 84, Theorem 111, it follows that

$$S_2^p x(x-1)+S_3^p x(x-1)(x-2)+\cdots+S_{p-1}^p x(x-1)\ldots(x-p+2)\equiv 0 \pmod p.$$

Divide first by $x(x-1)$ (see H-W, p. 82, Theorem 104) then set $x=2$; there follows $S_2^p\equiv 0 \pmod p$. Return now to the original congruence, divide by $x(x-1)(x-2)$, then set $x=3$; it follows that $S_3^p=0 \pmod p$, and so on.

247.2. The coefficients of $T_n(z)$ and $U_n(x)$ are integers, as we find if we compute them from the identity

$$\cos n\vartheta+i\sin n\vartheta=(\cos\vartheta+i\sin\vartheta)^n.$$

The assertion follows from

$$T_p'(x)=pU_{p-1}(x).$$

248. For $a\geqq 2$ (of course!) we have for

$$\begin{aligned} n \text{ even: } n^a=n\cdot n^{a-1}&=(n^{a-1}+1)+(n^{a-1}+3)+\cdots+(n^{a-1}+n-1)\\ &\quad+(n^{a-1}-1)+(n^{a-1}-3)+\cdots+(n^{a-1}-n+1);\\ n \text{ odd: } n^a&=n^{a-1}+(n^{a-1}+2)+(n^{a-1}+4)+\cdots+(n^{a-1}+n-1)\\ &\quad+(n^{a-1}-2)+(n^{a-1}-4)+\cdots+(n^{a-1}-n+1). \end{aligned}$$

249. A proper divisor of one of the numbers $2, 3, 4, \ldots, n$ is contained in the same sequence of numbers; hence the required number must be a prime number. If a number $m\leqq n/2$, then $2m$ is contained in the same sequence of numbers. Hence the required number must be $>n/2$.

250. Using Tchebychev's theorem: If $n>2$, p a prime number, $n\geqq p>n/2$, then we have

$$1+\frac{1}{2}+\cdots+\frac{1}{p}+\cdots+\frac{1}{n}=\frac{1}{p}+\frac{M}{N}=\frac{N+pM}{pN},$$

where $(M,N)=1$, $(p,N)=1$, from which it follows that $(pN, N+pM)=1$: no cancellation of factors in the denominator! Without using Tchebychev's theorem: Cf. **251**.

251. [J. Kürschák: Math. és phys. lapok, **27**, pp. 299–300 (1918).] We term α the "degree of parity of n" if n is divisible by 2^α and not divisible by $2^{\alpha+1}$.

$2^\alpha, 3\cdot 2^\alpha, 5\cdot 2^\alpha, 7\cdot 2^\alpha, \ldots$ are the numbers of degree of parity α. Between two consecutive numbers of this sequence lie the numbers $2\cdot 2^\alpha, 4\cdot 2^\alpha, 6\cdot 2^\alpha, \ldots$, respectively, and thus between any two numbers of the same degree of parity there lies one of higher degree of parity. Hence among the numbers $n, n+1, \ldots, m-1, m$ there is only *one* of maximal degree of parity μ: the factor 2^μ of the denominator cannot be cancelled.

252. [G. Pólya, problem: Arch. Math. Phys. Ser. 3, **23**, p. 289 (1915). Solution by S. Sidon: Arch. Math. Phys. Ser. 3, **24**, p. 284 (1916); present solution by A. Fleck.] It is sufficient to consider the case $\alpha = 1/k$, where k is an integer. Let n be divisible by $1, 2, 3, \ldots, [\sqrt[k]{n}]$, and hence also by the least common multiple V of these numbers. Denoting the νth prime number by p_ν, let

$$p_l \leqq \sqrt[k]{n} < p_{l+1}, \qquad V = p_1^{m_1} p_2^{m_2} \ldots p_l^{m_l}.$$

The exponent m_λ is determined by the inequality $p_\lambda^{m_\lambda} \leqq \sqrt[k]{n} < p_\lambda^{m_\lambda+1}$. In particular $\sqrt[k]{n} < p_\lambda^{2m_\lambda}$, $\lambda = 1, 2, \ldots, l$, whence by multiplication it follows that $n^{\frac{l}{k}} < V^2$. From the assumption it follows moreover that $V \leqq n$, and hence finally

$$\frac{l}{k} < 2, \qquad \sqrt[k]{n} < p_{2k}, \qquad n < p_{2k}^k.$$

E.g. if $k=2$, then $n<49$. We now arrive at the result for $n=24$ by direct computation. For $k=3$ the greatest permissible value of n is 420.

253. [L. Kollros.] Denote the least positive remainders of the numbers $P, 10P, 10^2P, \ldots$ (mod Q) by $p_0, p_1, p_2, \ldots$. If

$$\frac{P}{Q} = a.\ a_1 a_2 a_3 \ldots,$$

where $a_1, a_2, a_3 \ldots$ are the decimal digits, then we have

$$\frac{p_0}{Q} = 0.\ a_1 a_2 a_3 \ldots, \qquad \frac{p_1}{Q} = 0.\ a_2 a_3 a_4 \ldots, \qquad \frac{p_2}{Q} = 0.\ a_3 a_4 a_5 \ldots.$$

From this we deduce that the length of the shortest period l is the exponent of 10 (mod Q), i.e. $10^l \equiv 1$ (mod Q) and no smaller power of 10 is $\equiv 1$ (mod Q). If l is even, $=2\lambda$, and thus $(10^\lambda - 1)(10^\lambda + 1) \equiv 0$ (mod Q), then $10^\lambda \equiv -1$ (mod Q), and hence

$$\frac{p_0}{Q} + \frac{p_\lambda}{Q} = 0.\ a_1 a_2 a_3 \ldots a_l a_1 \ldots + 0.\ a_{\lambda+1} a_{\lambda+2} \ldots a_l a_1 \ldots a_\lambda = 0.999\ldots,$$

and thus $a_1 + a_{\lambda+1} = 9$, $a_2 + a_{\lambda+2} = 9, \ldots, a_\lambda + a_l = 9$. If on the other hand l is odd, then we have clearly

$$\frac{a_1 + a_2 + \cdots + a_l}{l} \neq \frac{9}{2}.$$

254. [E. Lucas; cf. A. Hurwitz: Interméd. des math. **3**, p. 214 (1896).]

(1) If $3^{2^{h-1}} \equiv -1$ (mod n), and hence $3^{2^h} \equiv 1$ (mod n), then 3 belongs mod n to the exponent $2^h = n-1$. The exponent is a divisor of $\varphi(n)$ and hence $\varphi(n) \geqq n-1$. It follows that $\varphi(n) = n-1$ and n is a prime number.

(2) If $n=2^h+1$ is a prime number, $h\geqq 2$, then $n\equiv 1$ (mod 4); moreover h must be even, $=2\nu$. (Otherwise $2^{2\nu+1}+1=4^\nu.2+1\equiv 0$, mod 3.) By the law of reciprocity we have

$$\left(\frac{3}{n}\right)=\left(\frac{n}{3}\right)=\left(\frac{4^\nu+1}{3}\right)=\left(\frac{2}{3}\right)=-1, \qquad 3^{\frac{n-1}{2}}\equiv\left(\frac{3}{n}\right) \quad (\text{mod } n).$$

255. [Euler: Opera Postuma, **1**, p. 220. Petropoli 1862; G. Pólya, problem: Arch. Math. Phys. Ser. 3, **24**, p. 84 (1916). Solution by G. Szegö: Arch. Math. Phys. Ser. 3, **25**, p. 340 (1917).] If x, y, z, t were a solution, then

$$\frac{4zt^2+1}{4yz-1}=4zx-1$$

would be integral, i.e.

$$(2zt)^2\equiv -z \quad (\text{mod } 4yz-1).$$

Let $z=2^\alpha z'$, α integral, $\alpha\geqq 0$, z' odd. We then have (Legendre-Jacobi symbols [V **45**])

$$\left(\frac{-z}{4yz-1}\right)=\left(\frac{-1}{4yz-1}\right)\left(\frac{2^\alpha}{4yz-1}\right)\left(\frac{z'}{4yz-1}\right).$$

The first factor is $=-1$. The third factor is

$$=\left(\frac{4yz-1}{z'}\right)(-1)^{\frac{z'-1}{2}}=\left(\frac{-1}{z'}\right)(-1)^{\frac{z'-1}{2}}=1.$$

The second factor, if $\alpha=2k+1$, k integral, $k\geqq 0$, $z=2z''$, is

$$=\left(\frac{2}{4yz-1}\right)=\left(\frac{2}{8yz''-1}\right)=1,$$

and, if $\alpha=2k$, k integral, $k\geqq 0$,

$$=\left(\frac{1}{4yz-1}\right)=1.$$

Thence we have

$$\left(\frac{-z}{4yz-1}\right)=-1. \quad \text{Contradiction!}$$

256. [G. Pólya, problem: Arch. Math. Phys. Ser. 3, **24**, p. 84 (1916). Cf. Gauss: Disquisitiones Arithmeticae, p. 125; Werke, **1**, pp. 94–95. Göttingen: Königliche Gesellschaft der Wissenschaften 1863. Solution by P. Bernays.]

(1) $q\equiv 1$ (mod 4). If p is a prime factor of $q-4$, then we have $\left(\frac{q}{p}\right)=\left(\frac{p}{q}\right)=1$; we have used $q>5$. Moreover, there must exist an odd prime number $p<q$, for which $\left(\frac{p}{q}\right)=\left(\frac{q}{p}\right)=-1$. For otherwise all odd numbers r less than q, and hence also the even numbers $q-r$ and finally all the numbers $1, 2, 3, \ldots, q-1$ would be quadratic residues of q.

(2) $q\equiv -1$ (mod 4). At least one prime factor p of $q-4$ is also $\equiv -1$ (mod 4). For this prime factor p we have $\left(\frac{q}{p}\right)=-\left(\frac{p}{q}\right)=1$.

(2a). $q \equiv 7 \pmod 8$. Of the four numbers $(q+1)/8, (q+9)/8, (q+25)/8, (q+49)/8$, one and only one is of the form $4n+3$. This number has a prime factor p of the same form, and $p<q$ if $q>7$. We have $\left(\frac{q}{p}\right)=\left(\frac{-1}{p}\right)=-1$, $\left(\frac{p}{q}\right)=1$.

(2b). $q \equiv 3 \pmod 8$. Of the two odd numbers $(q+1)/4$ and $(q+9)/4$, one and only one has the form $4n+3$. This number has a prime factor p of the same form, $p<q$, if $q>3$; $\left(\frac{q}{p}\right)=\left(\frac{-1}{p}\right)=-1=-\left(\frac{p}{q}\right)$.

257. [G. Pólya, problem: Arch. Math. Phys. Ser. 3, **21**, p. 288 (1913). Solution by O. Szász, G. Szegö, L. Neder: Arch. Math. Phys. Ser. 3, **22**, p. 366 (1914). Cf. W. H. Young and Grace Chisholm Young: The theory of sets of points, p. 3. Cambridge University Press, 1906.]

First solution: Let n be any positive integer. There are arbitrarily large prime numbers that, when expressed in decimal notation, contain at least n consecutive zeros. [The arithmetic progression $10^{n+1}x+1$ contains infinitely many prime numbers, **110**.] Hence the given decimal fraction cannot have a period of length n.

Second solution: By a result of Tchebychev [op. cit. **249**] there exists at least one prime number which, when expressed in decimal notation, has a prescribed number of digits. If the given decimal fraction were periodic with period $a_1, a_2, \ldots, a_k$, $k \geqq 2$, choose r so large that the prime numbers with kr digits follow the digit a_1 of the first period. Let x be the smallest of the prime numbers with kr digits. Then two cases are possible:

$$(1)\quad x = a_1a_2 \ldots \underbrace{a_k}_{1}a_1a_2\ldots\underbrace{a_k}_{2}\ldots a_1a_2\ldots\underbrace{a_k}_{r},$$

$$(2)\quad x = a_{l+1}a_{l+2}\ldots\underbrace{a_k}_{1}a_1a_2\ldots\underbrace{a_k}_{2}\ldots a_1a_2\ldots\underbrace{a_k}_{r}a_1a_2\ldots a_l, \qquad l>0.$$

In the first case the presumed prime number x would be divisible by the number $a_1a_2\ldots a_k$, while in the second case x would be divisible by $a_{l+1}a_{l+2}\ldots a_ka_1\ldots a_l$. Contradiction!

258. From Taylor's formula it follows that for $n=0, 1, 2, \ldots$

$$e = 1+\frac{1}{1!}+\frac{1}{2!}+\cdots+\frac{1}{n!}+\frac{e^{\vartheta_n}}{(n+1)!}, \qquad 0<\vartheta_n<1.$$

Assume $e=r/s$, (r, s positive integers, relatively prime, $s \geqq 2$). Set $n=s$. Then

$$s!\left(e-1-\frac{1}{1!}-\frac{1}{2!}-\cdots-\frac{1}{s!}\right)=\frac{e^{\vartheta_s}}{s+1}$$

would be integral. On the other hand we have that

$$0<\frac{e^{\vartheta_s}}{s+1}<\frac{e}{3}<1: \quad \text{Contradiction!}$$

259. If $ae+be^{-1}+c=0$ (a, b, c integral, $|a|+|b|>0$), then it would follow from Taylor's formula, applied to the function ae^x+be^{-x}, that

$$-c=\sum_{\nu=0}^{n}\frac{a+(-1)^\nu b}{\nu!}+\frac{ae^{\vartheta_n}-(-1)^n be^{-\vartheta_n}}{(n+1)!}, \qquad 0<\vartheta_n<1,$$

$n=0, 1, 2, \ldots$. Let $n \geqq 3|a|+|b|-1$, and further choose n such that the sign of $(-1)^n b$ coincides with the sign of $-a$. Then, [cf. **258**]

$$\frac{ae^{\vartheta_n}-(-1)^n be^{-\vartheta_n}}{n+1}$$

will be integral. On the other hand we have

$$0<\frac{|a|e^{\vartheta_n}+|b|e^{-\vartheta_n}}{n+1}=\left|\frac{ae^{\vartheta_n}-(-1)^n be^{-\vartheta_n}}{n+1}\right|<\frac{3|a|+|b|}{n+1}\leqq 1:$$

Contradiction!

260. [A. Hurwitz.] We have

$$\frac{\Gamma'(n)}{\Gamma(n)}=\frac{\Gamma'(1)}{\Gamma(1)}+\frac{1}{1}+\frac{1}{2}+\frac{1}{3}+\cdots+\frac{1}{n-1}$$

for integral n, and thus

$$\Gamma'(n+1)=n!\left(\frac{1}{1}+\frac{1}{2}+\frac{1}{3}+\cdots+\frac{1}{n}-C\right).$$

260.1. If λ were rational, $=m/n$, where m and n are positive integers, 3^n would equal 2^m which is impossible: The decomposition into prime factors is unique.

260.2. Eight examples are collected in the following table (use **260.1**):

a	b	a^b	
r	r	r	$2^1=2$
r	r	i	$2^{\frac{1}{2}}=\sqrt{2}$
i	r	r	$(\sqrt{2})^2=2$
i	r	i	$(\sqrt{2})^1=\sqrt{2}$
r	i	r	$2^\lambda=3$
r	i	i	$2^{\frac{\lambda}{2}}=\sqrt{3}$
i	i	r	$(\sqrt{2})^{2\lambda}=3$
i	i	i	$(\sqrt{2})^\lambda=\sqrt{3}.$

261. $\pi+(4-\pi)=3.9999\ldots.$

262. [G. Pólya, problem: Arch. Math. Phys. Ser. 3, **27**, p. 161 (1918).] The given expression is [**182, 82**]

$$=\left\{2^{n-[n]-\left[\frac{n}{2}\right]-\left[\frac{n}{4}\right]-\cdots}\prod_{p>2} p^{\left[\frac{2n}{p-1}\right]-\left[\frac{2n}{p}\right]-\left[\frac{2n}{p^2}\right]-\cdots}\right\}^{\frac{1}{n}}=2^{a_n}\prod\nolimits_1^{\frac{1}{n}}\prod\nolimits_2^{\frac{1}{n}} \quad ;$$

$\prod_{p>2}$ is extended over all odd prime numbers $\leqq 2n+1$. The product $\prod_1$ contains those factors for which $3 \leqq p \leqq \sqrt{2n}$, $\prod_2$ those for which $\sqrt{2n}<p\leqq 2n+1$. We have for $n \to \infty$

$$a_n=-\frac{1}{n}\left(\left[\frac{n}{2}\right]+\left[\frac{n}{4}\right]+\left[\frac{n}{8}\right]+\cdots\right) \to -1.$$

For $p>\sqrt{2n}$ we have $[2n/p^2]=[2n/p^3]=\cdots=0$. Since it is an alternating series with decreasing terms we have that

$$\sum_{p>\sqrt{2n}}\left(\left[\frac{p-1}{2n}\right]-\left[\frac{2n}{p}\right]\right)\leqq\left[\frac{2n}{[\sqrt{2n}]}\right]<\frac{2n}{\sqrt{2n}-1}.$$

We deduce that

$$1\leqq\prod\nolimits_2<(2n+1)^{\frac{2n}{\sqrt{2n-1}}},$$

and hence that $\prod_2^{\frac{1}{n}}\to 1$.

Let $m=a_0+a_1p+a_2p^2+\cdots+a_lp^l$ be a rational integer expressed in the p-adic system, that is

$$0\leqq a_\lambda\leqq p-1,\qquad \lambda=0,1,\ldots,l,\qquad a_l>0.$$

Then we have

$$\frac{m}{p}-\left[\frac{m}{p}\right]=\frac{a_0}{p},$$
$$\frac{m}{p^2}-\left[\frac{m}{p^2}\right]=\frac{a_0}{p^2}+\frac{a_1}{p},$$
$$\cdots\cdots\cdots\cdots\cdots$$

and if all the infinitely many equations are added we obtain

$$\frac{m}{p-1}-\left[\frac{m}{p}\right]-\left[\frac{m}{p^2}\right]-\cdots=\frac{a_0+a_1+\cdots+a_l}{p-1}\leqq l+1\leqq\frac{\log m}{\log p}+1,$$

where the last inequality follows from $l\log p\leqq\log m$. Hence we have for $3\leqq p<\sqrt{2n}$

$$p^{\left[\frac{2n}{p-1}\right]-\left[\frac{2n}{p}\right]-\left[\frac{2n}{p^2}\right]-\cdots}\leqq p^{\frac{\log 2n}{\log p}+1}<p^{\frac{2\log 2n}{\log p}}=(2n)^2,$$

$$1\leqq\prod\nolimits_1<(2n)^{2\sqrt{2n}},\qquad \prod\nolimits_1^{\frac{1}{n}}\to 1.$$

262.1. We must determine the limit of the sum of the first n rows of the double series

$$\frac{1}{2^2}+\frac{1}{3^2}+\frac{1}{4^2}+\cdots$$
$$+\frac{1}{2^3}+\frac{1}{3^3}+\frac{1}{4^3}+\cdots$$
$$+\frac{1}{2^4}+\frac{1}{3^4}+\frac{1}{4^4}+\cdots$$
$$+\cdots\cdots\cdots\cdots$$

The sum of the first n columns is

$$\frac{1}{2(2-1)}+\frac{1}{3(3-1)}+\cdots+\frac{1}{(n+1)n}=\left(1-\frac{1}{2}\right)+\left(\frac{1}{2}-\frac{1}{3}\right)+\cdots+\left(\frac{1}{n}-\frac{1}{n+1}\right)$$
$$=1-\frac{1}{n+1}.$$

262.2. Since

$$\log n = 1+\frac{1}{2}+\frac{1}{3}+\cdots+\frac{1}{n-1}-C+\varepsilon_n,$$

where $\varepsilon_n \to 0$ [II **19.1**], we must determine the limit of the sum of the first $n-1$ rows of the double series

$$\begin{aligned}&\frac{1}{2}\frac{1}{2^2}+\frac{1}{2}\frac{1}{3^2}+\frac{1}{2}\frac{1}{4^2}+\cdots\\+&\frac{1}{3}\frac{1}{2^3}+\frac{1}{3}\frac{1}{3^3}+\frac{1}{3}\frac{1}{4^3}+\cdots\\+&\frac{1}{4}\frac{1}{2^4}+\frac{1}{4}\frac{1}{3^4}+\frac{1}{4}\frac{1}{4^4}+\cdots\\+&\cdots\cdots\cdots\cdots\cdots\cdots\end{aligned}$$

The sum of the first $n-1$ columns is

$$-\log\left(1-\frac{1}{2}\right)-\frac{1}{2}-\log\left(1-\frac{1}{3}\right)-\frac{1}{3}-\cdots-\log\left(1-\frac{1}{n}\right)-\frac{1}{n}$$
$$=-\frac{1}{2}-\frac{1}{3}-\cdots-\frac{1}{n}+\log n \to 1-C \qquad \text{[II } \mathbf{19.1}.]$$

263. [G. Pólya: Gött. Nachr. p. 26 (1918).] By assumption there are only finitely many prime number powers $p'^{\alpha'}$ for which $|f(p'^{\alpha'})| \geqq 1$. Let the product of all such values $\Pi f(p'^{\alpha'})$ be $=C$. Let $0<\varepsilon<1$. Then there are also only finitely many prime number powers $p''^{\alpha''}$ such that

$$|f(p''^{\alpha''})| \geqq \varepsilon |C|^{-1}.$$

Thus, apart from finitely many, every natural number n has in its prime factor decomposition at least one prime number power P^A for which

$$|f(P^A)| < \varepsilon |C|^{-1}.$$

Then we have

$$\begin{aligned}|f(n)| &= |f(p^\alpha q^b \ldots P^A \ldots)| = |f(p^\alpha)||f(q^b)| \ldots |f(P^A)| \ldots\\ &< |C| \cdot \varepsilon |C|^{-1} = \varepsilon.\end{aligned}$$

264. Application of **263** [**25**, **44**]. Cf. **45**.

265. ax^2+bx+c is reducible if and only if b^2-4ac is the square of an integer, $=u^2$. We require an estimate for r_n. We can omit consideration of the coordinate planes ($a=0$, $b=0$, $c=0$), which can contribute at most $3(2n+1)^2$ units to r_n. For b there are thus $2n$ admissible values, $-n, \ldots, -1, 1, \ldots, n$. For fixed b we must have

$$b^2-u^2=4ac \geqq -4n^2, \qquad u^2 \leqq 4n^2+b^2 \leqq 5n^2.$$

For u there are thus at most $2\sqrt{5}n$ admissible values, $u^2=b^2$ is not permissible. For fixed b and u, $b^2-u^2=4ac$, there are a total of $2\tau(|b^2-u^2|/4)$ admissible values for a; then c is also determined. If then we denote by $T(n)$ the largest of the numbers $\tau(1), \tau(2), \ldots, \tau(n)$ we have

$$r_n < 3(2n+1)^2 + 2n \cdot 2\sqrt{5}n \cdot 2T(n^2).$$

We have $\tau(n) \cdot n^{-\varepsilon} \to 0$ for every fixed $\varepsilon > 0$ [**264**].

266. Let $k > 0$, $l > 0$, $k + l = h$, and

$$\alpha(x) = \alpha_0 + \alpha_1 x + \cdots + \alpha_k x^k,$$
$$\beta(x) = \beta_0 + \beta_1 x + \cdots + \beta_l x^l$$

two polynomials with integral coefficients, in whose product $\alpha(x)\beta(x) = A(x)$ all the coefficients are in absolute value $\leqq n$. Moreover let α_0, α_k, β_0, β_l be different from 0. Since the number of polynomials considered in the problem with $a_0 = 0$ or $a_h = 0$ is of the order of magnitude n^h, it is sufficient to show that the number of admissible systems $(\alpha_0, \alpha_1, \ldots, \alpha_k, \beta_0, \beta_1, \ldots, \beta_l)$ is equal to $0(n^h \log^2 n)$.

Let $x_1 < x_2 < \cdots < x_h$ be the h smallest natural numbers for which $A(x)$ does not vanish. Then clearly $x_h \leqq 2h$. Moreover $|\alpha(x_\nu)|$, $|\beta(x_\nu)|$ are positive integers and both divide $|A(x_\nu)|$; hence

$$|\alpha(x_\nu)| \leqq |A(x_\nu)| \leqq (1 + (2h) + (2h)^2 + \cdots + (2h)^h)n,$$
$$\nu = 1, 2, \ldots, h.$$

The same inequality holds for $\beta(x_\nu)$. From this it follows by the Lagrange interpolation formula (introduction to Part VI, § 9 p. 82) that all the coefficients of $\alpha(x)$ and $\beta(x)$ are in absolute value smaller than Cn where C depends only on h. Moreover $1 \leqq |\alpha_0\beta_0| \leqq n$, $1 \leqq |\alpha_k\beta_l| \leqq n$. The number of admissible systems of numbers (α_0, β_0) and (α_k, β_l) respectively is thus [**79**, II **46**] $= O(n \log n)$, the number of systems of numbers $(\alpha_0, \beta_0, \alpha_k, \beta_l)$ is hence $= O(n^2 \log {}^2 n)$, and finally the required number is

$$= O(n^{k-1} n^{l-1} n^2 \log^2 n) = O(n^h \log^2 n).$$

Part Nine. Geometric Problems

1. The points of the surface form a closed set of points $\mathfrak{O}$ from which P is at a certain minimum distance. This minimum distance can be attained only at a point M of the surface such that the direction of every curve element on $\mathfrak{O}$ emanating from the point M is perpendicular to the straight line MP. Thus M cannot be situated at a corner nor on an edge but must be in the interior of a face of the polyhedron.

1.1. [G. Pólya, problem: Elem. Math., **26**, p. 113 (1971).] Choose the system of rectangular coordinates x, y, z in such a way that the z-axis is the axis of the cone and the major axis of the ellipse in question lies in the xz-plane. By eliminating z from two equations, that of the cone and that of the intersecting plane, we obtain the equation of the projection of the ellipse onto the xy-plane:

$$\left[x-\frac{p-q}{2}\sin\alpha\right]^2+\left(\frac{p+q}{2\sqrt{pq}}\right)^2 y^2=\left(\frac{p+q}{2}\right)^2\sin^2\alpha,$$

which is again an ellipse with semi-axes

$$a=\frac{p+q}{2}\sin\alpha,\qquad b=\sqrt{pq}\sin\alpha.$$

Now

$$S\sin\alpha=\pi ab.$$

1.2. [G. Pólya, problem: Elem. Math., **16**, p. 92 (1961). Solution by H. Bieri, Elem. Math., **17**, p. 112 (1962).] Proof for V and S by elementary solid geometry. Observe that

$$S=4\pi r^2+4\pi R^2.$$

For the definition of M see the explanation preceding **7**.

1.3. [Cf. G. Pólya, problem: Amer. Math. Monthly, **54**, p. 340 (1947). Solution by Joseph Rosenbaum: Amer. Math. Monthly, **55**, pp. 162–164 (1948).] The "piece of paper" is bounded by two concentric circles, with radii r and $r+s$, respectively, and by rays emanating from the common center of the two circles and including an angle 2α,

$$r2\alpha=p,\qquad (r+s)2\alpha=P.$$

The assumption concerning the inclination i of the slant height implies that

$$\cos i=\frac{P-p}{2\pi s}\leq\frac{1}{2}$$

and therefore that

$$\alpha=\frac{P-p}{2s}\leqq\frac{\pi}{2}.$$

The piece of paper is obviously contained in a rectangle with sides

$$2(r+s)\sin\alpha \quad\text{and}\quad s+r(1-\cos\alpha).$$

That these sides are shorter than those mentioned in the problem follows from I **142**:

$$\sin\alpha<\alpha, \qquad 1-\cos\alpha<\frac{\alpha^2}{2}.$$

1.4. [G. Pólya, problem: Amer. Math. Monthly **54**, p. 479 (1947). Solution by N. J. Fine: Amer. Math. Monthly, **56**, p. 273 (1949).] Let

$$F(x)=2f(x)[1+(f'(x))^2]^{1/2}$$
$$G(x)=f(x)[x^2+(f(x))^2]^{1/2}.$$

It is sufficient to prove that

$$F(x)\leqq G'(x)$$

for $0\leqq x\leqq a$, but the case of equality is not attained throughout the whole interval. In fact we find that $G'\geqq 0$ and

$$(x^2+f^2)(G'^2-F^2)=x^2(x^2f'^2-f^2)+(2x^2f+4f^3)(xf'-f).$$

Since $f(0)=0$ and $f'(x)$ is not decreasing and not constant,

$$xf'(x)\geqq f(x),$$

and $>$ is valid instead of $\geqq$ at some points of the interval $0\leqq x\leqq a$.

2. [G. Pólya: Tôhoku Math. J., **19**, pp. 1–3 (1921).]

3. [II **121**.]

4. [Cf. H. Dellac: Interméd. des math., **1**, pp. 69–70 (1894); cf. also H. Poincaré: Interméd. des math., **1**, pp. 141–144 (1894)]. Let A and B be constants and set $A\cos x+B\sin x=u$. We have

$$(f''+f)u=(f''+f)u-f(u''+u)=\frac{d}{dx}(f'u-fu').$$

It follows that

$$(1)\qquad \int_a^{a+\pi}[f''(x)+f(x)]\sin(x-a)\,dx=f(a)+f(a+\pi)>0,$$

$$(2)\qquad \int_a^{b}[f''(x)+f(x)]\sin(x-a)\,dx=f'(b)\sin(b-a).$$

We have $f'(b)\leqq 0$ [solution V **10**]. If now we had $b-a\leqq\pi$ then the left-hand side would be >0 and the right-hand side $\leqq 0$.

5. [W. Blaschke, problem: Arch. Math. Phys. Ser. 3, **26**, p. 65 (1917). Solution by G. Szegö: Arch. Math. Phys. Ser. 3, **28**, pp. 183–184 (1920). Cf. also W. Süss: Jber. deutsch. Math. Verein. **33**, Section 2, pp. *32–33* (1924).] Let

$$h(\varphi)\sim\frac{h_0}{2}+\sum_{n=1}^{\infty}(h_n\cos n\varphi+k_n\sin n\varphi)$$

be the Fourier series of $h(\varphi)$. Then the Fourier series of $r(\varphi)$ is

$$r(\varphi)\sim\frac{h_0}{2}+\sum_{n=1}^{\infty}(1-n^2)(h_n\cos\varphi+k_n\sin n\varphi),$$

since because of the periodicity of $h(\varphi)$ and $h'(\varphi)$ we have

$$\int_0^{2\pi}r(\varphi)\,e^{in\varphi}\,d\varphi=(1-n^2)\int_0^{2\pi}h(\varphi)e^{in\varphi}\,d\varphi$$

$n=0, 1, 2, \ldots$. Moreover we have that

$$r(\varphi)-r(\varphi+\pi)\sim\sum_{n=1}^{\infty}(1-n^2)[1-(-1)^n](h_n\cos n\varphi+k_n\sin n\varphi)$$

$$\sim 2\sum_{\nu=1}^{\infty}[1-(2\nu+1)^2][h_{2\nu+1}\cos(2\nu+1)\varphi+k_{2\nu+1}\sin(2\nu+1)\varphi]$$

[II **141**].

6. [G. Pólya, problem: Arch. Math. Phys. Ser. 3, **27**, p. 162 (1918).] Let $H(\Phi)$ and $h(\varphi)$ be the support functions, and $R(\Phi)$ and $r(\varphi)$ the radii of curvature belonging to the two convex curves respectively. We then have

$$\int_0^{2\pi}\int_0^{2\pi}[H(\Phi)-h(\varphi)][R(\Phi)-r(\varphi)]\,d\Phi\,d\varphi$$

$$=\int_0^{2\pi}\int_0^{2\pi}[H(\Phi)R(\Phi)+h(\varphi)r(\varphi)-h(\varphi)R(\Phi)-H(\Phi)r(\varphi)]\,d\Phi\,d\varphi$$

$$=2\pi\cdot 2A+2\pi\cdot 2a-lL-Ll.$$

By assumption we have $H(\Phi)\geqq h(\varphi)$, $R(\Phi)\geqq r(\varphi)$ for all pairs of values of Φ and φ. The last expression above is hence non-negative.

7. [G. Pólya, problem: Arch. Math. Phys. Series 3, **27**, p. 162 (1918).] Let $H(\Omega)$ and $h(\omega)$ be the support functions, and $R(\Omega)$, $R'(\Omega)$ and $r(\omega)$, $r'(\omega)$ the principal radii of curvature belonging to the two convex surfaces, respectively. We then have

$$\iint[H(\Omega)-h(\omega)][R(\Omega)R'(\Omega)-r(\omega)r'(\omega)]\,d\Omega\,d\omega$$

$$=\iint[H(\Omega)R(\Omega)R'(\Omega)+h(\omega)r(\omega)r'(\omega)-h(\omega)R(\Omega)R'(\Omega)$$

$$-H(\Omega)r(\omega)r'(\omega)]\,d\Omega\,d\omega=4\pi\cdot 3V+4\pi\cdot 3v-mS-Ms,$$

$$\iint[H(\Omega)-h(\omega)]\{[R(\Omega)-r(\omega)[R'(\Omega)-r'(\omega)]$$

$$+[R(\Omega)-r'(\omega)][R'(\Omega)-r(\omega)]\}\,d\Omega\,d\omega$$

$$=\iint\{2H(\Omega)R(\Omega)R'(\Omega)-2h(\omega)r(\omega)r'(\omega)+2H(\Omega)r(\omega)r'(\omega)$$

$$+[R(\Omega)+R'(\Omega)]h(\omega)[r(\omega)+r'(\omega)]-2h(\omega)R(\Omega)R'(\Omega)$$

$$-[r(\omega)+r'(\omega)]H(\Omega)[R(\Omega)+R'(\Omega)]\}\,d\Omega\,d\omega$$

$$=6\cdot 4\pi V-6\cdot 4\pi v+2Ms+2M\cdot 2s-2mS-2m\cdot 2S.$$

By assumption we have

$$H(\Omega)\geqq h(\omega),\qquad \min[R(\Omega),\,R'(\Omega)]\geqq\max[r(\omega),\,r'(\omega)]$$

for all pairs of values of Ω and ω. The two expressions above are thus non-negative.

8. In Gauss's formula

$$\iint (X\cos\alpha+Y\cos\beta+Z\cos\gamma)\,dS=\iiint\left(\frac{\partial X}{\partial x}+\frac{\partial Y}{\partial y}+\frac{\partial Z}{\partial z}\right)dx\,dy\,dz,$$

set $X=1,\ Y=0,\ Z=0$, then $X=0,\ Y=-z,\ Z=y$.

9. Cf. **10.**

10. [Cf. Nouv. Annls Math. Ser. 4, **16**, p. 140 (1916).] Consider a surface parallel to the given surface at a distance ρ from it, and associate its points with those of the given surface in such a manner that corresponding points have the same normal. The coordinates, the direction cosines of the normal, the principal radii of curvature and the elements of area at corresponding points of the given surface and the parallel surface are given by

$$\begin{array}{cc} x,\ y,\ z & x+\rho\cos\alpha,\quad y+\rho\cos\beta,\quad z+\rho\cos\gamma \\ \cos\alpha,\quad \cos\beta,\quad \cos\gamma & \cos\alpha,\quad \cos\beta,\quad \cos\gamma \\ R_1,\quad R_2 & R_1+\rho,\quad R_2+\rho \\ dS & \dfrac{(R_1+\rho)(R_2+\rho)}{R_1R_2}\,dS, \end{array}$$

respectively. If we apply **8** to the *parallel surface* we obtain

$$\iint \cos\alpha\left(1+\frac{\rho}{R_1}\right)\left(1+\frac{\rho}{R_2}\right)dS=0,\ldots,$$

$$\iint (y\cos\gamma-z\cos\beta)\left(1+\frac{\rho}{R_1}\right)\left(1+\frac{\rho}{R_2}\right)dS=0,\ldots.$$

Now regard ρ as a variable and set the coefficients of ρ^2, ρ and 1 equal to 0. We then again obtain **10**, **9** and **8**, respectively.

11. By a limiting process from **9**, or by applying **8** to the "parallel surface" of the polyhedron and considering the coefficient of ρ, or as follows: Resolve the force F into two components that lie in the two faces of the polyhedron intersecting in the edge e. The new system of forces thus obtained maintains every face, taken *individually*, in equilibrium, by the analogy of **8** in the plane.

12. [G. Pólya, problem: Arch. Math. Phys. Ser. 3, **25**, p. 337 (1917). Solution by K. Scholl: Arch. Math. Phys. Ser. 3, **28**, p. 180 (1920).]

First Solution: Let the coordinates and the direction cosines of the principal normal be

$$x,\ y,\ z;\qquad l=r\frac{d^2x}{ds^2},\qquad m=r\frac{d^2y}{ds^2},\qquad n=r\frac{d^2z}{ds^2},$$

respectively; let the curve be described as s varies from 0 to L, and let x, y, z and their derivatives assume the same values for $s=0$ and $s=L$. We have

$$\int_0^L l\frac{ds}{r}=\int_0^L\frac{d^2x}{ds^2}\,ds=\left[\frac{dx}{ds}\right]_0^L=0,$$

$$\int_0^L (ny-mz)\frac{ds}{r}=\int_0^L\left(y\frac{d^2z}{ds^2}-z\frac{d^2y}{ds^2}\right)ds=\left[y\frac{dz}{ds}-z\frac{dy}{ds}\right]_0^L=0.$$

Second Solution: The family of spheres, whose center is a variable point of the curve and whose radius is fixed $=\rho$, has a "canal surface" as its envelope. Apply **8** to this surface and replace the system of forces acting at the surface by a system of forces acting at points of the curve.

13. [K. Löwner.] With the notation of the first solution of **12**, the coordinates of a variable point of the spherical image are given by

$$\xi=\frac{dx}{ds}, \qquad \eta=\frac{dy}{ds}, \qquad \zeta=\frac{dz}{ds}.$$

We have

$$\int_0^L \xi\, ds=\int_0^L \eta\, ds=\int_0^L \zeta\, ds=0,$$

and hence also for any set of real numbers α, β, γ

$$\int_0^L (\alpha\xi+\beta\eta+\gamma\zeta)\, ds=0.$$

There are always at least two values of s such that $\alpha\xi+\beta\eta+\gamma\zeta=0$.

14. [K. Löwner.] By (3), $F(x)$ assumes negative values both in the neighbourhood of a and in the neighbourhood of b. If then $F(x)$ were monotonic, we would have that $F(x)<0$, $f''(x)<0$ in the whole interval $a<x<b$. Let ξ, $a<\xi<b$, be the only zero of $f'(x)$, so that $f'(x)$ is positive from a to ξ and negative from ξ to b. The integral

$$(*) \qquad \int_a^x F(x)f(x)f'(x)\, dx=-\frac{1}{2}\frac{1}{1+[f'(x)]^2}$$

is convergent, and in view of (3) we have

$$\int_a^b F(x)f(x)f'(x)\, dx=\int_a^\xi F(x)f(x)f'(x)\, dx+\int_\xi^b F(x)f(x)f'(x)\, dx=0.$$

Since all the factors are of constant sign between a and ξ and also between ξ and b, it thus follows that

$$F(x_1)\int_a^\xi f(x)f'(x)\, dx+F(x_2)\int_\xi^b f(x)f'(x)\, dx=(F(x_1)-F(x_2))\frac{[f(\xi)]^2}{2}=0,$$

$$a<x_1<\xi<x_2<b,$$

i.e. $F(x_1)=F(x_2)=-c$, $c>0$. Thus $F(x)=-c$ for $x_1\leqq x\leqq x_2$. From the last two equations it follows that

$$\int_a^{x_1} (F(x)-F(x_1))f(x)f'(x)\, dx+\int_{x_2}^b (F(x)-F(x_2))f(x)f'(x)\, dx=0.$$

The two integrands are of the same constant sign, and so $F(x)=-c$ in the whole interval $a<x<b$.

From (*) it follows that

$$-c\frac{[f(x)]^2}{2}=-\frac{1}{2}\frac{1}{1+[f'(x)]^2}, \qquad f'(x)=\pm\frac{\sqrt{c^{-1}-[f(x)]^2}}{f(x)},$$

where the + sign applies for $a<x<\xi$ and the − sign for $\xi<x<b$. In the whole of the interval of integration we must have $[f(x)]^2 \leqq c^{-1}$. By integration we obtain

$$x-a=\sqrt{c^{-1}}-\sqrt{c^{-1}-[f(x)]^2}, \qquad a<x<\xi,$$
$$b-x=\sqrt{c^{-1}}-\sqrt{c^{-1}-[f(x)]^2}, \qquad \xi<x<b.$$

From the continuity of $f(x)$ it follows that $\xi-a=b-\xi$, $\xi=(a+b)/2$. Moreover we must have $[f(\xi)]^2=c^{-1}$, for otherwise differentiation with respect to x would give $1=-1$. Hence $\xi-a=\sqrt{c^{-1}}$ and thus $c^{-1}=(\xi-a)^2=\frac{1}{4}(b-a)^2$, $f(x)=\sqrt{(x-a)(b-x)}$.

15. [K. Löwner. Cf. also formulae (89) in H. Minkowski: Ges. Abhandlungen, **2**, p. 263. Leipzig and Berlin: B. G. Teubner, 1911.] Let the meridian curve be given by $y=f(x)$, $a \leqq x \leqq b$. To calculate the Gaussian curvature $K(x)=K_1(x)K_2(x)$ at points of the circle of latitude belonging to the value x, we note the following: The meridian section yields one of the principal curvatures

$$K_1(x)=\frac{f''(x)}{(1+[f'(x)]^2)^{3/2}}$$

The second principal curvature (by a theorem of Meusnier; cf. W. Blaschke: Vorlesungen über Differentialgeometrie, 2nd edition, **1**, pp. 57–58. Berlin: J. Springer, 1924) is obtained from the curvature of the circle of latitude by multiplication by the cosine of the angle between the plane of the circle of latitude and the normal to the surface:

$$K_2(x)=\frac{1}{f(x)}\cdot\frac{1}{\sqrt{1+[f'(x)]^2}}.$$

$f(x)$ satisfies the assumptions of **14**.

16. [Following J. Kürschák.]

$$(\beta-\gamma)(b-c)+(\gamma-\alpha)(c-a)+(\alpha-\beta)(a-b) \geqq 0,$$
$$\alpha(b+c-a)+\beta(c+a-b)+\gamma(a+b-c) \geqq 0.$$

17. [G. Pólya, problem: Arch. Math. Phys. Ser. 3, **27**, p. 162 (1918).] Let the edges of a closed convex polyhedron be denoted by $k_1, k_2, \ldots, k_m$, and the (interior) surface angles, formed by the faces of the polyhedron intersecting at the corresponding edge, by $\alpha_1, \alpha_2, \ldots, \alpha_m$. The *measure* of all the planes that intersect the polyhedron is given by

$$M=\tfrac{1}{2}[(\pi-\alpha_1)k_1+(\pi-\alpha_2)k_2+\cdots+(\pi-\alpha_m)k_m].$$

[G. Pólya: Wien. Ber. **126**, p. 319 (1917).] As a limiting case we have that the measure of all the planes in space that intersect a convex plane polygon is $=(\pi/2)\times$ perimeter of the polygon. We now consider a tetrahedron, $m=6$. Let the perimeters of the four faces be L_1, L_2, L_3, L_4. The measure of all the planes that intersect the tetrahedron without meeting its first face is $M-(\pi/2)L_1$, etc. Let the measure of all the planes that intersect the tetrahedron and meet all its four faces be T. We then have

$$M=\left(M-\frac{\pi}{2}L_1\right)+\left(M-\frac{\pi}{2}L_2\right)+\left(M-\frac{\pi}{2}L_3\right)+\left(M-\frac{\pi}{2}L_4\right)+T.$$

The desired inequality follows by rewriting the inequality

$$0<T<M.$$

The bounds are approached as the tetrahedron converges to a straight line segment, the upper bound as two corners of the tetrahedron tend to each end point of the segment, and the lower bound as three corners of the tetrahedron tend to one end of the segment.

17.1. The search for convenient parameters may lead (perhaps after various trials) to

$$x=\frac{b}{a}, \qquad y=\cos\gamma.$$

We are required to maximize the function of x and y

$$\frac{64A^6}{(abc)^4}=\frac{x^2(1-y^2)^3}{(1+x^2-2xy)^2}$$

in the domain D

$$x\geqq 0, \qquad -1\leqq y\leqq 1$$

on the boundary of which the function vanishes. (Define it by continuity for $x\to\infty$ and at the point $x=1$, $y=1$; observe that

$$1+x^2-2xy\geqq 1-y^2.)$$

Therefore the maximum is attained at an interior point of the domain D, where the first partial derivatives, with respect to x and to y, must vanish. The derivative with respect to x leads to $x=1$; the use of this value and the derivative with respect to y yield, appropriately combined, $y=\frac{1}{2}$. Thus

$$x=\frac{b}{a}=1, \qquad y=\cos\gamma=\tfrac{1}{2},$$

the triangle is equilateral.

17.2. Combine two facts: **17.1** and the inequality between the arithmetic and geometric means (Part II, chap. 2, § 1 p. 63). As p increases the inequality becomes steadily "weaker"; see II **82**. For the cases $p=1$ and $p=2$ see MPR, **1**, p. 133, ex. 16 and **2**, p. 212, ex. 8.63.4, respectively; for $p=3$ C. N. Mills, problem: Amer. Math. Monthly, **34**, pp. 382–4, 1927.

17.3. From

$$r=\frac{2A}{a+b+c}, \qquad R=\frac{abc}{4A}$$

$$\frac{2r}{R}=\frac{16A^2}{(a+b+c)abc}\leqq\frac{16A^2}{3(abc)^{4/3}}\leqq 1;$$

we have combined the same two facts as in solution **17.2**.

The inequality between r and R is well known; the novelty consists in using **17.1** to prove it.

17.4. Let V denote the volume of a tetrahedron and $k_1, k_2, \ldots, k_6$ the lengths of its six edges (as in **17**). Then

$$V\leqq\frac{\sqrt{2}}{12}(k_1k_2\ldots k_6)^{1/2},$$

with equality only for regular tetrahedra.

17.5. The proposition is conveniently verified in the following three cases:

(1) The base of the tetrahedron is an equilateral triangle whose center is the foot of the perpendicular drawn from the opposite vertex,

(2) MD, **1**, p. 45, ex. 2.10, and

(3) MD, **1**, p. 45, ex. 2.13.

18. [E. Steinitz, problem: Arch. Math. Phys. Ser. 3, **19**, p. 361 (1912). Solution by W. Gaedecke: Arch. Math. Phys. Ser. 3, **21**, p. 290 (1913).] Let P_2 lie at the origin, P_0 on the positive real axis, and P_1 on the positive imaginary axis. Let the length of the hypotenuse P_0P_1 be d, $d>0$, let the angle $P_2P_0P_1$ be α, $0<\alpha<\pi/2$, and let the vector $P_{n-1}P_n$ be represented by the complex number z_n, $n=0, 1, 2\ldots$; $P_{-1}=P_2$. Let $z_n=r_ne^{i\vartheta_n}$, $r_n>0$, $0\leqq\vartheta_n<2\pi$. The values of $r_0, r_1, r_2, \ldots$ are given by the sequence

$$d\cos\alpha,\quad \underbrace{d,\quad d\sin\alpha,\quad d\sin\alpha\cos\alpha,\quad d\sin^2\alpha\cos\alpha,}\quad$$
$$\underbrace{d\sin^2\alpha\cos^2\alpha,\quad d\sin^3\alpha\cos^2\alpha,}\quad \ldots;$$

the values of $\vartheta_0, \vartheta_1, \vartheta_2, \ldots$ are given by the sequence

$$0,\ \underbrace{\pi-\alpha,\ \frac{3\pi}{2},\ \frac{\pi}{2}-\alpha,}\ \underbrace{\pi,\ 2\pi-\alpha,\ \frac{\pi}{2},\ \frac{3\pi}{2}-\alpha,}\ 0,\ \pi-\alpha,\ \ldots.$$

We are required to find the sum of the infinite series

$$z_0+z_1+z_2+\cdots+z_n+\cdots=z_0+\frac{z_1+z_2+z_3+z_4}{1+\cos^2\alpha\sin^2\alpha}$$
$$=\frac{d\cos^3\alpha\sin^2\alpha}{1+\cos^2\alpha\sin^2\alpha}+i\frac{d\cos^2\alpha\sin\alpha}{1+\cos^2\alpha\sin^2\alpha}.$$

19. [Following A. Hirsch.] Let P_1, P_2, P_1', P_2' be the points of intersection of the line of intersection of the two given planes with the two given spheres. The equations of the line of intersection are

$$y=\frac{\begin{vmatrix}-D-Ax & C\\ -D'-A'x & C'\end{vmatrix}}{\begin{vmatrix}B & C\\ B' & C'\end{vmatrix}},\qquad z=\frac{\begin{vmatrix}B & -D-Ax\\ B' & -D'-A'x\end{vmatrix}}{\begin{vmatrix}B & C\\ B' & C'\end{vmatrix}}.$$

Substituting these expressions into the equations of the spheres, we obtain the quadratic equations

$$(1)\qquad A_0x^2+A_1x+A_2=0,\qquad A_0'x^2+A_1'x+A_2'=0,$$

which yield in turn the x-coordinates of P_1, P_2, P_1', P_2'. In a similar manner we obtain further equations that yield the y- and z-coordinates of these points:

$$(2)\qquad B_0y^2+B_1y+B_2=0,\qquad B_0'y^2+B_1'y+B_2'=0,$$
$$(3)\qquad C_0z^2+C_1z+C_2=0,\qquad C_0'z^2+C_1'z+C_2'=0.$$

Here we have

$$A_0=A_0'=B_0=B_0'=C_0=C_0'=\begin{vmatrix} B & C \\ B' & C' \end{vmatrix}^2+\begin{vmatrix} C & A \\ C' & A' \end{vmatrix}^2+\begin{vmatrix} A & B \\ A' & B' \end{vmatrix}^2.$$

This expression is $=0$ if and only if the two planes are parallel. If we now denote by $\mathfrak{A}$, $\mathfrak{B}$ and $\mathfrak{C}$ the resultants of the equations (1), (2) and (3) respectively (cf. solution V **194**), then $\mathfrak{A}+\mathfrak{B}+\mathfrak{C}$ satisfies the conditions for the function required by the problem.

(a) Assume that the two circles are interlocked. Then the pair of points P_1P_2 separates the pair of points $P_1'P_2'$. The same is therefore true for the projections onto each of the coordinate axes, with the possible exception of one or two axes, in case the straight line P_1P_2 is perpendicular to that axis; then the projections coincide into a single point. In any event, as can be seen from the relative positions of the roots of the quadratic equations (1), (2), (3), all the resultants $\mathfrak{A}$, $\mathfrak{B}$ and $\mathfrak{C}$ are negative (possibly one or two are equal to zero) [V **194**], and hence their sum is also negative.

(b) Conversely assume $\mathfrak{A}+\mathfrak{B}+\mathfrak{C}<0$. Then at least one of the three resultants is negative, $\mathfrak{A}<0$ say. Then $A_0>0$, $A_0'>0$, and hence the two planes are not parallel and the points of intersection P_1, P_2, P_1', P_2' are determined uniquely. The projection of the pair of points P_1P_2 on the x-axis separates that of $P_1'P_2'$ on the same axis [V **194**] and hence the same is true of the points themselves.

20. [A. Hirsch.] If (x_1, x_2, x_3) is a singular point of the conic section $\sum_{r=1}^3 \sum_{s=1}^3 a_{rs}X_rX_s=0$, then we have the equations

$$a_{\nu 1}x_1+a_{\nu 2}x_2+a_{\nu 3}x_3=0, \qquad \nu=1, 2, 3$$

[G. Kowalewski: Einführung in die analytische Geometrie, p. 202; Leipzig, Veit, 1910]. In our case, setting $x_1+x_2+x_3=-x_0$, we obtain $\lambda_0x_0=\lambda_1x_1=\lambda_2x_2=\lambda_3x_3$, and the equation of the conic section (pair of straight lines) in question has the form

$$x_1x_2x_3X_0^2+x_0x_2x_3X_1^2+x_0x_1x_3X_2^2+x_0x_1x_2X_3^2=0,$$

where $X_0+X_1+X_2+X_3=0$. Let dx_ν, $\nu=0, 1, 2, 3$, $dx_0+dx_1+dx_2+dx_3=0$, be a displacement along the integral curve. Then, setting $X_\nu=x_\nu+dx_\nu$, we have

$$\text{(1)} \qquad x_1x_2x_3(dx_0)^2+x_0x_2x_3(dx_1)^2+x_0x_1x_3(dx_2)^2+x_0x_1x_2(dx_3)^2=0.$$

This is the differential equation of the required family of curves. The change of variables $x_\nu=y_\nu^2$ yields the equation

$$\text{(1')} \qquad (y_0y_1y_2y_3)^2((dy_0)^2+(dy_1)^2+(dy_2)^2+(dy_3)^2)=0,$$

from which we obtain immediately the *singular solution* $x_\nu=0$ ($\nu=0, 1, 2, 3$). The *general solution* follows from the equations

$$(dy_0)^2+(dy_1)^2+(dy_2)^2+(dy_3)^2=0, \qquad y_0^2+y_1^2+y_2^2+y_3^2=0,$$
$$y_0\,dy_0+y_1\,dy_1+y_2\,dy_2+y_3\,dy_3=0.$$

Eliminating y_0 we obtain

$$(y_0\,dy_0)^2=(y_1^2+y_2^2+y_3^2)((dy_1)^2+(dy_2)^2+(dy_3)^2)$$
$$=(y_1\,dy_1+y_2\,dy_2+y_3\,dy_3)^2,$$

and thus

$$(y_2\,dy_3 - y_3\,dy_2)^2 + (y_3\,dy_1 - y_1\,dy_3)^2 + (y_1\,dy_2 - y_2\,dy_1)^2 = 0.$$

Setting

$$z_1 = y_2y_3' - y_3y_2', \qquad z_2 = y_3y_1' - y_1y_3', \qquad z_3 = y_1y_2' - y_2y_1',$$

where the differentiation is with respect to a parameter, it follows that

$$\begin{aligned} \sum y_r z_r &= 0, \qquad \sum y_r z_r' = 0, \\ \sum z_r z_r &= 0, \qquad \sum z_r z_r' = 0 \end{aligned} \tag{2}$$

where the summation is extended over $r = 1, 2, 3$. We deduce that $z_1' : z_2' : z_3' = z_1 : z_2 : z_3$, except in the case that $z_1 : z_2 : z_3 = y_1 : y_2 : y_3$. The latter assumption leads to $y_0 = 0$ and thus again to $x_0 = 0$. In the former case we obtain by integration $z_1 : z_2 : z_3 = \mu_1 : \mu_2 : \mu_3$, where $\mu_r = \text{constant}$, $\mu_1^2 + \mu_2^2 + \mu_3^2 = 0$, and hence by (2): $\mu_1 y_1 + \mu_2 y_2 + \mu_3 y_3 = 0$. Setting $\mu_r^2 = \chi_r$, $r = 1, 2, 3$, we obtain after rationalization:

$$(\chi_1 x_1 + \chi_2 x_2 + \chi_3 x_3)^2 - 2(\chi_1^2 x_1^2 + \chi_2^2 x_2^2 + \chi_3^2 x_3^2) = 0,$$
$$\chi_1 + \chi_2 + \chi_3 = 0.$$

This equation represents the family of those conic sections that are inscribed in the quadrilateral $X_0 = 0$, $X_1 = 0$, $X_2 = 0$, $X_3 = 0$. In fact the equation of the tangent at (x_1, x_2, x_3) is

$$(\chi_1 x_1 + \chi_2 x_2 + \chi_3 x_3)(\chi_1 X_1 + \chi_2 X_2 + \chi_3 X_3) - 2(\chi_1^2 x_1 X_1 + \chi_2^2 x_2 X_2 + \chi_3^2 x_3 X_3) = 0.$$

Let χ_1, χ_2, χ_3 be different from 0 and set (x_1, x_2, x_3) in succession equal to (χ_1, χ_2, χ_3), $(0, \chi_2^{-1}, \chi_3^{-1})$, $(\chi_1^{-1}, 0, \chi_3^{-1})$ and $(\chi_1^{-1}, \chi_2^{-1}, 0)$.

21. [A. Hirsch.]

$$\nu = \frac{x}{\sin\tau} + \frac{y}{\cos\tau},$$

$$\frac{dx}{d\tau} = \rho\cos\tau = 2n(x\cot\tau + y),$$

$$\frac{dy}{d\tau} = \rho\sin\tau = 2n(x + y\tan\tau).$$

Eliminating y we obtain

$$\frac{d^2x}{d\tau^2} - \frac{2n}{\sin\tau\cos\tau}\frac{dx}{d\tau} + \frac{2n}{\sin^2\tau}x = 0.$$

Now first introduce $u = -\tan^2\tau$ as the independent variable and then $\xi = xu^{-1/2}$ as the dependent variable. We obtain in this way the differential equation

$$u(1-u)\frac{d^2\xi}{du^2} + \left[\frac{3}{2} - n - \left(\frac{1}{2} + 1 - n + 1\right)u\right]\frac{d\xi}{du} - \frac{1}{2}(1-n)\xi = 0,$$

which is satisfied by the hypergeometric series [VIII **146**]

$$\xi = F(\tfrac{1}{2}, 1-n, \tfrac{3}{2}-n, u)$$

and thus the problem is solved in principle. On the basis of known results we can now discuss in particular the rational solutions. In order to obtain a simple expression for these, introduce the dependent variable $z=x(\tan\tau)^{-n}$ and then the independent variable $v=i\cot 2\tau$. We obtain the differential equation

$$(1-v^2)\frac{d^2z}{dv^2}-2v\frac{dz}{dv}+n(n-1)z=0,$$

one of whose particular solutions is $z=P_{n-1}(v)$ [VI **90**], if $n=1, 2, 3, \ldots$. We obtain in this way the rational expressions in $\tan\tau$

$$x=\frac{1}{i}(i\tan\tau)^nP_{n-1}(i\cot 2\tau),$$

$$y=(i\tan\tau)^nP_n(i\cot 2\tau).$$

For $n=1$ we obtain as a particular solution the parabola

$$x=\tan\tau,\qquad y=\frac{\tan^2\tau-1}{2},\qquad 2y=x^2-1.$$

22. [A. Hirsch.] (i) The equation for H follows by elimination of t from

$$F=0, \tag{1}$$

$$\frac{\partial F}{\partial t}\equiv\sum_{\nu=1}^{3}\frac{x_\nu^2}{(a_\nu-t)^2}-1=0. \tag{2}$$

If we write (1) in the form

$$\begin{aligned}&t^4-(\sum a_1)t^3+[\sum(a_2a_3+x_1^2)]t^2-[a_1a_2a_3+\sum(a_2+a_3)x_1^2]t+\sum a_2a_3x_1^2\\ &\equiv A_0t^4+4A_1t^3+6A_2t^2+4A_3t+A_4=0,\end{aligned}$$

where in the summations the indices are to be cyclically interchanged, and denote by Δ the discriminant of this polynomial in t, then $\Delta=0$ is the equation of H. Setting

$$I_2=A_0A_4-4A_1A_3+3A_2^2,$$

$$I_3=\begin{vmatrix}A_0 & A_1 & A_2\\ A_1 & A_2 & A_3\\ A_2 & A_3 & A_4\end{vmatrix}=A_0A_2A_4+2A_1A_2A_3-A_2^3-A_0A_3^2-A_1^2A_4$$

we now have [Cesàro, p. 387],

$$\begin{aligned}\Delta&=I_2^3-27I_3^2\\ &=27A_2^4(A_0A_4-4A_1A_3)+54A_2^3(A_0A_2A_4+2A_1A_2A_3-A_0A_3^2)+Q,\end{aligned}$$

where Q is of degree 8 at most. The leading term of Δ is thus equal to that of $27A_2^3(3A_2A_4-2A_3^2)$, i.e.

$$=27\left(\frac{\sum x_1^2}{6}\right)^3\left[\frac{\sum x_1^2}{2}\sum a_2a_3x_1^2-\frac{1}{8}\left(\sum(a_2+a_3)x_1^2\right)^2\right]$$

$$=-\frac{1}{64}\left(\sum x_1^2\right)^3\left[\sum(a_2-a_3)^2x_1^4+2\sum(a_1-a_2)(a_1-a_3)x_2^2x_3^2\right].$$

(ii) If dx_1, dx_2, dx_3 is a displacement on H, then we have [(2)]

$$\sum_{\nu=1}^{3} \frac{x_\nu \, dx_\nu}{a_\nu - t} = 0. \tag{3}$$

We deduce that the direction cosines X_1, X_2, X_3 of the appropriately oriented normal of H are given by

$$X_\nu = \frac{x_\nu}{a_\nu - t}, \qquad \nu = 1, 2, 3. \tag{4}$$

The coordinates of the tangent plane at the point x_1, x_2, x_3 are

$$u_\nu = \frac{X_\nu}{x_1 X_1 + x_2 X_2 + x_3 X_3}, \qquad \nu = 1, 2, 3.$$

By (1) we have $x_\nu = t(a_\nu - t)u_\nu$, $\nu = 1, 2, 3$. Substituting this in (1), (2) and eliminating t we obtain

$$\left(\sum_{\nu=1}^{3} a_\nu u_\nu^2\right)^2 - 4 \sum_{\nu=1}^{3} u_\nu^2 = 0.$$

(iii) If ρ is one principal radius of curvature of H at (x_1, x_2, x_3), $\rho \neq 0$, then the displacements on H along the corresponding line of curvature satisfy the following equations [W. Blaschke, op. cit. **15**, p. 63]:

$$dx_\nu + \rho \, dX_\nu = 0, \qquad \nu = 1, 2, 3$$

or by (ii)

$$dx_\nu + \rho \, dt \frac{x_\nu}{(a_\nu - t)(a_\nu - t + \rho)} = 0, \qquad \nu = 1, 2, 3. \tag{5}$$

In view of (3) it follows that

$$\sum_{\nu=1}^{3} \frac{x_\nu^2}{(a_\nu - t)^2 (a_\nu - t + \rho)} = 0. \tag{6}$$

This is the quadratic equation for the two principal radii of curvature. In view of the identity

$$\frac{1}{a^2(a+\rho)} = \frac{1}{\rho} \frac{1}{a^2} - \frac{1}{\rho^2} \frac{1}{a} + \frac{1}{\rho^2} \frac{1}{a+\rho},$$

and noting (1) and (2), it may be written as follows:

$$\sum_{\nu=1}^{3} \frac{x_\nu^2}{a_\nu - t + \rho} - (t - \rho) = 0. \tag{6'}$$

Combining the total differential of (6′) with (5) we now deduce

$$\left(\sum_{\nu=1}^{3} \frac{x_\nu^2}{(a_\nu - t + \rho)^2} - 1\right)(3 \, dt - d\rho) = 0. \tag{7}$$

The solution of (7) obtained by setting the first factor equal to zero is to be discarded, for it would mean that the equation $F(x, y, z, \lambda) = 0$ has two distinct double roots t and $t - \rho$, $(\rho \neq 0)$, which is impossible. For $F(\lambda)$ has an *odd* number of zeros in the interval a_1, a_2 as well as in the interval a_2, a_3. Thus $\rho = 3t + \text{const.}$ along the line of curvature corresponding to ρ. The other radius of curvature is equal to $t+\text{const.}$ [(6)].

The result can thus be interpreted as follows: If $w=u$, $w=v$ denote the two roots of

$$(8)\qquad \sum_{\nu=1}^{3}\frac{x_\nu^2}{(a_\nu-t)^2(a_\nu-w)}=0, \quad \text{or of} \quad \sum_{\nu=1}^{3}\frac{x_\nu^2}{a_\nu-w}=0,$$

then the equations of the lines of curvature are $u=\text{const.}$, $v=\text{const.}$ The parameters u, v are the two roots of the equation $F(x, y, z, w)=0$, which it has apart from the double root $w=t$.

The line of curvature $u=u_0$ lies on the surface $t=u_0$. The surfaces of the given family penetrate the enveloping surface H in its lines of curvature.

23. [A. Hirsch.] By **22**(8), (2) we have

$$\frac{x_1^2}{(a_1-t)^2}=\frac{(a_1-u)(a_1-v)}{(a_1-a_2)(a_1-a_3)}, \quad \ldots.$$

Combining this with **22**(1) we deduce that $2t+u+v=a_1+a_2+a_3$ and

$$x_1=\sqrt{\frac{(a_1-u)(a_1-v)}{(a_1-a_2)(a_1-a_3)}}\left(a_1+\frac{u+v-(a_1+a_2+a_3)}{2}\right), \ldots.$$

The determination of the points of intersection of the line of curvature $v=\text{const.}$ with a plane leads to an equation of the form:

$$c_0+\sum_{\nu=1}^{3}c_\nu\sqrt{a_\nu-u}(u+b_\nu)=0,$$

where c_0, a_ν, b_ν, c_ν are constants. Rationalization leads to an equation of 12th degree in u.

24. [A. Hirsch.] (1) If ξ_1, ξ_2, ξ_3 are the coordinates of the center of curvature corresponding to ρ, then we have [W. Blaschke, op. cit. **15**, p. 63 (47)]

$$\xi_\nu=x_\nu+\rho X_\nu=x_\nu\frac{a_\nu-t+\rho}{a_\nu-t}, \qquad \nu=1, 2, 3,$$

[**22** (4)], and hence [**22** (2), (6)]

$$\sum_{\nu=1}^{3}\frac{\xi_\nu^2}{(a_\nu-t+\rho)^2}=1, \qquad \sum_{\nu=1}^{3}\frac{\xi_\nu^2}{(a_\nu-t+\rho)^3}=0.$$

The central surface of H is thus the envelope of the family of ellipses

$$\sum_{\nu=1}^{3}\frac{\xi_\nu^2}{(a_\nu-s)^2}-1=0.$$

(2) Let h be a constant and set

$$\bar{x}_\nu=x_\nu+hX_\nu=x_\nu\frac{a_\nu-t+h}{a_\nu-t}, \qquad \nu=1, 2, 3.$$

We then have [**22** (1), (2)]

$$\sum_{\nu=1}^{3} \frac{\bar{x}_\nu^2}{(a_\nu - t + h)^2} = 1, \qquad \sum_{\nu=1}^{3} \frac{\bar{x}_\nu^2}{a_\nu - t + h} = t + h.$$

The parallel surface to H at a distance h is thus the envelope of the family of surfaces of second order

$$\sum_{\nu=1}^{3} \frac{\bar{x}_\nu^2}{a_\nu - s} = s + 2h.$$

25. [E. E. Levi: C.R. Acad. Sci. (Paris) **153**, p. 799 (1911).] If $x'' - x'$ and $y'' - y'$ are integers, then from the two points x', y' and x'', y'' there issue congruent and similarly situated integral curves, i.e. integral curves that differ only by a translation. Consider the integral curve $y = f(x)$ passing through the origin [i.e. for which $f(0) = 0$] and the two integral curves congruent to it passing through the lattice points $x = m$, $y = [f(m)]$ and $x = m$, $y = [f(m)] + 1$. Since two distinct integral curves do not intersect, we have for $m, n = 1, 2, 3, \ldots$

$$[f(m)] + f(n) \leqq f(m+n) < [f(m)] + 1 + f(n).$$

The sequence $f(1), f(2), \ldots, f(n), \ldots$ satisfies the conditions of I **99**. In addition we note that similar reasoning yields that

$$[f(-n)] + f(2n) \leqq f(n) < [f(-n)] + 1 + f(2n),$$

$$-\frac{1}{n} < \frac{f(n)}{n} + \frac{f(-n)}{-n} - 2\frac{f(2n)}{2n} < \frac{1}{n};$$

and moreover, denoting by M the maximum of $|F(x, y)|$ in the square $0 \leqq x \leqq 1$, $0 \leqq y \leqq 1$, we have $|f(x) - f([x])| < M$.

Appendix

Additional Problems to Part One

A I **9.1.** [Cf. Euler, loc. cit. I **9**, also Opera Omnia, Series 1, **2**, p. 267.] It will be sufficient (and it may even be more suggestive) to consider the concrete particular case of I **2**, where A_{nk} is the number of those non-negative integral solutions of

$$x+5y+10u+25v+50w=n$$

for which

$$x+y+u+v+w=k$$

(paying the sum of n cents with k coins). In this case

$$\begin{aligned}
&\frac{1}{(1-qz)(1-q^5z)(1-q^{10}z)(1-q^{25}z)(1-q^{50}z)} \\
&=(1+qz+q^2z^2+q^3z^3+\cdots+q^xz^x+\cdots) \\
&\quad(1+q^5z+q^{10}z^2+q^{15}z^3+\cdots+q^{5y}z^y+\cdots) \\
&\quad(1+q^{10}z+q^{20}z^2+q^{30}z^3+\cdots+q^{10u}z^u+\cdots) \\
&\quad(1+q^{25}z+q^{50}z^2+q^{75}z^3+\cdots+q^{25v}z^v+\cdots) \\
&\quad(1+q^{50}z+q^{100}z^2+q^{150}z^3+\cdots+q^{50w}z^w+\cdots) \\
&=\sum_{x=0}^{\infty}\sum_{y=0}^{\infty}\sum_{u=0}^{\infty}\sum_{v=0}^{\infty}\sum_{w=0}^{\infty} q^{x+5y+10u+25v+50w}z^{x+y+u+v+w} \\
&=\sum_{n=0}^{\infty}\sum_{k} A_{nk}q^nz^k,
\end{aligned}$$

which, in the general case,

$$=\frac{1}{(1-q^{\alpha_1}z)(1-q^{\alpha_2}z)\ldots(1-q^{\alpha_l}z)}.$$

A I **9.2.** We introduce the variables $z_1, z_2, \ldots, z_k$ (one variable for each equation of the system):

$$\begin{aligned}
&\sum_{n_1=0}^{\infty}\sum_{n_2=0}^{\infty}\cdots\sum_{n_k=0}^{\infty} A_{n_1n_2\ldots n_k}z_1^{n_1}\ldots z_k^{n_k} \\
&=\sum_{x_1=0}^{\infty}\sum_{x_2=0}^{\infty}\cdots\sum_{x_l=0}^{\infty} z_1^{a_{11}x_1+a_{12}x_2+\ldots+a_{1l}x_l}z_2^{a_{21}x_1+a_{22}x_2+\ldots+a_{2l}x_l} \\
&\quad\cdots\times z_k^{a_{k1}x_1+a_{k2}x_2+\ldots+a_{kl}x_l}
\end{aligned}$$

$$= \sum_{x_1=0}^{\infty} \sum_{x_2=0}^{\infty} \cdots \sum_{x_l=0}^{\infty} (z_1^{a_{11}} z_2^{a_{21}} \ldots z_k^{a_{k1}})^{x_1} (z_1^{a_{12}} z_2^{a_{22}} \ldots z_k^{a_{k2}})^{x_2}$$

$$\cdots \times (z_1^{a_{1l}} z_2^{a_{2l}} \ldots z_l^{a_{kl}})^{x_l}$$

$$= \frac{1}{(1 - z_1^{a_{11}} z_2^{a_{21}} \ldots z_k^{a_{k1}})(1 - z_1^{a_{12}} z_2^{a_{22}} \ldots z_k^{a_{k2}}) \ldots (1 - z_1^{a_{1l}} z_2^{a_{2l}} \ldots z_k^{a_{kl}})}$$

A I **20.1.** 1, 1, 2, 3, 5, 7, 11, 15, 22, 30, 42.

A I **20.2.** Limiting case ($l = \infty$) of the "change problem" I **9**:

$$a_1 = 1, \qquad a_2 = 2, \qquad a_3 = 3, \ldots .$$

Cf. I **54**.

A I **54.1.** In I **53**, substitute

$$q^{1/2} \text{ for } q, \qquad -q^{1/2}\zeta \text{ for } z.$$

$$\prod_{n=1}^{\infty} (1 - q^n \zeta)(1 - q^{n-1} \zeta^{-1})(1 - q^n)$$

$$= \frac{\zeta - 1}{\zeta} \prod_{n=1}^{\infty} (1 - q^n \zeta)(1 - q^n \zeta^{-1})(1 - q^n)$$

$$= \left(\sum_{n=0}^{\infty} + \sum_{n=-\infty}^{-1} \right) (-1)^n q^{n(n+1)/2} \zeta^n$$

$$= \sum_{n=0}^{\infty} (-1)^n q^{n(n+1)/2} \zeta^n + \sum_{k=0}^{\infty} (-1)^{k+1} q^{k(k+1)/2} \zeta^{-k-1}$$

$$= \sum_{n=0}^{\infty} (-1)^n q^{n(n+1)/2} (\zeta^{2n+1} - 1) \zeta^{-n-1}.$$

To obtain the penultimate line set $-n = k + 1$ in the preceding line. Equate the second line to the last line, divide both sides (i.e. each term on the right-hand side) by $\zeta - 1$; then set $\zeta = 1$.

A I **60.12.** By A I **9.1**

$$\sum_{\alpha=0}^{\infty} \sum_{r} c_{r+s,r,\alpha} \quad q^{\alpha} z^r = \frac{1}{(1-z)(1-qz)(1-q^2 z) \ldots (1-q^s z)},$$

which we set (cf. I **51**)

$$= \frac{1}{F(z)}$$

$$= Q_0 + Q_1 z + Q_2 z^2 + \cdots + Q_r z^r + \cdots,$$

and thus we require to find

$$Q_r = \sum_{\alpha} c_{r+s,\, r,\, \alpha} q^{\alpha}.$$

Yet

$$\frac{1}{(1-z)F(qz)} = \frac{1}{F(z)(1-q^{s+1}z)},$$

and thus

$$(1-q^{s+1}z)\sum_{r=0}^{\infty} Q_r q^r z^r=(1-z)\sum_{r=0}^{\infty} Q_r z^r,$$
$$Q_r q^r - Q_{r-1} q^{r+s} = Q_r - Q_{r-1},$$
$$Q_r = Q_{r-1}\frac{1-q^{r+s}}{1-q^r}.$$

Hence, since $Q_0=1$,

$$Q_r=\frac{1-q^{r+s}}{1-q^r}\frac{1-q^{r+s-1}}{1-q^{r-1}}\cdots\frac{1-q^{s+1}}{1-q}$$
$$=\begin{bmatrix} r+s \\ r \end{bmatrix}=\begin{bmatrix} n \\ r \end{bmatrix},$$

where the Gaussian binomial coefficient $\begin{bmatrix} n \\ r \end{bmatrix}$ is defined in Vol. 1, Pt. I, Chap. 1, § 5 p. 11. Compare this with the above expression for Q_r.

A I **191.1.** *First solution.* Consider the coefficient of x in the identity I **191**.

Second solution. From VIII **58.3** and VIII **58.2**. This argument presupposes knowledge of the fact that there are an infinity of primes (n is taken to be *any* positive integer greater than 1).

A I **191.2.** From VIII **58.3** and VIII **58.1**, by the second solution of A I **191.1**.

Is there a less roundabout solution?

A I **203.1.** The number of possible choices of partners is

$$\tilde{S}_{2n}^{4n}=1.3.5\ldots(4n-1)=\frac{(4n)!}{(2n)!\,2^{2n}},$$

see I **203**. Multiply this by the number of possible choices of opponent pairs

$$\tilde{S}_{n}^{2n}=1.3.5\ldots(2n-1)=\frac{(2n)!}{n!\,2^n}.$$

Without using I **203**, from the combinatorial interpretation of polynomial coefficients we have

$$\frac{1}{n!}\frac{(4n)!}{4!^n}3^n=\frac{(4n)!}{n!\,8^n}.$$

The first proof shows that this number is odd.

New Problems in English Edition

I *1, *2, *3, *13, 18.1, 18.2, *19, 27.1, 27.2, 31.1, 31.2, 34.1, 43.1, 60.1, 60.2, 60.3, 60.4, 60.5, 60.6, 60.7, 60.8, 60.9, 60.10, 60.11, 64.1, 64.2, 68.1, *99, 132.1, 132.2, *134, 181.1, 185.1, 185.2, *186, *187, *188, *189, *190, *191, *192, *193, *194, *195, *196, *197, *198, *199, *200, *201, *202, *203, *204, *205, *206, *207, *208, *209, *210.

II *5, 19.1, 19.2, 81.1, 81.2, 81.3, 81.4, 93.1, 94.1, 94.2, 94.3, 95.1, 95.2, 95.3, 95.4, 95.5, 114.1, 118.1, 122.1, 122.2, 122.3, 217.1, *223, *224, *225, *226.

III *20, *21, 55.1, 55.2, 55.3, 55.4, 55.5, 206.1, 206.2.

IV 49.1, *124, 124.1, 172.1, 174.1, 174.2, *206, *207, *208, *209, *210, *211, *212.

V 43.1, 44.1, 62.1, 65.1, 77.1, 83.1, 94.1, 171.1, 171.2, 171.3, 171.4, 171.5, 189.1, 189.2, 189.3, 196.1.

VII 1.1, 11.1, 11.2, 11.3, 43.1, 43.2, 46.1, 46.2, 54.1, 54.2, 54.3, 54.4.

VIII 21.1, 21.2, 22.1, 22.2, 22.3, 27.1, 27.2, 38.1, 58.1, 58.2, 58.3, 58.4, 58.5, 58.6, 58.7, 58.8, 68.1, 68.2, 68.3, 72.1, 74.1, *75, 75.1, 75.2, 75.3, 75.4, 75.5, 80.1, 80.2, 197.1, 197.2, 197.3, 197.4, 197.5, 197.6, 227.1, 227.2, 237.1, 237.2, 238.1, 238.2, 244.1, 244.2, 244.3, 244.4, 247.1, 247.2, 260.1, 260.2, 262.1, 262.2.

IX 1.1, 1.2, 1.3, 1.4, 17.1, 17.2, 17.3, 17.4, 17.5.

A I 9.1, I 9.2, I 20.1, I 20.2, I 54.1, I 60.12, I 191.1, I 191.2, I 203.1.

Author Index

Numbers refer to pages. Numbers in italics denote original contributions.

Subject Index

Roman numerals refer to parts, Arabic numerals in boldface to problems and Arabic numerals in plain type to pages. Thus, for example, VIII **204** 146 refers to problem **204** of Part VIII on page 146.

Topics

The topics listed below are not evident in the arrangement of problems in the book, but they are nevertheless represented by connected series of problems.

Roman numerals refer to parts, Arabic numerals in boldface to problems and Arabic numerals in plain type to pages in Volume I and in italics to pages in Volume II. Thus, for example, VIII **226** *148* refers to problem **226** of Part VIII on page 148 of Volume II.